THE MOLECULAR AND CELLULAR BIOLOGY OF FERTILIZATION

ADVANCES IN EXPERIMENTAL MEDICINE AND BIOLOGY

Recent Volumes in this Series

THE MOLECULAR AND CELLULAR BIOLOGY OF FERTILIZATION

Edited by

Jerry L. Hedrick

University of California, Davis
Davis, California

PLENUM PRESS • NEW YORK AND LONDON

Library of Congress Cataloging in Publication Data

Symposium on the Molecular and Cellular Biology of Fertilization (1984: University of
  California, Davis)
  The molecular and cellular biology of fertilization.

  (Advances in experimental medicine and biology; v. 207)
  "Proceedings of the Seventy-Fifth Annual Symposium on the Molecular and Cellular
Biology of Fertilization, held June 20–22, 1984, at the University of California, Davis,
in Davis, California"—T.p. verso.
  This 75th anniversary symposium was organized in celebration of the founding of the
Davis campus of the University of California and sponsored by its Division of
Biological Sciences.
  Includes bibliographies and index.
  1. Fertilization—Congresses. 2. Molecular biology—Congresses. 3. Cytology—
Congresses. I. Hedrick, Jerry L. II. University of California, Davis. Division of
Biological Sciences. III. Title. IV. Series. [DNLM: 1. Fertilization—congresses. 2.
Germ Cells—physiology—congresses. 3. Molecular Biology—congresses. W1 AD559
v.207/QH 485 S989m 1984]
QP273.S94   1984   596′.016   86-25380
ISBN-13: 978-1-4612-9320-0   e-ISBN-13: 978-1-4613-2255-9
DOI: 10.1007/978-1-4613-2255-9

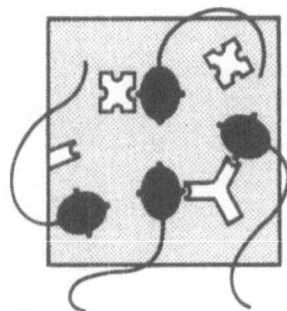

Proceedings of the Seventy-Fifth Annual Symposium on The Molecular and
Cellular Biology of Fertilization, held June 20–22, 1984, at the
University of California, Davis, in Davis, California

© 1986 Plenum Press, New York
 Softcover reprint of the hardcover 1st edition

A Division of Plenum Publishing Corporation
233 Spring Street, New York, N.Y. 10013

ACKNOWLEDGEMENTS

This book is derived from presentations made at the Symposium "The Molecular and Cellular Biology of Fertilization" held at the University of California, Davis, on June 20-22, 1984. This 75th Anniversary Symposium was sponsored by the Division of Biological Sciences, Donald L. McLean, Dean, and was organized in celebration of the founding of the Davis campus of the University of California. I am indebted to the members of the Organizing Committee, W. H. Clark, R. D. Grey, A. Lopo, and U. A. Urch, who ably assisted in the organization and execution of the Symposium. Financial support was provided by the 75th Anniversary Committee and the Division of Biological Sciences, both of the Davis campus; by a grant from the California Sea Grant College Program, University of California, La Jolla; by an NIH Training Grant/Reproductive Biology; and, by a contribution from Jack Daniels Distillery out of Sacramento. The members of my research laboratory, N. Wardrip, D. Goldhawk, M. Bakos, L. Lindsay, B. Shew, K. Clark provided essential logistical support critical to the smooth and efficient running of the meetings. A special thanks is due to Bert Urch, who so generously gave of his time, and of his scientific, epicurean, and enological abilities. I gratefully acknowledge the personal encouragement, patience, and involvement throughout the entire process from Symposium planning to volume publication of Karel J. Hedrick. Many thanks to the contributors to this volume and the Symposium participants who made the event both successful and enjoyable - it was fun!

Jerry L. Hedrick

J. L. Hedrick

C. B. Metz

A. Monroy

J. D. Biggers

S. Meizel

J. Dean

P. Wassarman

B. D. Shur

P. Saling

R. Grey

U. A. Urch

G. L. Gerton

C. Katagiri

D. C. Miceli

W. Clark, Jr.

C. Campanella

P. Jego

C. Cherr

M. Hoshi

A. Lopo

E. J. Carroll, Jr.

N. Ruiz-Bravo

D.L. Garbers

G. Ward

D. Katz

M. G. O'Rand

E. Goldberg

E. C. Yurewicz

D. P. Wolf

SPEAKERS AND SESSION CHAIRPERSONS*

CHIARA CAMPANELLA . . . . University of Naples, Naples, Italy

EDWARD J. CARROLL, JR. . . . University of California, Riverside, CA

GARY CHERR . . . . University of California, Davis, CA

WALLIS CLARK, JR.* . . . . University of California, Davis, CA

JURRIEN DEAN . . . . . National of Institute of Health, Bethesda, MD

DAVID L. GARBERS . . . . Vanderbilt University, Nashville, TN

GEORGE L. GERTON . . . . Harvard Medical School, Boston, MA

ERWIN GOLDBERG . . . . Northwestern University, Evanston, IL

ROBERT GREY* . . . . University of California, Davis, CA

JERRY L. HEDRICK* . . . . University of California, Davis, CA

MOTONORI HOSHI . . . . . . Nagoya University, Chikusa, Nagoya, Japan

PATRICK JEGO . . . . Université de Rennes, Rennes, France

CHIAKI KATAGIRI . . . . Hokkaido University, Sapporo, Japan

DAVID KATZ* . . . . University of California, Davis, CA

ALINA LOPO* . . . . University of California, Davis, CA

STANLEY MEIZEL* . . . . University of California, Davis, CA

CHARLES B. METZ . . . . University of Miami, Coral Gables, FL

DORA C. MICELI . Universidad Nacional de Tucuman, Tucuman, Argentina

ALBERTO MONROY . . . . Stazione Zoologica, Napoli, Italy

MICHAEL G. O'RAND . . University of North Carolina, Chapel Hill, NC

NORKA RUIZ-BRAVO . . . The University of Texas System Cancer Center, Houston, TX

PATRICIA SALING . . . . . Duke University Medical Center, Durham, NC

BARRY D. SHUR . . . . University of Connecticut Health Center, Farmington, CT

UMBERT A. URCH . . . . University of California, Davis, CA

GARY WARD . . . University of California at San Diego, La Jolla, CA

PAUL WASSARMAN . . . . Harvard Medical School, Boston, MA

DON P. WOLF . . . . The University of Texas, Houston, TX .

EDWARD C. YUREWICZ . . . . Wayne State University, Detroit, MI

CONTENTS

CONTENTS

Chapter 11
THE CORTICAL ENDOPLASMIC RETICULUM AND ITS POSSIBLE ROLE IN
ACTIVATION OF DISCOGLOSSUS PICTIS (ANURA) EGGS

Chiara Campanella, Riccardo Talevi, Umberto Atripaldi, and
Lucia Quaglia

Chapter 12
URODELE EGG JELLY AND FERTILIZATION

Patrick Jego, Hubert Lerivray, Amand Chesnel, and
Michel Charbonneau

Chapter 13
INDUCTION OF THE ACROSOMAL REACTION IN SPERM FROM THE WHITE
STURGEON, ACIPENSER TRANSMONTANUS

Gary Cherr and Wallis H. Clark, Jr.

Chapter 14
SPERM GLYCOSIDASE AS A PLAUSIBLE MEDIATOR OF SPERM BINDING
TO THE VITELLINE ENVELOPE IN ASCIDIANS

Motonori Hoshi

Chapter 15
STRUCTURE, ASSEMBLY AND FUNCTIONS OF THE SURFACE ENVELOPE
(FERTILIZATION ENVELOPE) FROM EGGS OF THE SEA URCHIN,
STRONGLYOCENTROTUS PURPURATUS

Edward J. Carroll, Jr., Mildred Acevedo-Duncan,
Robin Wilby Justice, and Lyric Santiago

Chapter 16
CHARACTERIZATION OF THE STRONGYLOCENTROTUS PURPURATUS EGG
CELL SURFACE RECEPTOR FOR SPERM

Norka Ruiz-Bravo, Daniel P. Rossignol, Glenn L. Decker,
Lawrence I. Rosenberg, and William J. Lennarz

Chapter 17
PEPTIDES ASSOCIATED WITH EGGS: MECHANISMS OF INTERACTION
WITH SPERMATOZOA

David L. Garbers, J. Kelley Bentley, Lawrence J. Dangott,
Chodavarapu S. Ramarao, Hiromi Shimomura, Norio Suzuki,
and David Thorpe

Chapter 18
DEPHOSPHORYLATION OF SEA URCHIN SPERM GUANYLATE CYCLASE
DURING FERTILIZATION

Gary E. Ward, Gary M. Moy, and Victor D. Vacquier

Chapter 19
STEPS IN THE FERTILIZATION PROCESS: UNDERSTANDING AND
CONTROL

Michael G. O'Rand

Chapter 20
CONTROL OF FERTILIZATION BY IMMUNIZATION WITH PEPTIDE
FRAGMENTS OF SPERM SPECIFIC LDH-C_4

Erwin Goldberg and Joyce A. Shelton

Chapter 21
PATHWAYS TO IMMUNOCONTRACEPTION: BIOCHEMICAL AND
IMMUNOLOGICAL PROPERTIES OF GLYCOPROTEIN ANTIGENS
OF THE PORCINE ZONA PELLUCIDA

Edward C. Yurewicz, Anthony G. Sacco,
and Marappa G. Subramanian

Chapter 22
RESEARCH FRONTIERS IN HUMAN IN VITRO FERTILIZATION

Don P. Wolf

PROLOGUE

Jerry L. Hedrick

Department of Biochemistry and Biophysics
University of California, Davis
Davis, CA 95616

Recent advances in the field of gamete interactions have produced new understandings of the mechanisms of cell-cell interactions at the molecular and supramolecular levels. This new knowledge, of course, brings pleasure and satisfaction to those of us
who are fulfilling our own sense of curiosity and wonder about this
fundamental biological process, but it also has significance for
mankind. Man's ability to control the fertilization process will
allow us to obtain fertilization where it is desired, but not
"naturally" obtainable, or to prevent fertility when it is not
desired. Also, it will allow us to increase fertility where it is
inadequate or limiting as in food producing animals, and with the
advent of in vitro fertilization in mammals, including humans, we
have the potential to produce and manipulate embryos, and to alter
the embryonic genome - events which, to a certain extent, are
already happening.

To the primary theme of the Symposium, advances in the
molecular and supramolecular mechanisms of sperm-egg interactions,
secondary themes were added. As the occasion for the Symposium was
the 75th Anniversary of the University California Davis campus, it
was appropriate to include a historical perspective to the
discussion of the current paradigms of gamete interactions. One
can question the utility of the history of science to scientists
themselves, but Joseph Needham expressed it well in his classical
work "A History of Embryology" when he wrote (1): "I suppose that
the history of science needs no apology. If at first sight the
discussion of what was thought in the past rather than what is
known now appears to be of merely antiquarian value, a deeper

---

1. Needham, J. (1959). "A History of Embryology."
    Abelherd-Shuman, New York.

consideration will admit ... that the history of science is the
guarantee of its freedom. The mistakes of our predecessors remind
us that we may be mistaken; their wisdom prevents us from assuming
that wisdom was born with us; and by studying the processes of
their thoughts, we may hope to have a better understanding, and
hence a better organization of our own."

Gamete interactions are studied using different methodologies
and different organismic systems. This often presents an added
degree of difficulty in understanding gamete interactions as the
electron microscopist must understand the language of the bio-
chemist, and vice versa, and the methodological and interpreta-
tional similarities and differences between various animal systems
such as sea urchins, amphibians, and mammals, must be recognized
and appreciated. Integrative and comparative approaches provide
additional perspectives in our attempts to comprehend gamete interactions. The utility of a comparative approach to understanding
biological processes was cogently stated by Albert Tyler (2):
"There are several reasons why the investigation of many diverse
kinds of organisms is important to the understanding of biological
processes. One is that some organisms may exhibit better than
others one or another of the special features of the biological
process under investigation ...Another reason that studies with
diverse organisms are of value is that conditions for successful
experimental manipulation with respect to particular processes may
differ, and some organisms may prove more suitable than others ...A
further, and perhaps the most cogent reason for adopting a compara-
tiave approach is that it permits broad generalizations ...It is
clear that the most important features of any biological process
are those which are common to diverse organisms."

The program of the Symposium and the organization of this
volume were primarily structured along organismal lines. Attempts
to organize the meeting using common molecular or cellular
phenomena as the schema was not as satisfactory as an organization
based on the organisms studied. This may be reflective of the
relatively incomplete state of our understanding of the molecular
and supramolecular basis of sperm-egg interactions, and cell-cell
interactions in general, or it may reflect the true diversity of
gamete interaction mechanisms in different species. If the former
is the case, then perhaps at future meetings devoted to the topic
of gamete interactions, the ability to organize our knowledge in
terms of molecular mechanisms of cellular phenomenon e.g.,
structure-function properties of recognition and binding molecules
and the molecular nature of the acrosome reaction, will reflect a
deeper and more comprehensive understanding of gamete interactions.

---

2. Tyler, A. (1967). in "Fertilization" Vol. 1, C. B. Metz and
    A. Monroy, eds., Academic Press, New York, p. 7.

FERTILIZATION AND IMMUNITY - REVISITED

Charles B. Metz

Institute for Molecular and Cellular Evolution
University of Miami
Coral Gables, Florida 33134

SUMMARY

This article provides an historical account and comparison of immune reactions and fertilization phenomena, particularly as they relate to sperm and egg receptors, and the considerable effect developments in immunology have had upon our conceptual and practical understanding of fertilization.

Why the title: Fertilization and Immunity - Revisited? Fertilization and Immunity is the title of an article by Albert Tyler published in 1948. Here he compared characteristics of immune reactions with fertilization phenomena as understood at that time. In keeping with the title, this article attempts to provide an historical account of these two subjects as they relate to sperm and egg receptors and the considerable effect immunology has had upon our understanding of fertilization - at both the conceptual and practical levels.

The two systems have many characteristics in common. First consider the origins and maturations of the germ cells and immune system effector cells. The similarities are striking. Both the primordial germ cells and the lymphocyte stem cells originate in or near the embryonic yolk sac. The primordial germ cells (mouse) originate in the allantoic rudiment and migrate into the yolk sac endoderm at a very early stage (44,61,62,67 for review). The lymphocytes are derived from the yolk sac mesoderm (37; see also authors in 4). The primordial cells of both systems migrate from their yolk sac sites of origin to distant organs where they differentiate to functional maturity: the primordial germ cells migrate to the embryonic genital ridges which, in the course of develop-

3

ment, differentiate into the gonads and at sexual maturity further
produce functional gametes. Like the primordial germ cells, the
primordial lymphocytes migrate to special organs where they differentiate. In birds those primordial lymphocytes that migrate to the
Bursa of Fabricius differentiate there into B lymphocytes, the
immuno-globulin producing or humoral components of the immune
system. Lymphocytes that enter the thymus gland differentiate into
T lymphocytes, the effector cells of the cellular immune system.
During their functional differentiation, both the germ cells and
lymphocytes share another strikingly similar property - namely,
development of exquisitely specific cell surface receptors that are
specific for complementary receptors of foreign origin (see 22).
Among gametes these receptors include the specific complementary
surface components of eggs and sperm that participate in sperm-egg
interactions at fertilization. The sum of these interactions leads
to the acrosome reaction, sperm-egg attachment, egg activation,
membrane fusion, in some cases apparently the symmetry of the egg,
and ultimately the mitotic cleavages of development. Likewise,
lymphocyte differentiation results in surface receptors specific
for a single or a few similar foreign antigenic receptors and, like
sperm-egg interaction, lymphocyte-antigen interaction can result in
"activation" and cell multiplication of the lymphocytes. Finally,
macrophage engulfment of foreign particulate matter can be mediated
by specific receptors, notably the Fc components of specific IgG
antibodies attached to the foreign particulate; a phenomenon with
properties including specificity, in common with sperm-egg fusion.

     The two systems do differ strikingly in one important respect,
namely, the time of functional maturity during ontogeny. For example, in mammals the immune system is fully developed and functional
at birth whereas the reproductive system does not attain full functional maturity - notably gametogenesis - until puberty. The potential immunological consequences of this - and of placentation -
have very important conceptual and practical significance. This
significance is based on the fact that the gametes and products of
conception including the placenta contains cell-specific antigens.
These should be recognized as foreign by the immune system which,
in turn, should mount an immune response to destroy these "foreign"
cells. This subject is now treated as a new science - Immunorepro-
duction - complete with at least two journals (see 58 for
review).

     The very early history of gametology, fertilization and immun-
ity has been reviewed by Lillie (36), Morgan (64), Tyler (81) and
many others. Suffice it to record that by 1900 the species and
tissue specificity of fertilization was throughly established. In
fact, Castle (2) and Morgan (63) had demonstrated self-sterility in
the hermaphroditic ascidian Ciona intestinalis. A thread-like
structure extending between the starfish sperm, through the egg
jelly, and to the egg was described by Fol (17), later investigated

in greater detail (Fig. 1) by R. Chambers (3), and finally recognized as the acrosomal filament (Fig. 2) following Jean Dan's (8,9,
10) discovery of the acrosome reaction. Natural parthenogenesis
in bees was apparently recognized by the early Greeks and Egyp-
tians, and recognized in silk worms in the 18th and 19th centuries.
More importantly, artificial parthenogenesis had been achieved in a
variety of marine organisms – with varying degrees of perfection –
by 1900 (see 38, 64 for review of early literature). This
achievement and its subsequent refinement was of great significance, obviously, because it showed that the actual activation of
the egg could be a relatively non-specific, physicochemical event –
and by implication, that the specificity of fertilization might
reside in such preliminary events as sperm penetration of egg envelopes and sperm-egg attachment.

At the turn of the century the concepts of macromolecular
specificity were perhaps best developed in the fields of immunology. For example, Paul Ehrlich (16) demonstrated that diptheria
toxin possessed at least two receptors. This was done by showing
that conversion to the toxoid form destroyed the toxic properties
but not all of the immunological properties of the toxin molecule.
From these studies Ehrlich developed his "side chain" theory and
extended it to include complement-dependent lytic effects of
specific antibody on erythrocytes. As seen in Figures 3 and 4, the
concepts were remarkably modern implying structural specificity of
combining sites or receptors. Although Ehrlich was aware of the

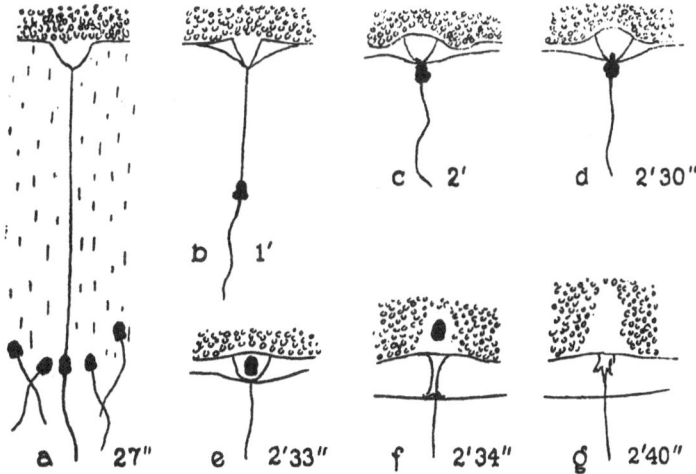

Figure 1. Reprinted from: Chambers, R. 1923. The mechanism of
          the entrance of sperm into the starfish egg. J. Gen.
          Physiol. 5:821-829.

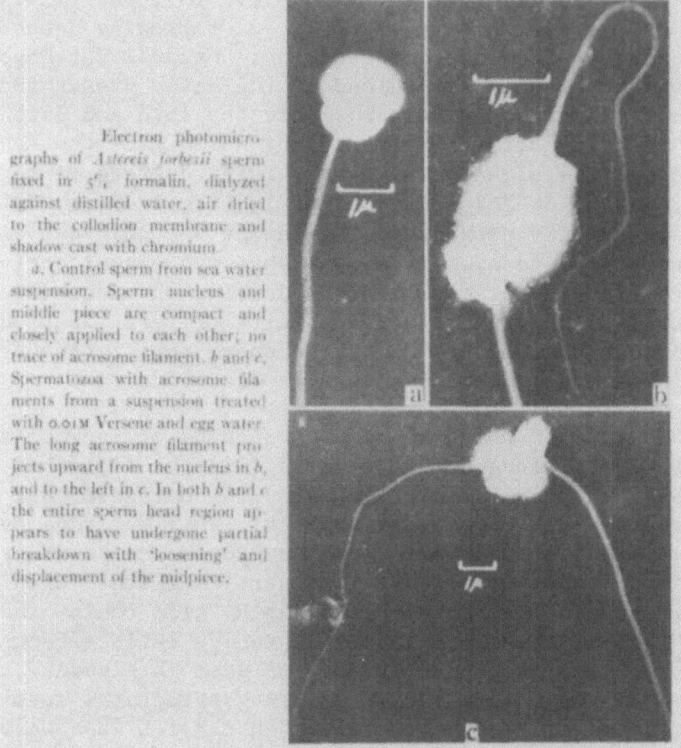

Figure 2. Reprinted from: Metz, D.B. 1957b. Mechanisms in fertili-
         zation, in: "Physiological Triggers," T.H. Bullock, ed.
         Am. Physiol. Soc., Washington, DC., pp. 17-45.

principal of optimal proportions in immune reactions such as agglutination and precipitation, he apparently did not appreciate the
necessity of antigen and antibody multivalency for such reactions.
At this time too the antigenic properties of gametes were recognized, e.g., Metchnikoff, (45; see 82 for comprehensive reference
list) including agglutination of mammalian sperm by antiserum; von
Dunger (12) described starfish sperm agglutination by rabbit antisera and aggregation of sea urchin sperm when mixed with egg water
(12) e.g., sea water in which eggs of the species had been suspended. This observation was subsequently confirmed by several others
including De Meyer (11). Two other historically interesting
reports appeared in 1911: (1) utilization of Ehrlich's side chain
theory - complete with diagrams by Godlewski (21) in what T. H.
Morgan (64) called a "carefully guarded comparison between the
results of the antagonistic action of sperm of different species
and serological work on blood;" and (2) at least partially successful attempts to produce infertility in guinea pigs by immunizing
females with sperm of the species (73,74).

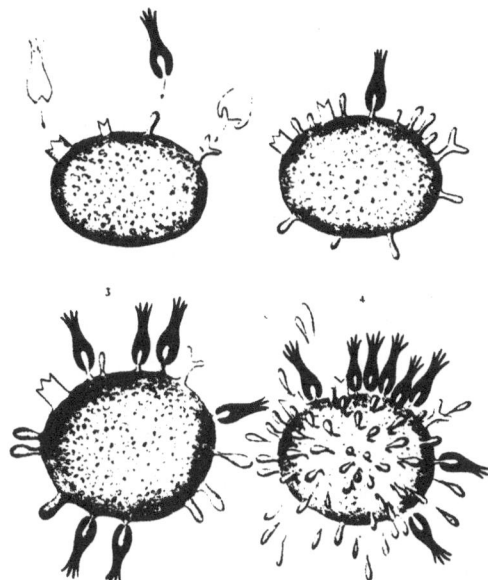

Figure 3. Reprinted from: Ehrlich, P. 1900. On immunity with
          special reference to cell life [Croonian Lecture].
          Proc. R. Soc. 66:424-448.

        Virtually simultaneously F. R. Lillie (34) began his studies
on sperm agglutination by egg water using primarily the sea urchin,
Arbacia punctulata, at the Marine Biological Laboratory, Woods
Hole, Massachusetts. Lillie called the agglutinin "fertilizin"
because of the central role he believed it played in fertilization.
He concluded from his experiments that:

        (1) Fertilizin agglutination was specific for sperm of the
species. (Failure to obtain interspecific cross reactions.)
        (2) Fertilizin was actively secreted by the unfertilized, but
not the fertilized egg. (Now known to be an egg jelly component
which dissolves in sea water.)
        (3) Fertilizin was essential for fertilization. (Probably an
aid to fertilization but not an absolute requirement.)
        (4) The egg contained an antifertilizin which reacted with and
neutralized fertilizin. (Egg extracts have such properties.)
        (5) Fertilization was inhibited by a factor in blood of the
species and that it acted upon the egg. (Probably a secretion from
the animal's epidermis e.g., 56.)

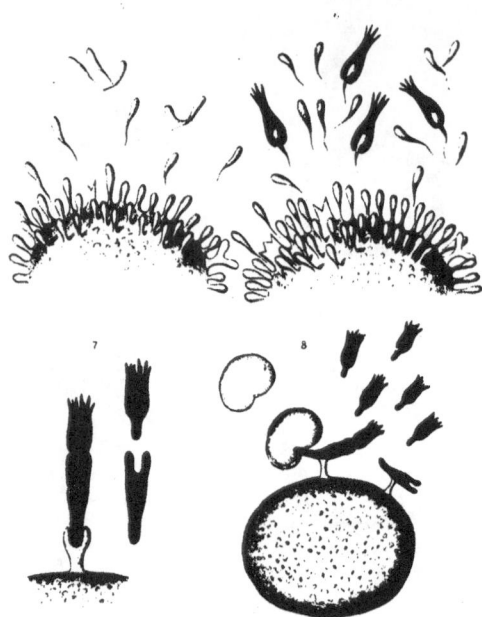

Figure 4. Reprinted from: Ehrlich, P. 1900. On immunity with
          special reference to cell life [Croonian Lecture]. Proc.
          R. Soc. 66:424-448.

From these conclusions and again utilizing Ehrlich's side
chain theory of specific receptor interactions, Lillie constructed
his Fertilizin Theory of fertilization (35,36). The specific sperm
agglutination phenomenon played a central role in this theory, in
fact as indicated above, Lillie called the agglutinin "Fertilizin"
for this reason. The theory was presented in dramatic diagrammatic
form (Fig. 5) and had enough support from experimental data to com-
mand recognition but not complete acceptance. In short it immediately generated controversy - especially from Jacques Loeb who at
first (39) denied the reality of the agglutination reaction[1], but
eventually admitted its reality - if not its significance (Fig. 6).
The theory was defended hotly by Lillie, his student E. E. Just and
others into the 1920s, after which interest diminished only to be
revived, especially by Albert Tyler in 1939.

At this time (1939) Linus Pauling, Albert Tyler, and others at
Cal Tech decided that immunology was the "wave of the future."
Naturally, many of their students (including this writer) were
swept up in this wave. The interests of individuals in this group
varied widely. Tyler was especially interested in immune phenomena
as model systems and tools to study reproduction and development.

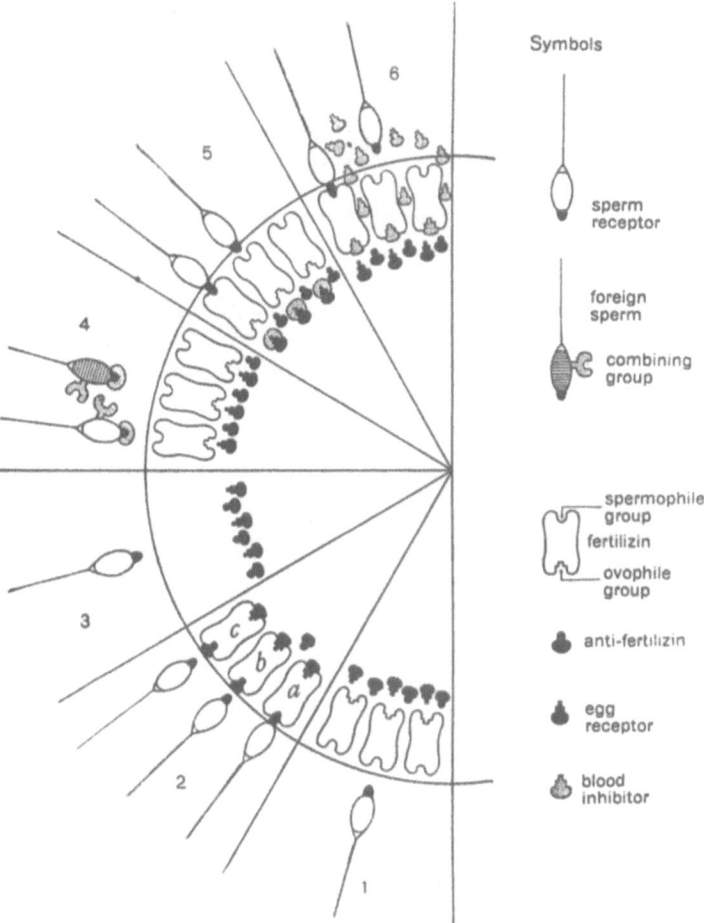

Figure 5. In successive sectors of the egg there are represented
the mechanism of fertilization and the blocks to the
mechanism, as follows:

Sector 1. The arrangement of substances in the unfertilized
egg and in the spermatozoon that are active in fertilization. See
explanation of symbols.

Sector 2. The mechanism of normal fertilization. The sperm
receptor unites with the spermophile group of the fertilizin and
the egg-receptors with the ovophile group of the fertilizin owing
to activation of the latter by the sperm (a). Molecules of the
antifertilizin (b and c) and thus block the way for supernumerary
spermatozoa. This is the postulated mechanism for prevention of
polyspermy. At the same time molecules b and c of the fertilizin
have also united with the egg-receptors.

Sector 3. Inhibition of fertilization by loss of the active body, fertilizin.

Sector 4. Theory of antagonistic action of spermatozoa of different phyla. The sperm receptors are occupied by combining groups cast off by the antagonistic spermatozoa.

Sector 5. Fertilization is blocked by occupancy of the egg-receptors. Purely hypothetical.

Sector 6. Theory of inhibitory action of blood of the same species. The ovophile group of the fertilizin is occupied by molecules in the blood (inhibitor) possessing the same combining group as the egg-receptors. Molecules of the blood inhibitor also shown in the medium.

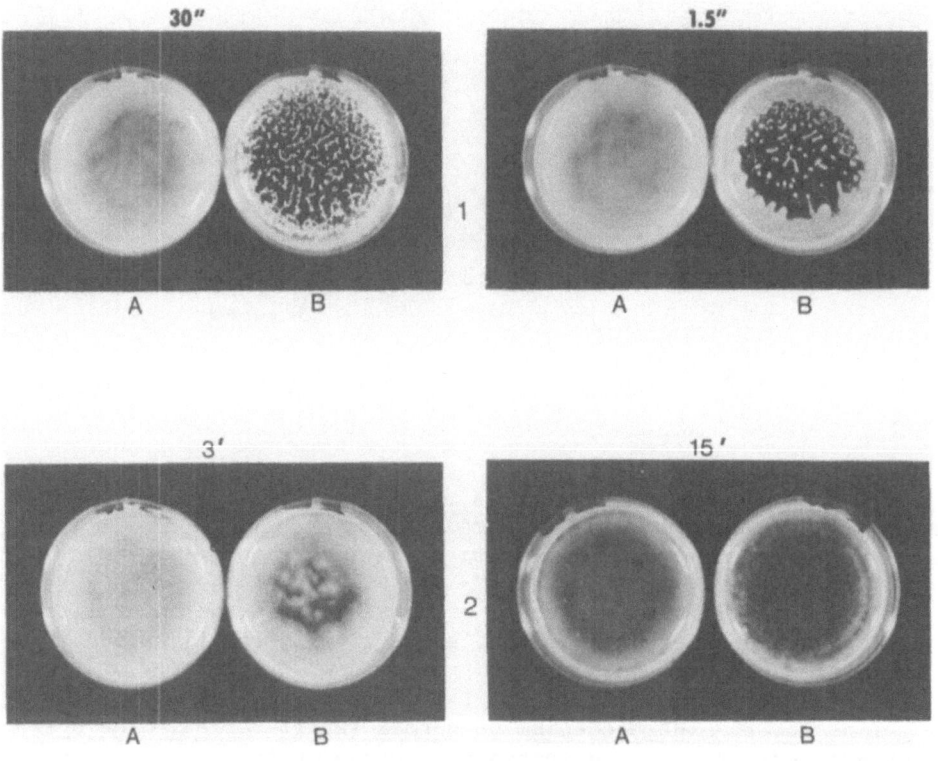

Figure 6. Egg water agglutination and spontaneous reversal of _Mellita quinquiesperferata_ sperm. Sperm suspension and egg water prepared in CA^{++}-free sea water. A. – Control: sperm in Ca^{++}-free sea water. B. – Sperm mixed with egg water prepared in Ca^{++}-free sea water. Time after mixing: 1) 90 s; 2) 120 s. From: J.B. Morrill, J.M. Shea, and C.B. Metz, unpublished.

In this context he naturally undertook a reinvestigation of Lillie's studies on fertilizin agglutination of sperm and its implications for fertilization. He confirmed the high order of fertilizin agglutination specificity, its spontaneous reversal in sea urchins but not in other species notably the mollusc, <u>Megathura crenulata</u> (Fig. 7), and its source, namely, the egg jelly coat in agreement with Loeb (39)[2]. In addition Tyler showed that the agglutinin was absorbed from solution by the sperm, that "reversed" sea urchin sperm could not be reagglutinated, and that the fertilizing capacity of reversed sperm was severely reduced, especially in <u>Strongylocentrotus purpuratus</u> (see 80 for review and references). Although such marked reduction in fertilizing capacity is less evident in <u>Arbacia punctulata</u> and probably some other sea urchins, it can be demonstrated using dejellied eggs (84).

The fertilizin is represented in the diagram as occurring only in the cortex of the egg, but it also occurs in high concentration in the jelly surrounding the egg. The spermatozoon must thus normally arrive at the egg-membrane loaded with combined fertilizin. This fact, however, makes no essential difference in the theory, and its representation would complicate the diagram (35). Reprinted from: Lillie, F. R., 1914. Studies on fertilization in <u>Arbacia</u>. J. Exp. Zool. 16:523-590.

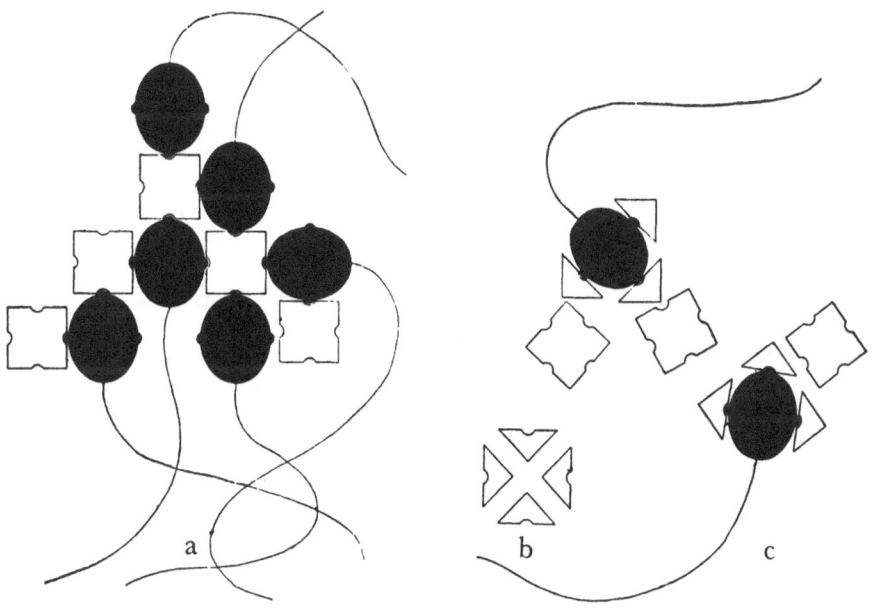

Figure 7. Reprinted from: Rothchild, Lord. 1951. Sea urchin spermatozoa. Biol. Rev. 26:1-27.

In addition Tyler confirmed that acid dejellied eggs had
reduced fertilizability and that even trypsin-treated eggs retained
some fertilizability except in cases of extreme exposure (see 84).

Tyler also undertook an investigation of the chemical nature
of the fertilizin from the sea urchin. Lillie (35) had already
demonstrated that the agglutinin was a macromolecule because it
failed to diffuse through a semi-permeable (collodion) membrane.
Tests for protein (20,88) were negative except for a weak xantho-
proteic test. Tyler and Fox (83) then began a chemical analysis
which was shortly joined by others (41,75,87) and which has continued sporadically to current times as interest has been revived
and new methods of analysis have been developed. The agglutinin is
an acid (20-25% sulfate) polysaccharide complex (see 57 for
review). Tyler (76) examined for the receptor of sperm that
reacted with fertilizin and found that supernatants from standing
sperm suspensions, and extracts prepared by freeze thawing, pH 3
extraction, or boiling (18) sperm had "antifertilizin activity"
e.g., the preparations agglutinated eggs, precipitated egg jelly,
and neutralized the sperm agglutinating action of fertilizin. The
fertilizin neutralizing and egg jelly precipitating agent in supernatants of sperm suspensions and weak acid extracts of sperm may be
the authentic sperm receptor involved in the agglutination reaction. Which, if any, of the products of the various extraction
procedures are "authentic" antifertilizin, e.g., the actual
specific sperm surface receptor for fertilizin, remains in doubt in
the writer's mind because of the relatively low order of specificities of the various preparations (see 80), sizable molecular weight
ranges, and the reactivity of fertilizin with a variety of
unrelated basic substances (e.g., protamines and histones, 50;
ribonuclease, 33). However, all agree that the material is predominantly protein. Tyler also obtained egg extracts with
properties similar to the "antifertilizin" from sperm including a
relatively low level of specificity (see 57, 80 for review and
references). This "egg antifertilizin" neutralizes the sperm
agglutinating action of fertilizin. Fertilizin may also be the
natural sperm acrosome initiating agent in sea urchin egg jelly
since purified acrosome initiating preparations (28,75) have apparently not been tested for sperm agglutinating activity. Acrosome
initiating and sperm agglutinating activities of sea urchin egg
jellies apparently have not been separated.

Three features of the fertilizin agglutination phenomena have
interested, if not troubled, many investigators. These are: (1)
the spontaneous reversal of sperm agglutination in echinoid echinoderms, (2) the apparent lack of universality of the agglutination
phenomenon, and (3) apparent requirement that sperm be living and
motile for fertilizin agglutination to occur.

The spontaneous reversal of agglutination in echinoid echinoderms is a striking and rare, if not unique, phenomenon (Fig. 6).
It is certainly not shared with antibody agglutination or precipitation reactions or with egg water agglutination in other forms
including asteroid echinoderms (48). To explain the spontaneous
reversal Tyler (77) took recourse to then current immunological
theory; more specifically the "lattice theory" of Heidelberger (25)
and of Marrack (43). This theory was based on the fact that
antigen-antibody precipitates formed maximally at certain optimal
proportions of the two reactants and that the precipitates dissolved in excess of either antigen or antibody. The only adequate
explanation of this phenomenon was that both the antigen and antibody molecules must be multivalent e.g., possess two or more
specific receptors such that each antibody and each antigen molecule combined with two or more of the complementary macromolecules
to form a "lattice" which then came out of solution.

With respect to reversal of fertilizin agglutination of sperm
Tyler (77) proposed that the agglutinin split after combination
with sperm thereby reversing the agglutination and leaving "univalent" fragments attached to and blocking the sperm surface antifertilizin receptor sites[3]. This interpretation was consistent
with failure of reversed sperm to reagglutinate on addition of more
fertilizin. When sea urchin egg water was heated, treated with
proteolytic enzymes (77,78), ultraviolet, X-irradiation (46), or
H_2O_2 (51) to the point where the egg water no longer agglutinated
sperm, this inactivated material still reacted with sperm as evidenced by failure of the treated sperm to reagglutinate when subsequently mixed with untreated, native egg water. Thus the spontaneous reversal of sea urchin agglutination was explained (see 54, 80
for reviews) as illustrated in Figure 7 from Rothschild (71).
Failure of the spontaneous reversal in the mollusc, Megathura
crenulata (Fig. 8) and other forms was attributed to greater sta-
bility of the agglutinin and/or absence of the physical or chemical
forces (e.g., enzymes) required to split the agglutinin. Apart
from explaining the reversal of agglutination the reversal itself
and the methods for obtaining univalent fertilizin provided valuable research tools because "reversed" sperm or univalent
fertilizin-treated sperm were freely motile e.g., any effect of the
treatment cannot be attributed to mechanical trapping of sperm by
permanent agglutination. The fertilization inhibiting action,
then, must be attributed to blocking sperm surface antifertilizin
or by producing some cellular purturbation - such as initiating a
"premature" acrosomal reaction (28). Availability of univalent
fertilizin, then, provided a "quantum" improvement in the technical
and conceptual levels of the research at that time.

Thus the univalent concept offered an explanation for the
failure of egg waters from many species to agglutinate sperm e.g.,

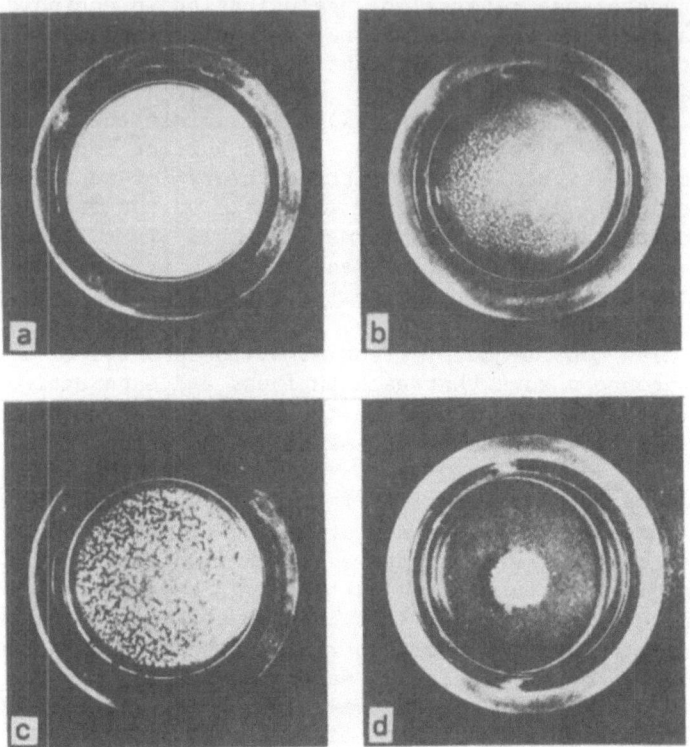

Figure 8. Reprinted from: Tyler, A. 1940. Sperm agglutination in
the keyhole limpet, _Megathura crenulata_. Biol. Bull.
78:159-178.

in these cases the fertilizin would be released from the egg (or
jelly) in a univalent, non-agglutinating form (77).

However, interesting conditions in starfish show that a
natural univalent fertilizin is not the only possible explanation
for failure of fertilizin agglutination. Starfish sperm do not
agglutinate when mixed with homologous egg water alone. However,
when any one of a variety of metal chelating agents is included in
the mixture of sperm and egg water, strong, permanent sperm agglutination results (Fig. 9) (48,52). This phenomenon has been demonstrated in over ten starfish species where the species specificity
of the reaction between sperm and egg water was virtually absolute.
The function of the metal chelating agent evidently is to expose
receptors on the sperm surface (48). The agglutinin is apparently
multivalent because ultraviolet irradiation renders the starfish
sperm agglutinin univalent e.g., non-agglutinating but capable of
blocking agglutination by unirradiated material (48).

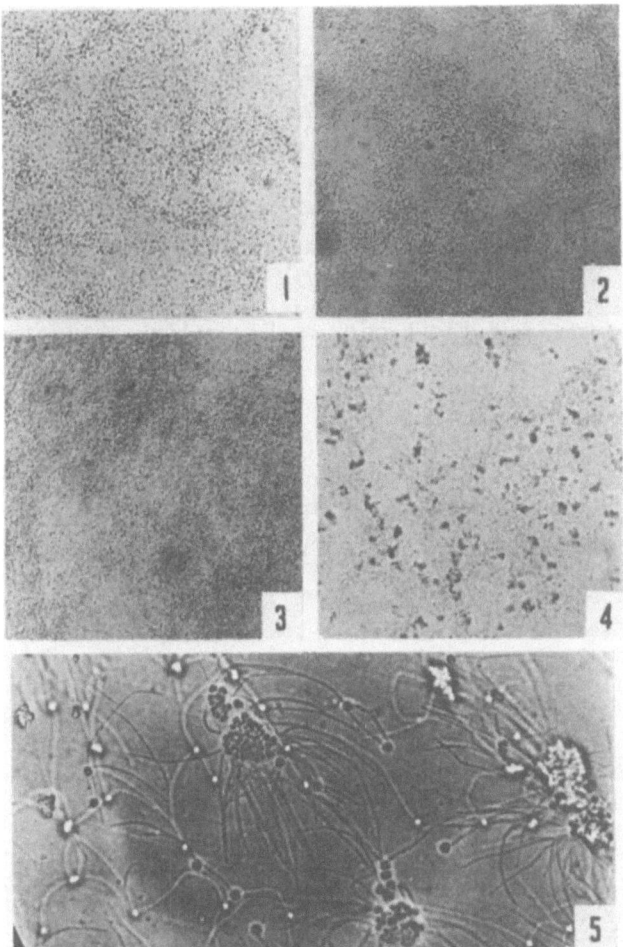

Figure 9. Metz, C.B. 1957a. Specific sperm and egg substances and
the activation of the egg. Pp. 23-69 in The Beginnings
of Embryonic Development. Am. Assoc. Adv. Sci.,
Washington, DC.

The starfish system, then, is a clear demonstration of the
sperm agglutinating system by special methods. A final unusual
feature of the starfish system is the chemical nature of the agglutinin. Uno and Hoshi (85) find that the agglutinin is an astro-
saponin e.g., a saponin core with carbohydrate side chains, not a
mucopolysaccharide as in sea urchins (75; see 57 for review).

Univalent antibodies

The earlier immunological literature contains reports of non-agglutinating and non-precipitating antibodies which nevertheless bind to antigen e.g., neutralize toxins (see 80 for reference). From our (e.g., 84) experiences with non-agglutinating, univalent fertilizin, it was obvious that univalent antibodies would be very valuable tools e.g., specific "made-to-order" blocking agents or probes for specific receptors. Accordingly, efforts were made to produce such material (e.g., 47,79). Tyler (79) did succeed in obtaining non-agglutinating antibody preparations by intense photodynamic oxidation, but these were highly viscous solutions of limited value.

Immunochemistry and knowledge about antibody structure advanced rapidly in the 1950s, 60s, and 70s. In the present context the most significant advance was development of enzymatic digestion procedures for fragmenting IgG. The first of these was the papain digestion procedure of R. R. Porter (68,69) which produce univalent antibody (Fab) fragments of relatively small size (M_r = 55,000) but with intact specific combining sites for antigen. The procedure removed the Fc component of the IgG molecule, thus eliminating the possibility of complement-dependent reactions. The second, or pepsin digestion procedure (65,66) removed the Fc component but left a bivalent Fc-free antibody. This could then be split by a reducing agent into Fab_1 fragments. The latter could then be oxidized to permanent Fab_1 univalent fragments or combined with reduced Fab_1 fragments with different specificity to produce "hybrid" bivalent, Fc-free antibodies (19).

With this simple technology available we first showed (59,60) that treatment with Fab antisperm antibody rendered Arbacia punctulata sperm incapable of agglutinating with egg water and incapable of fertilizing eggs. It is of special interest that certain of the Fab antisera inhibited the fertilizing capacity of the sperm but did not inhibit fertilizin agglutination of the sperm. This strongly suggests that reaction of the sperm with fertilizin is not an essential event in the fertilization process.

Attempts to isolate the "fertilization antigen" from Arbacia sperm extracts prepared by freeze thawing sperm yielded a single glycoprotein which neutralized the fertilization inhibiting action of Fab antisperm antibody (6,14,15,42). Evidently this glycoprotein is an essential fertilization antigen. Labeling studies (42) indicated that the antigen was located in the acrosome region. Examination of the behavior of Fab antibody treated sperm showed that these sperm failed to attach to the egg surface of dejellied or demembranated egg (72) and failed to undergo the acrosome reaction (72). Evidently the univalent antibody blocked the Arbacia sperm surface receptor involved in initiating the acrosome reaction

as also occurs in <u>Strongylocentrotus</u> <u>purpuratus</u> (40). Consequently, the glycoprotein described above must be the specific sperm
surface receptor involved in initiating the acrosome reaction.
Such univalent antibody blocked sperm did undergo acrosome reactions when treated with ionophore A23187 (72).

The unreacted sea urchin sperm acrosome contains a dense granule. The contents of this granule are released and coat the tip of
the acrosome projection following the acrosome reaction. This
adhesive material is a glycoprotein called "bindin" (86) which
attaches the sperm to the vitelline membrane of the egg. The
receptor for bindin, a component of the vitelline envelope, is a
very interesting high molecular weight protease-sensitive glycoconjugate complex (70; see Ruiz-Bravo et al. and Carroll, this
volume).

In summary, the mechanisms of sperm egg interaction and
attachment up to and including sperm acrosome attachment to the
vitelline envelope of the egg are reasonably well described at the
molecular level for the sea urchin. Each of these steps includes a
high order of species specificity. The final step(s) may be a
relatively simple, non-specific physical process such as spontaneous fusion of opposing hydrophobic surfaces of the sperm and egg
plasma membrane. Perhaps perturbations during such fusion result
in transitory local increase in permeability accompanied by egg
membrane depolarization. However, this interpretation is not
entirely consistent with fertilization events in at least one lower
organism.

Paramecium: a model system

The critical final events of the activation process in terms
of receptors are perhaps best understood in the protozoan, <u>Para-</u>
<u>mecium</u>. In this "model system" the complementary sexually competent
cells are structurally indentical (isogamous), fertilization is
reciprocal, and no diffusing ("action at a distance") sex substances confuse the events. In addition, defective mutant stocks
with mating [e.g., the CM block in the can't mate mutant, Fig. 10
(55)] abnormalities can provide valuable information. In the mating
process, sexually competent individuals of complementary sex adhere
by the tips of their ventral cilia upon random contact. After
approximately an hour (<u>P. aurelia</u>) this primary attachment of cilia
is replaced by secondary ("hold fast") and paroral unions. However, the initial (ciliary) mating reaction union is the critical
event because it triggers the entire series of preprogrammed fertilization events (26,31,49,53). These conclusions are based on
the fact that dead mating reactive cells or isolated cilia retain
specific activity e.g., specifically attach to ventral cilia of
living paramecia of complementary sex and initiate the full

sequence of sexual responses (Fig. 10). This appears to be a
strictly "antigen-antibody like" reaction between complementary
mating type substances. No proteolytic enzymes appear to be involved since protease inhibitors have no effect on the mating
events (29).

     Attempts to solubilize and purify the active mating type sub-
stance from ventral cilia have failed. However, inactivation
studies with enzymes and other agents show that the mating type
substances on the tips of cilia are highly specific, strongly
hydrophobic proteins. They appear when the ventral cilia become
hydrophobic e.g., when the paramecia become sexually reactive.
When reactive cilia membranes are solubilized (2 M urea - EDTA or
LIS), centrifuged, and the supernatant dialyzed to remove the sol-
ubilizing agent, membrane vesicles can be recovered (50 - 100 nm
diameter in the case of LIS vesicles). These vesicles bind specifically to living mating reactive paramecia of opposite sex and
specifically initiated the chain of fertilization events e.g.,
activate the animals (30). Hiwatashi and Kitamura (26) propose
that the mating type substances are inserted into the ciliary membrane tip as it becomes more hydrophobic. This, then, could provide a transmembrane receptor. Interaction with the complementary
receptor of the opposite mating type could produce perturbations
e.g., conformational transductions through the membrane to initiate
the internal events of fertilization activation (see 1 for review
of transduction mechanisms). This concept is at least consistent
with the absence of electrical (32) and/or pronounced ionic changes
at activation except for a possible internal release of Ca^{++} (7).
Unfortunately, antisera to paramecia have almost universally failed
to show mating type specificity in spite of extensive trials in

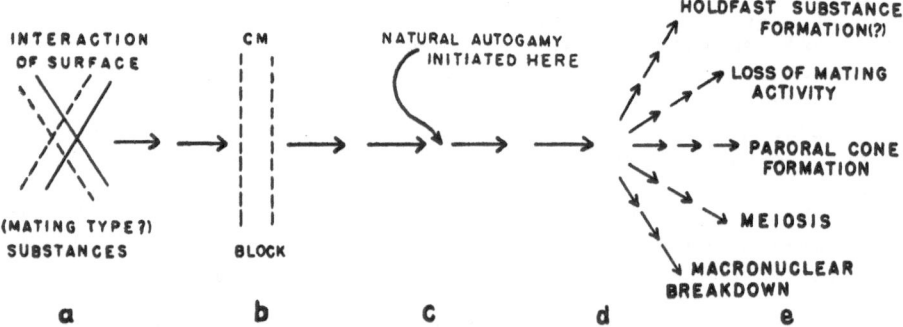

Figure 10. Metz, C.B. 1957b. Mechanisms in fertilization. Pp. 17-45
           in Physiological Triggers, T.H. Bullock, ed. Am.
           Physiol. Soc., Washington, DC.

several laboratories. Use of monoclonal antibodies may be a solution to this problem, as Hiwatashi and Kitamura (26) suggest.
These important studies by Hiwatashi, Kitamura, and associates are
a major advance in our understanding of the critical final step(s)
in fertilization, namely the actual activation process.

ACKNOWLEDGEMENTS

The author is grateful to Dr. John B. Morrill and Ms.
Jacqueline Shea for collaboration and for the S.E.M. illustrations
in Figure 6. The study was performed in Dr. Morrill's laboratory
at New College, Sarasota, FL. The author also appreciates many
helpful suggestions and references from Dr. William R. Eckberg,
Howard University, Washington, DC, and Dr. Michael G. O'Rand,
University of North Carolina, Chapel Hill.

FOOTNOTES

[1] Loeb (39) originally held that the phenomenon described by
Lillie was not "real" agglutination but a "cluster formation"
related to surface tension effects. Loeb also raised some problems
which still have not been solved, notably why non-motile sperm
ordinarily fail to agglutinate when treated with egg-water. This
question of the "reality" of fertilizin agglutination of sperm has
been resurrected recently by Collins (5). The confusion is largely
attributed to the spontaneous reversal of the agglutination
reaction in echinoids. This reversal is evidently a special
adaptation to permit the enchinoid sperm with its short acrosomal
filament to attach to but still penetrate the egg jelly (27,57).
In fact, were it not for this reversal phenomenon, the
investigations of Lillie and the concepts derived from them could
not have occurred. Figure 6 shows the sperm agglutination and
spontaneous reversal phenomenon in the sanddollar, Mellita
quinquiespurpurata, the echinoid that gives the most dramatic
reaction in the author's experience. Sperm suspension and egg
water were prepared in Ca^{++}-free sea water to prevent acrosome
reactions and confirmed by scanning electron microscopy.

[2] A material, cytofertilizin, with properties identical to egg
jelly fertilizin is contained in the egg cortical granules and is
released with cortical granule discharge at fertilization (23).

[3] The "splitting" is evidently not a simple "halving" of the
fertilizin molecule. ^{35}S is removed from labeled fertilizin
solutions by sperm. However, half of the sperm-bound label is
released following reversal of agglutination (24).

REFERENCES

1. Carraway, K.L., and Carraway, C.A.C. 1984. Cell surface receptors and transduction mechanisms, in: "Biology of Fertilization," Vol. I, pp. 23-55, C.B. Metz and A. Monroy, eds, Academic Press, New York.

2. Castle, W.E. 1896. The early embryology of Ciona intestinalis. Bull. Mus. Com. Zool. Harvard Coll. Vol. 27.

3. Chambers, R. 1923. The mechanism of the entrance of sperm into the starfish egg. J. Gen. Physiol. 5:821-829.

4. Clarkson, B., Marks, P.A., and Till, J.E. 1978. Differentiation of normal and neoplastic hematopoitic cells, A. Book, ed. Cold Spring Harbor Conference on Cell Proliferation, Vol. 5. Cold Spring Harbor Laboratory.

5. Collins, F. 1976. A reevaluation of the fertilizin hypothesis of sperm agglutination and the description of a novel form of sperm adhesion. Dev. Biol. 49:381-394.

6. Cordle, C.T., and Metz, C.B. 1973. Isolation of Arbacia sperm antigens important in fertilization. Biol. Bull. 145:430.

7. Cronkite, D.L. 1979. The genetics of swimming and mating behavior in Paramecium, in: "Biochemistry and Physiology of Protozoa," Vol. 3, pp. 221-273, M. Levandowsky and S.H. Hunter, eds, Academic Press, New York.

8. Dan, J.C. 1952. Studies on the acrosome. I. Reaction to egg-water and other stimuli. Biol. Bull. 103:54-66.

9. Dan, J.C. 1954. Studies on the acrosome. II. Acrosome reaction in starfish spermatozoa. Biol. Bull. 107:203-218.

10. Dan, J.C. 1954. Studies on the acrosome. III. Effect of Ca^{++} deficiency. Biol. Bull. 107:335-349.

11. De Meyer, J. 1911. Observations et expériences relatives a l'action exercée par les extracts d'ocufs et d'autre substances sur les spermatozoïdes. Arch. Biol. 26:65-101.

12. von Dungern, E. 1901. Dic Ursachen der Specietat der Befrüchtung. Centralol. F. Physiol. 15:1-4.

13. von Dungern, E. 1902. Neue Versuche zur Physiologie der Befrüchtung. Z. Allgemeine. Physiol. 1:34-55.

14. Eckberg, W.R., and Metz, C.B. 1974. Effects of univalent (Fab) antibody fragments on the fertilizing capacity of sperm of the sea urchin, Arbacia punctulata. Biol. Bull. 147:475.

15. Eckberg, W.R., and Metz, C.B. 1982. Isolation of an Arbacia sperm fertilization antigen. J. Exp. Zool. 221:101-105.

16. Ehrlich, P. 1900. On immunity with special reference to cell life. (Croonian Lecture) Proc. R. Soc. 66:424-448.

17. Fol, H. 1877. Sur le premier développement d'une étoile de mer. Compt. Rendu. Acad. Sci. 84:357.

18. Frank, J.A. 1939. Some properties of sperm extracts and their relationship to the fertilization reaction in Arbacia punctulata. Biol. Bull. 76:190-216.

19. Fundenberg, H.H., Drews, G., and Nisonoff, A. 1964. Serological demonstration of direct specificity of rabbit bivalent hybrid antibody. J. Exp. Med. 119:151-166.

20. Glaser, O. 1914. A qualitative analysis of the egg-secretions and extracts of Arbacia and Asterias. Biol. Bull. 26:367-386.
21. Godlewski, E. 1911. Studien uber die Entwicklungserregung. Wilhelm Roux Arch. Entwicklungs. Organismen 33:196-254.
22. Goldschneider, I. 1980. Early stages in lymphocyte development. Curr. Top. Dev. Biol. 14(pt. II):33-53.
23. Gregg, K.W. 1969. Cortical response antigens released at fertilization from sea urchin eggs and the relation to antigens of the jelly coat. Biol. Bull. 137:146-154.
24. Hathaway, R.R., and Metz, C.B. 1961. Interactions between Arbacia sperm and S^{35}-labelled fertilizin. Biol. Bull. 120:360-369.
25. Heidelberger, M. 1939. Chemical aspect of the precipitin and agglutination reactions. Chem. Rev. 24:323-343.
26. Hiwatashi, K., and Kitamura, A. 1984. Fertilization in Paramecium, in: "Biology of Fertilization," Vol. 1, pp. 57-85, C.B. Metz and A. Monroy, eds, Academic Press, New York.
27. Hoshi, M., and Nagai, Y. 1977. Glycoconjugates and fertilization, in: "Biology of Fertilization," Jpn. Soc. Dev. Biol., Iwanomi - Shoten, Tkyo, in Japanese.
28. Kinsey, W.H., Rubin, J.A., and Lennarz, W.J. 1980. Studies on the specificity of sperm binding in echinoderm fertilization. Dev. Biol. 74:245-250.
29. Ditamura, A. 1983. Effects of protease inhibitors on early events in the conjugation of Paramecium caudatum. J. Exp. Zool. 225:501-503.
30. Kitamura, A., and Hiwatashi, K. 1980. Reconstitution of mating reactive vesicles in paramecium. Exp. Cell Res. 155:486-489.
31. Kitamura, A., and Hiwatashi, K. 1984. Cell contact and the activation of conjugation in paramecium. Zool. Sci. 1:161-168.
32. Kitamura, A., Onimaru, H., Naitoh, Y., and Hiwatashi, K. 1979. Relation between swimming and sexual behavior in Paramecium caudatum. Dubutsugaka Zasshi 88:528.
33. Ledoux, L., and Metz, C.B. 1960. Inhibition of sea urchin egg cleavage by ribonuclease. Experentia 6:1-3.
34. Lillie, F.R. 1912. The production of sperm iso-agglutinins by ova. Science 36:527-530.
35. Lillie, F.R. 1914. Studies on fertilization in Arbacia. J. Exp. Zool. 16:523-590.
36. Lillie, F.R. 1919. Problems of Fertilization, University of Chicago Press, Chicago.
37. Linman, J.L. 1966. Origin of hemotopoic cells, in: "Principles of Hematology," MacMillan Co., New York.
38. Loeb, J. 1913. Artificial Parthenogenesis and Fertilization, University of Chicago Press, Chicago.
39. Loeb, J. 1914. Cluster formation of spermatozoa caused by specific substances from eggs. J. Exp. Zool. 17:123-140.
40. Lopo, A.C., and Vacquier, V.D. 1980. Antibody to a sperm surface glycoprotein inhibits the jelly-induced acrosome reaction of sea urchin sperm. Dev. Biol. 79:325-333.

41. Lorenzi, M., and Hedrick, J.L. 1973. On the macromolecular composition of the jelly coat from S. purpuratus eggs. Exp. Cell
    Res. 79:417-422.

42. Maitra, U.S., and Metz, C.B. 1974. Purification and properties of
    an *Arbacia* sperm fertilization antigen. Biol. Bull. 147:490.

43. Marrack, J.R. 1938. The chemistry of antigens and antibodies. MRC
    Special Rep., series #230, London.

44. McLaren, A. 1983. Germ Cells and Soma: a New Look at an Old
    Problem, Yale University Press, New Haven.

45. Metchnikoff., F. 1899. Etudes sur la résorption des cellules.
    Ann. Inst. Pasteur 13:737-769.

46. Metz, C.B. 1942. The inactivation of fertilizin and its conversion to the "univalent" form by x-rays and ultraviolet light.
    Biol. Bull. 82:446-454.

47. Metz, C.B. 1942. Doctoral Thesis, California Institute of
    Technology, Pasedena.

48. Metz, C.B. 1945. The agglutination of starfish sperm by
    fertilizin. Biol. Bull. 89:84-94.

49. Metz, C.B. 1948. The nature and mode of action of the mating type
    substances. Am. Nat. 82:85-95.

50. Metz, C.B. 1949. Agglutination of sea urchin eggs and sperm by
    basic proteins. Proc. Exp. Biol. Med. 70:422-424.

51. Metz, C.B. 1954. The effect of some protein group reagents on
    the sperm agglutinating action of *Arbacia* fertilizin. Anat.
    Record 120:713.

52. Metz, C.B. 1954. The adjuvant action of chelating agents in
    fertilizin agglutination of starfish sperm. Biol. Bull.
    107:317.

53. Metz, C.B. 1954. Mating type substances and the physiology of
    fertilization in ciliates, in: "Sex and Microorganism," D.H.
    Wenrich, ed., Am. Assoc. Adv. Sci., Washington, DC., pp. 284-
    334.

54. Metz, C.B. 1957. Specific sperm and egg substances and the
    activation of the egg, in: "The beginnings of Embryonic
    Development," Am. Assoc. Adv. Sci., Washington, DC., pp. 23-
    69.

55. Metz, C.B. 1957. Mechanisms in fertilization, in: "Physiological Triggers," T.H. Bullock, ed., Am. Physiol. Soc.,
    Washington, DC., pp. 17-45.

56. Metz, C.B. 1961. Use of inhibiting agents in studies of fertilization mechanisms. Int. Rev. Cytol. 11:219-253.

57. Metz, C.B. 1978. Sperm and egg receptors involved in fertilization. Curr. Top. Dev. Biol. 12:107-147.

58. Metz, C.B. 1979. Immunological inhibition of sperm and egg function. Prospects for an antifertility vaccine from gametes,
    in: "Contraception: Science, Technology, and Applications,"
    E.V. Jenner, ed., Natl. Acad. Sci., Washington, DC., pp. 180-
    220.

59. Metz, C.B., and Schuel, H. 1961. Inhibition of agglutinability
    and fertilizing capacity of sea urchin sperm by papain
    digested antibody. Am. Zool. 1:92.

60. Metz, C.B., Schuel, H., and Bischoff, W.E. 1964. Inhibition of the fertilizing capacity of sea urchin sperm by papain digested, non-agglutinating antibody. J. Exp. Zool. 155: 261-272.
61. Mintz, B. 1961. Formation and early development of germ cells. In Symp. Germ Cells Dev. 1:24.   Embryological Division, I.U.B.S.
62. Mintz, G., and Russell, E.S. 1957. Gene-induced embryological modifications of primordial germ cells in the mouse. J. Exp. Zool. 134:207-238.
63. Morgan, T.H. 1904. Self-fertilization induced by artificial means. J. Exp. Zool. 1:135-178.
64. Morgan, T.H. 1927. Experimental Embryology, Columbia University Press, New York.
65. Nisonoff, A., and Inman, F.P. 1965. Structural basis of the specificity of antibodies, in: "Reproduction: Molecular, Subcellular and Cellular," M. Locke, ed. 24th Symposium of the Soc. Dev. Biol. Academic Press, New York, pp. 39-64.
66. Nisonoff, A., Wissler, F.C., Lipman, L.H., and Woernley, D.L. 1960. Separation of univalent fragments from the bivalent rabbit antibody molecule by reduction of disulfide bonds. Arch. Biochem. Biophys. 89:230-244.
67. Ozdzenski, W. 1967. Observations on the origin of primordial germ cells in the mouse. Zool. Pol. 17:367-379.
68. Porter, R.R. 1958. Separation and isolation of fractions of rabbit gamma-globulin containing the antibody and antigenic combining sites. Nature 182:670-671.
69. Porter, R.R. 1959. The hydrolysis of rabbit gamma-globulin and antibodies with crystalline papain. Biochem. J. 73:119-126.
70. Rossignol, D.P., Earles, B.J., Decker, G.L., and Lennarz, W.J. 1984. Characerization of the sperm receptor on the surface of eggs of Strongylocentrotus purpuratus. Dev. Biol. 104:308-321.
71. Rothchild, Lord. 1951. Sea urchin spermatozoa. Biol. Rev. 26: 1-27.
72. Saling, P.M., Eckberg, W.R., and Metz, C.B. 1982. Mechanism of univalent antisperm antibody inhibition of fertilization in the sea urchin, Arbacia punctulata. J. Exp. Zool. 221:93-99.
73. Savini, E., and Savini-Castan, Th. 1911a. Immunité spermatoxique et fécundation. Compt. Rendu. Soc. Biol. 71:22-23.
74. Savini, E., and Savini-Castano, Th. 1911b. Contribution à l'étude des spermatoxines. Compt. Rendu. Soc. Biol. 71:106-108.
75. SeGall, G.K., and Lennarz, W.J. 1979. Chemical characterization of the components of the egg jelly coat from sea urchin eggs responsible for the acrosome reaction. Dev. Biol. 71:33-48.
76. Tyler, A. 1939. Extraction of an egg membrane - lysin from sperm of the giant keyhole limpet (Megathura crenulata). Proc. Natl. Acad. Sci. USA 25:317-323.
76a.Tyler, A. 1940. Sperm agglutination in the keyhole limpet, Megathura crenulata. Biol. Bull. 78:159-178.

77. Tyler, A. 1941. The role of fertilizin in the fertilization of eggs of the sea urchin and other animals. Biol. Bull. 81:190-204.
78. Tyler, A. 1942. Specific interacting substances of eggs and sperm. West. J. Surg. 50:126-138.
79. Tyler, A. 1945. Conversion of agglutinins into "univalent" (non-agglutinating or non-precipitating) antibodies by photodynamic irradiation of rabbit antisera vs. pneumonocci, sheep-red-cells and sea urchin sperm. J. Immunol. 51:157-172.
80. Tyler, A. 1948. Fertilization and immunity. Phys. Rev. 28: 180-219.
81. Tyler, A. 1967. Problems and procedures of comparative gametology and syngamy, in: "Fertilization," Vol. I, C.B. Metz and A. Monroy, eds, Academic Press, New York, pp. 1-26.
82. Tyler, A., and Bishop, D.W. 1963. Immunological phenomena, in: "Conference on Physiological Mechanisms Concerned with Conception," Pergammon Press, pp. 397-482.
83. Tyler, A., and Fox, S.W. 1940. Evidence for the protein nature of the sperm agglutinins of the keyhole limpet and the sea urchin. Biol. Bull. 79:153-165.
84. Tyler, A., and Metz, C.B. 1955. Effects of fertilizin - treatment of sperm and trypsin - treatment of egg of homologous and cross-fertilization in sea urchins. Pubbl. Staz. Zool. Napoli 27:128-145.
85. Uno, Y., and Hoshi, M. 1978. Separation of the sperm agglutinin and the acrosome-inducing substance from egg jelly of star-fish. Science 200:58-59.
86. Vacquier, V.D., and Moy, G.W. 1977. Isolation of bindin: the protein responsible for adhesion of sperm to sea urchin eggs. Proc. Natl. Acad. Sci. USA 74:2456-2460.
87. Vasseur, E. 1952. Chemistry and physiology of the egg jelly coat of the sea urchin. Kihlstromstryck, A.B., Stockholm.
88. Woodward, A.E. 1918. Studies on the physiological significance of certain precipitates from the egg secretions of Arbacia and Asterias. J. Exp. Zool 26:459-501.

SPERM-EGG INTERACTIONS PREPARATORY TO FERTILIZATION

Alberto Monroy

Stazione Zoologica
Villa Comunale
80121 Napoli, Italy

SUMMARY

In 1902, Boveri introduced the important concept that for the success of fertilization the gametes must activate one another. Based primarily on studies on the sea urchin and ascidian fertilization the suggestion is presented here that "activation" of the spermatozoon actually involves a switching off of its metabolic machinery as a result of its interaction with the sperm receptors of the egg envelopes and prior to its fusion with the egg.

Concerning the activation of the egg, there is a fairly large body of old and new experimental evidence that activation per se does not require sperm incorporation. Indeed, the chain of reactions culminating in the activation of the egg is initiated upon attachment of the spermatozoon to the egg plasma membrane.

INTRODUCTION

The essence of the problems of fertilization was clearly outlined by Theodore Boveri in 1902. Boveri indeed pointed out that the main property of the gametes is that they should be in a repressed condition, namely that "they should not be able to develop by themselves and in fact they should be blocked in such a way that inhibition is released by the other cell." The second condition for fertilization to occur is that "the two cells must be able to find each other." The third condition is a corollary of what Boveri defines "the equivalence of the germ cells." This led him to propose that "in the same way that the ability of the egg to develop is completed by the spermatozoon, the same is true for the spermatozoon by the egg. In the same way as we say that the sper-

matozoon fertilizes the egg we may say that the egg fertilizes the
spermatozoon." This important concept introduced a new dynamic
view of the gamete encounter: in fertilization neither gamete
behaves passively in the sense that one is the activator and the
other the activated one: <u>for the success of fertilization the</u>
<u>gametes must activate one another.</u>

        A few years later the same view was expressed by F.R. Lillie
(22) namely that in fertilization the spermatozoon "needs to be
fertilized." What he specifically meant was that the role of "fertilizin" with which the spermatozoon reacts before fusing with the
egg is "to prepare the spermatozoon for exercising fertilization
effect in the interior of the egg." This "preparation" of the
spermatozoon we interpret nowadays to be the changes that the
spermatozoon undergoes in preparation for its fusion with the egg,
primarily the acrosome reaction. Again in 1963 the same concept
was proposed by the Colwins in their classical studies of the
acrosome reaction and of the sperm-egg fusion (6). In their words,
"The proposed concept of sperm activation offers a framework within
which the changing acrosomal region can be related to its precise
roles in the initial stages of fertilization, much as the concept
of egg activation has facilitated analysis of the initial stages of
fertilization." In the Colwins' view the "major role of the acrosome is <u>not</u> to activate the egg but to convey the sperm plasma
membrane to the egg plasma membrane, as a consequence of which the
egg subsequently is activated."

        In later years most of the attention of the students of ferti-
lization has focused on the events that take place in the egg as a
result of fertilization while interest in the spermatozoon has been
confined primarily to the molecular mechanisms leading to the acrosome reaction. The acrosome reaction is, however, but the final
event of a series of physiological and morphological changes that
the spermatozoon undergoes upon encountering the egg and which are
now spoken of as <u>sperm activation</u>. This term needs qualification.
Sperm activation is an expression commonly used to indicate <u>all</u> the
changes the spermatozoa undergo from the time of ejaculation, and
in fact irrespective of whether they will ever meet an egg, to
their actual encounter with the egg. In the present context this
expression will be used in a more restricted sense, namely limited
to those alterations that are triggered in the spermatozoon at the
time it meets the egg.

        From the side of the egg, the question may be asked, "At which
stage of gamete encounter is egg activation triggered?" I will
review the evidence that suggests that egg activation <u>per se</u> -
independently of whether development will ensue - does <u>not</u> require
actual fusion with the spermatozoon.

THE "ACTIVATION" OF THE SPERMATOZOON

Most of the work on "sperm activation" has been carried out on the sea urchin. As is well known, exposure of sea urchin spermatozoa to a solution of egg jelly immediately triggers all the events described as sperm activation including the acrosome reaction. Indeed, the jelly coat in solution is the most potent trigger of the acrosome reaction (9). The acrosome reaction is associated with ionic movements (32,33, and 35 for a review), increased intracellular pH (34) uncoupling of the mitochondrion which becomes spherical and with a decrease of O_2 consumption of the spermatozoa (4). The massive Ca-uptake that accompanies the acrosome reaction (32,33) may be responsible for the uncoupling of oxidative phosphorylation. Moreover, Verapamil, a drug that blocks Ca-uptake inhibits the jelly coat-triggered acrosome reaction as well as the decrease of O_2 consumption and the uncoupling of oxidative phosphorylation (4). Since, however, it is essentially impossible to remove the jelly coat from the sea urchin egg completely, it is very difficult to decide whether the actual physiological site of the acrosome reaction is the jelly coat or the underlying vitelline coat (see 27, for a discussion).

The Ascidian egg on the other hand appears to be a more favorable system than the sea urchin egg for the analysis of the sequence of the events taking place in the spermatozoon upon binding to the vitelline coat (the chorion) and prior to the acrosome reaction. In the Ascidians the vitelline coat is separated from the egg surface by a wide perivitelline space in which the test cells are floating (whose role in fertilization is unknown). The vitelline coat can be readily separated from the egg manually; also, procedures have been worked out for the en masse preparation of clean vitelline coats (29). The vitelline coat is the site of the species-specific sperm recognition and of the acrosome reaction (10). An important breakthrough in this work was the discovery (30) that the treatment of the eggs with cold glycerol results in the shedding of the follicle cells while the egg and the test cells cytolyze within the unbroken vitelline coat. The vitelline coat of these "ghost eggs" retains its ability to bind spermatozoa species-specifically while rejecting the spermatozoa of the same animal (Fig. 1). "Self sterility" is a striking peculiarity of the Ascidian eggs (28) and in particular of Ciona which is the species with which we have done most of our work. A noteworthy property of the "ghost eggs" is that the vitelline coat, though allowing sperm binding, seems to have lost the ability to trigger the acrosome reaction. This is in contrast to the condition in the living eggs deprived of their follicle cells (e.g. by shaking) in which some of the attached spermatozoa undergo the acrosome reaction (10,27). "Ghost eggs" thus appear to be an excellent system for investigating the events of sperm binding independently of the acrosome reaction. At the same time the possibility of obtaining relatively

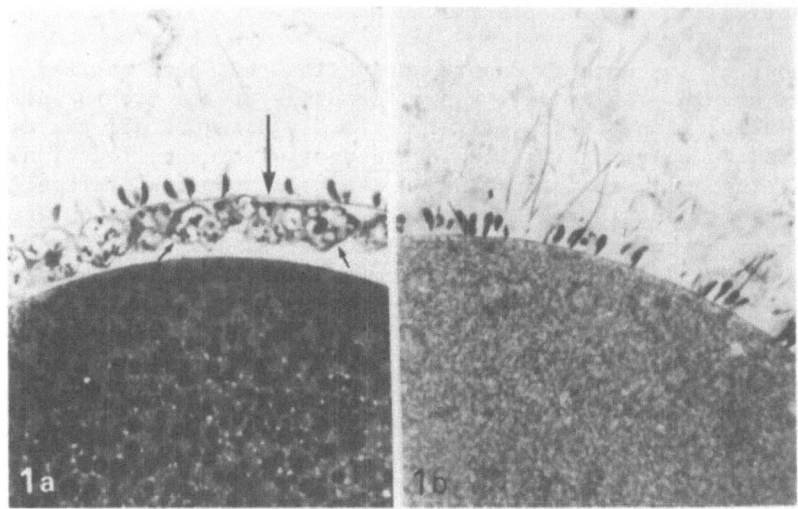

Figure 1. (a) Spermatozoa attached to the vitelline coat (arrow) of
an egg of Ciona after mechanical removal of the follicle
cells: small arrows: test cells x 1500.   (b) Spermatozoa
attached to a "ghost egg" x 1500.

clean preparations of vitelline coats offers an opportunity for the
isolation of the sperm receptors. The results so far obtained in
work in progress indicate the glycoprotein nature of the receptors
and that their function depends on the exposed fucosyl residues
(10,11,29,31).

At the morphological level, the most spectacular change of the
Ascidian spermatozoon upon binding to the vitelline coat of living
eggs is a rounding up and a tailward translocation of the mitochondrion which in the most extreme cases may even be cast off (19).
When binding to the ghost eggs, translocation of the mitochondrion
is in general not so pronounced.

Our microcalorimetric studies have shown that sperm binding to
the ghost eggs (13) is accompanied by the production of heat in
excess of that of spermatozoa in the absence of eggs (Fig. 2).
Also, the excess heat production is not accompanied by O_2 consumption.

Sea urchin spermatozoa lose high energy phosphate compounds
upon being shed in sea water (5). Ciona spermatozoa upon dilution
in sea water lose about 70% of their ATP with respect to their ATP
content in the dry semen within a few seconds; then a slight recovery of ATP content occurs over a period of about one hour. On the
other hand, when the spermatozoa attach to the ghost eggs there is
an immediate loss of about 90% of their ATP and no recovery phase

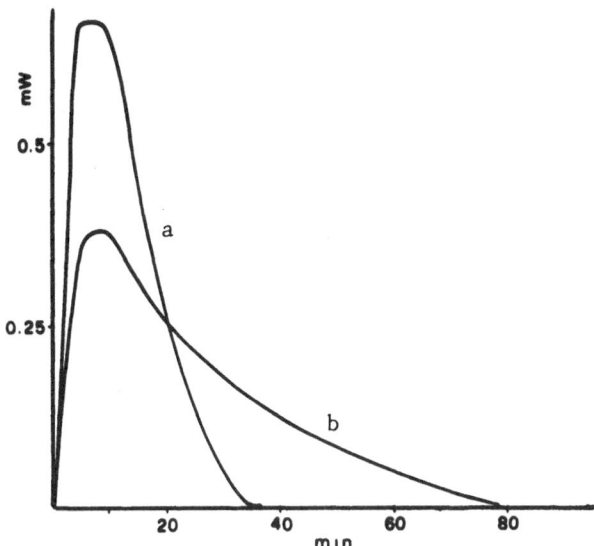

Figure 2. The thermograms show the excess heat, i.e., the heat
          evolved by <u>Ciona</u> spermatozoa when binding to ghost eggs <u>in</u>
          <u>excess</u> to that produced by the same amount of spermatozoa
          in the absence of eggs. The thermograms describe the time
          course of heat production (in m Watts, mW) when ghost eggs
          at two different concentrations, (5 x 10^3 (a) and 2 x 10^3
          (b)/ml) were challenged with the same amount of spermatozoa
          (5 x 10^7/ml). (Experiments by F. Rosati, V. Elia and G.
          Barone, with permission.)

ensues (Rosati & Tosti, personal communication). Whether a Ca-uptake
occurs in the <u>Ciona</u> spermatozoa upon binding to the ghost eggs is not
known; by analogy with the events in the sea urchin Ca-uptake may be
suggested to be the cause of the morphological changes that take
place in the mitochondrion. Our interpretation of the observations I
have just described is that when the spermatozoa are bound to the
vitelline coat, the oxidative phosphorylation is uncoupled whereby
most of the energy generated from ATP breakdown goes largely into
heat rather than to provide energy for motility; a small part of the
energy released may be used for the polymerization of actin that is
likely to be instrumental in the translocation of the mitochondrion
(20).

     Our observations together with those from other laboratories on
different materials lead us to a re-appraisal of the so-called sperm
activation. Indeed, upon interacting but prior to fusion with the
egg the spermatozoa appear to burn out their energy sources.

This is most clearly shown by the futile heat production associated with a massive ATP breakdown and uncoupling of the sperm
oxidative phosphorylation. Thus the so-called activation of the
spermatozoon is actually a switching off of its metabolic
machinery.

THE ACTIVATION OF THE EGG

The next question that I would like to address is whether the
activation of the egg requires fusion with the spermatozoon or
whether the processes leading to activation are started before
actual fusion takes place. F.R. Lillie (21) was the first to try
to answer this question in the egg of Nereis. Nereis eggs are a
favorable material for this kind of experiments as attachment of
the spermatozoon is followed by a prominent cortical reaction which
consists in the extrusion of the contents of the cortical vesicles
and in the formation of a thick jelly layer around the egg. However, the fertilizing spermatozoon does not enter the egg until 40
to 50 minutes later "in coincidence with the late anaphase of the
first maturation division." The extruded jelly material carries
away the spermatozoa that have not succeeeded in attaching to the
egg.

Lillie took advantage of the long delay between sperm attachment and its penetration into the egg by attempting to remove the
"activating" spermatozoon by submitting the egg to centrifugation.
In a number of cases he succeeded in detaching the spermatozoon
from the egg: these eggs were able to complete meiosis but failed
to cleave. From these observations Lillie concluded that for the
cortical reaction to take place "mere contact of the spermatozoon
with the membrane is apparently not sufficient; but actual attachment of at least a single spermatozoon is required." And he drew
the more general conclusion that "we would at least have to distinguish two stages in the fertilization action of the spermatozoon,
one before and the other after penetration."

Similar results were obtained by Goodrich (14) who succeeded
in removing the spermatozon attached to the oocyte surface with a
microneedle at a much earlier time after attachment than in
Lillie's experiments, i.e., even after only 3 minutes. The same
experimental approach was used by Hiramoto (16) with the sea
urchin egg. He was able to show that removal of the spermatozoon
immediately after its attachment to the egg did not prevent elevation of the fertilization membrane and secretion of the hyaline
layer. A faint aster also developed around the egg pronucleus but
development of the egg did not proceed any farther. Occasionally,
a few monastral cycles followed as in parthenogenetically activated
eggs. Removal of the spermatozoon after it had penetrated the egg
but before its fusion with the egg pronucleus had the same result.

These results are confirmed by experiments in which fertilization was accomplished in the presence of Cytochalasin B which prevents incorporation of the spermatozoon that has attached to the egg; and yet neither cortical reaction nor egg activation were inhibited (2,15, 25).

Also some recent experiments of Lynn and Chambers (26) have shown that in sea urchin eggs which were voltage clamped at a negative potential of −80 mV the spermatozoa attached to, but did not penetrate the egg. Nevertheless a cortical reaction occurred and in fact the elevating fertilization membrane detached and carried away the spermatozoon from the egg. It thus seems that the mere fusion of the sperm and egg plasma membranes is sufficient to elicit a propagated response that triggers activation.

Elucidating the nature and the mechanism(s) of the sperm-triggered membrane changes would be of the utmost importance for understanding the molecular events underlying egg activation.

The first possibility that comes to mind is that Ca^{2+} is the trigger of the membrane changes. This idea has been championed primarily by Schackmann et al. (32), Shapiro & Eddy (35), and Jaffe (17,18) and defined as the hypothesis of the "calcium bomb." In particular, Shapiro & Eddy (35) have proposed that "a Ca^{2+} - calmodulin complex of sperm is the active component that initiates the cortical reaction of the egg." According to the hypothesis, the Ca^{2+} - calmodulin complex acts as a mediator of the Ca^{2+} release from the endoplasmic reticulum which in its turn binds to the calmodulin of the egg thus propagating the reaction. However, the experiments of Chambers and Hinkley (3) show that localized applications of the Ca-ionophore A23187 to sea urchin eggs result in a non-propagated response, i.e., membrane elevation and cortical granule breakdown are restricted to the area acted upon by the ionophore. Thus we are confronted with the recurrent idea of the spermatozoon injecting an "activating factor" in the egg, an idea that goes back to J. Loeb (24) who suggested that a lysin carried by the spermatozoon breaks down a calcium-lipid complex in the egg cortex thus releasing a catalyst which in its turn sets in motion the metabolic machinery of the egg. This would find support in the experiments of Sugiyama (36,37) who showed that exocytosis of cortical granules is not the trigger of a propagated exocytosis and in fact propagation of the "activation wave" does not appear to require the cortical granules (38). With some difference in wording, F.R. Lillie (23) expressed a similar idea when he thought of "fertilizin" as "the effective agent which is transformed from an inactive to an active state by some substance in the spermatozoon." We do not seem to have progressed much along the way to understanding the molecular mechanisms of the activation of the egg!

In this connection it should be mentioned that sperm attachment to the sea urchin egg is marked by a step depolarization which
is accompanied by a transient increase in the voltage noise (7).
This was interpreted as indicating the activation or the insertion
in the membrane of high conductance (~ 33pS) unspecific channels.
The same has been observed in the Ascidian egg by a patch clamp
technique; the conductance of the channels in this egg is, however,
much higher (~ 400 pS, 8).

By the single channel recording technique, Dale and De Felice
(8) have further observed sperm induced channel currents from
patches of membrane of the Ascidian oocyte at some distance from
the site of sperm attachment. The investigators suggest that a
chemical factor released by the spermatozoon may diffuse through
the oocyte, thus activating channels from inside. This interpretation is supported by the activation of sea urchin eggs in response
to the injection of minute amounts of fractions prepared from spermatozoa: the active component is possibly a low molecular weight
organic molecule (12). These lines offer promise to solving the
problem of the "sparking events" of the activation of the egg.

CONCLUDING REMARKS

In this presentation I have attempted to review the evolution
of our ideas on the interactions between sperm and egg at the time
of their encounter but prior to their fusion.

As long as the gametes are in the gonad, their metabolism is
repressed. Upon release from the gonad, the egg starts aging and
this process manifests itself as a progressive loss of its fertilizability and as a general decrease of its metabolism. Fusion with
the spermatozoon triggers the metabolic derepression which turns
the dormant egg into a most metabolically active system. On the
contrary, the spermatozoon upon release into the outer medium
undergoes a frantic metabolic activation related to the large
energy requirements to sustain its motility in anticipation to its
meeting with the egg. This is, however, a short-lived process as
upon binding to the egg the sperm metabolism is turned off or at
least is drastically reduced. I submit that this may be a prerequisite for sperm-egg fusion. However, this does not mean that from
that point on the spermatozoon behaves passively in the sense that
it simply "waits" to be incorporated into the egg. Indeed, there
is evidence that the chain of reactions leading to the activation
of the egg is initiated before the spermatozoon has actually
entered the egg and in fact upon its attachment to the egg plasma
membrane. The key question that still awaits an answer is whether
the reaction is initiated by Ca^{2+}, either conveyed or locally
released by the spermatozoon, or by the injection of some sperm
factor which in its turn triggers the Ca^{2+} release.

ACKNOWLEDGMENTS

    The work of the author and of his colleagues described in this
report has been supported by Grants from the Consiglio Nazionale
delle Ricerche, Progetti Finalizzati "Biologia della Riproduzione"
& "Chimica Fine e Secondaria", and from the Ministry of Education.
The author is indebted to his colleagues, Drs. G. Casazza, B. Dale,
L. Nelson, M.R. Pinto, and R. Rosati for discussions and for their
comments to the manuscript.

REFERENCES

1. Boveri, Th. 1902. Das Problem der Befruchtung. G. Fischer
   Verl., Jena.
2. Byrd, W. and Perry, G. 1980. Cytochalasin B blocks sperm
   incorporation but allows activation of the sea urchin egg.
   Exp. Cell Res. 126:333–342.
3. Chambers, E.L. and Hinkley, R.E. 1979. Non-propagated cortical
   reactions induced by the divalent calcium ionophore A 23187
   in eggs of the sea urchin Lytechinus variegatus. Exp. Cell
   Res. 124:441–446.
4. Christen, R., Schackmann, R.W., and Shapiro, B.M. 1983.
   Interactions between sperm and sea urchin egg jelly. Dev.
   Biol., 98:1–14.
5. Christen, R., Schackmann, R.W., Dahlquist, F.W., and Shapiro,
   B.M. 1983. ^{31}P-NMR analysis of sea urchin sperm activation.
   Reversible formation of high energy phosphate compounds by
   changes in intracellular pH. Exp. Cell Res. 149:289–294.
6. Colwin, L.H. and Colwin, A.L. 1963. Role of the gamete
   membranes in fertilization of Saccoglossus kowalewski
   (Enteropneusta) II. Zygote formation by gamete membrane
   fusion. J. Cell Biol. 19:501–518.
7. Dale, B., De Felice, J. L., and Taglietti, V. 1978. Membrane
   noise and conductance increase during single
   spermatozoon–egg interactions. Nature 275:217–219.
8. Dale, B. and De Felice, J. L. 1984. Sperm–activated
   channels in Ascidian oocytes. Dev. Biol. 101:235–239.
9. Dan, J.C. 1952. Studies on the acrosome. 1. Reaction to egg
   water and other stimuli. Biol. Bull. 103:54–66.
10. De Santis, R., Jamunno, G. and Rosati, F. 1980. A study of the
    chorion and of the follicle cells in relation to sperm
    egg interaction in the Ascidian, Ciona intestinalis.
    Dev. Biol. 74:490–499.
11. De Santis, R., Pinto, M.R., Cotelli, F., Rosati, F., Monroy,
    A., and D'Alessio, G. 1983. A fucosyl glycoprotein
    component with sperm-receptor and sperm-activating
    activities from the vitelline coat of Ciona intestinalis
    eggs. Exp. Cell Res. 148:508–513.

12. Ehrenstein, G., Dale, B. and De Felice, J.L. 1984. A soluble
    fraction of sperm triggers cortical granule exocytosis in
    sea urchin eggs. Bioph. J. 45:23a.
13. Elia, V., Rosati, F., Barone, G., Monroy, A., and Liquori, A.M.
    1983. A thermodynamic study of sperm-egg interaction. The
    EMBO J. 2:2053-2058.
14. Goodrich, H.B. 1920. Rapidity of activation in the fertiliza-
    tion of Nereis. Biol. Bull. 38:106-201.
15. Gould-Somero, M., Holland, L., and Paul, M. 1977. Cytochalasin
    B inhibits sperm penetration into eggs of Urechis caupo
    (Echiura). Dev. Biol. 58:11-22.
16. Hiramoto, Y. 1962. Microinjection of the live spermatozoa into
    sea urchin eggs. Exp. Cell Res. 27:416-426.
17. Jaffe, L.F. 1983. Sources of calcium in egg activation: a
    review and hypothesis. Dev. Biol. 99:265-276.
18. Jaffe, L.F. 1980. Calcium explosions as triggers of develop-
    ment, in: "Growth Regulation by Ion Fluxes." Ann. N.Y. Ac.
    Sci. 339:86-101.
19. Lambert, C.C. and Epel, D. 1979. Calcium-mediated mitochondrial
    movement in Ascidian sperm during fertilization. Dev. Biol.
    69:296-304.
20. Lambert, C.C. and Lambert, G. 1983. Mitochondrial movement
    during the Ascidian sperm reaction. Gamete Res. 8:295-
    307.
21. Lillie, F.R. 1911. Studies of fertilization in Nereis. 1. The
    cortical changes in the egg. 2. Partial fertilization. J.
    Morphol. 22:3610393.
22. Lillie, F.R. 1914. Studies on fertilization. 6. The mechanism
    of fertilization in Arbacia. J. Exp. Zool. 16:523-590.
23. Lillie, F.R. 1919. Problems of fertilization, Univ. of Chicago
    Press.
24. Loeb, J. 1913. Artificial parthenogenesis and fertilization,
    Chicago Univ. Press.
25. Longo, F.J. 1978. Effects of Cytochalasin B on sperm-egg inter-
    actions. Dev. Biol. 67:249-265.
26. Lynn, J.W. and Chambers, E. L. 1983. Voltage clamp studies of
    fertilization in sea urchin eggs. 1. Effect of clamped mem-
    brane potential in sperm entry, activation and development.
    Dev. Biol. 102:98-109.
27. Monroy, A. and Rosati, F. 1983. A comparative analysis of
    sperm-egg interactions. Gamete Res. 7:85-102.
28. Morgan, T.H. 1923. Removal of the block to self-fertilization
    in the Ascidian Ciona. Proc. Natl. Ac. Sci. USA. 9:170-
    171.
29. Pinto, M.R., De Santis, R., D'Alessio, G., and Rosati, F. 1981.
    Studies on fertilization in the Ascidians Fucosyl sites on
    the vitelline coat of Ciona intestinalis. Exp. Cell Res.
    132:289-295.
30. Rosati, F. and De Santis, R. 1978. Studies on fertilization in
    the Ascidians. 1. Self-sterility and specific recognition

between gametes of Ciona intestinalis. Exp. Cell Res.
112:111-119.

31. Rosati, F. and De Santis, R. 1980. The role of the surface
    carbohydrates in sperm-egg interaction in Ciona intesti-
    nalis. Nature 283:762-764.

32. Schackmann, R.W., Eddy, E.M. and Shapiro, B.M. 1978. The acro-
    some reaction of Strongylocentrotus purpuratus sperm. Ion
    requirement and movements. Dev. Biol. 65:483-495.

33. Schackmann, R.W. and Shapiro, B.M. 1981. A partial sequence of
    ionic changes associated with the acrosome reaction of
    Strongylocentrotus purpuratus. Dev. Biol. 81:145-154.

34. Schackmann, R.W., Christen, R., and Shapiro, B.M. 1981. Mem-
    brane potential depolarization and increased intracellular
    pH accompany the acrosome reaction of sea urchin sperm.
    Proc. Natl. Acad. Sci. 78:6066-6070.

35. Shapiro, B.M. and Eddy, E.M. 1980. When sperm meets egg: bio-
    chemical mechanisms of gamete interactions. Intern. Rev.
    Cytol. 66:257-302.

36. Sugiyama, M. 1953. Physiological analysis of the cortical res-
    ponse of the sea urchin egg to stimulating reagents. 1. Res-
    ponse to sodium choleinate and wasp venom. Biol. Bull.
    104:210-215.

37. Sugiyama, M. 1953. Physiological analysis of the cortical
    response of the sea urchin egg to stimulating reagents. 2.
    The propagating or non-propagating nature of the cortical
    changes induced by various reagents. Biol. Bull. 104:216-
    223.

38. Uehara, T. and Katou, K. 1972. Changes of the membrane poten-
    tial at the time of fertilization in the sea urchin egg with
    special reference to the fertilization wave. Dev. Growth &
    Differ. 14:175-184.

EFFECTS OF ANTI-ZONA PELLUCIDA MONOCLONAL ANTIBODIES

ON FERTILIZATION AND EARLY DEVELOPMENT

Jurrien Dean[1] and Iain J. East[2]

Laboratory of Cellular and Developmental Biology,[1]
Laboratory of Chemical Biology,[2]
NIADDK, NIH, Bethesda, MD 20205

SUMMARY

The murine zona pellucida surrounds the growing oocyte, ovulated egg and dividing embryo. It is comprised of three sulfated glycoproteins designated ZP-1, ZP-2, and ZP-3 which have molecular weights of 185,000, 140,000 and 83,000 daltons respectively. We have isolated a series of cell lines that produce monoclonal antibodies to ZP-2 and to ZP-3. These immunogical probes indicate that the extracellular matrix proteins of the zona pellucida are found uniquely in the ovary where they surround maturing oocytes. We have demonstrated that passive immunization with antibodies specific either for ZP-2 or ZP-3 inhibit in vivo and in vitro fertilization. This effect is observed with ng/ml quantities of antibody. It appears that the antibodies do not preclude sperm binding but rather prevent sperm penetration of the zona by steric hinderance. Although long-term, the contraceptive effect is fully reversible and this reversibility is associated with loss of antibody from the zona pellucida surrounding intra-ovarian oocytes. Antibodies to ZP-2 or ZP-3 had no other adverse effect on in vivo or in vitro preimplantation development.

INTRODUCTION

The maternal genome appears to have an important influence on the growing embryo during mammalian pre-implanation development (18). Many gene products are synthesized during this period of development; few are stage specific and still fewer have known biological functions (6-8,28). In contrast to this situation, the three sulfated glycoproteins which make up the zona pellucida derive from developmentally regulated genes and serve known functions during fertiliza-

tion and early embryogenesis. These proteins are designated ZP-1,
ZP-2 and ZP-3 and have molecular weights of 185,000, 140,000 and
83,000 respectively (3,23). The zona pellucida genes are coordin-
ately expressed during oogenesis and their gene products are synthesized only in the oocyte. Thus, they are not only tissue specific
but germline specific. The zona proteins are secreted to form a
glycocalyx which surrounds the oocyte and this extracellular matrix
has known biological functions (19,30). ZP-3, which makes up 19% of
the mass of the zona pellucida, is known to trigger the sperm acrosome reaction and has been reported to have sperm binding activity
(4) associated with its carbohydrate side chains (15). ZP-2 makes up
more than 50 percent of the zona and is biochemically modified in
conjunction with the post-fertilization block to polyspermy (5). No
function has yet been ascribed to ZP-1.

Until recently, little was known about the tissue distribution
of the zona proteins or about the localization of individual proteins
within the zona pellucida. Previous studies had shown that parenteral administration of anti-ovarian or anti-zona can inhibit fertilization (1,2,19,22,24,27). Polyclonal antibodies injected after fertilization will bind to the zona pellucida but do not affect development
up to the formation of the blastocyst (26). Although an anti-zona
antiserum has been reported to inhibit the hatching reaction (25),
more recent data suggests that this effect may be due to the presence
of antibodies to non-zona proteins (26). In these earlier studies it
was not known to which of the three zona proteins these antisera were
specific. To address these and other questions, we have developed a
series of cell lines which produce monoclonal antibodies to ZP-2 and
to ZP-3 (10,11). These immunological probes have been used to demonstrate that the zona proteins are found uniquely in the ovary where
they form a diffuse coating on the surface of the zona (10). Using
monoclonal antibodies specific to the major protein of the zona pellucida or to the putative sperm receptor, we have ascertained their
effects on in vivo and in vitro fertilization and early development
(11-13).

METHODS

Monoclonal Antibodies

Ovaries were obtained from DBA/2 or NIH random-bred mice and
homogenized with a flattened glass rod. After seiving through a
series of screens, the eggs and torn zonae were freeze-thawed in the
presence of 0.2% NP-40 and centrifuged through 50% sucrose as previously described (10). For immunization, zonae were diluted in 3.0 ml
of modified Brinster's media (14) and collected by micropipette.
Otherwise the pellet was washed five times with PBS and solubilized
by heating at 60°C for 20 min. The solubilized zonae were labeled
with [^{125}I]-Bolton-Hunter reagent as described (10).

Osborn-Mendel male rats were each immunized with a total of 80 μg of particulate zonae. Three days after the last 50 μg was given by intravenous injection, the rat was sacrificed and the spleenocytes fused to the cells of SP2/0 mouse myeloma line. Cell lines secreting anti-zona antibody were selected in HAT media and detected by radioimmunoassay (10,11). Positive cell lines were then cloned twice by limiting dilution. After characterization by immuno-precipitation and analysis by SDS-PAGE (10), cell lines of interest were propagated as ascites tumors in NIH nu/nu mice. The IgG fractions were purified by Sephacryl S-200 column chromatography and characterized (10,11).

Binding studies with ovulated eggs were performed with the IgG fractions labeled with [^{125}I]-Bolton-Hunter reagent (6.0 x 10^8 cpm/ μg). Three week old female mice were injected intra-peritoneally with 2.7 μCi of iodinated antibody to determine the tissue distribution of the zona proteins (12). Immunofluorescence of eggs, two-cell embryos and ovarian sections were obtained with either rhodamine- or fluorescein-conjugated second antibody (10,13).

Passive Immunization of Mice

Either NIH random-bred or B6C3 F_1 hybrid female mice were injected parenterally with 250 μg of monoclonal antibody in 0.5 ml of saline or 0.5 ml of saline alone at the time of pregnant mare serum gonadotropin (PMSG) administration and then induced to ovulate with human chorionic gonadotrophin (hCG) (11,13). The mice were mated overnight and sacrificed either 40 h after hCG injection to recover 2-cell embryos or maintained until the 19th day of pregnancy. Other mice, similarly immunized were mated continuously until obviously pregnant at which time the males were removed (11,13).

In Vitro Fertilization and Development

Eggs in cumulus mass were recovered from superovulated B6C3 F_1 hybrid mice and incubated with 1.0-1.5 x 10^6/ml epididymal sperm from ZBW F_1 males. Fertilization was determined at 5-6 h by the presence of two pronuclei and two polar bodies (11). To titrate the effect of monoclonal antibodies on fertilization, eggs in cumulus mass were incubated for 15 min with increasing concentrations of antibody (0-90 μg/ml) and washed three times in T6 medium before the addition of sperm.

Cumulus-free eggs were obtained by treatment with hyaluronidase (125 units/ml). Half of these cumulus-free eggs were subsequently exposed to acid Tyrode's media (pH 2.5) for 15 sec to remove their zonae. Oocytes from each group were incubated for 15 min with one of the following: a) T6 medium; b) control ascites fluid (diluted 1:10); c) monoclonal antibody IE-3 (2 μg/ml); or d) monoclonal antibody IE-10 (200 μg/ml). Oocytes were then washed and fertilized with capacitated sperm (11).

Two-cell embryos were collected from super-ovulated mice and incubated in vitro in medium No. 16 (29) for 5 days at 37°C in 5% CO_2-air in the presence of one of the following: a) T6 medium; b) control ascites fluid; c) monoclonal antibody IE-3 (2 μg/ml); or d) monoclonal antibody IE-10 (200 μg/ml) (11).

RESULTS

Production of Monoclonal Antibodies

Male rats were immunized with 30 μg of particulate zonae in divided doses and then boosted intravenously with 50 μg of zonae three days prior to being sacrificed. Their spleen cells were fused with cells of the SP2/0 mouse myeloma line and screened for the production of anti-zona antibodies by radioimmunoassay. Five fusion experiments were performed. The first three resulted in no cell lines producing anti-zona antibodies, a single positive cell line (IE-1) was found in the fourth fusion but more than fifty cell lines secreting anti-zona antibodies were isolated from the fifth fusion.

Positive cell lines were sub-cloned twice by limiting dilution and propagated as ascites tumors in NIH nu/nu mice. Their ability to immunoprecipitate [^{125}I]-zona protein was determined by SDS-PAGE (see Fig. 1). Of twenty-three clones assayed, sixteen secreted antibodies that bound to ZP-2, five made antibodies that bound to both ZP-2 and ZP-3 and two had antibodies that immunoprecipitated only ZP-3. No cell lines produced antibodies capable of binding to ZP-1; nor did the rat polyclonal antisera immunoprecipitate ZP-1. As a criterion for homogeneity the IgG fraction from the ascites fluid of each cell line was chromatograhpically purified and shown to have a unique pattern by O'Farrell two-dimensional electrophoresis. Furthermore, each cell line had a single light chain on the 2-D gel and a single IgG heavy chain isotype (10,11).

Binding studies utilizing super-ovulated eggs and [^{125}I]-monoclonal antibodies were used to determine dissociation constants for each of the monoclonal antibodies; these ranged from 3×10^{-9} to 6×10^{-10} mol 1^{-1} (see Table 1). Scatchard analysis of the binding data indicate that there are 1.3×10^8 binding sites for anti-ZP-2 antibodies on the surface of the zona pellucida which represents approximately 2% of the ZP-2 molecules in the zona (10). The arbitrarily defined epitopes shown in Table 1 were obtained by competitive binding studies (10). Thus, there are at least three antigenic determinants on ZP-2 and two on ZP-3 that are accessible on the surface of the zona pellucida.

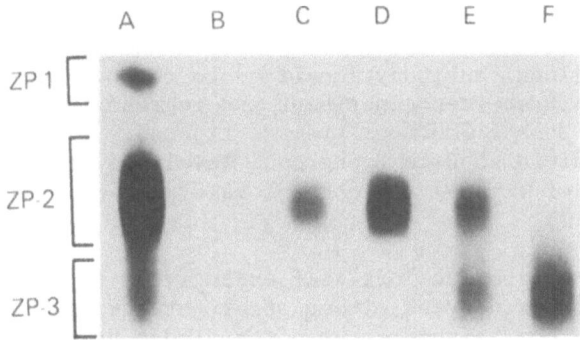

Figure 1. Autoradiograph of SDS-PAGE of Immunoprecipitates. (A) Isolated [^{125}I]-zona pellucida (3 x 10^4 cpm). The lanes following show immunoprecipitation with: (B) control ascites fluid; (C) monoclonal antibody IE-1; (D) IE-3; (E) IE-4; and (F) IE-10.

Distribution of Zona Proteins

The zona pellucida is not unlike other extracellular matrixes. It is comprised of three sulfated glycoproteins and plays an impor- tant role in cell-cell interactions. We have demonstrated that the zona proteins are immunologically distinct from laminin, fibronectin, collagen IV and entactin (10). To determine the tissue distribution of the zona proteins we parenterally administered [^{125}I]-anti-ZP-2 monoclonal antibody into fifteen, three-week old female mice. Groups

Table 1.

Hybridoma Cell Line	Isotype	Protein A Binding	K_{diss} mol l^{-1}	Target Antigen	Epitope
IE-I	IgG$_{2a}$	+/-	6 x 10^{-10}	ZP-2	A
IE-3	IgG$_{2a}$	-	2 x 10^{-10}	ZP-2	B
IE-4	IgG$_1$	+	3 x 10^{-9}	ZP-2/3	C
IE-10	IgG$_{2a}$	-	n.d.	ZP-3	D

of three mice were sacrificed on days 4, 7, 10, 14 and 22, and the
tissue distribution of radioactivity determined (Fig. 2). The amount
of [^{125}I]-monoclonal antibody localized to the ovaries remained high
throughout the three-week experiment and represented 2.6% of the
total injected dose (1300% per gram of tissue). Radioactivity in
other tissues diminished to background levels (e.g. the liver had
0.11% per gram of tissue) and control male mice had no such tissue
localization (12).

Anti-zona antibodies localized exclusively to the zona pellucida
surrounding growing oocytes either after parenteral administration of
unlabeled anti-zona antibodies or after direct staining of ovarian
tissue sections with antibodies (see Fig. 9B). No localization was

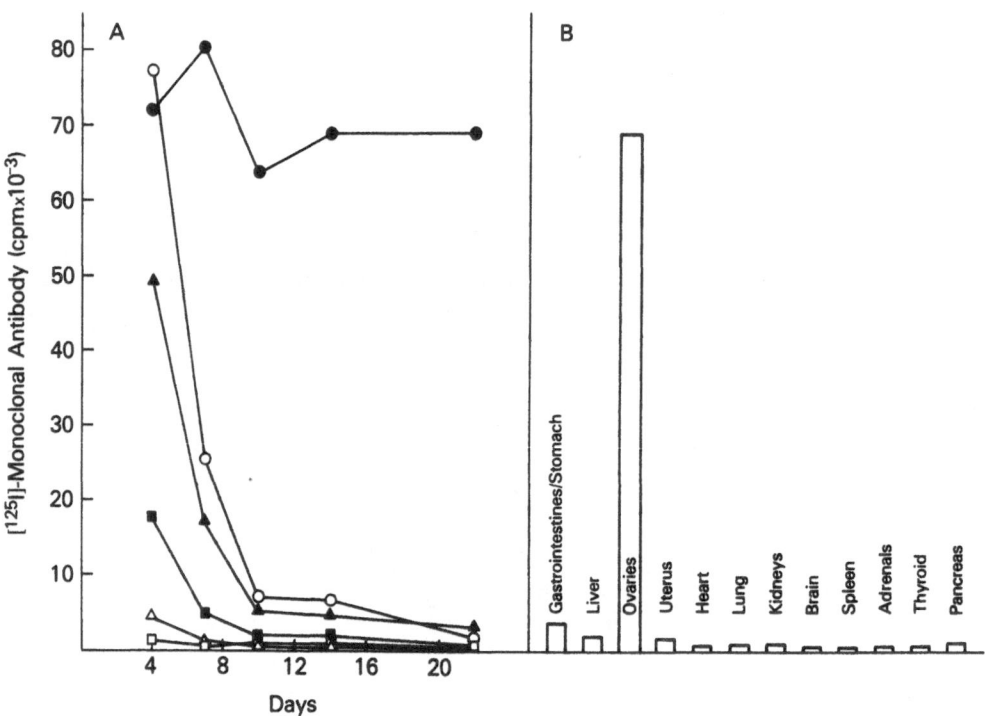

Figure 2. Tissue Distribution of Radioactivity in Mice Immunized with
           [^{125}I]-Monoclonal Antibody. (A) Fifteen mice were immun-
           ized with 2.7 Ci (6.0 x 10^8 cpm/μg) of [^{125}I]-anti-ZP-2
           antibody (IE-3) and three were sacrificed at each time
           point: ● —— ● ovary; o --- o liver; Δ --- Δ gastrointes-
           tinal tract; ■ --- ■ kidney; Δ --- Δ heart; and ☐ --- ☐
           - uterus. (B) Distribution of radioactivity in mice after
           extensive dissection on day 22.

observed in the basement membrane surrounding follicles (10). The distribution within the zona pellucida of individual components was determined by indirect immunofluorescence of ovulated eggs and 2-cell embryos. Monoclonal antibodies to ZP-2 (Fig. 3) or to ZP-3 (data not shown) completely stain the zona and demonstrate a fenestrated staining pattern not unlike that reported with scanning electron microscopy (20,21). No differences in the staining patterns were seen between the zonae surrounding eggs and two-cell embryos.

Passive Immnization of Mice

To determine if the diffuse coating of anti-ZP-2 or anti-ZP-3 monoclonal antibodies perturbed fertilization and early development, female mice were passively immunized with 250 µg of anti-zona monoclonal antibody (Fig. 4). After nineteen days, 50% of control mice were pregnant and each had an average of 20 pups which weighed 0.9 gm each. Animals treated with antibody to ZP-2 and sacrificed on day 19 had no pregnancies. However, if the monoclonal antibody was administered after fertilization at the 2-cell embryo stage or after implantation, there was no difference in pregnancy rate or litter characteristics from control animals (13).

This study was repeated with four groups of animals each of which had been immunized with monoclonal antibodies directed against one of the following: One of two different epitopes on ZP-2 (monoclones IE-1 and IE-3), one epitope present both on ZP-2 and ZP-3 (monoclone IE-4), or one epitope on ZP-3 (monoclone IE-10). Females with post-coital vaginal plugs were sacrificed and the number of

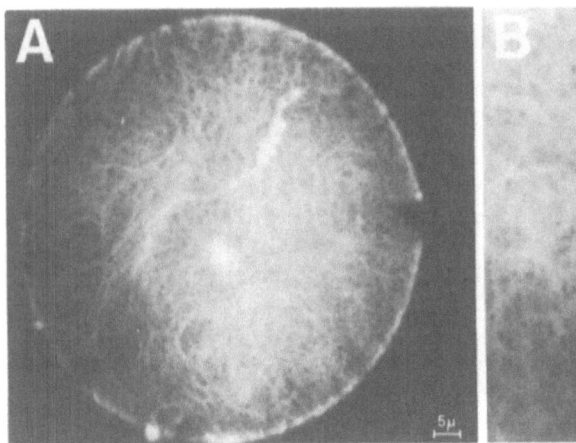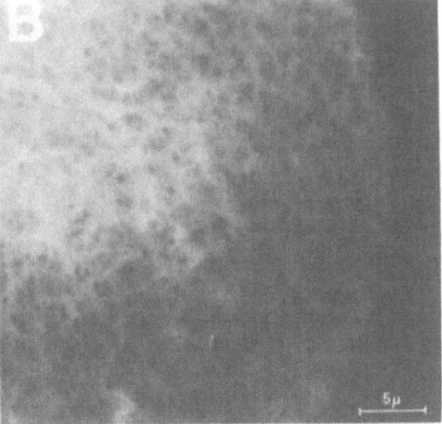

Figure 3. Indirect Immunofluorescence of Zona Pellucida. (A) Ovulated egg stained with monoclonal antibody IE-3 (specific for ZP-2) and detected with fluorescein-conjugated sheep anti-rat IgG. (B) Higher magnification of (A).

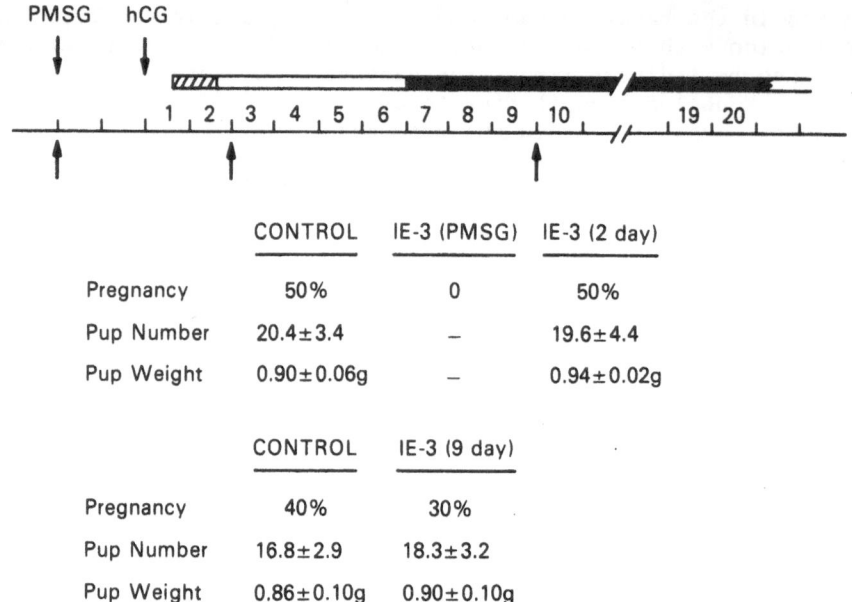

Figure 4. Passive Immunization with Anti-Zona Antibodies. Bar repre-
     sents mouse development. Hatched area is development to
     2-cell embryos, clear areas is pre-implantation development
     and darkened area is post-implantation development. Admin-
     istration of PMSG and hCG are indicated by arrows above the
     bar. Arrows (↑) below the line indicate administration of
     anti-ZP-2 monoclonal antibody (IE-3) or saline. Days are
     determined from time of hCG injection.

intra-oviductal 2-cell embryos and eggs was determined (Fig. 5).
Control animals injected with pre-immune serum (or saline) had 85%
two-cell embryos. Mice treated with any one of the four monoclonal
antibodies had fewer than 5% two-cell embryos. This block to the
formation of two-cell embryos was associated with the diffuse coating
of the ovulated eggs by monoclonal antibody as demonstrated by indirect immunofluorescence (Fig. 5). Two-cell embryos from control
animals had no such coating. The total number of eggs plus 2-cell
embryos from the treated and the untreated animals was the same (11,
13).

In Vitro Fertilization

To examine further the molecular mechanism by which two cell
embryo formation was blocked, we examined the effect of the monoclonal antibodies on in vitro fertilization (Fig. 6). After exposure

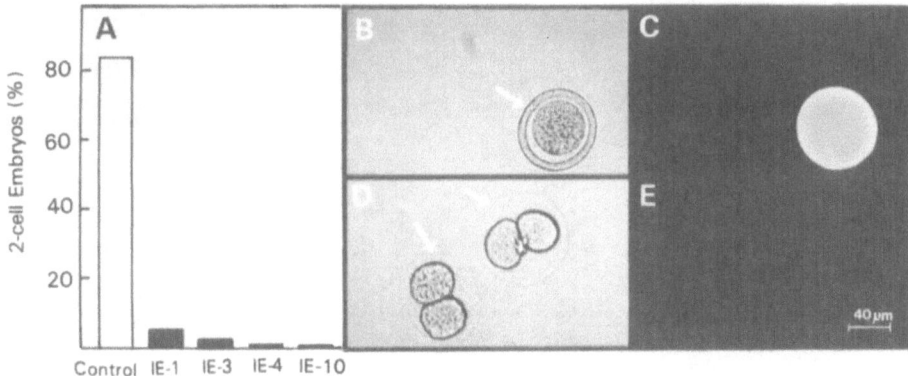

Figure 5. Production of 2-cell Embryos after Administration of Antibody. (A) The percentage of 2-cell embryos formed after
passive immunization with preimmune serum or monoclonal
antibody. (B) Light microscopy of egg from mouse after
passive immunization with IE-3 (anti-ZP-2) and stained with
FITC-conjugated sheep anti-rat Ig antibodies. Arrow indicates outer diameter of zona pellucida. (C) Fluorescence
microscopy of egg from (B). (D) Light microscopy of 2-cell
embryo from mouse treated with preimmune rat serum and
stained with FITC-conjugated sheep anti-rat Ig antibodies.
Arrow indicates outer diameter of translucent zona pellucida. (E) Fluorescence microscopy of embryo from (D).

to increasing concentrations of monoclonal antibodies to ZP-2 or ZP-
3, ovulated eggs in cumulus mass were incubated with epididymal sperm
and fertilization was determined by the presence of two pronuclei and
two polar bodies. Each data point represents 33 ± 12 eggs. The contraceptive effect was dose dependent and 50% inhibition of fertilization was obtained with 20 ng/ml of anti-ZP-2 or 185 ng/ml of anti-
ZP-3 monoclonal antibody.

This contraceptive effect was clearly dependent on the presence
of the zona. In vitro fertilization was effectively blocked with
antibodies specific for ZP-2 or for ZP-3 when tested with zona-intact
eggs (fewer than 10% eggs fertilized). However, even with concentrations of antibody 2.7-5.0 fold in excess of that amount needed to
totally block fertilization, zona-free eggs were 100% fertilized
(Fig. 7). Although effective in blocking fertilization, the monoclonal antibodies did not dramatically alter sperm binding (more than
90% of the eggs bound greater than 100 sperm), even after attempts to
remove loosely bound sperm by pipetting the eggs up and down three
times in a 100 µm bore pipette. In contrast, after similar treatment, fewer than 15 sperm were found associated with the zona-free
eggs. The presence of the antibody did not affect sperm morphology
or motility and sperm exposed to antibody and then washed in media
were fully capable of in vitro fertilization (11).

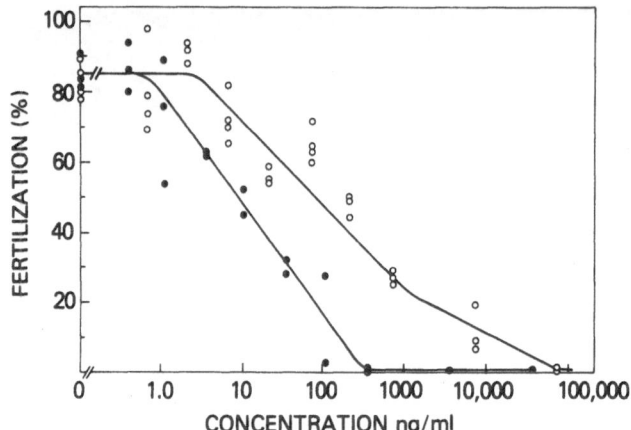

Figure 6. In Vitro Fertilization. The percent of superovulated eggs
from B6C3 F_1 hybrid females fertilized in vitro with sperm
from ZBW F_1 males in the presence of increasing concentrations of ● --- ● monoclonal antibody IE-3 (specific for
ZP-2) or o --- o, monoclonal antibody IE-10 (specific for
ZP-3).

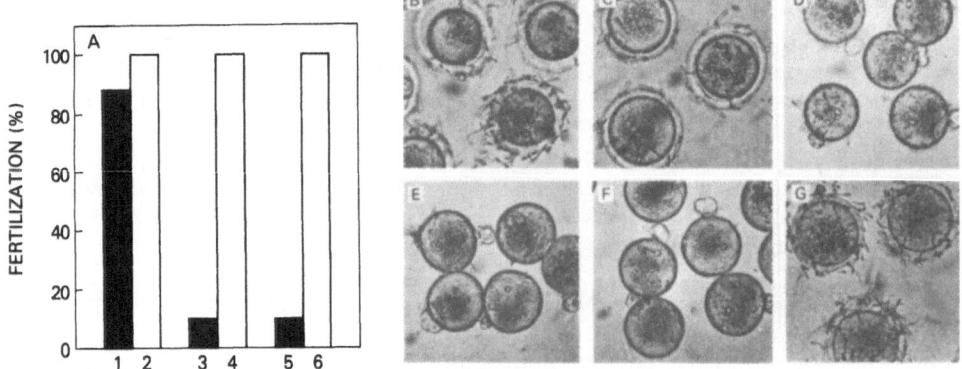

Figure 7. Fertilization of Zona-intact and Zona-free Eggs. (A) The
percent of 1-cell zygotes with two pronuclei and two polar
bodies formed after in vitro fertilization of zona-intact
(1,3,5) and zona-free (2,4,6) superovulated eggs treated
with: control ascites fluid (1,2); monoclonal antibody
IE-3 (3,4) or monoclonal antibody IE-10 (5,6). (B-G)
Photomicrographs of eggs/1-cell zygotes after in vitro
fertilization of zona-intact (B,C,D) or zona-free (E,F,G)
eggs and treatment with ascites fluid (B,E); IE-3 (C,F); or
IE-10 (D,G).

Reversibility of Contraception

The contraceptive effect induced by monoclonal antibodies to the zona pellucida was fully reversible. Female mice, parenterally injected either with saline or 250 μg of monoclonal antibody to ZP-2 were continuously mated and their cumulative litters tabulated (Fig. 8). Fifty percent of the control animals had given birth to litters by 23 days. In contrast, fifty percent of the antibody-treated animals did not give birth until 80 days, although all of the treated mice eventually gave birth. Nine of the treated mice were again mated and 50% of their litters were born by 29 days (13).

The reversibility of contraception was associated with a loss of monoclonal antibodies from growing intra-ovarian oocytes. Frozen sections of ovaries were obtained from mice three days after immunization with anti-ZP-2 monoclonal antibody. In each case, the monoclonal antibody could be localized to the zonae pellucidae surrounding growing oocytes (Fig. 9B). Several mice treated with anti-zona monoclonal antibody were sacrificed after they eventually gave birth (13). Frozen sections of their ovaries were histologically similar to control animals injected with saline and showed no immunofluorescence of their zonae (Fig. 9D).

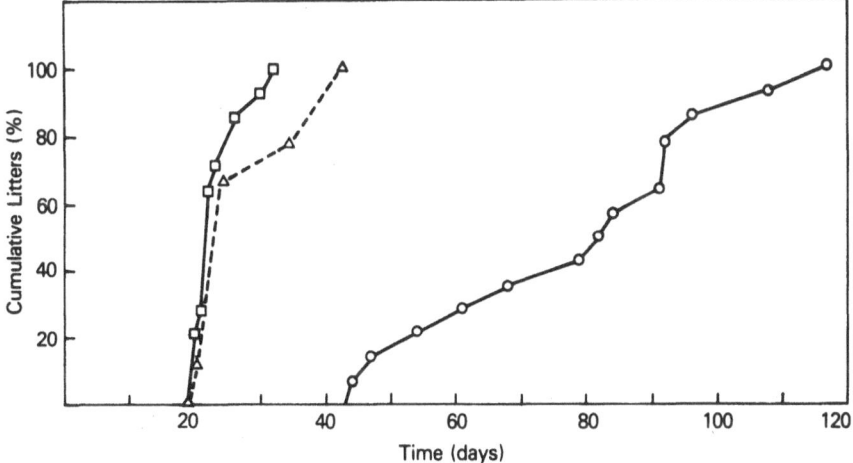

Figure 8. Reversibility of Block to Fertility. Cumulative litters (%) over time of groups of 14 mice mated 2 days after administration of saline (□) or monoclonal antibody IE-3 (0). Nine mice from the antibody-treated group remated after the resumption of fertility (△). Days are numbered from time that males were caged with females.

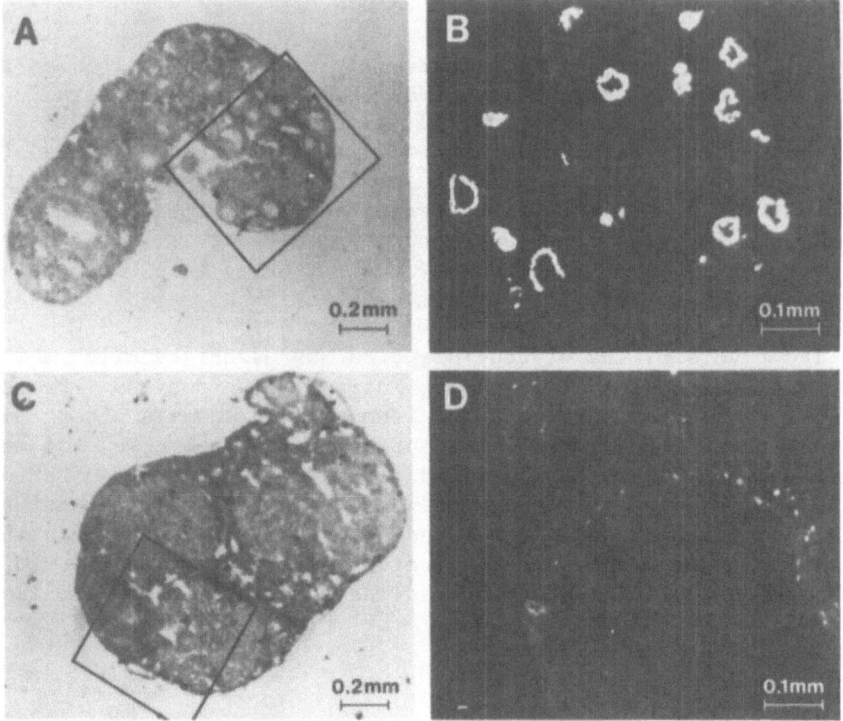

Figure 9. Indirect Immunofluorescence of Ovarian Sections. Light
microscopy of frozen ovarian sections stained with methy-
lene blue isolated, (A) 3 days after passive administration
with monoclonal antibody IE-3 and (C) 79 days after antibody administration from a mouse that had resumed fertility. (B) Fluorescence microscopy of area indicated in (A).
(D) Fluorescence microscopy of area indicated in (C).

Effect on Development

The initial in vivo studies on control and antibody treated mice
as well as those demonstrating the reversibility of contraception
show that the antibodies have no effect on the litter size or pup
weight (Fig. 4). However, because of the possible important role
that the zona may play during pre-implantation development and the
hatching reaction, the effect of monoclonal antibodies on in vitro
development was studied. Two-cell embryos were incubated for five
days in the continuous presence of monoclonal antibodies to either
ZP-2 or ZP-3 in concentrations 5-10 fold in excess of that necessary
to block fertilization. Compared to the controls we observed no
effect on the number of embryos that progressed to the blastocyst
stage or that eventually hatched from their surrounding zonae pellucidae (see Table 2).

Table 2. In vitro Effect of Monoclonal Antibodies on Early Development.

	2-Cell Embryos	Blastocyst (%)	Hatched Blastocyst (%)
Control	60	59/60 (98.3)	50/60 (83.3)
Ascites fluid	60	55/60 (93.3)	50/60 (83.3)
IE-3 (2 µg/ml)	60	59/60 (98.3)	48/60 (80.0)
IE-10 (200 µg/ml	60	57/60 (95.0)	45/60 (75.0)

Thus, the anti-zona monoclonal antibodies do not appear to have an
adverse affect on early development (13).

DISCUSSION

Monoclonal antibodies to individual zona proteins have provided
us with exquisitely specific immunological probes with which to
examine the structure and biological function of the murine zona
pellucida. We have demonstrated that ZP-2 and ZP-3 are found
uniquely in the ovary where they surround growing oocytes. These
sulfated glycoproteins are immunologically distinct from other extracellular matrix proteins and do not participate in basement membrane
structures (10). The monoclonal antibodies bind to ovulated eggs as
well as to 2-cell embryos. The reticulated fibrous pattern seen on
the surface of the eggs does not appear to change during early development. It is not clear how the zona pellucida, which has a very
long half-life (23), accomodates the five-fold increase in the
surface area of the ooctye as the ooctye's diameter increases from 30
µm to 70 µm during oogenesis. Although laser photo-bleaching studies
indicate no lateral diffusion of ZP-2 (10), the fenestrated network
of the zona proteins surrounding ovulated eggs may indicate the
flexibility of the zona proteins which would be necessary for this
growth.

Monoclonal antibodies both to ZP-2 as well as to ZP-3 have
roughly 10^8 binding sites, diffusely located on the surface of the
zona pellucida. We estimate that only about 2% of the zona proteins
are available to react with antibodies. Thus, it was of interest to
determine what effect the presence of antibody would have on known
functions of the zona. Following parenteral administration of monoclonal antibody, the only perturbation observed was an effective,

longterm block to fertilzation. Although most studies were performed
with an injection of 250 μg of antibody to ZP-2 or ZP-3, as little as
10 μg of antibody was equally effective. Furthermore, in vitro fertilization was 50% inhibited by 20 ng/ml of monoclonal antibody to
ZP-2 or 185 ng/ml of antibody to ZP-3. The effectiveness of the
in vivo block lasted an average of 57 days or approximately 14 mouse
menstrual cycles. Although long-lasting, the effect was completely
reversible and the treated mice were again as fertile as the
untreated control animals. This reversibility was associated with a
loss of intra-ovarian monoclonal antibodies. Thus, it appears that
parenterally administered antibodies localize to the ovary where they
coat the zonae pelucidae surrounding the many growing oocytes. When
ovulated, these eggs cannot be fertilized and this block continues
until all of the coated oocytes are ovulated. The primordial oocytes
in the intra-ovarian resting pool which have not yet initiated zona
production are not coated. As they grow and mature in the absence of
circulating anti-zona antibody, their zonae are not coated and, thus,
when ovulated they can be fertilized normally.

The most likely explanation for the molecular mechanism of this
effect is that the monoclonal antibodies block fertilization by
steric hinderance. The antibodies do not affect ovulatory rates nor
do they affect sperm morphology or function (11,13). There is no
induction of the biochemical changes in the zona proteins associated
with the block to polyspermy (5,13). However, the antibody block is
clearly dependent on the presence of the zona pellucida and is
associated with the presence of the antibody on the surface of the
zona (13). The absence of perivitelline sperm or of a male pronucleus
further suggests that the antibody blocks penetration of the zona by
the sperm.

Nevertheless, it appears unlikely that the block is caused by
specific antibody-masking of the sperm binding site on the surface of
the zona. Although ZP-3 has been reported to have sperm binding
properties (4), antibodies to three different epitopes on ZP-2 were
equally effective in blocking fertilization. Recently an O-linked
carbohydrate side chain on ZP-3 has been suggested as the sperm
binding site (15). However, the anti-ZP-3 monoclonal antibody is
directed against the protein backbone (unpublished data) and, thus,
should not directly block sperm binding. This hypothesis is further
supported by our studies which show that in vitro fertilization is
blocked by antibodies to ZP-2 or ZP-3 but the presence of either
antibody does not dramatically decrease sperm binding to the zona
pellucida. The polyclonal antiserum from the rat whose spleen cells
were used to make the cell lines secreting monoclonal antibodies
reacts with the zonae from a number of mammals including rat, guinea
pig, rabbit, hamster, dog and cat. However, the tested monoclonal
antibodies reacted only with mouse and rat zonae. It will be of
interest to determine if additional cell lines have greater cross-
reactivity as others have reported (9,17).

We have also examined the effect of the monoclonal antibodies on post-fertilization development. In vivo studies have shown that antibody given at the 2-cell embryo stage or later had no effect on litter size or pup weight. In vitro development to the blastocyst stage was also unaffected by antibody to either ZP-2 or ZP-3 present in amounts 5-10 fold in excess of that needed to block fertilization and neither antibody inhibited the hatching reaction. Thus, the only observed perturbation of the anti-zona monoclonal antibodies on early development was an effective, long-lasting but fully reversible block to fertilization.

The development of monoclonal antibodies to the zona pellucida has opened up several exciting areas of research. Using these protein-specific probes one can begin to dissect the supramolecular structure of the zona pellucida during oogenesis and early embryo-genesis. Differentially labeled fluorescent monoclonal antibodies can be used to determine the spatial relationship of the ZP-2 to ZP-3 in the zona and to detect changes which may occur during development. These probes have also been instrumental in cloning and characteriz-ing the zona genes utilizing the lambda gt11 expression vector (31) to express proteins coded for by a cDNA library derived from ovarian tissue (Ringuette, M., Sobieski, D., Chamow, S. and Dean, J.; unpub-lished data). It will be of great interest to determine the struc-ture of these genes and the mechanism(s) that control their expres-sion during development.

The monoclonal antibodies have also suggested a powerful immuno-logical aproach to birth control. These monoclonal antibodies can be purified, isolated in pharmacological amounts and used to passively immunize female mice. Furthermore, our ability to express cloned zona genes as β-galatosidase fusion proteins will permit us to begin to test fusion products containing zona protein sequences as contra-ceptive vaccines. At present these are model systems, but such strategies may in the future have an important role to play in animal husbandry and the control of world over-population.

ACKNOWLEDGEMENTS

We appreciate the critical reading of the manuscript by Dr. Robert Simpson and we thank Ms. Dorothy Neely for expert typing.

REFERENCES

1. Aitken, R.J., Holme, E., Richardson, D.W. and Hulme, M. 1982. Properties of intact and univalent (Fab) antibodies raised against isolated, solubilized, mouse zonae pellucida, J. Reprod. Fertil. 66:327-334.

2. Aitken, R.J., Rudak, E.A., Richardson, D.W., Dor, J., Djahan-
   bahkch, O., and Templeton, A.A. 1981. The influence of anti-
   zona and anti-sperm antibodies on sperm-egg interactions. J.
   Reprod. Fertil. 62:597-606.
3. Bleil, J.D. and Wassarman, P.M. 1980. Structure and function of
   the zona pellucida: Identification and characterization of the
   proteins of the mouse oocyte's zona pellucida. Dev. Biol.
   76:185-202.
4. Bleil, J.D. and Wassarman, P.M. 1980. Mammalian sperm-egg
   interaction: Identification of a glycoprotein in mouse egg
   zonae pellucidae possessing receptor activity for sperm. Cell
   20:873-882.
5. Bleil, J.D., Beall, C.F. and Wassarman, P.M. 1981. Mammalian
   sperm-egg interaction: Fertilization of mouse egg triggers
   modification of the major zona pellucida glycoprotein, ZP-2.
   Dev. Biol. 86:189-197.
6. Braude, P., Pelham, H., Flach, G. and Lobatto, R. 1979. Post-
   transcriptional control in the early mouse embryo. Nature
   (London) 282:102-105.
7. Cascio, S.M. and Wassarman, P.M. 1982. Program of early development in the mammal: Post-transcriptional control of a class of
   proteins synthesized by mouse oocytes and early embryos. Dev.
   Biol. 89:397-408.
8. Cullen, B., Emigholz, K. and Monahan, J. 1980. The transient
   appearance of specific proteins in one-cell mouse embryos.
   Dev. Biol. 76:215-21_.
9. Drell, D.W. and Dunbar, B.S. 1984. Monoclonal antibodies to
   rabbit and pig zonae pellucidae distinguish species-specific
   and shared antigenic determinants. Biol. Reprod. 30:445-447.
10. East, I.J. and Dean, J. 1984. Monoclonal antibodies as probes of
   the distribution of ZP-2, the major sulfated glycoprotein of
   the murine zona pellucida. J. Cell Biol. 98:795-800.
11. East, I.J., Gulyas, B. and Dean, J. 1984. Monoclonal antibodies
   to the murine zona pellucida protein with sperm receptor
   activity: Effects on fertilization and early development.
   Dev. Biol. 109:268-273.
12. East, I.J., Keenan, A.M., Larson, A.M. and Dean, J. 1984. Scinti-
   graphy of normal mouse ovaries with monoclonal antibodies to
   ZP-2, the major zona pellucida protein. Science 225:938-941.
13. East, I.J., Mattison, D.R. and Dean, J. 1984. Monoclonal antibodies to the major protein of the murine zona pellucida:
   Effects on fertilization and early development. Dev. Biol.
   104:49-56.
14. Eppig, J.J. 1977. Mouse oocyte development in vitro with various
   culture systems. Dev. Biol. 60:371-388.
15. Florman, H. and Wassarman, P.M. 1985. O-linked oligosaccharides
   of mouse egg ZP-3 account for its sperm receptor activity.
   Cell 41:313-324.
16. Greve, J.M., Salzmann, G.S., Roller, R.J. and Wassarman, P.M.
   1983. Biosynthesis of the major zona pellucida glycoproteins

secreted by oocytes during mammalian oogenesis. Cell 31:749-759.

17. Isojima, S., Koyama, K., Hasegawa, A., Tsunoda, Y. and Hanada, A. 1984. Monoclonal antibodies to porcine zona pellucida antigens and their inhibitory effects on fertilization. J. Reprod. Immunol. 6:77-87.

18. Johnson, M.H. 1981. The molecular and cellular basis of pre-implantation mouse development. Biol. Rev. 56:463-498.

19. Jones, W.R. 1982. Immunization against the oocyte, in: "Immuno-logical Fertility Regulation." pp. 147-174, Blackwell, Oxford.

20. Longo, F.J. 1981. Changes in the zonae pellucidae and plasma-lemmae of aging mouse eggs. Biol. Reprod. 25:339-411.

21. Phillips, D.M. and Shalgi, R. 1980. Surface architecture of the mouse and hamster zona pellucida and ooctye. J. Ultrastruct. Res. 72:1-12.

22. Sacco, A.G. 1979. Inhibition of fertility in mice by passive immunization with antibodies to isolated zonae pellucidae. J. Reprod. Fertil. 56:533-537.

23. Shimizu, S., Tsuji, M. and Dean, J. 1983. In vitro biosynthesis of three sulfated glycoproteins of murine zonae pellucidae by oocytes grown in follicle culture. J. Biol. Chem. 258:5858-5863.

24. Tsunoda, Y. and Chang, M.C. 1976. The effect of passive immuniza-tion with hetero- and iso-immune anti-ovary antiserum on the fertilization of mouse, rat and hamster eggs. Biol. of Reprod. 15:361-365.

25. Tsunoda, Y. and Chang, M.C. 1978. Effect of antisera to eggs and zonae pellicudae on fertilization and development of mouse eggs in vivo and in culture. J. Reprod. Fertil. 54:233-237.

26. Tsunoda, Y. and Whittingham, D.G. 1982. Lack of effect of zona antibody on the development of mouse embryos in vivo and in vitro. J. Reprod. Fertil. 66:585-589.

27. Tsunoda, Y., Sugie, T., Mori, J., Isojima, S. and Koyama, K. 1981. Effect of purified zona antibody on fertilization in the mouse. J. Exp. Zool. 217:103-108.

28. Van Blerkom, J. 1981. Structural relationship and posttransla-tional modification of stage specific proteins synthesized during preimplantation development in the mouse. Proc. Natl. Acad. Sci. USA 78:7629-7633.

29. Whittingham, D.G. 1971. Culture of mouse ova. J. Reprod. Fertil. Suppl. 14:7-21.

30. Yanagimachi, R. 1981. Mechanism of fertilization in mammals, in: "Fertilization and Embryonic Development in vitro." (L. Mastroianni, Jr. and J. D. Biggers, eds.), pp. 81-182, Plenum, New York.

31. Young, R.A. and Davis, R.W. 1983. Efficient isolation of genes using antibody probes. Proc. Natl. Acad. Sci. USA 80:1194-1198.

NATURE OF THE MOUSE EGG'S RECEPTOR FOR SPERM

Paul M. Wassarman, Jeffrey D. Bleil, Harvey M. Florman,
Jeffrey M. Greve, Richard J. Roller and George S. Salzmann

Department of Biological Chemistry, Harvard Medical School
Boston, MA 02115

SUMMARY

The zona pellucida is an extracellular coat that surrounds all
mammalian eggs. Sperm must penetrate the zona pellucida in order to
reach and fuse with the plasma membrane of unfertilized eggs. Penetration is accomplished by a sequence of events involving both egg
and sperm. First, sperm must bind to the outer margin of the zona
pellucida. Such binding is mediated in a relatively species-specific
manner by "sperm receptors" in the zona pellucida. Second, sperm
must undergo the "acrosome reaction", a membrane fusion event, in
order to traverse the zona pellucida.

Here we review results from our own laboratory which demonstrate
that, during the course of sperm-egg interaction in mice, zona pellucida glycoprotein ZP3 serves as both receptor for sperm and inducer
of the acrosome reaction. Furthermore, we review evidence from our
laboratory indicating that the sperm receptor activity of ZP3 is
dependent only on its O-linked carbohydrate components, whereas acrosome reaction-inducing activity is dependent on the polypeptide
portion of ZP3 as well.

INTRODUCTION

Of the many receptor mediated cell-cell interactions that occur
during animal development, none has a more extensive history of
investigation than the interaction between sperm and eggs (43; see
chapter by C.B. Metz, this volume). This can be attributed to many
factors, among which is the ability to work with relatively large
numbers of gametes as relatively homogeneous populations of cells.
Additionally, sperm-egg interaction is of inherent interest since it

represents the first instance of cell-cell interaction during development of a new individual and, therein, has profound ramifications.

A variety of experimental evidence has led to the current belief
that extracellular coats of both mammalian (zona pellucida) and
nonmammalian (vitelline envelope) eggs possess receptors for sperm
that mediate species-specific fertilization (27,29,67,69,72,73).
(See appropriate chapters in this volume.) In recent years, substan-
tial progress has been made toward the goal of isolating and characterizing sperm receptors. Here, we review work from our own laboratory on the mouse egg's receptor for sperm.

Overall, we have identified, isolated, and characterized a
unique zona pellucida glycoprotein, called ZP3 (83 kD molecular
weight), that exhibits properties expected of a mammalian sperm
receptor (4,5,7,8,66,67,69). Along with the two other zona pellucida
glycoproteins, ZP1 (200 kD molecular weight) and ZP2 (120 kD molecular weight), ZP3 is synthesized and secreted exclusively by mouse
oocytes during their growth phase (6,23,49,53,66,68,70). Sperm
receptor activity has been traced to the O-linked oligosaccharides of
ZP3 (15,16; H.M. Florman and P.M. Wassarman, unpublished results)
suggesting involvement of "lectin-like" interactions during binding
of sperm to eggs. Furthermore, we have found that ZP3 induces mouse
sperm to undergo the acrosome reaction (8,15,16,66,67,69), completion
of which enables bound sperm to traverse the zona pellucida and fuse
with the egg's plasma membrane. Thus, only one zona pellucida component, ZP3, appears to be involved in bringing about those events
that precede fertilization in mice.

RESULTS

Composition of the Mouse Zona Pellucida

Mouse zonae pellucidae isolated from fully-grown oocytes (Fig.
1) each contain approximately 3 ng of protein, or about 10% of the
oocyte's total protein. In order to determine the number and assess
the nature of protein species that constitute zonae pellucidae,
several experimental approaches have been employed: (i) radiolabeling of isolated zonae pellucidae (both intact and solubilized
preparations) with reagents such as ^{125}I-iodosulfanilic acid and the
N-hydroxysuccinimide ester of iodinated (^{125}I) p-hydroxyphenylpro-
pionic acid (Bolton-Hunter reagent), followed by electrophoretic
separation (SDS-PAGE) and autoradiography; (ii) electrophoresis of
isolated zonae pellucidae, followed by staining for protein with
either Coomassie brilliant-blue or silver; (iii) electrophoresis of
isolated zonae pellucidae, followed by staining for glycoprotein with
either periodic acid-Schiff or dansyl hydrazine. Surprisingly, in
all cases, we found that the 3 ng or so of zona pellucida protein is
distributed among only three different glycoprotein species having

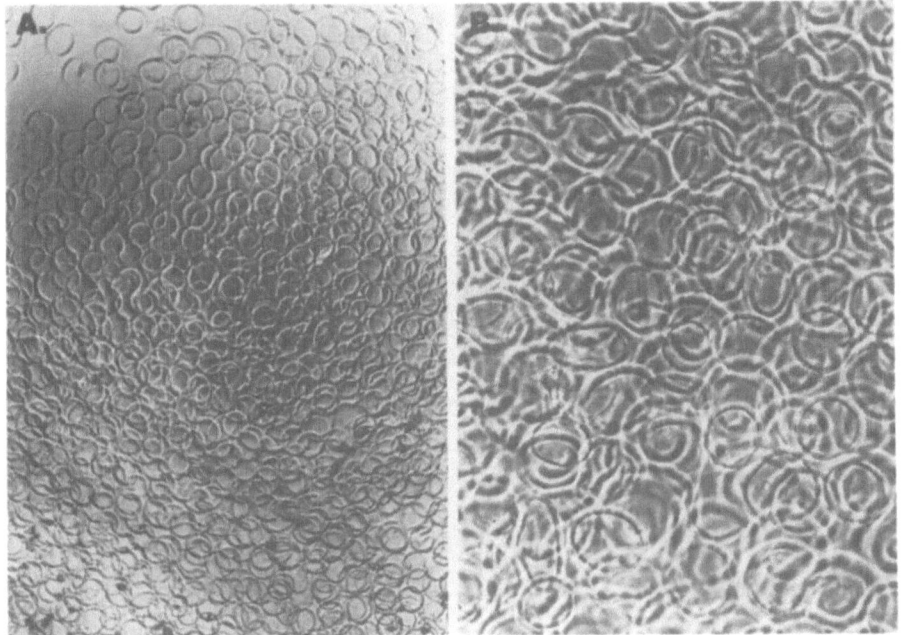

Figure 1. Photomicrographs of zonae pellucidae isolated from mouse
          oocytes. (A) 65x; (B) 150x.

molecular weights of 200, 120, and 83 kD, and referred to these as
ZP1, ZP2, and ZP3, respectively (4,5; J.D. Bleil and P.M. Wassarman,
unpublished results) (Fig. 2). These three glycoproteins together
account for virtually all of the mass of the zona pellucida. Results
of amino acid analysis, high-resolution two-dimensional gel electrophoresis, electrophoretic analysis of proteolytic digests, and immunological analysis, all strongly suggest that each glycoprotein
represents a unique polypeptide chain. All three zona pellucida
glycoproteins are acidic (isoelectric points less than 5.5), in part
due to the presence of sialic acid, and only ZP1 consists of more
than one polypeptide chain (linked by intermolecular disulfides).
These are the "mature" forms of the glycoproteins found in the zona
pellucida itself.

Identification of the Mouse Egg's Sperm Receptor: Experimental
Approach

     At least three distinct events take place prior to penetration
of the mouse egg's zona pellucida by sperm: (i) "attachment" of
sperm to the zona pellucida, (ii) "binding" of sperm to the zona
pellucida, and (iii) induction of the acrosome reaction (66,67,69).
The initial step in sperm-egg interaction in the mouse is attachment,
a loose association between the sperm head and zona pellucida, and

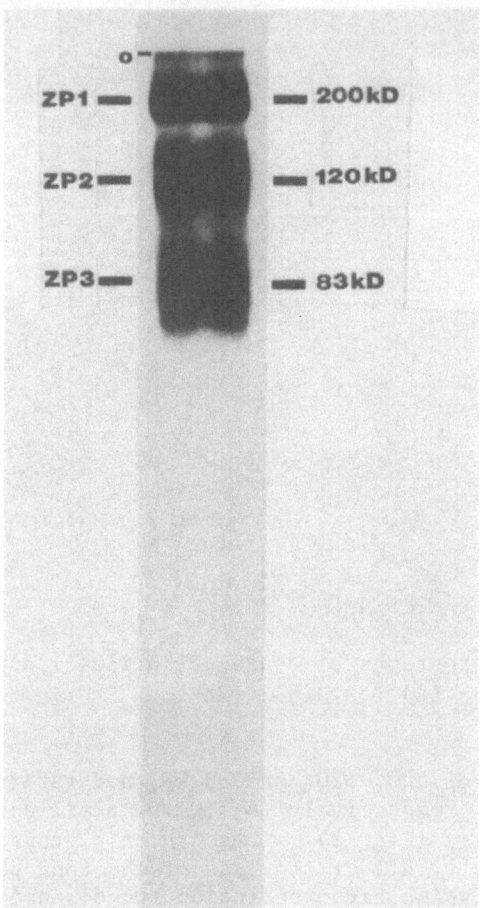

Figure 2. Autoradiograph of ^{125}I-labeled zonae pellucidae subjected
    to SDS-PAGE.   o, origin; kD, kilodaltons.

this precedes binding, a much tighter association between the sperm
head and zona pellucida (7,8,14,16,30,35,51,52). We have demonstrated that sperm attached to zonae pellucidae of unfertilized mouse
eggs during a brief incubation period ("pulse") in vitro, can be
successfully "chased" into bound sperm during a subsequent incubation
period carried out in the absence of free-swimming sperm (8). This
is consistent with the observation that attached sperm removed from
one group of unfertilized eggs could subsequently bind to zonae
pellucidae of another group of eggs (30). It seems likely that
attachment of mouse sperm to unfertilized eggs is analogous to the
association between noncomplementary gametes seen in lower species
(2), and to the reversible step preceding bacteriophage adsorption to

<u>Escherichia</u> <u>coli</u> (9). Furthermore, several lines of evidence suggest
that binding of sperm to zonae pellucidae is relatively species
specific and is attributable to the presence of sperm receptors in
zonae pellucidae of unfertilized mouse eggs (43,67,69,72,73). Since
the mouse egg's zona pellucida is relatively simple in composition
(see above), we attempted to identify the nature of the sperm
receptor.

     Each of the zona pellucida glycoproteins from unfertilized eggs,
as well as from 2-cell embryos, was tested for sperm receptor
activity in an in vitro "competition assay." In all experiments,
sperm were first "capacitated" (27,29,72,73) and then cultured for an
additional period in either culture medium alone or in the presence
of substances to be tested for sperm receptor activity. This experimental design precluded any indirect effect of agents to be tested on
gamete interaction through an inhibition of capacitation. Ovulated
eggs and 2-cell embryos were then added to the sperm cultures and,
within seconds, zonae pellucidae of both eggs and embryos were
covered with motile sperm (Fig. 3). Such sperm were loosely associated with zonae pellucidae and could be removed by gentle pipetting
with a broad bore micropipette; this state of adhesion has been
referred to as "attachment". During the next 10 min or so, contact
between sperm and zonae pellucidae of unfertilized eggs became more
tenacious, such that gentle pipetting no longer dissociated the
gametes; this state of adhesion has been referred to as "binding".
Although the initial, reversible attachment of sperm to embryo zonae
pellucidae was indistinguishable from that observed with unfertilized
eggs, in the former case attachment did <u>not</u> proceed to the binding
stage. This difference in behavior provided an operational definition of "bound sperm" as those adhering to the egg's zona pellucida
under conditions that resulted in complete removal of sperm from the
embryo's zona pellucida (Fig. 3). When gametes were cocultured for
about 6 hr under these conditions, more than 60% of the eggs were
fertilized, as indicated by the presence of a sperm tail and either a
decondensing sperm head or, more typically, two pronuclei within the
ooplasm. Sperm binding levels observed under experimental conditions
(e.g., in the presence of either solubilized zonae pellucidae or
purified zona pellucida glycoproteins) were expressed as a percentage
of binding levels observed in the presence of culture medium alone,
and these values were used for comparisons between individual experiments. This procedure of analysis resulted in highly reproducible
patterns.

<u>Identification of the Mouse Egg's Sperm Receptor: Observations</u>

     As anticipated from previous reports (28), we found that sperm
exposed to solubilized zonae pellucidae from 2-cell embryos bound to
ovulated eggs to the same extent as untreated sperm, but that sperm
exposed to solubilized zonae pellucidae from unfertilized eggs exhibited greatly diminished binding to eggs as compared to untreated

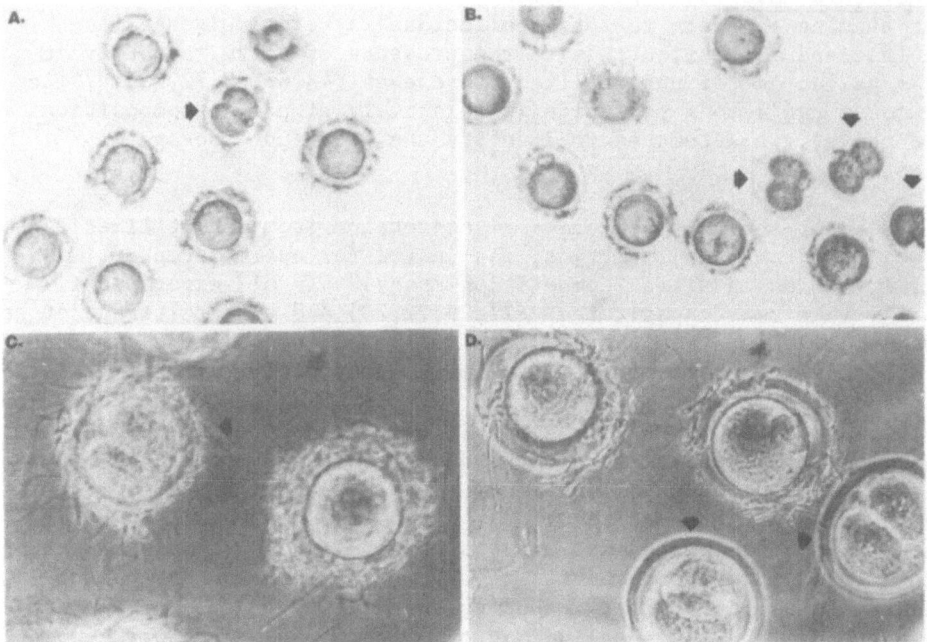

Figure 3. Interaction of mouse sperm and eggs in vitro. Shown are
photomicrographs comparing "attachment" (A,C) and "binding"
(B,D) of sperm to zonae pellucidae of unfertilized eggs and
2-cell embryos. Arrows indicate 2-cell embryos. (A,B)
170x; (C,D) 340x.

sperm. Titration of sperm with egg zonae pellucidae revealed that
binding of sperm could be reduced to as little as 10% of control
values. Solubilized zonae pellucidae from unfertilized eggs were
shown to affect sperm rather than eggs, since mouse eggs which failed
to bind sperm exposed to solubilized zonae pellucidae could be
removed from these cultures and were shown to bind untreated sperm to
the same extent as control eggs. It was also noted that solubilized
zonae pellucidae from unfertilized eggs and embryos appeared to have
no significant effect on sperm motility during the incubation period.
These results suggest that zonae pellucidae from unfertilized mouse
eggs, but not from embryos, contain a component that behaves in the
manner expected for a sperm receptor in the competition assay.

Each of the mouse egg's zona pellucida glycoproteins was purified to homogeneity in order to identify the component(s) possessing
sperm receptor activity in vitro (5,7,8,15,16). The purification
procedure included: (i) removal of zonae pellucidae from unfertilized eggs individually, using mouth-operated micropipettes; (ii) SDS-
polyacrylamide gel electrophoresis (SDS-PAGE) to separate the three

glycoproteins; (iii) electroelution of the individual glycoproteins
from the gel; (iv) extensive dialysis of the glycoproteins, first
against urea to remove SDS, and then against distilled water to
remove urea, and; (v) lyophilization to dryness. For each preparation, a small portion of the starting material (about 10%) was radiolabeled with either iodosulfanilic acid or Bolton-Hunter reagent in
order to be able to locate the glycoproteins on gels by autoradiography, monitor the effectiveness of the purification procedure, and
quantitate the recovery of each glycoprotein.

     Whereas purified ZP1 and ZP2 did not have a significant effect
on the extent of sperm binding in the competition assay, purified ZP3
was nearly as effective as solubilized zonae pellucidae from unfertilized eggs in reducing the number of sperm bound to eggs. Several
other purified glycoproteins (including bovine serum albumin, bovine
submaxillary mucin, bovine transferrin, and human chorionic gonadotropin) were also tested in the competition assay and found to have
no significant effect on binding of sperm to eggs; even at ten-times
the highest concentration of purified zona pellucida glycoprotein
tested. These results suggest that ZP3 is the component of zonae
pellucidae from unfertilized eggs that possesses sperm receptor
activity. Comparisons of results obtained with purified material and
total solubilized zonae pellucidae indicate that ZP3 alone accounts
for all sperm receptor activity present in zonae pellucidae of unfertilized eggs.

     ZP1, ZP2, and ZP3 are present in zonae pellucidae isolated from
embryos, as well as from unfertilized eggs. Since sperm do not bind
to fertilized eggs or embryos, and solubilized zonae pellucidae from
fertilized eggs or embryos have no effect on sperm binding in the
competition assay, zona pellucida glycoproteins were isolated from
2-cell embryos (as above) and tested for their ability to prevent
sperm binding to unfertilized eggs. If ZP3, indeed, possessed the
unfertilized egg's sperm receptor activity, it was anticipated that
it would have lost the ability to prevent sperm binding in the competition assay as a consequence of fertilization. This proved to be
the case, since none of the three glycoproteins purified from 2-cell
embryo zonae pellucidae had any effect on the number of sperm bound
to eggs in vitro. In the same experiments, solubilized zonae pelucidae from unfertilized eggs and ZP3 purified from egg zonae pellucidae
had a marked inhibitory effect on the binding of sperm to eggs. It
would appear, therefore, that ZP3 is altered as a result of fertilization such that it no longer exhibits sperm receptor activity.

Identification of the Mouse Egg's Inducer of the Acrosome Reaction:
Experimental Approach

     In order for sperm to penetrate the extracellular layers of eggs
and reach the plasma membrane, they must undergo the acrosome reaction. This reaction involves the fusion and extensive vesiculation

of the sperm's plasma and outer acrosomal membranes, and results in
exposure of the inner acrosomal membrane with its associated enzymatic activities (13,27,29,57,67,69,72,73). Mouse sperm that have
completed the acrosome reaction prior to exposure to unfertilized
eggs in vitro fail to bind to the zona pellucida (8,14,51,52,67,69)
suggesting that it is the sperm's plasma membrane that interacts with
zonae pellucidae and that a zona pellucida component induces bound
sperm to undergo the acrosome reaction. Using an experimental
approach similar to that described above, we attempted to identify
the nature of the inducer of the acrosome reaction present in zonae
pellucidae of unfertilized mouse eggs.

In order to determine whether or not one or more zona pellucida
glycoproteins had acrosome reaction-inducing activity, sperm were
exposed to test material and then scored for the presence of intact
acrosomes by either transmission electron microscopy or indirect
immunofluorescence (8,15,16,66). The latter approach was made
possible by the availability of a monoclonal antibody, HS19, that is
directed against the acrosomal cap region of mammalian sperm; only
sperm with intact acrosomes display localized head fluorescence after
treatment with HS19 (16; K.B. Bechtol, unpublished results). In
these experiments, the lower and upper limits of the extent of acrosome reaction within a population of sperm were determined by using
sperm exposed to either culture medium alone (lower limit; "background") or to medium containing ionophore A23187, an inducer of the
acrosome reaction in vitro (upper limit). These results were then
compared with the extent of acrosome reaction in populations of sperm
exposed to solubilized zonae pellucidae from either unfertilized eggs
or 2-cell embryos, as well as to purified ZP1, ZP2, or ZP3 from
either unfertilized eggs or 2-cell embryos.

Identification of the Mouse Egg's Inducer of the Acrosome Reaction: Observations

Solubilized zonae pellucidae isolated from unfertilized mouse
eggs were found to be as effective as ionophore A23187 in inducing
capacitated sperm to undergo the acrosome reaction in vitro. On
average, about 25% of the sperm incubated in the absence of either
zonae pellucidae or ionophore A23187 underwent the acrosome reaction;
this was the case even though only motile sperm were used in these
experiments. Incubation of sperm in the presence of ionophore A23187
resulted, on average, in a two to three-fold increase in the percentage of the sperm population that underwent the acrosome reaction as
compared to the control samples. Incubation of sperm in the presence
of solubilized egg zonae pellucidae was as effective as ionophore
A23187 in inducing the acrosome reaction and, whereas sperm exposed
to ionophore became immotile during the incubation period, sperm
incubated in the presence of solubilized zonae pellucidae could not
be distinguished from control samples on the basis of motility.

The results just described suggest that solubilized zonae pellucidae from unfertilized eggs contain a component that has acrosome reaction-inducing activity. In order to identify such a component, each of the zona pellucida glycoproteins was purified (described above) and tested for the ability to induce the acrosome reaction in vitro. As before, samples of sperm were incubated alone and in the presence of ionophore A23187 in order to determine the background and maximal levels of acrosome reaction, respectively. Whereas ZP1 and ZP2 from unfertilized eggs had little effect on the extent of acrosome reaction, as compared to control samples, ZP3 was as effective as ionophore A23187 and solubilized zonae pellucidae in inducing the reaction. On the other hand, ZP1, ZP2, and ZP3 purified from 2-cell embryo zonae pellucidae were all much less effective than ionophore A23187, solubilized egg zonae pellucidae, or purified egg ZP3 in inducing sperm to undergo the acrosome reaction. These results strongly suggest that ZP3 is the glycoprotein component of egg zonae pellucidae that is responsible for inducing the acrosome reaction, and that ZP3 is altered as a result of fertilization such that it is no longer able to induce the reaction.

Structure of ZP3: The Mouse Egg's Receptor for Sperm

In view of the important biological functions of ZP3 described above, it is appropriate to review what we have learned about the structure of ZP3. Much of this information has come from biosynthetic studies carried out with isolated, growing mouse oocytes cultured in vitro in the presence of radioactively labeled precursors (e.g., ^{35}S-methionine and ^{3}H-fucose), and was made possible by the availability of polyclonal antisera directed specifically against ZP3 (23,49,53,70).

In order to identify intracellular precursors of ZP3 (83 kD molecular weight), growing mouse oocytes were metabolically radiolabeled by culturing isolated oocytes, freed of follicle cells, in the presence of ^{35}S-methionine. At the end of the culture period, one group of oocytes was processed immediately for immunoprecipitation ("pulsed" group), while another was placed in fresh medium in the presence of a large molar excess of cold methionine, and cultured an additional period ("chased" group) prior to immunoprecipitation. Fluorograms of high-resolution two-dimenstional gels of the immunoprecipitates from the two groups of oocytes, from which zonae pellucidae had been removed following the culture period (i.e., only intracellular material was examined), were then developed. Two radiolabeled species, with apparent molecular weights of 56 and 53 kD (pI 6.7), were specifically immunoprecipitated from the pulsed oocytes. On the other hand, immunoprecipitates from the chased oocytes contained only the mature (83 kD molecular weight; pI 4.6) form of ZP3. These observations strongly suggest that the 56 and 53 кD molecular weight species are intracellular precursors of mature, 83 kD molecular weight ZP3 (53,70).

Since mature ZP2 (120 kD molecular weight), its precursor (91 kD
molecular weight), and many other secreted glycoproteins contain N-
linked (i.e., asparagine-linked) oligosaccharides, it seemed likely
that biosynthesis of ZP3 would include cotranslational "core"-
glycosylation of the polypeptide chain (23,33,70). If true, ZP3
intracellular precursors (56 and 53 kD molecular weight) would be
expected to be sensitive to the enzyme endo-beta-N-acetylglucosamini-
dase H (Endo H) that cleaves between the two proximal GlcNAc residues
of high mannose-type oligosaccharide, leaving a GlcNAc attached to an
asparagine residue of the polypeptide chain (33,47,61,62).

As in the case of mature ZP2 (23,70) mature, 83 kD molecular
weight ZP3 proved to be completely insensitive to Endo H, indicating
that most or all of the high mannose oligosaccharide present on the
precursor are processed to complex-type oligosaccharides (53). On
the other hand, digestion of the 56 and 53 kD molecular weight precursors of mature ZP3 with Endo H resulted in their conversion into a
single species of 44 kD molecular weight. Two-dimensional gel
analyses of intermediates in Endo H partial digests of the ZP3 precursors revealed a total of five discrete species at about 44, 47,
50, 53, and 56 kD molecular weight, all with isoelectric points of
6.7. These results suggest that the 56 kD molecular weight ZP3
precursor contains four high mannose-type oligosaccharides, and the
53 kD molecular weight species is simply an underglycosylated form
containing only three such oligosaccharides. Whether or not the
extent of glycosylation of ZP3 varies in a similar manner in vivo is
not known. Of course, it is possible that the presence of two ZP3
precursors is not the result of variable glycosylation of the same
polypeptide chain, but instead results from complete glycosylation of
two different polypeptide chains, of virtually identical molecular
weights, but with different numbers (3 vs 4) of "core"-glycosylation
sites.

As an alternative experimental approach to identifying the nascent polypeptide chain of ZP3, growing oocytes were cultured in the
presence of tunicamycin (53). This drug prevents the addition of
GlcNAc to dolichol phosphate, the first step in formation of "core"
oligosaccharides, and, thereby, results in synthesis of glycoproteins
deficient in N-linked oligosaccharides (33,60,63). Under these conditions, immunoprecipitates contained a 44 kD molecular weight
species, rather than the 56 and 53 kD molecular weight precursors
found in the absence of the drug. This result is completely consistent with those of the Endo H experiments in which the 56 and 53 kD
molecular weight precursors of ZP3, found within oocytes, are,
indeed, formed by "core"-glycosylation of a 44 kD molecular weight
polypeptide chain.

In summary, (Fig. 4), ZP3, the mouse egg's sperm receptor, is
synthesized by growing oocytes as a 44 kD molecular weight polypeptide chain to which either three or four high mannose-type oligo-

saccharides ("core" structure, $Glc_3Man_9GlcNAc_2$) are added cotranslationally, giving rise to 53 and 56 kD molecular weight species, respectively; these events presumably occur in the rough endoplasmic reticulum (33). The oligosaccharides of these two precursors are then processed (removal of outer mannose residues and addition of terminal sugars, including GlcNAc, galactose, fucose, and sialic acid), presumably in the Golgi apparatus (33), to complex-type oligosaccharides, giving rise to the mature, 83 kD molecular weight form of ZP3 which is then secreted by the oocyte (Fig. 4). The N-linked oligosaccharides, rather than the polypeptide chain, appear to be partly responsible for the molecular heterogeneity of mature ZP3. Other results from our laboratory strongly suggest that ZP3 is also glycosylated at serine/threonine residues (O-linked) (15,16; H.M. Florman, F.G. Samuels, and P.M. Wassarman, unpublished results; see below) and other modifications as well (e.g., phosphorylation and/or sulfation) may contribute to the molecular heterogeneity of ZP3.

Identification of the Molecular Nature of the Sperm Receptor Activity of ZP3

We have examined the relative roles of polypeptide and carbohydrate domains of ZP3 in sperm binding as part of a systematic analysis of the sperm receptor's structure-function relationships (15,16,69; H.M. Florman and P.M. Wassarman, unpublished results). Since carbohydrate recognition figures centrally in many current models of cell-cell interaction, it was anticipated that the oligosaccharide portion of ZP3, not the polypeptide chain, might serve as the ligand to which sperm bind. Such a situation would be compatible with the suggestion that the binding of sperm to eggs in invertebrates is mediated via a sperm lectin:egg glycoprotein ("receptor") interaction (18,19,22,54,65).

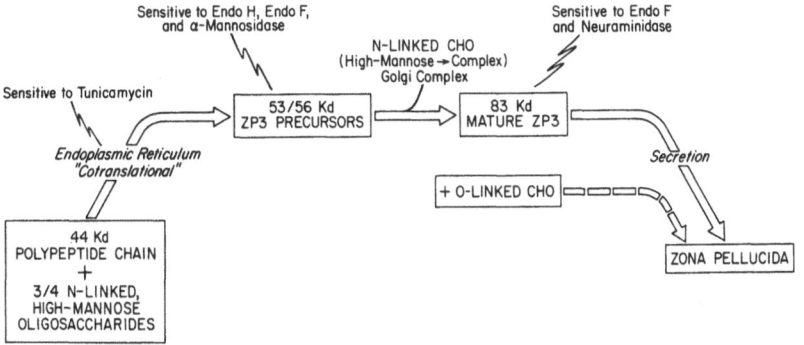

Figure 4. Pathway for the biosynthesis and secretion of the mouse egg's receptor for sperm. For details see references 6,23, 53,66,68,70.

In order to determine the molecular nature of the receptor
activity of ZP3, solubilized egg zonae pellucidae and purified egg
ZP3 (prepared as described above) were modified either enzymatically
or chemically and then tested in the competition binding assay
(described above) with sperm. The results of these experiments are
summarized in Fig. 5. Extensive digestion of either solubilized
zonae pellucidae or purified ZP3 by carboxymethyl-cellulose-conju-
gated Pronase had no significant effect on sperm receptor activity;
glycopeptides (1.5-6 kD molecular weight) were as effective as undi-
gested material in the competition assay. On the other hand,
complete deglycosylation of either solubilized zonae pellucidae or
purified ZP3 by trifluoromethanesulfonic acid (11,37) resulted in
loss of receptor activity. These data implicate carbohydrate, rather
than the polypeptide backbone, as responsible for the receptor
activity of ZP3. This conclusion is consistent with the finding that
the receptor activity of ZP3 is extremely stable (see Discussion).
Furthermore, we found that removal of O-linked carbohydrate from
either solubilized zonae pellucidae or purified ZP3 by mild alkaline
hydrolysis (beta-elimination) in the absence of a reducing agent,
under conditions that maintained the integrity of the polypeptide

RECEPTOR ACTIVITY

TREATMENT	TOTAL ZP	PURIFIED ZP3
PRONASE	PRESENT	PRESENT
TRIFLUOROMETHANE SULFONIC ACID	ABSENT	ABSENT
ENDO-N-ACETYLGLUCOS-AMINIDASE F	PRESENT	PRESENT
SODIUM HYDROXIDE	ABSENT	ABSENT
SODIUM HYDROXIDE, SODIUM BOROHYDRIDE	PRESENT	PRESENT

Figure 5. Enzymatic and chemical modification of the mouse egg's
          receptor for sperm. In each case, receptor activity was
          assessed in the "competition assay" following treatment of
          either solubilized egg zonae pellucidae (Total ZP) or ZP3
          purified from egg zonae pellucidae (Purified ZP3).   In the
          case of sodium hydroxide-sodium borohydride treated
          material, released oligosaccharides were tested for recep-
          tor activity; in all other cases, the modified glycoprotein
          was tested for receptor activity. For details see refer-
          ences 15 and 16. Endo-beta-N-acetylglucosaminidase F (Endo
          F) was generously provided by Dr. John Elder.

chain, resulted in loss of receptor activity. Removal of N-linked oligosaccharide, on the other hand, by digestion with endo-beta-N-acetylglycosidase F (Endo F; 12) did not have a significant effect on receptor activity. Finally, we have been able to isolate the O-linked carbohydrate released by mild alkaline hydrolysis of solubilized zonae pellucidae and purified ZP3 in the presence of sodium borohydride (to prevent the "peeling reaction"), and have found that this material retains full sperm receptor activity. Overall, these data strongly suggest that it is O-linked carbohydrate, not N-linked oligosaccharide nor the polypeptide portion of ZP3, that is responsible for its sperm receptor activity.

Identification of the Molecular Nature of the Acrosome Reaction-Inducing Activity of ZP3

Recently (16), we found that, while sperm receptor activity of egg ZP3 is unaffected by extensive proteolysis of the glycoprotein (see above), its acrosome reaction-inducing activity is destroyed by such treatment (16,69). Sperm receptor activity was retained completely by small (1.5-6 kD molecular weight) glycopeptides present in Pronase digests of either solubilized egg zonae pellucidae or purified ZP3, but these glycopeptides were unable to induce sperm to undergo the acrosome reaction. These results strongly suggest that, although the polypeptide chain of ZP3 is not directly involved in sperm receptor function, it probably plays a role in the acrosome reaction-inducing activity of ZP3. It is not possible as yet to distinguish between the alternate possibilities of a direct effect of polypeptide chain on sperm through a secretagogue domain, as has been described for certain hormones, or an indirect effect simply by serving as the backbone for a polyvalent carbohydrate structure.

DISCUSSION

Common to nearly all eggs are one or more extracellular layers that completely surround the plasma membrane. In echinoderms these are the vitelline envelope and jelly coat, whereas in mammals there is the zona pellucida. Since sperm must bind to and penetrate these extracellular layers in a relatively species-specific manner, they are intimately involved in regulating sperm-egg interactions both before and after fertilization (Fig. 6).

The results of experiments reviewed here strongly suggest that ZP3, a glycoprotein having an apparent molecular weight of 83 kD, an isoelectric point of about 4.6, and representing about 30-40% of the mass of zonae pellucidae, possesses the receptor activity responsible for sperm binding to unfertilized eggs. Since the purification of zona pellucida glycoproteins involved the use of denaturants, it is possible, of course, that any sperm receptor activity associated with ZP1 and/or ZP2 in intact zonae pellucidae was lost. However, the

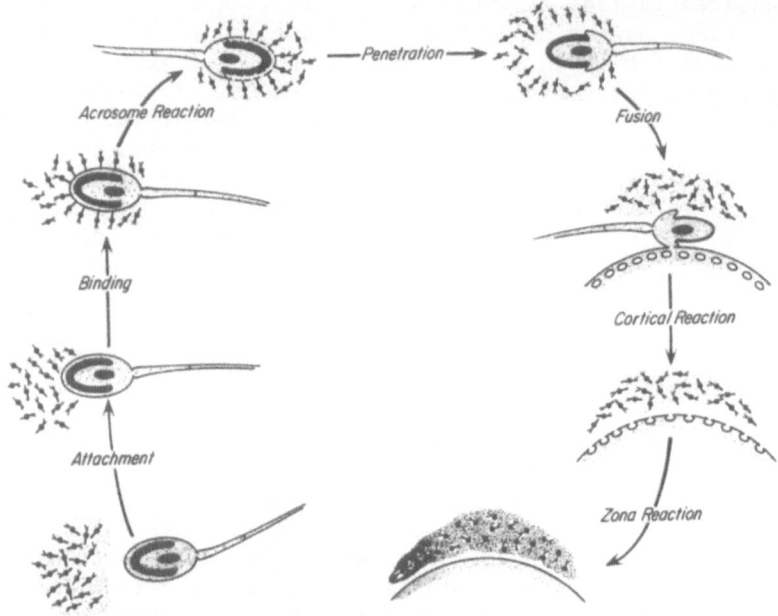

Figure 6. Diagrammatic presentation of various aspects of sperm-egg
          interaction in the mouse. Depicted are: (i) nonspecific
          attachment of sperm to the unfertilized egg's zona pellu-
          cida; (ii) specific binding of acrosome-intact sperm to
          receptors (ZP3; ball-and-stick model, with O-linked oligo-
          saccharides serving as the recognition site) in the zona
          pellucida; (iii) induction of the acrosome reaction
          mediated by binding of sperm to receptors (ZP3); (iv)
          passage of acrosome-reacted sperm through the zona pellu-
          cida; (v) fusion of an acrosome-reacted sperm with the
          egg's plasma membrane; (vi) induction of the cortical
          reaction (fusion of cortical granules with the egg's plasma
          membrane); and (vii) inactivation of sperm receptors by
          cortical granule exudate (presumably by modification of
          O-linked oligosaccharides of ZP3).

likelihood of this possibility is diminished in view of our finding
that the concentration dependence of purified ZP3 in reducing the
binding of sperm to eggs in vitro closely resembled that of solubilized zonae pellucidae from unfertilized eggs. In fact, there are
numerous reports of the isolation of proteins involved in cell-cell
recognition that retained biological activity even after exposure to
denaturing conditions. These include: an avian hepatic receptor
which binds to glycoproteins possessing a terminal galactose that was

solubilized in the presence of Triton X-100 (41); contact sites A
from Dictyostelium discoideum that were extracted from membranes
with 1% sodium deoxycholate (34); the sexual agglutinin from
flagellae of Chlamydomonas reinhardii that was extracted with octyl
glucoside (1,42); and the sperm binding factor from Hemicentrotus
pulcherrimus that was isolated by extraction of eggs with 1 M urea
(64). Finally, bacteriophage receptors that are, for the most
part, specific oligosaccharide components of the bacterial cell
wall, have been isolated under denaturing conditions without
altering their receptor activity (39). For example, Micrococcus
lysodeikticus cell walls treated with 2% SDS did not lose their
capacity to bind the tail tips of bacteriophage N1 (40) and,
similarly, a lipoprotein fraction extracted from Escherichia coli B
with 2 M urea inactivated bacteriophages T2 and T6, and lost
receptor activity only after heat or protease treatment (44).
Therefore, on the basis of these analogous observations, it should
not be surprising that ZP3 retained receptor activity even after
exposure to SDS and urea.

On the basis of electrophoretic analysis, ZP3 is present in
zonae pellucidae isolated from 2-cell embryos. However, unlike ZP3
purified from egg zonae pellucidae, ZP3 purified from embryo zonae
pellucidae does not exhibit sperm receptor activity. These
observations are consistent with both the inability of sperm to
bind to fertilized mouse eggs and embryos and the inability of
solubilized zonae pellucidae isolated from 2-cell embryos to reduce
the binding of sperm to eggs in vitro. It seems probable that ZP3
undergoes a structural change at the time of fertilization that is
sufficient to abolish its receptor activity, but insufficient to
alter significantly its electrophoretic mobility. Since certain
lectins bind to zonae pellucidae and prevent fertilization in vitro
(45,46,69), it is possible that this structural change involves the
carbohydrate moieties of ZP3 rather than the polypeptide chain
itself (see below). Such a situation would be compatible with the
suggestion that the binding of sperm to eggs in sea urchins is
mediated via a lectin-glycoprotein receptor interaction
(18,19,22,54,65).

We demonstrated that the sperm receptor activity of egg ZP3 is
unaffected by extensive proteolysis of the glycoprotein. Sperm
receptor activity was retained completely by small (1.5-6 kD
molecular weight) glycopeptides present in Pronase digests of
either egg zonae pellucidae or purified ZP3. These results
strongly suggest that the polypeptide chain of ZP3 is not directly
involved in sperm receptor function and that carbohydrate alone is
involved in this aspect of ZP3 function. This conclusion is
consistent with the finding that the receptor activity of ZP3 is
extremely stable, being virtually unaffected by lyophilization,
freezing and thawing, boiling, or by exposure to reducing agents
(dithiothreitol or 2-mercaptoethanol), detergents (0.1% SDS or 1%

Triton X-100), or denaturants (7 M urea). Furthermore, although
extensive proteolysis of ZP3 does not affect sperm receptor activity, other results from our laboratory suggest that selective
removal of O-linked, but not N-linked oligosaccharide from ZP3, by
treatment with mild alkali, inactivates the receptor. In fact, it
has been suggested previously that carbohydrate plays a role in the
binding of sperm to zonae pelludicae, since various lectins, monosaccharides, and glyconconjugates inhibit the binding of sperm to
mammalian eggs (32,45,58,59). Similarly, as discussed above,
oligosaccharides have been implicated as playing a role in cell-
cell recognition in a variety of other biological systems (10,17,
24). However, in nearly all cases, it has not been possible as yet
to completely exclude a role for polypeptide chain in receptor
function. In sea urchins, several lines of evidence suggest that
gamete adhesion is mediated by "bindin", a lectin-like sperm
protein associated with acrosomes and a carbohydrate containing
sperm receptor in the egg's vitelline envelope (18,2). Although
sperm receptor activity has been found associated with glycopeptides (about 6 kD molecular weight) derived from vitelline envelopes (38), the glycopeptides did not exhibit the species specificity observed with high molecular weight glycoconjugates (greater
than 350 kD molecular weight) found in vitelline envelopes or with
intact gametes (21,50). The small zona pellucida glycopeptides we
have examined act in a cell-specific manner, in that they inhibit
the tenacious binding of sperm to eggs without affecting the reversible attachment of sperm to embryos. Whether or not mouse zona
pellucida glycopeptides inhibit sperm binding in a species-specific
manner is not known as yet. In this context, it should be noted
that, although in vitro mammalian fertilization is, with few
exceptions, species-specific, it is not clear that this specificity
is always achieved at the level of sperm binding (69).

In order for sperm to penetrate the extracellular layers of
eggs and reach the plasma membrane, they must undergo the acrosome
reaction (Fig. 6). This reaction involves the fusion and extensive
vesiculation of the sperm's plasma and outer acrosomal membranes,
and results in exposure of the inner acrosomal membrane (13,27,29,
57,72,73). Mouse sperm that have completed the acrosome reaction
prior to exposure to unfertilized eggs in vitro fail to bind to the
zona pellucida (8,14,51,52,66,67,69), suggesting that it is the
sperm's plasma membrane that interacts with zonae pellucidae. This
is unlike the situation in sea urchins, since species-specific
interaction of the protein bindin (associated with the inner acrosomal membrane) with sperm receptors in the vitelline envelope can
only occur after completion of the acrosome reaction (13,18,57,65).

The acrosome reaction is characterized by calcium influx and
proton efflux through the plasma membrane surrounding the sperm
head (13,57). Presumably, a change in permeability of the plasma
membrane allows movement of ions to occur when sperm interact with

either the jelly coat (echinoderms) or zona pellucida (mammals).
Extracellular calcium is required in order for the acrosome reaction to take place, and the calcium ionophore A23187 acts as an
inducer of the acrosome reaction with both mammalian and nonmamm-
malian sperm. Recently, it has been proposed that calmodulin found
in the anterior region of the sperm head may serve as a link
between calcium influx and calcium-dependent fusion of the plasma
and outer acrosomal membranes (36). Concomitant with fusion of the
sperm's plasma and outer acrosomal membranes acid is released, and
it seems likely that an increase in intracellular pH is required
for the calcium-dependent acrosome reaction to take place (13,57).

In sea urchins it would appear that a fucose sulfate polysaccharide component of the egg's jelly coat is responsible for
inducing the acrosome reaction (55,56). Here, we have reviewed
evidence indicating that zona pellucida glycoprotein ZP3 is responsible for inducing mouse sperm to undergo the acrosome reaction.
We have found that ZP3 purified from unfertilized egg zonae pellucidae, but not from 2-cell embryo zonae pellucidae, can induce the
acrosome reaction to the same extent as calcium ionophore A23187;
neither ZP1 or ZP2, from eggs or embryos, was able to affect the
extent of acrosome reaction significantly above background levels.
The difference in effectiveness of ZP3 from egg and embryo zonae
pellucidae is, perhaps, the clearest indication that this glycoprotein is the component of egg zonae pellucidae responsible for
inducing the acrosome reaction in vitro. It is of interest to note
that, while both egg ZP3 and ionophore A23187 induce the acrosome
reaction in vitro, the ionophore drastically alters sperm motility,
whereas ZP3 does not. This difference may reflect the site-
specific interaction of ZP3 with plasma membrane surrounding only
the sperm head, while ionophore A23187 presumably interacts with
plasma membrane surrounding the entire sperm. Although the nature
of the role of the polypeptide chain of ZP3 in inducing the acrosome reaction is not clear, our observations using ZP3 glycopeptides suggest that the polypeptide chain does play a role. It is
not possible as yet to distinguish between the alternate possibilities of a direct effect of the polypeptide chain on sperm through
a secretagogue domain, as has been described for certain hormones
(31,48), or an indirect effect simply by serving as the backbone
for a polyvalent carbohydrate structure. In the latter context, it
should be noted the fucose-sulfate polysaccharide in sea urchin egg
jelly coat induces sea urchin sperm to undergo the acrosome
reaction; based on available compositional data, this polysaccharide is polyvalent with respect to both fucosyl and sulfate components (55,56). The sequence of events, initiated by binding of
sperm to ZP3 and culminating in changes in permeability of the
sperm's plasma membrane to various ions that lead directly to the
acrosome reaction, needs to be further defined.

Both binding of sperm to the mouse egg's zona pellucida and induction of the acrosome reaction appear to be mediated by ZP3 alone (Fig. 6). These two events can be distinguished from one another in vitro because of their distinctive time-courses (8,14), sensitivity to inhibitors (14), and dependence on ZP3 concentration (8), and can be biochemically dissected from one another by proteolysis of the polypeptide chain of ZP3 (16). Therefore, in the mouse, a single glycoprotein alone performs the functions ascribed to fucose sulfate polysaccharide of the jelly coat (acrosome reaction-inducing activity) and (a) glycoprotein(s) of the vitelline envelope (sperm receptor activity) of sea urchin eggs. Following fertilization, ZP3 exhibits neither sperm receptor nor acrosome reaction-inducing activities. In this connection, it has been reported that exposure of mammalian egg zonae pellucidae to material exuded from activated eggs, presumably cortical granule contents (Fig. 6), results in a dramatic decrease in sperm binding (3,28,71). Furthermore, since the effect on sperm binding was prevented by trypsin inhibitors, it was concluded that a protease, originating from cortical granules, inactivated sperm receptors in zonae pellucidae. While there is no evidence as yet for proteolysis of ZP3 following either fertilization (7) or parthenogenetic activation (71) of mouse eggs, it is possible that release of a small glycopeptide, possessing sperm receptor activity, has simply not been detected due to limited resolution of the analytical procedures used thus far. Since a thorough understanding of the effects of cortical granule material on zonae pellucidae is lacking at present (25), the potential involvement of specific glycosidases in inactivation of sperm receptors following fertilization should also be considered.

ACKNOWLEDGEMENTS

This research was supported by grants from the National Institute of Child Health and Human Development and the National Science Foundation. H.M.F. and J.M.G. were National Institutes of Health and Rockefeller Foundation postdoctoral fellows, respectively, and J.D.B., R.J.R., and G.S.S. were predoctoral fellows supported by National Research Service Awards.

REFERENCES

1. Adair, W.S., Monk, B.C., Cohen, R., Hwang, C., and Goodenough, U. W. 1982. Sexual agglutinins from the Chlamydomonas flagellar membrane. Partial purification and characterization. J. Biol. Chem. 257:4593-4602.
2. Austin, C.R. 1965. "Fertilization", Prentice-Hall, New Jersey.
3. Barros, C., and Yanagimachi, R. 1971. Induction of the zona reaction in golden hamster eggs by cortical granule material. Nature 233:268-269.

4. Bleil, J.D., and Wassarman, P.M. 1978. Identification and characterization of the proteins of the zona pellucida. J. Cell Biol. 79:173a.

5. Bleil, J.D., and Wassarman, P.M. 1980. Structure and function of the zona pellucida: identification and characterization of the proteins of the mouse oocyte's zona pellucida. Devel. Biol. 76:185-203.

6. Bleil, J.D., and Wassarman, P.M. 1980. Synthesis of zona pellucida proteins by denuded and follicle-enclosed mouse oocytes during culture in vitro. Proc. Nat. Acad. Sci. USA 77:1029-1033.

7. Bleil, J.D., and Wassarman, P.M. 1980. Mamalian sperm-egg interaction: identification of a glycoprotein in mouse egg zonae pellucidae possessing receptor activity for sperm. Cell 20:873-882.

7a. Bleil, J.D., Beall, C.F., and Wassarman, P.M. 1981. Mammalian sperm-egg interaction: fertilization of mouse eggs triggers modification of the major zona pellucida glycoprotein. Devel. Biol. 86:189-197.

8. Bleil, J.D., and Wassarman, P.M. 1983. Sperm-egg interactions in the mouse: sequence of events and induction of the acrosome reaction by a zona pellucida glycoprotein. Devel. Biol. 95:317-324.

9. Christiensen, J.R. 1975. The kinetics of reversible and irreversible attachment of bacteriophage T1. Virol. 26:727-737.

10. Culp, L.A. 1978. Biochemical determinants of cell adhesion. Curr. Topics Memb. Trans. 11:327-396.

11. Edge, A.S.B., Faltynek, C.R., Hof, L., Reichert, L.E., and Weber, P. 1981. Deglycosylation of glycoproteins by trifluoromethane-sulfonic acid. Anal. Biochem. 118:131-137.

12. Elder, J.H., and Alexander, S. 1982. Endo-beta-N-acetyl-glucosaminidase F: endoglycosidase from Flavobacterium meningosepticum that cleaves both high-mannose and complex glycoproteins. Proc. Nat. Acad. Sci. USA 79:4540-4544.

13. Epel, D., and Vacquier, V.C. 1978. Membrane fusion events during invertebrate fertilization. In "Cell Surface Reviews" (G. Poste and G. L. Nicolson, eds.), 5:1-63, Elsevier: North Holland, Amsterdam.

14. Florman, H.M., and Storey, B.T. 1982. Mouse gamete inter-actions: the zona pellucida is the site of the acrosome reaction leading to fertilization in vitro. Devel. Biol. 91:121-130.

15. Florman, H.M., and Wassarman, P.M. 1985. O-linked oligosaccha-rides of mouse egg ZP3 account for its sperm receptor activity. Cell 41:313-324.

16. Florman, H.M., Bechtol, K.B., and Wassarman, P.M. 1984. Enzy-matic dissection of the functions of the mouse egg's recep-tor for sperm. Devel. Biol. 106:243-255.

17. Frazier, W.A., and Glaser, L. 1979. Surfact components and
    cell recognition. Ann. Rev. Biochem. 48:491-523.
18. Glabe, C.G., and Vacquier, V.D. 1978. Egg surface glycoprotein
    receptor for sea urchin sperm bindin. Proc. Nat. Acad. Sci.
    USA 75:881-885.
19. Glabe, C.G. 1979. A supramolecular theory for specificity in
    intercellular adhesion. J. Theoret. Biol. 78:417-432.
20. Glabe, C.G., and Lennarz, W.J. 1979. Species-specific sperm
    adhesion in sea urchins: a quantitative investigation of
    bindin mediated egg agglutination. J. Cell Biol. 83:595-
    604.
21. Glabe, C.G., and Lennarz, W.J. 1981. Isolation of a high
    molecular weight glycoconjugate derived from the surface of
    Strongylocentrotus purpuratus eggs that is implicated in
    cell adhesion. J. Supramol. Struc. Cell Biochem.
    15:387-394.
22. Glabe, C.G., Grabel, L.B., Vacquier, V.D., and Rosen, S.D.
    1982. Carbohydrate specificity of sea urchin sperm bindin:
    a cell surface lectin mediating sperm-egg adhesion. J. Cell
    Biol. 94:123-128.
23. Greve, J.M., Salzmann, G.S., Roller, R.J., and Wassarman, P.M.
    1982. Biosynthesis of the major zona pellucida glycoprotein
    secreted by oocytes during mammalian oogenesis. Cell
    31:749-759.
24. Grinnel, F. 1978. Cellular adhesiveness and extracellular
    substrata. Inter. Rev. Cytol. 53:65-144.
25. Gulyas, B.J. 1980. Cortical granules of mammalian eggs.
    Inter. Rev. Cytol. 63:357-392.
26. Gwatkin, R.B.L., Williams, D.T., Hartman, J.F., and Kniazuk,
    M. 1973. The zona reaction of hamster and mouse eggs:
    production in vitro by a trypsin-like protease from cortical
    granules. J. Reprod. Fert. 32:259-265.
27. Gwatkin, R.B.L. 1976. Fertilization. In "Cell Surface Reviews"
    (G. Poste and G. L. Nicholson, eds.), 1:1-52, Elsevier:
    North Holland, Amsterdam.
28. Gwatkin, R.B.L., and Williams, D.T. 1976. Receptor activity of
    the solubilized hamster and mouse zona pellucida before and
    after the zona reaction. J. Reprod. Fert. 49:55-59.
29. Gwatkin, R.B.L. 1977. "Fertilization Mechanisms in Man and
    Mammals", Plenum Press, New York.
30. Hartmann, J.F., Gwatkin, R.B.L., and Hutchinson, C.F. 1972.
    Early contact interactions between mammalian gametes in
    vitro: evidence that the vitellus influences adherence
    between sperm and zona pellucida. Proc. Nat. Acad. Sci. USA
    69:2767-2769.
31. Hoffmann, K., Wingender, W., and Finn, F.M. 1970. Correlations
    of adrenocorticotropic activity with ACTH analogs with
    degree of binding to an adrenal gland particulate
    preparation. Proc. Nat. Acad. Sci. USA 67:820-833.

32. Huang, T.T.F., Ohzu, E., and Yanagimachi, R. 1982. Evidence suggesting that L-fucose is part of the recognition signal for sperm-zona pellucida attachment in mammals. Gamete Res. 5:355-361.

33. Hubbard, S.C., and Ivatt, R.J. 1981. Synthesis and processing of asparagine-linked oligosaccharides. Ann. Rev. Biochem. 50:555-583.

34. Huesgen, A., and Gerisch, G. 1975. Solubilized contact sites A from cell membranes of Dictyostelium discoideum. FEBS Lett. 56:46-49.

35. Inoue, M., and Wolf, D.P. 1975. Sperm binding characteristics of the murine zona pellucida. Biol. Reprod. 13:340-346.

36. Jones, H.P., Lenz, R.W., Palevitz, B.A., and Cormier, M.J. 1980. Calmodulin localization in mammalian spermatozoa. Proc. Natl. Acad. Sci. USA 77:2772-2776.

37. Kalyan, N.K., and Bahl, O.P. 1983. Role of carbohydrate in human chorionic gonadotropin. Effect of deglycosylation on the subunit interaction and on its in vitro and in vivo biological properties. J. Biol. Chem. 258:67-74.

38. Kinsey, W.J., and Lennarz, W.J. 1981. Isolation of a glycopeptide fraction from the surface of the sea urchin egg that inhibits sperm-egg binding and fertilization. J. Cell Biol. 91:325-331.

39. Linberg, A.A. 1973. Bacteriophage receptors. Ann. Rev. Microbiol. 27:205-239.

40. Lovett, P.S., and Shockman, G.D. 1970. Interaction of bacteriophage N1 with cell walls of Micrococcus lysodeikticus. J. Virol. 6:135-144.

41. Lunney, J., and Ashwell, G. 1976. A hepatic receptor of avian origin capable of binding specifically modified glycoproteins. Proc. Nat. Acad. Sci. USA 73:341-343.

42. Mesland, D.A.M., Hoffman, J.L., Caligor, E., and Goodenough, U.W. 1980. Flagellar tip activation stimulated by membrane adhesions in Chlamydomonas gametes. J. Cell Biol. 84:599-617.

43. Metz, C.B., and Monroy, A. 1985. "Biology of Fertilization", Academic Press, New York.

44. Michael, J.G. 1968. The surface antigens and phage receptors on Escherichia coli B. Proc. Soc. Exp. Biol. Med. 128:434-438.

45. Oikawa, T., Nicolson, G.L., and Yanagimachi, R. 1973. Wheat germ agglutinin blocks mammalian fertilization. Nature 241:256-259.

46. Oikawa, T., and Yanagimachi, R. 1975. Block of hamster fertilization by anti-ovary antibody. J. Reprod. Fert. 45:487-494.

47. Robbins, P.W., Hubbard, S.C., Turco, S.J., and Wirth, D.F. 1977. Proposal for a common oligosaccharide intermediate in the synthesis of membrane glycoproteins. Cell 12:893-900.

48. Rodbell, M., Birnbaumer, L., Pohl, S.L., and Sunby, F. 1971. The reaction of glucagon with its receptor: evidence for discrete regions of activity and binding in the glucagon molecule. Proc. Nat. Acad. Sci. USA 68:909-913.

49. Roller, R.J., and Wassarman, P.M. 1983. Role of asparagine-linked oligosaccharides in secretion of glycoproteins of the mouse egg's extracellular coat. J. Biol. Chem. 258:13243-13249.

50. Rossignol, D.P., Roschelle, A.J., and Lennarz, W.J. 1981. Sperm-egg binding: identification of specific sperm receptor from eggs of Strongylocentrotus purpuratus. J. Supramol. Struct. Cell. Biochem. 15:347-358.

51. Saling, P.M., Wosinski, J., and Storey, B.T. 1979. An ultra-structural study of epididymal mouse spermatozoa binding to zonae pellucidae in vitro: sequential relationship to the acrosome reaction. J. Exp. Zool. 109:229-238.

52. Saling, P.M., and Storey, B.T. 1979. Mouse gamete interactions during fertilization in vitro. Chlortetracycline as a fluorescent probe for the mouse acrosome reaction. J. Cell Biol. 83:544-555.

53. Salzmann, G.S., Greve, J.M., Roller, R.J., and Wassarman, P.M. 1983. Biosynthesis of the sperm receptor during oogenesis in the mouse. EMBO J. 2:1451-1456.

54. Schmell, E., Earles, B.J., Breaux, C., and Lennarz, W.J. 1977. Identification of a sperm receptor on the surface of eggs of the sea urchin, Arbacia punctulata. J. Cell Biol. 72:35-46.

55. SeGall, G.K., and Lennarz, W.J. 1979. Chemical characterization of the component of the jelly coat from sea urchin eggs responsible for induction of the acrosome reaction. Devel. Biol. 71:33-48.

56. SeGall, G.K., and Lennarz, W.J. 1981. Jelly coat and induction of the acrosome reaction in echinoderm sperm. Devel. Biol. 86:87-93.

57. Shapiro, B.D., Schackmann, R.W., and Gabel, C.A. 1981. Molecular approaches to the study of fertilization. Ann. Rev. Biochem. 50:815-843.

58. Shur, B.D., and Hall, N.G. 1982. Sperm surface glycosyltrans-ferase activities during in vitro capacitation. J. Cell Biol. 95:567-573.

59. Shur, B.D., and Hall, N.G. 1982. A role for mouse sperm surface galactosyltransferase in sperm binding to the egg zona pellucida. J. Cell Biol. 95:574-579.

60. Takatsuki, A., Fukui, Y., and Tamura, A. 1975. Inhibition of biosynthesis of polyisoprenol sugars in chick embryo micro-somes of tunicamycin. Agric. Biol. Chem. 39:2089-2091.

61. Tarentino, A.L., and Maley, F. 1974. Purification and properties of an endo-beta-N-acetylgucosaminidase from Streptomyces griseus. J. Biol. Chem. 249:811-817.

62. Tarentino, A.L., Plummer, T.H., and Maley, F. 1974. The release
    of intact oligosaccharides from specific glycoproteins by
    endo-beta-N-acetylglucosaminidase H. J. Biol. Chem.
    249:818-824.
63. Tkacz, J.S., and Lampen, J.O. 1975. Tunicamycin inhibition of
    polyisoprenyl N-acetylglucosaminyl pyrophosphate formation
    in calf liver microsomes. Biochem. Biophys. Res. Commun.
    65:248-257.
64. Tsuzuki, H., Yoshida, M., Onitake, K., and Aketa, K. 1977.
    Purification of the sperm-binding factor from the egg of the
    sea urchin Hemicentrotus pulcherrimus. Biochem. Biophys.
    Res. Commun. 76:502-511.
65. Vacquier, V.D., and Moy, G. 1977. Isolation of bindin: the
    protein responsible for adhesion of sperm to sea urchin
    eggs. Proc. Nat. Acad. Sci. USA 74:2456-2460.
66. Wassarman, P.M., and Bleil, J.D. 1982. The role of zona pellu-
    cida glycoproteins as regulators of sperm-egg interaction in
    the mouse. In "Cellular Recognition" (W.A. Frazier, L.
    Glaser, and D.I. Gottlieb, eds.), pp. 845-863, Alan R. Liss,
    New York.
67. Wassarman, P.M. 1983. Fertilization. In "Cell Interactions and
    Development: Molecular Mechanisms" (K. Yamada, ed.) pp.
    1-27, John Wiley and Sons, New York.
68. Wassarman, P.M. 1983. Oogenesis: synthetic events in the
    developing mammalian egg. In "Mechanism and Control of
    Animal Fertilization", (J. F. Hartmann, ed.), pp. 1-54,
    Academic Press, New York.
69. Wassarman, P.M., Florman, H.M., and Greve, J.M. 1985. Receptor
    mediated sperm-egg interactions in mammals. In "Biology of
    Fertilization" (C.B. Metz and A. Monroy, eds.), 2:341-360,
    Academic Press, New York.
70. Wassarman, P.M., Greve, J.M., Perona, R.M., Roller, R.J., and
    Salzmann, G.S. 1984. How mouse eggs put on and take off
    their extracellular coat. In "Molecular Biology of
    Development" (E.H. Davidson and R. A. Firtel, eds.), pp.
    213-225, Alan R. Liss, New York.
71. Wolf, D.P., and Hamada, M. 1977. Induction of zonal and
    oolemmal blocks to sperm penetration in mouse eggs with
    cortical granule exudate. Biol. Reprod. 21:205-211.
72. Yanagimachi, R. 1978. Sperm-egg association in mammals. In
    "Current Topics in Developmental Biology" (A. A. Moscona and
    A. Monroy, eds.), 12:83-105, Academic Press, New York.
73. Yanagimachi, R. 1981. Mechanisms of fertilization in mammals.
    In "Fertilization and Embryonic Development In Vitro" (L.
    Mastroianni and J. D. Biggers, eds.) pp. 81-182, Plenum
    Press, New York.

THE RECEPTOR FUNCTION OF GALACTOSYLTRANSFERASE DURING MAMMALIAN FERTILIZATION

Barry D. Shur

University of Connecticut Health Center
Department of Anatomy
Farmington, Connecticut 06032

SUMMARY

The molecular mechanisms underlying mouse sperm binding to the egg zona pellucida are being examined. A variety of studies suggest that galactosyltransferase (GalTase) on the sperm surface at least partly mediates gamete adhesion by binding to its appropriate carbohydrate substrate in the egg zona pellucida. The first indication that GalTase serves as a sperm receptor came from a biochemical analysis of sperm bearing mutant alleles of the t-complex. t-bearing sperm, which are transmitted perferentially during fertilization relative to normal sperm, have four times the surface GalTase activity of wild-type, while eight other enzyme activities are indistinguishable between normal and t-sperm populations. Interestingly, mutant t-sperm that have lost their fertilizing superiority due to genetic recombination, no longer show elevated GalTase activity. Before sperm are capable of binding to the eggs, the GalTase active site becomes exposed on the sperm surface, due to the release of a competitive substrate. When added back to sperm/zona binding assays, this GalTase substrate inhibits binding by competing for the GalTase active site. Inhibition and/or modification of the sperm GalTase with either substrate analogues or modifier proteins, produces a parallel inhibition of sperm GalTase activity and sperm binding to the zona pellucida. The presence of the second GalTase substrate, UDPGal, causes the dissociation of sperm bound to the zona pellucida. UDPGal forces catalysis to occur, thus releasing the sperm GalTase from its zona pellucida substrate. Purified GalTase inhibits sperm/-zona binding, as does galactosylation of the zona pellucida. Monospecific anti-GalTase IgG, and its Fab fragments, inhibit sperm/zona binding and GalTase catalytic activity. The sperm surface GalTase is

localized by indirect immunofluorescence to the plasma membrane overlying the intact acrosome. Finally, the zona pellucida possess substrates for the sperm GalTase, and enzymatic modification of these
substrates inhibits subsequent sperm binding. These studies, as well
as others, all suggest that sperm GalTase serves as at least one, and
possibly the principal, sperm surface receptor for binding to the
zona pellucida during fertilization. Studies are in progress to
identify the complementary GalTase substrate in the zona pellucida.

INTRODUCTION

     The union of sperm and eggs is one of the most intensely studied
examples of cellular recognition that occurs during development.
Over seventy years ago, Lillie proposed that gamete recognition is
mediated by complementary cell surface receptors, and that the specificity of sperm-egg binding is dictated by species-specific receptors
(10; see Metz, this volume). The nature of these receptors still
remains unknown, even though the last ten years has seen considerable
progress towards their identification.

     According to present knowledge, sperm-egg binding is thought to
be mediated by interactions between complementary cell surface proteins and glycoconjugates, a theme that has been popular throughout
studies of cell social behavior in other developmental systems. For
example, a protein has been isolated from sea urchin sperm, "bindin",
that is capable of agglutinating eggs of homologous species (31).
The bindin receptor isolated from eggs has been characterized as a
large molecular weight glycoconjugate (9). Similarly, studies of
Ascidian (11), and horseshoe crab (3) fertilization, support the
notion that gamete recognition is mediated by interactions between
cell surface glycoconjugates and presumably, these interactions are
indeed responsible for species-specific fertilization.

     Studies described in this paper, as well as those from other
laboratories (1,5) suggest that mammalian fertilization is also
dictated by interactions between complementary cell surface proteins
and carbohydrates. However, unlike sperm from lower species, mammalian sperm must first be "capacitated" within the female reproductive
tract before they are able to bind, penetrate and fuse with the egg.
Capacitation is a poorly understood process that involves extensive
alterations in the sperm plasma membrane, including surface glycoconjugates, antigens, lipids, intramembrane particles and ion permeability, as well as dramatic changes in sperm intermediary metabolism (6)
Besides these changes to the sperm per se, the sperm must traverse
through the female reproductive tract, encountering changes in the
fluid and cellular milieu, as well as physical obstacles to forward
transport, including the cervix and uterotubal junction (13). It
remains unclear how any of these sperm-associated changes that occur

within the female reproductive tract affect the sperm receptor for
the egg zona pellucida.

In addition, there are genetic factors that influence the
efficacy of mammalian fertilization. It is generally assumed that
the union of gametes within a species is random, and independent of
their genotypes. This is not necessarily true, since mouse sperm
that carry particular rearrangements of the 17th chromosome appear to
be physiologically superior to normal sperm, and are therefore
transmitted preferentially during fertilization relative to normal
sperm (4,28).

Thus, there are a multitude of factors - biochemical, physiological and genetic - that influence mammalian fertilization, and
sperm-egg binding in particular. The work discussed in this paper
will briefly review studies in our laboratory that have helped to
identify the receptor on the mouse sperm surface that participates in
binding to the egg zona pellucida. Through the use of conventional
biochemical, genetic and serological studies, we now realize that a
sperm surface enzyme, galactosyltransferase (GalTase), which has a
strict binding specificity for particular carbohydrate residues,
appears to be at least one of the components on the sperm surface
that mediates gamete recognition (Fig. 1). In the following discussion, GalTase will first be defined, and then the effect of genetic
mutation and in vitro capacitation on the sperm surface GalTase
activity will be examined. Finally, the biochemical and antibody
studies that have directly examined the sperm receptor function of
surface GalTase will be reviewed.

Cell Surface GalTase

GalTase is one member of a family of enzymes called glycosyltransferases that are responsible for the synthesis of all complex
glycoconjugates, including glycoproteins, glycolipids, and glycosaminoglycans. Most glycosyltransferases are membrane bound, but some
activities have been found in soluble form in a variety of body
fluids (17).

Glycosyltransferases synthesize complex carbohydrates by the
addition of specific monosaccharide residues donated from sugar
nucleotide substrates to the non-reducing terminus of the growing
polysaccharide chain. [An exception to this rule, however, is the
synthesis of the core oligosaccharide of the asparagine-linked glycoproteins, which is synthesized on lipid soluble intermediates and
transferred en bloc to nascent polypeptides (30)]. The glycosyltransferases are named according to their sugar nucleotide substrates, such that GalTase specifically transfers galactose donated
from its UDPGal substrate to a particular sugar acceptor substrate.
In the case of glycoprotein biosynthesis, the sugar acceptor sub-

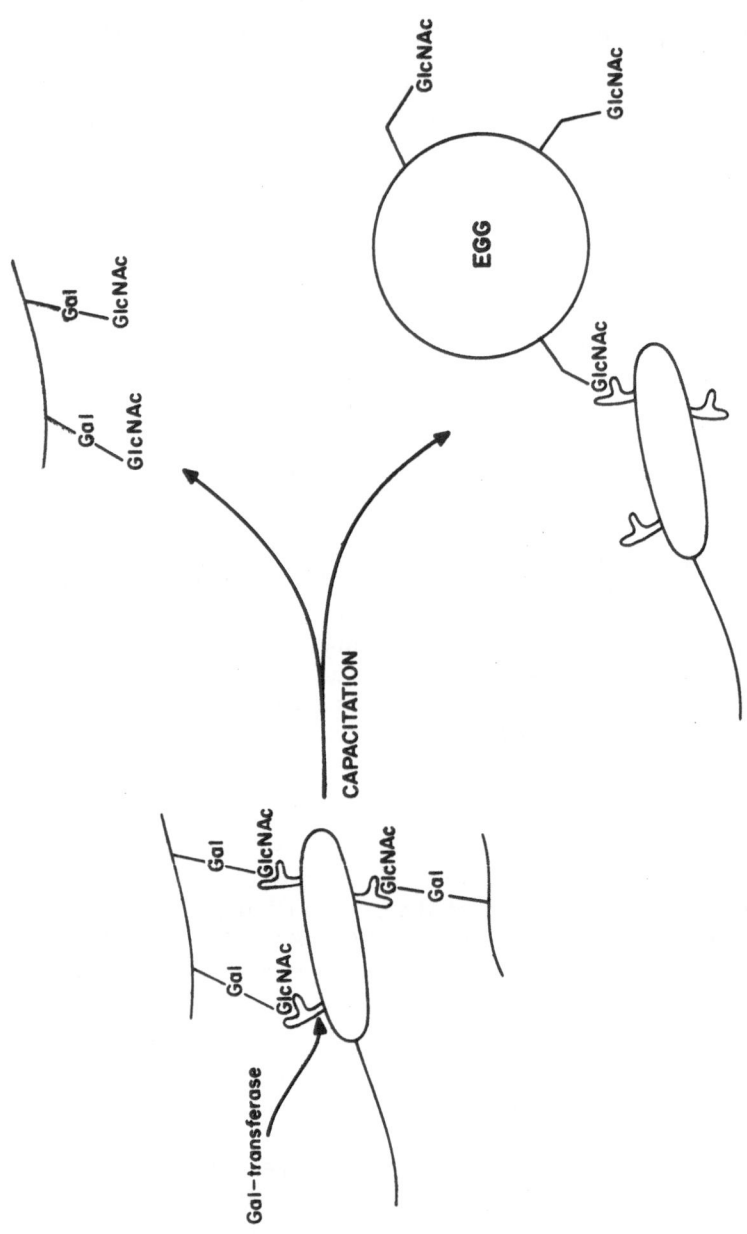

Figure 1. A proposed model for mouse sperm binding to the egg zona pellucida. Studies suggest that sperm surface galactosyltransferase (Gal-transferase) participates during fertilization by binding to its carbohydrate substrate in the zona pellucida. Within the epididymis, the sperm Gal-transferase active site is occupied by a large molecular weight lactosaminoglycan substrate, of which only the terminal dissaccharides are pictured. As a result of capacitation, the lactosaminoglycan substrate is released, thus exposing the Gal-transferase active site for binding to terminal N-acetylglucosamine (GlcNAc) residues in the zona pellucida. Reprinted from (24).

strate usually terminates in an N-acetylglucosamine (GlcNAc) residue. Thus, polysaccharide chains are synthesized by the sequential addition of monosaccharides catalyzed by specific glycosyltransferases. The composition of the resulting glyconconjugate will therefore reflect the spatial and temporal specificities of glycosyltransferases active within each cell.

It has long been thought that membrane-bound glycosyltransferases were localized exclusively in the intracellular membranes of the endoplasmic reticulum and Golgi apparatus, where they synthesize glycoconjugates for secretion and transport to the cell surface. However, it is now clear that glycosyl transferases exist on the cell surface as well, with their active sites exposed to the extracellular milieu (for review, see 26,27). On the cell surface, glycosyltransferases have been suggested to participate in cell-cell adhesions by recognizing, and binding to, their specific sugar acceptor substrates on adjacent cell surfaces. Since sugar nucleotide substrates are not present in the extracellular fluids, glycosyltransferase-mediated cell adhesions presumably remain stable in a "dead-end" enzyme-substrate complex. If desired, cell dissociation could result by providing the second glycosyltransferase substrate, (i.e., sugar nucleotide), which would force catalysis to occur.

Usually, glycosyltransferases are detected on the cell surface by direct enzymatic assay of intact cells, which requires the addition of radiolabeled sugar nucleotide substrates to cell suspensions (26,27). The surface glycosyltransferase is then able to transfer the labeled sugar motiety to sugar acceptor substrates either on the cell surface (endogenous acceptors) or to substrates added exogenously to the incubation medium. In these experiments, it is critical to control for the intracellular utilization of sugar nucleotide breakdown products, such as free sugar, which can be incorporated into glycoconjugates intracellularly. One can effectively eliminate this potential problem either through the use of inhibitors of sugar nucleotide hydrolysis and/or inhibitors of free sugar transport. One must also consider the possible contribution of glycosyltransferases secreted from the cells into the incubation medium. Additionally, the exposure of intracellular glycosyltransferase due to cell death or lysis must be controlled for.

Recently, assays of surface GalTase have circumvented these problems through the use of isolated plasma membrane fractions rather than whole cells (7) and by the use of monospecific anti-GalTase antisera raised against purified GalTase, which have been used to detect surface GalTase by immunofluorescence (18,25,27) and immuno-precipitation (29).

The T/t-Complex of the Mouse

Our present focus on GalTase originated from biochemical studies on the T/t-complex of the mouse. The T/t-complex is located near the centromeric end of chromosome 17 and contains over 100 known dominant (T) and recessive (t) mutations. The mutant T/t alleles have pronounced effects on spermatogenesis, fertilization, embryonic morphogenesis and meiotic recombination, and consequently, have captured the attention of reproductive, developmental and molecular biologists (4,28).

Numerous studies have suggested that mutant T/t alleles modify the cell surface, which leads to abnormalities in cellular interactions characteristic of mutant T/t tissues. Defects in intermediary metabolism have also been reported in some T/t mutant tissues. Many of the diverse effects of the T/t-complex on fertilization and development can be dissociated from one another by rare recombinational events. In this way, mouse strains can be constructed that demonstrate only selected phenotypes of the complete T/t-complex. Such recombinants have proved invaluable in further characterizing the molecular defects associated with particular phenotypic abnormalities.

Surface GalTase on Mutant t-Sperm

Sperm bearing mutant t alleles demonstrate a phenomenon called "transmission-distortion", during which t-sperm are preferentially transmitted during fertilization, relative to normal sperm. In order to approach the biochemical defect associated with sperm transmission-distortion, the specific activities of nine enzymes were compared between normal and t-sperm populations. Of the nine enzymes assayed, only surface GalTase proved to be different, having four times the specific activity on transmission-distorting t-sperm populations, relative to normal sperm (19,22). The GalTase assay was linear with time, cell number, and included saturating concentrations of all required substrates and cofactors. The eight other enzymes that were assayed, and which proved to be indistinguishable between normal and mutant sperm, included two other glycosyltransferases, three glycosidases, and three phosphatases.

In order to determine which portions of the T/t-complex were responsible for the elevated surface GalTase activity, GalTase was assayed on non-transmission-distorting sperm bearing one of a number of recombinant t alleles or dominant T alleles. Results showed that without exception, all non-transmission-distorting t- and T-sperm no longer showed elevated GalTase activity, whereas seven different complementation groups of t-sperm that show strong transmission distortion during fertilization, all produced a four-fold increase in enzyme activity (22). Therefore, the elevated sperm GalTase activity strictly correlated with the fertilizing ability of these sperm, suggesting that sperm surface GalTase plays some fundamental role in mammalian fertilization.

The Effects of In Vitro Capacitation on Surface GalTase Activity

In light of these results, we examined the consequences of sperm capacitation on surface GalTase activity, since it is believed that one result of capacitation is the release of sperm surface glycoconjugates, which, when reintroduced to the sperm, can inhibit binding to the egg zona pellucida (6). Specifically, we examined whether in vitro capacitation involves the release of GalTase substrates from the sperm surface, thereby exposing the enzyme active site for binding to the zona pellucida.

It was found that surface GalTase on uncapacitated sperm is occupied by a novel, large molecular weight glycoconjugate substrate shown to be composed of repeating N-acetyllactosamine (galactose --→ GlcNAc) residues (24). The lactosaminoglycan composition of the substrate was further established by immunoprecipitation with antiserum raised against F9 embryonal carcinoma cells (2). Anti-F9 antiserum recognizes a class of lactosaminoglycans on embryonal carcinoma (12,21) and embryonic (20) cells, and immunoprecipitates 87% of the sperm GalTase reaction product. Under identical conditions, normal mouse serum precipitates only 9.2% of the product.

Sperm capacitation is thought to be Ca^{++} dependent (16), and when sperm are capacitated in Ca^{++}-containing medium, the lactosaminoglycan substrates are spontaneously released from the sperm surface GalTase, thereby exposing the enzyme's active site for binding competitive exogenous substrates, or for binding to the zona pellucida (24). Sperm capacitation can be mimicked, in the absence of Ca^{++}, by either washing sperm in Ca^{++}-free medium, or by pretreating sperm with anti-F9 antiserum to displace the GalTase substrate. In both instances, sperm galactosylation of endogenous substrates is reduced, while galactosylation of competitive exogenous substrates is increased, as is binding to the zona pellucida. Normal mouse serum has no effect on either sperm GalTase activity or on binding to the zona pellucida. In addition, anti-F9 antiserum has no effect on four other sperm enzymes including sialyltransferase, acid phosphatase, alkaline phosphatase, and nucleotide pyrophosphatase (19,24).

Binding of capacitated sperm to the egg can be inhibited by purified deproteinized, delipidated high molecular weight lactosaminoglycans extracted from epididymal fluids or from embryonal carcinoma cells (24). Thus, these glycosides function as competitive inhibitors when added back to in vitro fertilization assays. These glycosides inhibit sperm egg binding by competing for the sperm surface GalTase since they are galactosylated by sperm in the presence of radiolabeled UDPGal, and enzymatic removal of terminal GlcNAc residues reduces their competition. On the other hand, conventional

glycosides extracted from epididymal fluids fail to inhibit capacitated sperm binding to the zona pellucida, nor do extracts from
differentiated embryonal carcinoma cells that no longer synthesize
lactosaminoglycans. Collectively, these results suggest a molecular
mechanism for one aspect of sperm capacitation, and help define why
removal of epididymal glycoconjugates is a necessary prerequisite for
sperm binding to the zona pellucida.

Perturbing Sperm GalTase Inhibits Sperm/Egg Binding

The effects of in vitro capacitation on sperm GalTase directly
suggested that the surface enzyme may participate during fertilization by binding to the zona pellucida. To test this possibility, the
effects of three different reagents that perturb surface GalTase were
examined on sperm/zona binding (23,25).

First, the effect of UDP-dialdehyde, a substrate analogue that
inhibits GalTase activity, was examined on sperm binding. When UDP-
dialdehyde was added to sperm/zona binding assays, it inhibited sperm
binding to the zona pellucida coincident with inhibition of sperm
surface GalTase (23). Of five other sperm enzymes assayed, four were
totally unaffected by UDP-dialdehyde, while one was inhibited only
slightly. Covalent linkage of UDP-dialdehyde to sperm GalTase dramatically inhibited binding to eggs, while treatment of eggs with UDP-
dialdehyde had no effect on sperm binding.

Second, the effect of the milk protein, α-lactalbumin (α-LA), on
sperm/zona binding was examined. α-LA binds to glycoprotein GalTase
and changes its substrate specificity away from GlcNAc and towards
glucose (8). Thus, milk sugar, or lactose, is synthesized rather
than an N-acetyllactosamine linkage. According to present knowledge,
the only known binding affinity of α-LA is for GalTase (8). α-LA
simultaneously inhibits sperm surface GalTase towards GlcNAc, stimulates activity towards glucose, and inhibits sperm binding to the
zona pellucida in a dose-dependent manner (23). Under identical
conditions, bovine serum albumin does not interfere with sperm/zona
binding, and α-LA loses its ability to inhibit GalTase activity and
sperm/zona binding when denatured by boiling (Fig. 2).

Third, the effects of adding the GalTase sugar donor substrate,
UDPGal, was examined on sperm/zona binding (25). If the surface
GalTase is mediating sperm/zona binding in a "dead-end" enzyme-
substrate complex, then we reasoned that addition of the missing
second GalTase substrate, i.e. UDPGal, would force catalysis to completion, thereby allowing the sperm GalTase to dissociate from its
zona pellucida glycoside substrate. The net result would be an
inhibition of sperm/zona binding. Indeed, the addition of UDPGal
produced a dose-dependent inhibition of binding, while identical
concentrations of the related sugar nucleotide, UDPglucose, had no
effect on gamete binding.

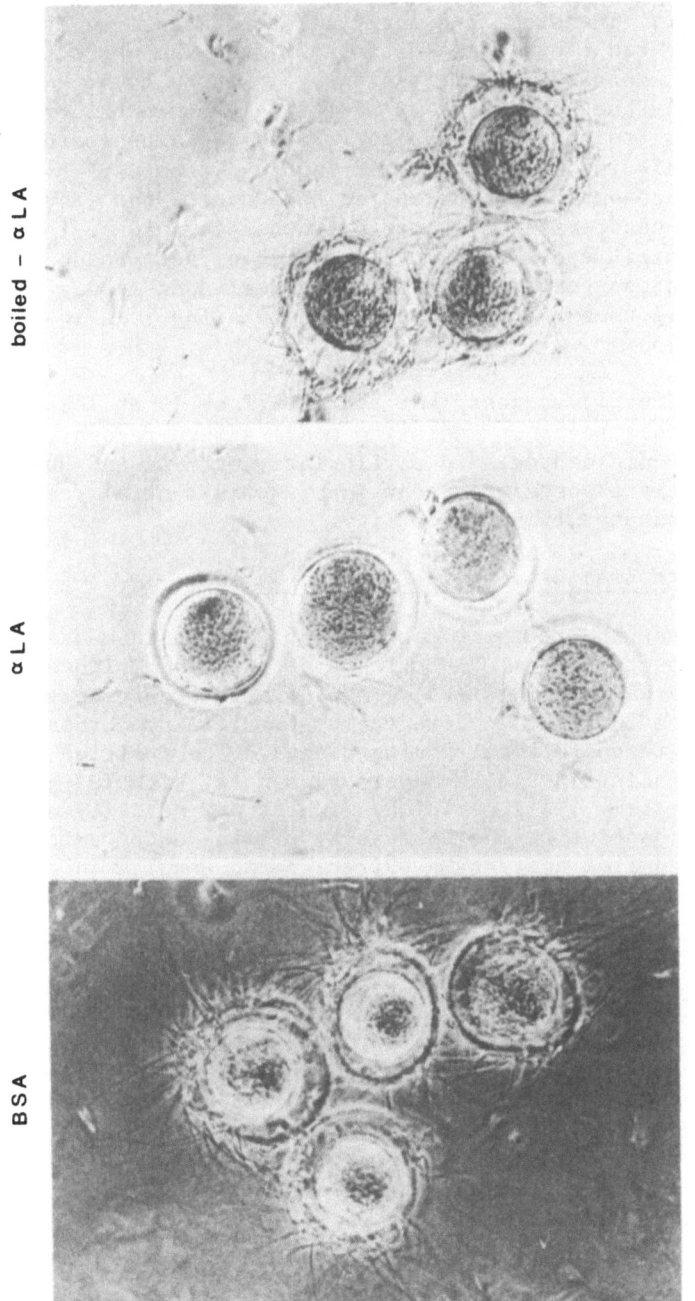

Figure 2. The effect of 2% α-lactalbumin (α-LA) on mouse sperm binding to the zona pellucida. α-LA is a GalTase substrate modifier protein, which alters the specificity of the sperm GalTase away from its traditional GlcNAc substrate, and which produces a coincident inhibition of sperm/zona binding. Control incubations contain either 2% bovine serum albumin or 2% heat-denatured α-LA. Photographs were taken of the sperm/zona binding assay prior to fixation and washing away of the unbound sperm.

Furthermore, when sperm bound to eggs were transferred into medium containing UDPGal, the sperm dissociated from the eggs (25).
UDPGal dissociation of initial sperm/zona adhesions was temperature
and time-dependent. Transfer into identical concentrations of
UDPglucose did not cause any release of sperm from the egg surface,
relative to medium controls. Interestingly, if sperm bound to eggs
were allowed to further adhere for an additional 25 min before transferring into UDPGal solutions, those sperm/zona adhesions were now
resistant to UDPGal dissociation, suggesting that secondary binding
mechanisms, the acrosome reaction, and/or penetration had occurred.
Adding similar concentrations of competitive hexosaminidase inhibitors failed to inhibit sperm/zona binding, and in fact produced a
slight stimulation, suggesting that sperm hexosaminadases may be
hydrolyzing a small number of the zona glycosides involved in sperm
GalTase binding (25).

These data directly suggest that sperm GalTase is at least one,
if not the principal, receptor mediating mouse gamete recognition.
The ability of UDP-dialdehyde, and particularly of α-LA and UDPGal to
inhibit and dissociate sperm/zona adhesions, while control reagents
do not, directly support this possibility.

Inhibition of Sperm/Zona Binding by Purified GalTase

To further test the receptor function of GalTase in mouse sperm
binding to the zona pellucida, the effects of purified GalTase on
sperm/zona binding was examined (25). A naturally occurring soluble
form of GalTase was purified to homogeneity by affinity chromatography, and added to sperm/zona binding assays. The purified enzyme
inhibited sperm binding in a dose-dependent manner, with 10 ug/ml
producing approximately 90% inhibition. Inhibitory activity could be
removed either by denaturing the GalTase by boiling or by omitting
the required cation from the enzyme buffer, which results in an inactive GalTase.

To control for the possibility that the GalTase was inhibiting
sperm binding by sterically blocking the appropriate receptors on the
zona pellucida, the enzyme was catalytically eluted off of the zona
by co-incubation with UDPGal (25). In this way, the bound GalTase is
replaced with a covalently linked galactose residue (Fig. 3). Following exhaustive washing to remove unused UDPGal and soluble Gal
Tase, the galactosylated eggs only supported 33% of control levels of
sperm binding. Parallel incubations with GalTase and an inappropriate substrate, UDPglucose, resulted in normal levels of sperm/zona
binding. Thus, the zona pellucida receptor is a GalTase substrate,
which when galactosylated becomes incompetent to bind sperm.

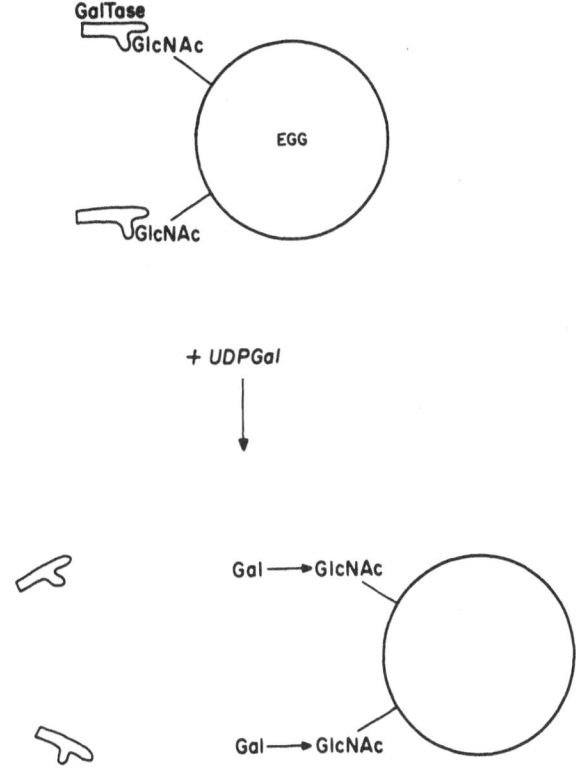

Figure 3. The experimental protocol is pictured for inhibiting sperm
        binding to the zona pellucida by adding purified GalTase,
        and by galactosylation of the zona with GalTase and its
        sugar donor substrate, UDPGal. After incubation, the
        galactosylated eggs are exhaustively washed to remove
        soluble GalTase and unused UDPGal, and then fresh sperm are
        added. Zonae pellucidae galactosylated in this way support
        sperm binding to only 33% of control levels.

Inhibition of Sperm/Zona Binding by Monospecific Anti-GalTase IgG

        Monospecific IgG was prepared against affinity purified GalTase,
and its effect was examined on sperm binding to the zona pellucida
and on sperm GalTase activity (25). Anti-GalTase IgG, as well as its
Fab fragments, produced a dose-dependent inhibition of sperm/zona
binding, and of GalTase activity, with no apparent inhibition of
sperm motility. Pre-immune IgG and anti-mouse brain IgG, which also

binds to the sperm surface, had no effect on sperm binding or GalTase
activity.

The GalTase was localized by indirect immunofluorescence using
biotinylated goat-anti-rabbit IgG and fluoresceinated-avidin to the
plasma membrane overlying the dorsal aspect of the anterior head
(25). Such a localization is consistent with the inability of
acrosome-reacted sperm to bind to the zona pellucida, since the
plasma membrane bearing the GalTase is lost following the acrosome
reaction.

Sperm GalTase Substrates in the Zona Pellucida

Finally, substrates for sperm surface GalTase can be directly
demonstrated in the zona pellucida. Both intact and solubilized zona
pellucida can be glycosylated by sperm GalTase in the presence of
UDPGal (23).

If sperm GalTase is responsible for binding to zona pellucida
glycoside substrates, than enzymatic modification of these zona glycoconjugates should perturb subsequent sperm binding. As discussed
above, galactosylation of the zona surface with purified GalTase and
UDPGal results in an inhibition of sperm/zona binding (25). Since
this enzyme is specific for terminal GlcNAc residues, it is consistent with the notion that the active moiety on the zona receptor for
sperm is GlcNAc, which is recognized by sperm surface GalTase during
gamete recognition. Alternatively, digestion of the zona with N-
acetyglucosaminidase, which hydrolyzes the terminal GlcNAc residues,
also inhibits subsequent sperm binding, while parallel treatment with
another glycosidase does not (23).

CONCLUSIONS

Our interests have been to define the surface receptors that
participate in mammalian gamete recognition. Results accumulated
over the past few years have focused our attention on the receptor
function of GalTase on the sperm surface, which appears to participate during fertilization by binding specific GlcNAc-terminating substrates in the zona pellucida (Fig. 1). Support for this model of
gemete binding comes from a variety of studies.

First, sperm that have genetic predisposition for increased fertilizing ability, have a specific four-fold increase in their surface
GalTase activity (19,22).

Second, as sperm acquire the ability to bind eggs following
capacitation, the surface GalTase active site becomes exposed, due to
the release of a large molecular weight competitive substrate. This

released substrate is able to inhibit sperm/zona binding by competing for the active site of the GalTase (24).

Third, inhibitors and modifiers of surface GalTase produce a coincident inhibition of sperm/zona binding, and sperm bound to the zona pellucida can be dissociated by completion of the enzymatic reaction with UDPGal (23,25).

Fourth, purified GalTase inhibits sperm/zona binding, and does so by recognizing the sperm receptor, since galactosylation of the zona results in a similar inhibition of binding (25).

Fifth, monospecific anti-GalTase IgG, and its Fab fragments, produce a dose-dependent inhibition of sperm/zona binding and concommitantly block sperm GalTase catalytic activity. Pre-immune IgG or anti-mouse brain IgG, which also binds to the sperm surface, has no effect. The sperm GalTase has been localized by indirect immunofluorescence to a discrete plasma membrane domain on the dorsal surface of the anterior head overlying the intact acrosome (25).

Finally, intact and soluble zona pellucida possess substrates for sperm surface GalTase, and enzymatic modification of these GalTase substrates inhibits subsequent binding (23,25).

We are now in a position to determine by SDS-PAGE fluorography which zona pellucida glycoprotein serves as the substrate for the sperm GalTase. One zona pellucida component in particular, ZP3, has been suggested to be the zona receptor for sperm binding by competition experiments (5; see Wassarman et al., this volume). Whether ZP3 is the sperm GalTase substrate is of obvious interest. It is also of interest to determine whether surface GalTase serves a similar receptor function during gamete binding in other species.

It is likely that other sperm components, such as trypsin-like enzymes (15) also participate in gamete binding, and it will be of interest to compare them, and their biological activities, with sperm surface GalTase.

ACKNOWLEDGMENTS

The author is indebted to Evelyn Bayna, Jonathan Reichner and Raymond Runyan for their suggestions regarding this manuscript. Studies performed in the author's laboratory were supported by grants from the National Institutes of Health (HD 15856) and the National Science Foundation (PCM 8400817).

REFERENCES

1. Ahuja, K.K. 1982. Fertilization studies in the hamster. The role
   of cell-surface carbohydrates. Exp. Cell Res. 140:353-362.
2. Artzt, K., Dubois, D., Bennett, D., Condamine, H., Babinet, C.,
   and Jacob, F. 1973. Surface antigens common to mouse cleavage
   embryos and primitive teratocarcinoma cells in culture. Proc.
   Natl. Acad. Sci. USA 70:2988-2992.
3. Barnum, S.R., and Brown, G.G. 1983. Effect of lectins and sugars
   on primary sperm attachment in the horseshoe crab, Limulus
   polyphemus L. Devel. Biol. 95:352-359.
4. Bennett, D. 1975. The T-locus of the mouse: a review. Cell
   6:441-454.
5. Bleil, J.O., and Wassarman, P.M. 1980. Mammalian sperm-egg interaction: identification of a glycoprotein in mouse egg zonae
   pellucidae possessing receptor activity for sperm. Cell
   20:873-882.
6. Clegg, E.D. 1983. Mechanisms of mammalian sperm capacitation, in:
   "Mechanism and Control of Animal Fertilization," J. F.
   Hartman, ed., Academic Press, N.Y. pp. 177-212.
7. Cummings, R.D., Cebula, T.A., and Roth, S. 1979. Characterization
   of a galactosyltransferase in plasma membrane-enriched fractions from BALB/c 3T12 cells. J. Biol. Chem. 254:1233-1240.
8. Ebner, K.E., and Magee, S.C. 1975. Lactose synthetase: α-
   lactalbumin and β $(1 \longrightarrow 4)$ galactosyltransferase, in: "Subunit Enzymes," Biochemistry and Function, K. Ebner, ed.,
   Marcel Dekker, N.Y. pp. 137-179.
9. Glabe, C.G., and Lennarz, W.J. 1981. Isolation of a high molecular weight glycoconjugate derived from the surface of S.
   purpuratus eggs that is implicated in sperm adhesion. J.
   Supramol. Struct. Cell Biochem. 15:387-394.
10. Lillie, F.R. 1913. The mechanism of fertilization. Science
    38:524-528.
11. Monroy, A., and Rosati, F. 1983. A comparative analysis of sperm-
    egg interaction. Gamete Res. 7:85-102.
12. Muramatsu, T., Gachelin, G., Damonneville, M., Delarbre, C., and
    Jacob, F. 1979. Cell surface carbohydrates of embryonal carci-
    noma cells: polysaccharidic side chains of F9 antigens and of
    receptors to two lectins, FBP and PNA. Cell 18:183-191.
13. Overstreet, J.W. 1983. Transport of gametes in the reproductive
    tract of the female mammal, in: "Mechanism and Control of
    Animal Fertilization," J. F. Hartman, ed. Academic Press, N.Y.
    pp. 499-543.
14. Pierce, M., Turley, E.A., and Roth, S. 1980. Cell surface
    glycosyltransferase activities. Int. Rev. Cytol. 65:1-47.
15. Saling, P.M. 1981. Involvement of trypsin-like activity in bind-
    ing of mouse spermatozoa to zonae pellucidae. Proc. Natl.
    Acad. Sci. USA 78:6231-6233.

16. Saling, P.M., Storey, B.T., and Wolf, D.P. 1978. Calcium-dependent binding of mouse epididymal spermatozoa to the zona pellucida. Devel. Biol. 65:515-525.

17. Schacter, H., and Roseman, S. 1980. Mammalian glycosyltrans-ferases, in: "The Biohemistry of Glycoproteins and Proteo-glycans," W. J. Lennarz, ed., Plenum Press, N.Y., pp. 85-160.

18. Shaper, J.H., and Mann, P.L. 1981. The demonstration of a cell surface UDP-galactosyltransferase on mammalian cells by indirect immunofluorescence. J. Supra. Struct. Suppl. 5:272.

19. Shur, B.D., and Bennett, D. 1979. A specific defect in galactosyltransferase regulation on sperm bearing mutant alleles of the T/t locus. Devel. Biol. 71:243-259.

20. Shur, B.D. 1982. Cell surface glycosyltransferase activities during normal and mutant (T/T) mesenchyme migration. Devel. Biol. 91:149-162.

21. Shur, B.D. 1982. Evidence that galactosyltransferase is a surface receptor for poly (N)-acetyllactosamine glycoconjugates on embryonal carcinoma cells. J. Biol. Chem. 257:6871-6878.

22. Shur, B.D. 1981. Galactosyltransferase activities on mouse sperm bearing multiple t lethal and t viable haplotypes of the T/t-complex. Genet. Res. 38:225-236.

23. Shur, B.D., and Hall, N.G. 1982. A role for mouse sperm surface galactosyltransferase in sperm binding to the egg zona pellu-cida. J. Cell Biol. 95:574-579.

24. Shur, B.D., and Hall, N.G. 1982. Sperm surface galactosyltrans-ferase activities during in vitro capacitation. J. Cell Biol. 95:567-573.

25. Shur, B.D., Litoff, D., and Bayna, E. 1985. The receptor function of mouse sperm surface galactosyltransferase during fertiliza-tion. J. Cell Biol., in press October.

26. Shur, B.D., and Roth, S. 1975. Cell surface glycosyltransferases. Biochim. Biophys. Acta 415:473-512.

27. Shur, B.D. 1984. The receptor function of galactosyltransferase during cellular interactions. Molec. Cell. Biochem. 61:143-158.

28. Silver, L.M. 1981. Genetic organization of the mouse t complex. Cell 27:239-240.

29. Strous, G.J.A.M., and Berger, E.G. 1982. Biosynthesis, intra-cellular transport, and release of the Golgi enzyme galacto-syltransferase (Lactose synthetase A protein) in HeLa cells. J. Biol. Chem. 257:7623-7628.

30. Struck, D.K., and Lennarz, W.J. 1980. The function of saccharide-lipids in synthesis of glycoproteins, in: "The Biochemistry of Glycoproteins and Proteoglycans," W. J. Lennarz, ed., Plenum Press, N.Y. pp. 35-83.

31. Vacquier, V.D., and Moy, G.W. 1977. Isolation of bindin: the protein responsible for adhesion of sperm to sea urchin eggs. Proc. Nat. Acad. Sci. USA 74:2456-2460.

IMMUNOLOGICAL IDENTIFICATION OF SPERM ANTIGENS

THAT PARTICIPATE IN FERTILIZATION

P. M. Saling, R. Waibel, and K. A. Lakoski

Duke University Medical Center
Department of Obstetrics and Gynecology
Box 3143
Durham, North Carolina 27710

SUMMARY

Using monoclonal antibodies (mAbs) as probes for specific components of the spermatozoon, we have initiated an investigation aimed
at identifying those sperm components that participate in the events
of gamete interaction. We intend to exploit these mAbs not only for
identifying functional sperm components, but also for defining the
constituents of discrete domains that comprise the structure of this
highly diferentiated cell.

Among the large group of anti-sperm mAbs that we have generated,
we have focused to date upon two categories. The first category consists of mAbs that localize to the acrosomal crescent, a restricted
region of plasma membrane overlying the acrosome. Within this category, the mAbs share many similarities with regard to subclass,
species and tissue cross-reactivity, and antigen solubility, in addition to cellular distribution. Nevertheless, despite these similarities, some mAbs in this category (e.g., M42) inhibit fertilization,
whereas others (e.g., M41) are non-inhibitory. The block to fertilization observed in the presence of M42 is dependent upon the zona
pellucida surrounding the egg. The specific event prevented by M42
appears to be the induction of the acrosome reaction at its physiological site, the surface of the zona pellucida. The sperm components
recognized specifically by M42 are a cluster of high molecular weight
moieties, ranging from approximately 220,000 to 240,000.

The mAbs described in the second category display common localization at the equatorial segment of the sperm head. The pair of mAbs
discussed from this category, M2 and M29, again bear considerable

similarity to each other, yet differ significantly in their ability
to inhibit fertilization. M2 does not inhibit, whereas M29 causes a
marked inhibition of fertilization. With M29, however, the block to
fertilization is independent of the zona pellucida. The M29 mAb
interferes with sperm interaction with the egg plasma membrane subsequent to sperm attachment; since M29 does not prevent sperm binding
to the egg plasma membrane, the specific event affacted, in all likelihood, is gamete membrane fusion. M29 recognizes a single sperm
component, with subunit molecular weight of approx. 40,000.

A variety of experiments are underway currently, both to characterize the antigens recognized by these mAbs further as well as to
identify additional sperm components that participate in the fertilization process.

INTRODUCTION

The interactions between sperm and egg that lead to fertilization are complex. Investigation during the last few years has delin-
eated many details concerning the sequence of events that the spermatozoon undergoes during this process (in the mouse, for instance, see
refs. 2,3,5,20,22,23). Results of experiments using both cumulus-
intact and cumulus-denuded eggs from different species (e.g., mouse
(23,24), rabbit (25)), indicate the emergence of a common pattern for
this sequence. Sperm traverse the cumulus layer and arrive at the
surface of the zona pellucida with a morphologically intact acrosome;
recognition of and binding to zonae occurs via the sperm's plasma
membrane. Once bound to the zona surface, the acrosome reaction is
induced in a time-dependent fashion. Following the induction of the
acrosome reaction, sperm penetrate the zona matrix and enter the
perivitelline space. There, the acrosome-reacted sperm associate
rapidly with the egg plasma membrane using the plasma membrane in the
equatorial segment region as the sperm's starting point. Membrane
fusion commences without significant delay, initiating fertilization.

Prominent in this description of the events of the fertilization
process is the absence of information concerning the identity of
sperm components that participate in the events. In large part, this
has been due to a lack of probes specific for individual components
of the sperm cell. The availability of hybridoma technology has
rectified this situation considerably. We have produced a large
family of monoclonal antibodies (mAbs) against mouse sperm, and are
using these as immunological tools to identify and characterize sperm
components involved in each of the events of gamete interaction.

Syngeneic mouse testis was used as the immunogen for all of the
mAbs described here. Our primary screening procedure for initial
hybridoma selection is indirect immunofluorescence (IIF) on mature
mouse sperm. This combination of immunization and screening proce-

dures has proved to be very valuable for several reasons: 1) Immunization with testis and screening on mature sperm permits selection for components that arise in the testis and persist throughout spermatogenesis. Thus, non-germ cell components as well as transient differentiation antigens (8,12,13) are avoided at the outset. 2) Use of testicular material as the immunogen, in contrast to the use of epididymal sperm, prevents another complicating factor, the generation of mAbs against epididymal secretory products adsorbed to the sperm surface (4,27), rather than mAbs against actual membrane constituents. Use of mature epididymal sperm as immunogen will almost certainly be necessary to define the sperm components that participate in interaction with zonae, however, since the sperm's ability to recognize and bind to zonae is acquired during epididymal maturation (6,14,16). 3) The tendency toward generation of autoantibodies is enhanced with syngeneic immunization (although cannot be presumed due to the immunological background of the immunized animal); autoantigens have been implicated in infertility in a variety of species (11,18,26). 4) Use of immunofluorescence as a primary screen provides immediate information upon mAb localization. Thus, attention can be focused from the start upon those mAbs with potentially useful distribution patterns, such as restriction to discrete regions of the sperm head. Furthermore, categories of mAbs may be defined on the basis of mAb distribution. Comparison of the antigens recognized by the members within such categories will identify the constituents of each cellular domain. This strategy will also provide a method to enrich for the identification of mAbs that interact with the same molecule, but at different antigenic determinants, thereby permitting the construction of a sub-molecular map for those components that function in fertilization.

So far in our investigation of the molecular basis of mammalian sperm interaction with the egg, we have concentrated upon two particular categories of mAbs: one category consists of mAbs that localize at the equatorial segment of the sperm head, whereas mAbs in the second category are distributed in a restricted region of plasma membrane overlying the acrosome which we term the acrosomal crescent. A pair of mAbs, one which does and one which does not inhibit fertilization, from each of these categories will be compared below.

ACROSOMAL CRESCENT LOCALIZATION: M42 AND M41

Based upon their appearance following IIF analysis, these two mAbs, M41 and M42, are indistinguishable (Fig. 1). Several other characteristics of these mAbs are shown in Table I. It was concluded that these mAbs label the sperm's plasma membrane based on the following results: a) both fixed sperm and unfixed, motile sperm display the same IIF pattern; b) within a fresh population of sperm, nearly all cells display the IIF pattern shown in Figure 1, whereas after incubation in acrosome reaction-inducing conditions, less than

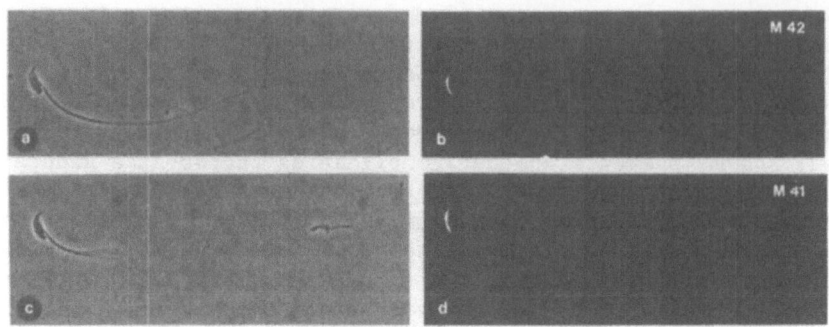

Figure 1. Paired phase contrast (a,c) and epifluorescence (b,d)
          photomicrographs demonstrating the localization patterns of
          the two mAbs indicated using indirect immunofluorescence on
          formaldehyde-fixed cauda epididymal mouse sperm. Both M42
          and M41 display indistinguishable distribution patterns
          confined to a restricted region of plasma membrane
          overlying the acrosome. Control specimens (myeloma cell
          line culture supernatants or PBS) display very weak, if
          any, fluorescence which appears totally dark under the
          photographic conditions used.

40% of the sperm display any fluorescence (consistent with the extent
of acrosome reaction induction under these conditions); c) M42 is
cytotoxic (immobilizing) in the presence of guinea pig complement
with a titer of 1/192 (M41 has not yet been assayed); and d) hamster
sperm, which cross react with these mAbs, display fluorescence confined to their readily visible acrosomal cap. After the acrosome
reaction, no fluorescence is associated with the sperm head (17).

    Not only do these observations indicate the plasma location of
these mAbs, but also suggest that these mAbs may be used as probes
for the occurrence of the acrosome reaction in mouse sperm. This is
particularly useful since it is not possible to distinguish the acrosome of the mouse sperm with standard light optics (brightfield,
phase contrast or differential interference contrast).

    For accurate comparison of mAbs within a particular category, a
similar immunoglobulin subclass for the category members is desirable
in addition to similar localization pattern. Once a mAb that inhibits fertilization is identified, efforts are made (when possible) to
restrict other members of that category to mAbs of the same subclass
to control for non-specific effects. Thus, although M41 and M42 are
not identical, they are both IgG's (Table 1).

Table 1. Monoclonal Antibody Characteristics

mAb	Localization Pattern	Immunoglobulin Subclass	Tissue Cross-Reactivity				Species Cross-Reactivity			
			Liver	Kidney	Ovary	Testis	Mouse	Hamster	Human	Rabbit
M41	AC*	IgG_{2a}	-	-	-	+	+	+	-	N.T.
M42	AC	IgG_1	-	-	-	+	+	+	-	-
M2	ES	IgM	-	-	-	+	+	+	+	+
M29	ES	IgM	-	-	-	+	+	+	+	+

*AC, acrosomal crescent (i.e., restricted region of plasma membrane overlying the acrosome)
ES, equatorial segment

-, no cross reactivity demonstrated
+, cross reactivity present

Both of these mAbs appear to be tissue, but not species, specific (Table 1). While the mAbs do not cross-react with fixed, paraffin-embedded sections of mouse liver, kidney and ovary, both M41 and M42 cross-react with antigens present in the mouse testis. Since testis was used as the immunogen, this result is not surprising, but preliminary evidence indicates that the antigens recognized by these mAbs appear asynchronously during spermatogenesis despite their ultimate coincident location. Studies are in progress currently to pursue this topic.

With regard to species cross-reactivity, these mAbs will cross-react with hamster, but not human or rabbit sperm. When cross-reaction occurs, the distribution of the mAb is confined to the acrosomal crescent. Whether this restricted cross-reactivity reflects a rodent specific parameter or the site of origin of the sperm sample (mouse, hamster: epididymis vs. human, rabbit: ejaculate) has not yet been addressed.

Although the majority of characteristics described for these mAbs are similar, a fundamental difference between M41 and M42 was apparent when the mAbs were tested for an effect upon mouse fertilization in vitro (Fig. 2). When mouse eggs possessing intact zonae pellucidae (the presence of the cumulus layer was irrelevant) were challenged with capacitated mouse sperm, a marked difference was observed depending upon mAb treatment. M41-treated sperm penetrated zonae and fertilized eggs at levels indistinguishable from control sperm. M42-treated sperm, in contrast, significantly blocked both penetration and fertilization when zona-intact eggs were used. However, a dramatic alteration in the inhibitory effect due to M42 was found upon removal of zonae from mouse eggs prior to sperm addition. Under these conditions, no block to fertilization was observed. Thus, M42 mAb blocks fertilization by a specific inhibition of sperm-zona interaction.

In the mouse, at least three discrete events occur between sperm and zonae: sperm first recognize and bind to zonae, the acrosome reaction is then induced, and finally, sperm penetration through the zona occurs (3,5,15,20,23). Our preliminary experiments to define the specific locus of inhibition indicate that sperm binding to zonae is unaffected by M42. The next event, the induction of the acrosome reaction, appears to be the step that is inhibited specifically by this mAb. The results of these preliminary experiments suggest that the mAb is capable of discriminating between physiological and unphysiological modes of acrosome reaction induction. Thus, M42 does not inhibit A23187-induced (20) mouse sperm acrosome reactions. In contrast, at the same concentration, the mAb will prevent the induction of the acrosome reaction in sperm bound to the surface of the zona pellucida. Further characterization of the concentration and temporal requirements of these effects are necessary, but the available evidence indicates that M42 may provide a valuable probe for

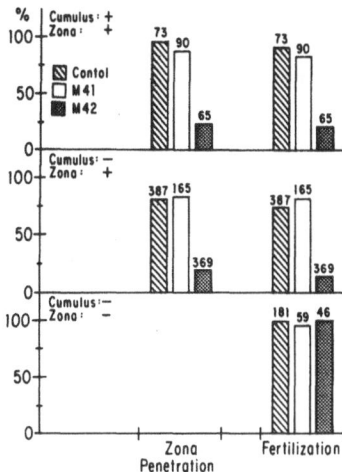

Figure 2. Effect of ascites fluid containing M41 vs. M42 on mouse
          fertilization in vitro. Superovulated tubal eggs were pre-
          pared in three different ways to determine the event of the
          gamete interaction process affected by the mAb under
          examination. In the upper panel, eggs retained both the
          cumulus layer and the zona pellucida. In the middle
          panel, the cumulus layer around the eggs was removed by
          brief incubation (<10 min) in 0.1% hyaluronidase (Sigma) in
          CM (16,20,22), the culture medium used for capacitation and
          fertilization in vitro. In the lower panel, both the
          cumulus layer and zonae pellucidae were removed from the
          eggs. Zona removal was accomplished manually using narrow-
          bore micropipettes. For all of the egg preparations, cauda
          epididymal sperm were capacitated by incubation in CM at a
          concentration of approx. 10^6 cells/ml at 37°C. After 90
          min, 20 μl of capacitated sperm suspension were deposited
          into 160 μl of CM plus 20 μl of ascites fluid. The final
          concentration of each mAb tested was approx. 0.4 mg/10^5
          sperm/ml. Ten to fifteen minutes later, the eggs were
          added; the gametes were incubated together for 4-5 hrs at
          37°C. After thorough washing, the eggs were fixed and
          stained to determine the occurrence of fertilization: eggs
          were scored as penetrated if sperm were found within the
          perivitelline space and/or vitellus, and as fertilized only
          if both sperm head (or pronucleus) and sperm tail were
          identified within the vitellus.

          The results were obtained by averaging the fertilization
          levels and the zona penetration levels from 3 replicate
          experiments using cumulus-+-zona-intact eggs, and 8 repli-
          cate experiments each using zona-intact and zona-free eggs.
          The numbers at the top of each bar indicate the total
          number of eggs examined.

understanding the physiological mechanism(s) underlying the mammalian sperm acrosome reaction. It is of considerable interest that Lopo and Vacquier (7) have described an anti-84 Kd antibody directed against sea urchin sperm that prevents the induction of the acrosome reaction under physiological (egg jelly), but not under unphysiological (high pH), conditions.

Identification of the antigens recognized by these mAbs has been accomplished immunologically following transfer of sperm components from SDS-gels to nitrocellulose sheets ("Western blots"). M42 recognizes a high molecular weight cluster (approx. 220,000 to 240,000) (Fig. 3), whereas M41 does not interact with any components in the high MW range, but, instead, with components of 60, 35, and 21 Kd (Fig. 4). Experiments are in progress to determine what relationship exists for these multiple components.

EQUATORIAL SEGMENT LOCALIZATION: M2 AND M29

Examination of the immunofluorescent distribution of M2 and M29 indicates that they share localization at the equatorial segment of the sperm head (Fig. 5). In addition, M2 binds in a patchy, variable distribution along the principal piece of the sperm tail. These two mAbs are considered here as an equatorial segment-staining pair, despite the accessory tail labelling by M2, due to a variety of other comparable characteristics which are described below (see Table 1).

Both M2 and M29 are of the same immunoglobulin class, IgM, a parameter considered particularly important in this case to control for any non-specific effects due to a molecule as large as an IgM. Like the two acrosomal crescent-staining mAbs, these two equatorial segment-staining mAbs appear tissue-specific, staining only mouse testis sections, but not mouse liver, kidney or ovary sections. Unlike the former category of mAbs, these mAbs demonstrate a wider pattern of species cross-reactivity. Cross reaction was apparent when hamster, human or rabbit sperm were incubated with either M2 or M29. The cross-reaction was confined to the equatorial segment in these species, despite the varied configurations that homologous region assumes in different species (Fig. 6).

The plasma membrane in the equatorial segment region of the sperm head does not appear to be the most likely site of localization for either M2 or M29. This conclusion is based upon the following findings: a) essentially all fixed sperm label with these mAbs, but extremely few fresh, motile sperm label; and b) in unfixed hamster sperm, which have readily apparent acrosomes and which cross-react with these mAbs, mAb labelling is seen only in the absence of the acrosome. It is likely, therefore, that the antigens recognized by these mAbs are sequestered in the inner and outer acrosomal membrane crypt that forms the equatorial segment; these membrane surfaces are

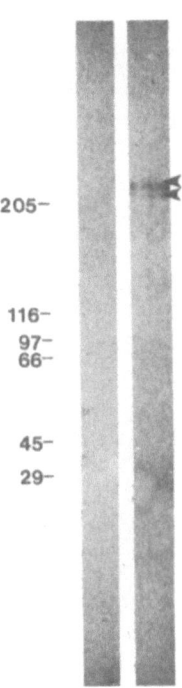

205—

116—
97—
66—

45—
29—

Figure 3. Identification of sperm antigens recognized by M42 mAb.
SDS-extracted mouse sperm components were separated
electrophoretically in a 12% SDS-polyacrylamide gel in the
presence of β-mercaptoethanol, followed by transfer of the
proteins to nitrocellulose. The strips were treated equi-
valently except for the primary antibody incubation: Lane
1: Control parent myeloma cell line culture supernatant;
and lane 2: Culture supernatant from M42 hybridoma cell
line. After thorough washing, the strips were incubated in
peroxidase-conjugated goat anti-mouse IgG + IgM. Peroxidase activity was demonstrated using Bio-Rad's HPR color
development reagent containing 4-chloro-1-napthol after the
strips had been washed thoroughly. The arrows point to the
sperm components recognized specifically by M42. Molecular
weight standards (x 10^{-3}) are indicated to the left of
lane 1.

accessible to the exterior only after loss of the acrosome from the
cell. Consistent, although not confirming, of this location is the
finding that M29 is not cytotoxic in the presence of guinea pig
complement (M2 has not been tested).

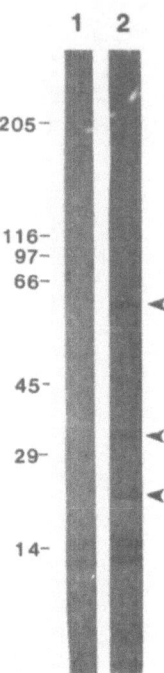

Figure 4. Identification of sperm antigens recognized by M41 mAb.
        SDS-extracted mouse sperm components were separated elec-
        trophoretically in a gradient (7-24%) SDS-polyacrylamide
        gel in the presence of β-mercaptoethanol, followed by
        transfer of the proteins to nitrocellulose. The strips
        were treated equivalently except for the primary antibody
        incubation: Lane 1: culture supernatant from parent
        myeloma cell line; lane 2: culture supernatant from M41
        hybridoma cell line. Secondary antibody incubation and
        color development was accomplished as described in Figure
        3. The arrowheads point to the bands in lane 2 that are
        recognized specifically by M41. Molecular weight standards
        (x 10^{-3}) are indicated to the left of lane 1.

        When tested for an effect upon fertilization in vitro, M2 and
M29 demonstrate dramatically different effects (Fig. 7). Again,
different types of egg preparations were used to define inhibition at
a specific stage of gamete interaction. M2 did not provoke any sub-
stantial reduction in zona penetration or fertilization, regardless
of the presence or absence of either cumulus cells or zonae. In
marked contrast, M29 caused a significant reduction in fertilization
with all of the types of egg preparations used. We conclude, therefore, that the block to fertilization caused by M29-treated sperm

occurs at the level of sperm interaction with the egg plasma
membrane.

Qualitative assessment of sperm binding to the egg plasma membrane in the presence of M29 indicated that this event is unaffected.
The next recognized event between sperm and egg plasma membrane is
fusion of their membranes; experiments are in progress to determine
whether this is the event specifically blocked by M29. It would not
be unlikely that gamete membrane fusion is the affected step since
ultrastructural studies (1,9,10,28) have established that the area of
the sperm cell that initiates fusion with the egg is the equatorial
segment.

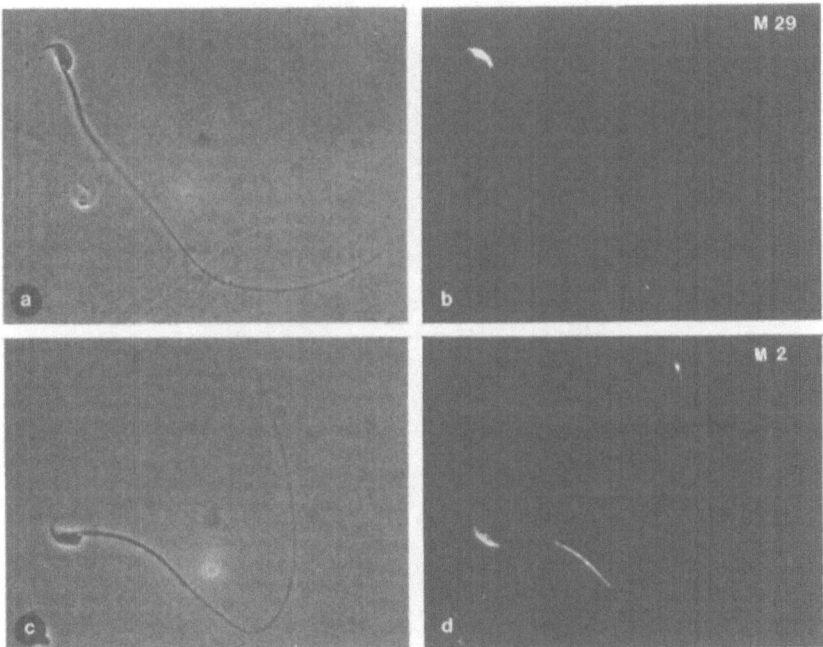

Figure 5. Paired phase contrast (a,c) and epi-fluorescent (b,d)
photomicrographs demonstrating the indirect immunofluorescent localization patterns of the 2 mAbs indicated on
formaldehyde-fixed cauda epididymal mouse sperm. One of the
mAbs (M29) is distributed on the equatorial segment alone,
whereas the other mAb (M2) displays additional variable
staining on the sperm tail. Control specimens (parental
myeloma cell line culture supernatant or PBS) display very
faint, if any, fluorescence over the entire sperm cell
which appears totally dark in the epi-fluorescent micrographs under the photographic conditions used.

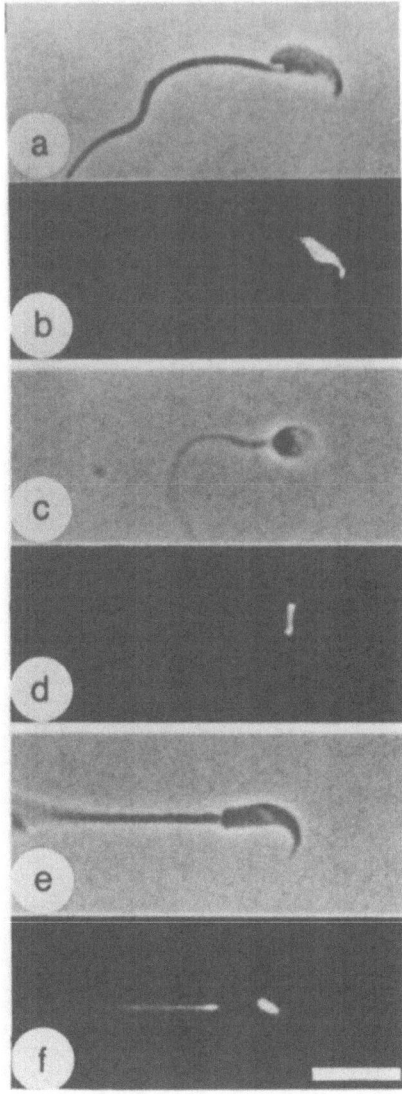

Figure 6. Paired phase contrast (a,c,e) and epifluorescent (b,d,f)
photomicrographs demonstrating the indirect immunofluorescent localization pattern of M29 on formaldehyde-fixed
sperm: (a,b) cauda epididymal mouse sperm; (c,d) ejaculated human sperm; and (e,f) cauda epididymal hamster
sperm. Regardless of the species examined, the antibody
is distributed in the region of the equatorial segments
(see ref. 19).

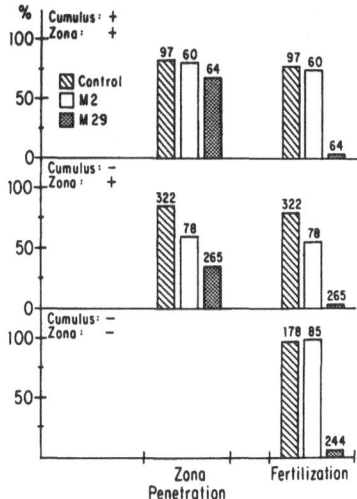

Figure 7. Effect of ascites fluid containing M29 vs. M2 on mouse
          fertilization in vitro. The experimental details are the
          same as those found in the legend to Figure 2. Again, the
          numbers at the top of each bar indicate the number of mouse
          eggs examined to obtain the result indicated.

Immuno-blotting methods have been used with these mAbs as well
to identify the specific sperm antigens recognized. Figure 8 demon-
strates that M29 mAb recognizes a single sperm component with a subunit molecular weight of approx. 40,000 (lane 4, arrowhead); M2, in
contrast, recognizes a band running at approx. 44,000 as well as a
cluster of bands centered at approx. 36,000 (lane 6, arrowheads).

ACKNOWLEDGEMENTS

This work was supported by grants from the N.I.H. (HD 18201)
and the Ford Foundation (#820-1096).

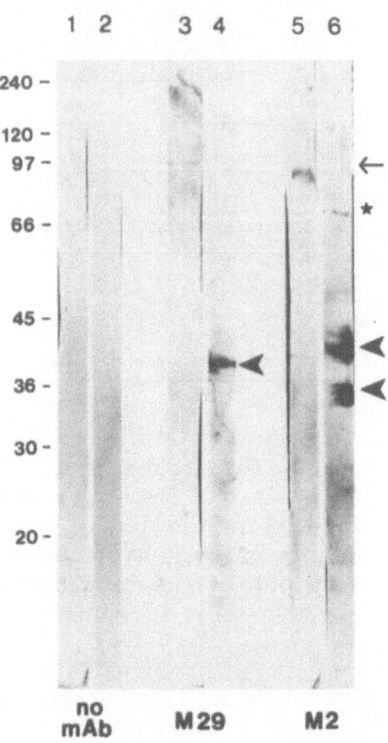

Figure 8. Immunological detection of the antigens recognized by M29
          mAb and M2 mAb. SDS-extracts of either mouse liver (lanes
          1,3, and 5) or mouse sperm (lanes 2,4, and 6) were elec-
          trophoresed in a 12% polyacrylamide gel in the presence of
          β-mercaptoethanol, followed by transfer of the proteins to
          nitrocellulose. Lanes 1,2: Nitrocellulose strips incu-
          bated in buffer without primary (i.e., monoclonal) anti-
          body. Lanes 3,4: Nitrocellulose strips incubated in M29
          mAb. Lanes 5,6: Nitrocellulose strips incubated in M2
          mAb. Secondary antibody incubation and color development
          were accomplished as described in Figure 3. The arrowheads
          indicate the sperm components recognized specifically by
          the respective mAbs (lanes 4 and 6); the arrow indicates a
          liver component recognized mutually by both mAbs, but not
          by sperm; and the star indicates a tear in the nitrocel-
          lulose strip due to serrated forceps. Molecular weight
          standards (x 10^{-3}) are indicated to the left of lane 1.

REFERENCES

1. Bedford, J.M. and Cooper, B.W. 1978. Membrane fusion events in the fertilization of vertebrate eggs. Cell Surface Rev. 55:65-125.

2. Bleil, J.D. and Wasserman, P.M. 1980. Mammalian sperm-egg interaction: Identification of a glycoprotein in mouse egg zonae pellucidae possessing receptor activity for sperm. Cell 20:873-882.

3. Bleil, J.D. and Wasserman, P.M. 1983. Sperm-egg interactions in the mouse: Sequence of events and induction of the acrosome reaction by a zona pellucida glycoprotein. Develop. Biol. 95:317-324.

4. Feuchter, F.A., Vernon, R.B., and Eddy, E.M. 1981. Analysis of the sperm surface with monoclonal antibodies: Topographically restricted antigens appearing in the epididymis. Biol. Reprod. 24:1099-1110.

5. Florman, H.M. and Storey, B.T. 1982. Mouse gamete interactions: The zona pellucida is the site of the acrosome reaction leading to fertilization in vitro. Develop. Biol. 91:121-130.

6. Fournier-Delpech, S., Courtens, J.C., Pisselet, C.L., Delateu, B. and Courot, M. 1982. Acquisition of zona binding by ram spermatozoa during epididymal passage, as revealed by interaction with rat oocytes. Gamete Res. 5:403-408.

7. Lopo, A.C. and Vacquier, V.D. 1980. Antibody to a sperm surface glycoprotein inhibits the egg jelly-induced acrosome reaction of sea urchin sperm. Develop. Biol. 79:325-333.

8. Millette, C.F. and Moulding, C.T. 1981. Cell surface marker proteins during mouse spermatogenesis: two-dimensional electrophoretic analysis. J. Cell Sci. 48:367-382.

9. Moore, H.D.M. and Bedford, J.M. 1978. Ultrastructure of the equatorial segment of hamster during penetration of oocytes. J. Ultrastruct. Res. 62:110-117.

10. Noda, Y.D. and Yanagimachi, R. 1976. Electron microscopic observations of guinea pig spermatozoa penetrating eggs in vitro. Dev. Growth Diff. 18:15-23 (1976).

11. O'Rand, M.G. 1980. Antigens of spermatozoa and their environment, in: "Immunological Aspects of Infertility and Fertility Regulation," D. S. Dhindsa and G. F. B. Schumacher, eds., Elsevier-North Holland, New York, pp. 155-171.

12. O'Rand, M.G. and Romrell, L.J. 1980. Appearance of regional surface autoantigens during spermatogenesis: Comparison of anti-testis and anti-sperm autoantisera. Develop. Biol. 75:431-441.

13. O'Rand, M.G. and Romrell, L.J. 1981. Localization of a single sperm membrane autoantigen (RSA-1) on spermatogenic cells and spermatozoa. Develop. Biol. 84:322-332.

14. Orgebin-Crist, M.-C. 1967. Maturation of spermatozoa in the rabbit epididymis: fertilizing ability and embryonic mortality in does inseminated with epididymal spermatozoa. Ann. Biol. Anim. Biochem. Biophys. 7:373-389.

15. Phillips, D.M. and Shalgi, R.M. 1980. Surface properties of the zona pellucida. J. Exp. Zool. 213:1-8.

16. Saling, P.M. 1982. Development of the ability to bind to zonae pellucidae during epididymal maturation: Reversible immobilization of mouse spermatozoa by lanthanum. Biol. Reprod. 26:429-436.

17. Saling, P.M. and Lakoski, K. 1985. Mouse sperm antigens that participate in fertilization, II. Inhibition of sperm penetration through the zona pellucida using monoclonal antibodies. Biol. Reprod. 33:527-536.

18. Saling, P.M. and O'Rand, M.G. 1982. Anti-mouse sperm antiserum: Fertility inhibition in vitro and preliminary antigen identification. J. Androl. 3:434-439.

19. Saling, P.M., Raines, L. and O'Rand, M.G. 1983. Monoclonal antibody against mouse sperm blocks a specific event in the fertilization process. J. Exp. Zool. 227:481-486.

20. Saling, P.M. and Storey, B.T. 1979. Mouse gamete interactions during fertilization in vitro. Chlortetracycline as a fluorescent probe for the mouse sperm acrosome reaction. J. Cell Biol. 83:544-555.

21. Saling, P.M., Irons, G. and Waibel, R. 1985. Mouse sperm antigens that participate in fertilization, I. Inhibition of sperm fusion with the egg plasma membrane using monoclonal antibodies. Biol. Reprod. 33:515-526.

22. Saling, P.M., Sowinski, J. and Storey, B.T. 1979. An ultra-structural study of epididymal mouse spermatozoa binding to zonae pellucidae in vitro: Sequential relationship to the acrosome reaction. J. Exp. Zool. 209:229-238.

23. Saling, P.M., Storey, B.T. and Wolf, D.P. 1978. Calcium-dependent binding of mouse epididymal spermatozoa to the zona pellucida. Develop. Biol. 65:515-525.

24. Storey, B.T., Lee, M.A., Muller, C., Ward, C. and Wirtshafter, D.G. 1984. Mouse spermatozoa bind to the zonae pellucidae of mouse eggs in cumulus with acrosomes intact. Biol. Reprod. 31:1119-1128.

25. Suarez, S., Katz, D.F. and Overstreet, J.W. 1983. Movement characteristics and acrosomal status of rabbit spermatozoa recovered at the site and time of fertilization. Biol. Reprod. 29:1277-1287.

26. Tung, K.S.K., Goldberg, E.H. and Goldberg, E. 1979. Immunobio-logical consequence of immunization of female mice with homologous spermatozoa: Induction of infertility. J. Reprod. Immunol. 1:145-158.

27. Vernon, R.B., Muller, C.H., Herr, J.C., Feuchter, F.A. and
     Eddy, E.M. 1982. Epididymal secretion of a mouse sperm
     surface component recognized by a monoclonal antibody.
     Biol. Reprod. 26:523-535.
28. Yanagimachi, R. and Noda, Y. 1970. Ultrastructural changes in
     the hamster sperm head during fertilization. J. Ultra-
     struct. Res. 31:465-485.

22.

THE ACTION OF ACROSIN ON THE ZONA PELLUCIDA

Umbert A. Urch

Department of Biochemistry and Biophysics
University of California
Davis, CA 95616

SUMMARY

The presence of hydrolytic enzymes in and associated with the sperm head has long argued for their functioning in fertilization. Several observations led investigators to propose that the acrosomal trypsin-like enzyme, acrosin in mammals, functioned in fertilization in aiding the sperm to penetrate the zona pellucida. While many have raised significant objections to this role, the action of acrosin on its presumed physiological substrate has not been characterized in a biochemical fashion. The intent of this study was to examine the effect of sperm proteases on the innermost egg envelopes in a parallel study, with the pig, Sous scrofa and the South African clawed frog, Xenopus laevis. With the pig, a great deal of information exists concerning the boar enzyme, acrosin, but little is known about the chemical structure of the zona pellucida. The opposite situation exists in X. laevis where the vitelline envelope is well characterized chemically, but little is known about the putative sperm lysins.

INTRODUCTION

The hypothesis that acrosin facilitates sperm penetration evolved from observations that sperm and sperm extracts effected visual dissolution of egg envelopes. Hibbard (19) found that Discoglossus pictus sperm could liquify isolated jelly plugs of D. pictus eggs. The exact physiological significance of this observation was cast in doubt by Wintrebert (44), when he showed artificially activated oocytes in the absence of sperm caused similar liquification. Subsequently, Yamane (46) observed the dispersion of

113

the rabbit follicle cell mass with the addition of rabbit sperm suspensions. Hypothesizing that a protease must be functioning in this
dispersion, he added a crude pancreatic extract to the rabbit
follicle cell mass and observed the same dispersion. Oddly, this is
often cited as the first evidence for an acrosin in cell dispersion,
when most clearly the effect is one of the sperm enzyme hyaluronidase. Indeed, McLean and Rowlands (26) demonstrated that sperm
hyaluronidase caused dispersion of the cumulous oophorus. Later,
Yamane (47) removed both the follicle cell mass and the zona
pellucida with a toluene extract of sperm. Tyler (39) found that an
extract of Megathura crenulata sperm dissolved the egg membranes
without any other effect on the egg. This observation and others led
him to hypothesize in 1948 that sperm lysins must be instrumental in
sperm penetration of egg membranes.

The presence of zona lysins was also postulated in rodent spermatozoa by Austin and Bishop (21). However, since they observed that
the acrosome was lost before sperm penetration of the zona, they
suggested the lysins might originate in the perforatorium of the
sperm. The first direct evidence to suggest that proteolytic enzymes
of sperm origin contribute to denudation of rabbit eggs and probably
facilitate penetration of the zona pellucida by sperm was put forward
by Srivastava, Adams and Hartree (38). They measured hyaluronidase
and protease activities in acrosomal preparations of ram, bull, boar
and rabbit sperm and were able to demonstrate the removal of the
cumulous oophorus, corona radiata and zona pellucida from rabbit
eggs. The nature of this acrosomal enzyme in bull sperm was probed
by Waldschmidt et al. (41) who found that soybean trypsin inhibitor
was an effective inhibitor of this protease activity.

These observations formed the backdrop for two sets of experiments that led investigators to further embellish the hypothesis that
acrosin was important in aiding sperm penetration of the zona pellucida. The first set of observations dealt with the use of natural and
synthetic inhibitors of trypsin to prevent in vivo (37) and in vitro
(49,50) fertilization, and the prevention of zona penetration (12).
The second set of experiments dealt with visual dissolution of the
homologous zona pellicuda by acrosin or acrosin-rich acrosomal
extracts (rabbit, 36; hamster, 25; mouse, 5).

In direct conflict with these observations are the reports that
neither inhibition of fertilization by synthetic and natural inhibitors of trypsin nor complete visual dissolution of the zona pellucida
are satisfactory as criteria for acrosin's function. With trypsin
inhibitors, several reports indicate these inhibitors do block fertilization, however this inhibition approaches but never reaches 100%
and is often much lower. Hartmann and Hutchison (15) reported that
trypsin inhibitors no longer inhibited fertilization in the golden
hamster once the gametes had become tightly bound. Also Schill
et al. (34) and Müller-Esterl et al. (29) have reported that some

inhibitors of trypsin were not permeable across acrosomal membranes, and that diffusion across the acrosomal membrane of these chemicals varied from one species to the next. In addition, the target of inhibition may be acrosin in the enzymatic assay for esterase or amidase, but it is not clear where these inhibitors are functioning in the physiological milieu. Both Saling (33) and Frazer (11) have demonstrated inhibition of sperm binding to the zona pellucida in the mouse when trypsin inhibitors were present. With regard to visual zona pellucida dissolution, many acrosins have no visual effect on the homologous zona pellucida. For example, the boar enzyme has been reported by Dudkiewicz and Garrison (10) to have little or no effect on the pig zona. Nor would complete visual dissolution appear to be a physiologically significant criterion from the standpoint that only a sperm penetration slit is formed in the zona (3), never total dissolution. The zona pellucida serves a number of roles in fertilization and early development and total dissolution may at best be a physicochemical criteria with little or no physiological significance.

In addition to the problems associated with the above criteria, several conceptual problems have been raised about the role of acrosin, notably by Bedford (3) and by Yanagimachi (48). These problems include: 1) unequivocal localization of acrosin on the inner acrosomal membrane or in the acrosomal matrix (21); 2) location of the acrosome reaction in the fertilization process; and 3) the exact turn-on mechanism for and timing of the proacrosin to acrosin conversion. In addition, Bedford and Cross (4) showed wheat germ agglutinin could make the rabbit zona resistant to trypsin and acrosin dissolution, yet sperm still penetrated the zona and fertilized the egg. To further this hypothesis for purely mechanical penetration of the zona by sperm, Green and Purves (13) calculated the force a sperm generated and based on assumptions of zona physicochemical structure, they suggested that while a swimming sperm does not generate enough force to break covalent bonds, it may however be able to swim through a semi-solid liquid, as the zona had been described.

In response to the localization studies is the work of Brown and Harrison (6) and Hartree (16). They were able to measure membrane bound acrosin activity after disrupting the sperm membranes with a cell disruptor. The membrane bound form had radically different inhibition kinetics vis-a-vis the soluble form of acrosin.

Lastly, several investigators have argued for other functions for acrosin. Meizel (24) has implicated acrosin in the acrosome reaction. Others have implicated acrosin in gamete binding and acrosomal granule dehiscence and acrosomal matrix dispersion.

With these controversies and associated problems with the proposed role for acrosin in zona penetration, unequivocal biochemical

evidence is lacking. To this end the porcine zona pellucida was
challenged with pure boar acrosin to determine whether it was a
substrate for acrosin, and which, if any, of the components of the
zona pellucida were hydrolyzed. With this information, experiments
to inhibit this interaction specifically, such as monoclonal antibodies to the substrate site, may determine what specific role
acrosin plays in fertilization. In a parallel study, partial characterization of the sperm lysin(s) from the frog, X. laevis, was
initiated and its action on the vitelline envelope examined.

MATERIALS AND METHODS

    Biological   Porcine ZP[1] were isolated following the procedures
of Dunbar et al. (9). The ZP ghosts were suspended in 5 mM NH_4 HCO_3
and either sonicated to form a heterogeneous suspension (particulate
ZP), solubilized by moderate heating at 72°C for 20 minutes (heat
solubilized ZP), or dissociated in 2% SDS (SDS-ZP). Xenopus laevis
VE were isolated after the procedures of Wolf et al. (45), and the VE
ghosts were solubilized by suspending them in 20 mM sodium borate
buffer at pH 9.0 and heating at 70°C for 5 minutes. Boar acrosin was
purified to homogeneity following the methods of Müller-Esterl et al.
(28). X. laevis sperm were suspended in distilled water and put
through five freeze-thaw cycles using liquid nitrogen and a room
temperature water bath. The supernatant solution after centrifugation
was used without further purification.

    Chemical   Protein concentrations were determined by a modified
Lowry procedure developed by Peterson (31).

    The ZP oligosaccharides were removed by TFMS treatment after the
method of Karp et al. (22).

    Enzyme assays for trypsin and acrosin were carried out using an
amidase assay involving BANA and following the release of β-naphthy-
lamine fluorimetrically (excitation λ = 335 nm, emission λ = 410 nm)
as described by Greenberg (14). Enzyme molarities were determined
by active site titration with NPGB (7). Inhibition of acrosin by
DFP was performed at pH 8.0 in 0.1M Tris with a 10-fold molar excess
of DFP added in isopropyl alcohol.

ZP[1]: Abbreviations used: BANA, benzoyl-DL-arginine-β-naphthylamide;
BSA, bovine serum albumin; 2D-PAGE, two-dimensional polyacrylamide
gel electrophoresis; DFP, diisopropylfluorophosphate; NPGB,
p-nitro-phenyl-p'-guanidinobenzoate; SDS, sodium dodecyl sulfate;
SDS-PAGE, sodium dodecyl sulfate polyacrylamide gel electrophoresis;
TFMS, tri-fluoro-methane sulfonic acid; VE, vitelline envelopes; ZP,
zona pellucida.

For the assay of X. laevis sperm extracts, peptidyl-4-methyl-
coumaryl-4-amide substrates purchased from Chemical Dynamics Corp.,
were used following the procedure of Zimmerman et al. (51). A
standard curve for the product, amino-4-methylcoumarin was generated at excitation λ = 380 nm and emission λ = 460 nm.

For kinetic estimation of inhibition, Lineweaver and Burke
inverse plots (1/v vs 1/[S]) were used to determine K_m's in the presence and absence of the investigated substance. Kinetic analysis was
carried out as per Segel (35) and K_i's were calculated from the
equation:

$$K_{mapp} = \frac{K_m}{K_i} [I] + K_m$$

Heat solubilized ZP, acrosin, and the isolated macromolecular
components of the VE were radioiodinated by the procedure of
Salacinski et al. (32) using Iodogen to a specific activity of
1.11×10^7 dpm/μg, 1.32×10^6 dpm/μg and 1.59×10^9 dpm/mg respectively. The release of ^{125}I-hydrolysis products from ^{125}I-ZP was
followed through a modification of the procedure of Urch et al. (40).
This modification involved the use of 66% acetone to precipitate
unreacted radioactive proteins.

Hydrolysis reactions measured by SDS-PAGE contained 583 μg of
heat solubilized ZP with 27 μg acrosin per time point. The reactions
were stopped at the appropriate time interval and totally deglycosylated by TFMS prior to SDS-PAGE. For 2D PAGE (isoelectric focusing
in the first direction and SDS-PAGE in the second), the individual
time points for hydrolysis contained 175 μg of ZP, 2.6×10^{-8} M to
7.1×10^{-6} M concentrations of acrosin or trypsin and pH 8.0 Tris
buffer to a final concentration of 0.1 M. For Sephadex chromatography, the individual time points contained 500 μg ZP and 5 μg
acrosin or trypsin. In the peptide mapping gel systems, 100 μg of ZP
electrophoretically isolated family component was run on isoelectric
focusing gel and then either incubated with 49 μg acrosin, or 7 to
49 μg acrosin was added on top of the IEF gel when run in the second
dimension.

All other chemicals used were purchased from commercially available sources and were of the highest purity available.

Physical Ultracentrifugation clearing times for ^{125}I-acrosin
and ^{125}I-ZP were generated following the procedure of Howlett et al.
(20) developed for the air driven, high speed airfuge.

SDS-PAGE was carried out according to Laemmli (23). Two dimensional electrophoresis was performed after the methods of O'Farrell
(30) and of Ames and Nikaido (1). Peptide maps utilizing acrosin as

the protolytic agent were performed after the method of Cleveland et al. (8).

Sephadex G-75 chromatography was performed in 0.1 M Tris pH 8.0 with 1% SDS added.

RESULTS AND DISCUSSION

To investigate whether the ZP was a substrate for acrosin, we wanted to know if the enzyme, a protease, would bind to the ZP, a complex network of glycoprotein components. This would be a minimum requirement for enzymatic activity and was investigated using the air-driven, table-top ultracentrifuge, the Airfuge. Clearing times were generated for radioiodinated acrosin and ZP. ^{125}I-Acrosin centrifuged at 148,000 x g for two hours was sedimented 21%; 75% of the ^{125}I-ZP was cleared in this time. It is not clear why so much acrosin sedimented, since by its molecular size (38 K), it should have required longer sedimentation times. It has been reported that acrosin is both hydrophobic and behaves as an extrinsic membrane protein, and that acrosin forms higher molecular weight aggregates (Muller-Esterl et al., 27). It is also not clear why the heat solubilized ^{125}I-ZP did not completely sediment since it is believed that heat solubilization causes the formation of supramolecular islands that are several million in molecular weight (Wardrip, unpublished observations). When a 60 fold excess of heat solubilized ZP was added to ^{125}I-acrosin, 50% of the labelled material was sedimented, indicating binding by acrosin to ZP. This binding was mediated through the active site, since active site inhibited acrosin was not removed from solution by sedimenting heat solubilized ZP. These experiments indicated acrosin bound to ZP and did so at its active site.

To further examine this binding, questions raised as to the specificity of ZP binding by acrosin were investigated using heat solubilized ZP as an inhibitor of acrosin's amidase activity on BANA. The K_m for BANA is 0.25 mM. Lineweaver and Burke inverse plots of substrate versus initial velocity in the presence and absence of heat solubilized ZP indicated that ZP acted as a competitive inhibitor, with a calculated K_i of 0.12 mg/ml. A control protein, BSA, also behaved as a competitive inhibitor, but with a 11.6 fold higher K_i. These results indicated a specificity of binding above that of a simple protease-protein interaction.

Since acrosin bound to ZP specifically when compared to a control protein, BSA, we investigated whether acrosin would affect hydrolysis of the ZP. We challenged ^{125}I-ZP with acrosin and equimolar concentrations of trypsin and assayed this reaction with time by precipitating the large molecular weight species with acetone, and measuring the acetone-soluble supernatant solution (Fig. 1). With time, there

was the release of small, acetone soluble pieces from ^{125}I-ZP by both
acrosin and trypsin, but by markedly different kinetics. The ZP was
a better substrate for trypsin, since more of it was hydrolyzed and
in a faster fashion.

Since acrosin bound to and hydrolyzed ZP, the nature of the
hydrolysis was examined with 2D-PAGE. The porcine ZP when electro-
phoresed in one dimension in the absence of β–mercaptoethanol, yielded two broad smears, one at roughly 90K daltons and the other at 55K.
With β–mercaptoethanol, the smears merge and yield little information
about macromolecular components. With 2D-PAGE the ZP is still very

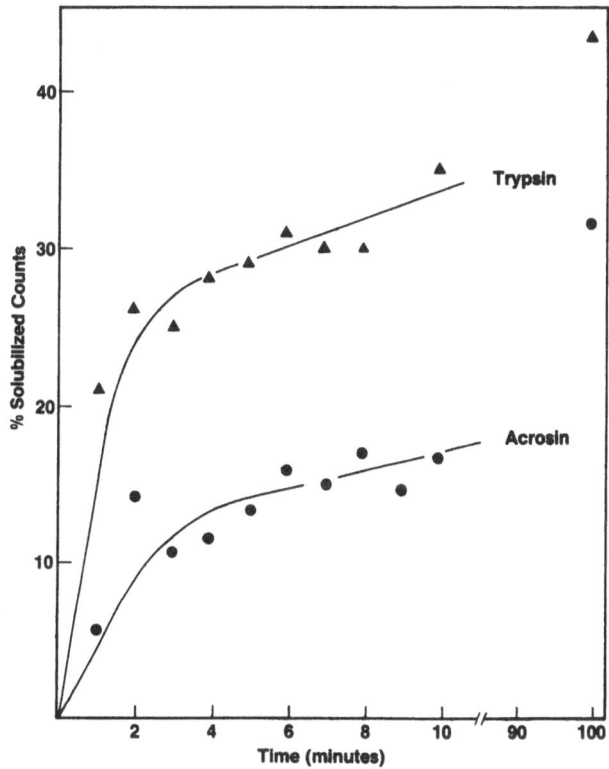

Figure 1. Time course of ^{125}I-heat solubilized ZP hydrolysis by
equimolar amounts of trypsin and acrosin. Individual
reaction time points contained: 143 μmole of acrosin or
trypsin, 62 μg ^{125}I-heat solubilized ZP, and 0.1M Tris-HCl
buffer at pH 8.0 with 0.5 M CaCl$_2$ and 0.1% Triton X-100.
The reactions were stopped by adding two volumes of acetone
and measuring soluble counts released. The ordinate
represents percent counts released over total cpm added.

complex, but can be described in terms of families of similar molecular weight species with perhaps as many as thirty family members, each differing by their isoelectric point (42) (Fig. 2a). It has been shown that this microheterogeneity is largely due to differences in carbohydrate present and in the number of acidic groups present on that carbohydrate (42). Recently it was shown that this charge heterogeneity was not due to the presence of differing amounts of sialic acid (43). When acrosin was added in a 1 to 100 weight ratio to heat solubilized ZP, and the resultant hydrolyses assayed with 2D-PAGE, the ZP was hydrolyzed rapidly to small pieces no longer retained in the gel (Fig. 2b,c). Equimolar concentrations of trypsin hydrolyzed the heat solubilized ZP faster, consistent with hydrolysis described above. When a sonicated suspension of ZP was used for these hydrolyses, a different kinetic hydrolysis sequence was obtained with acrosin (Fig. 3) and with trypsin (Fig. 4). In both hydrolyses, the 90K family was hydrolyzed first, similar to the heat solubilized ZP hydrolyzed, with the 65K hydrolysed next but only after extended hydrolyses times. With both enzymes the 55K family was resistant to hydrolysis even after very long hydrolyses times, unlike the heat solubilized ZP experiments. Heat solubilized ZP consists of supramolecular islands of all component ZP families, and in all ZP forms, all family members and their component macromolecules were found to be available for iodination by both Iodogen and lactoperoxidase (18). A crude acid extract of boar sperm when incubated with the particulate ZP for 24 hr caused no further hydrolysis than that seen with pure acrosin (Fig. 5). Therefore both acrosin and trypsin were limited in their hydrolysis of the ZP by conformational restriction of the solution state of the ZP, and the only hydrolytic agent active against ZP in the sperm appeared to be acrosin.

In an attempt to quantitate this hydrolysis, SDS-PAGE of the ZP hydrolysis products with acrosin was stained with Coomassie blue and the resulting gel scanned with a soft laser scanner. There was a reduction in the general area of the 90K family, but the gels did not resolve the components well enough for quantitation. When the acrosin hydrolyses of ZP were deglycosylated by TFMS treatment to completely remove both N- and O-linked oligosaccharide side chains of the zona glycoproteins, the resultant polypeptide backbones can be resolved well on SDS-PAGE. Scanning of the hydrolyses after TFMS allowed for integration of the stained materials (Fig. 6). With 2D-PAGE, the 90K and 65K families were seen to disappear rapidly. With TFMS treatment and one dimensional SDS-PAGE, there was agreement with both the order and extent of acrosin hydrolysis seen with 2D-PAGE. There was a rapid loss of the 90K and 65K family peptide backbones. After five hours when 70% of the 90K has been hydrolyzed to smaller components, the 55K family was still intact.

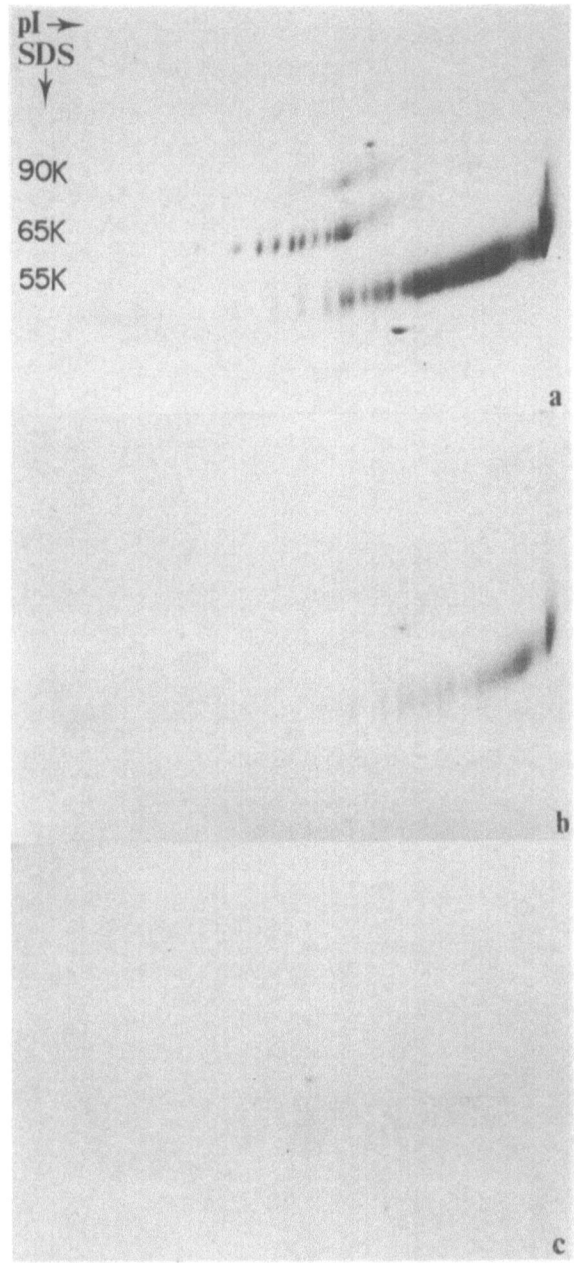

Figure 2. Hydrolysis action patterns of acrosin on heat solubilized
         ZP. Figure 2a: 0 time hydrolysis; 2b: 30 min hydro-
         lysis; and 2c: 5 h hydrolysis. Experimental details
         for this electrophoresis and for all other gels are given
         in the text.

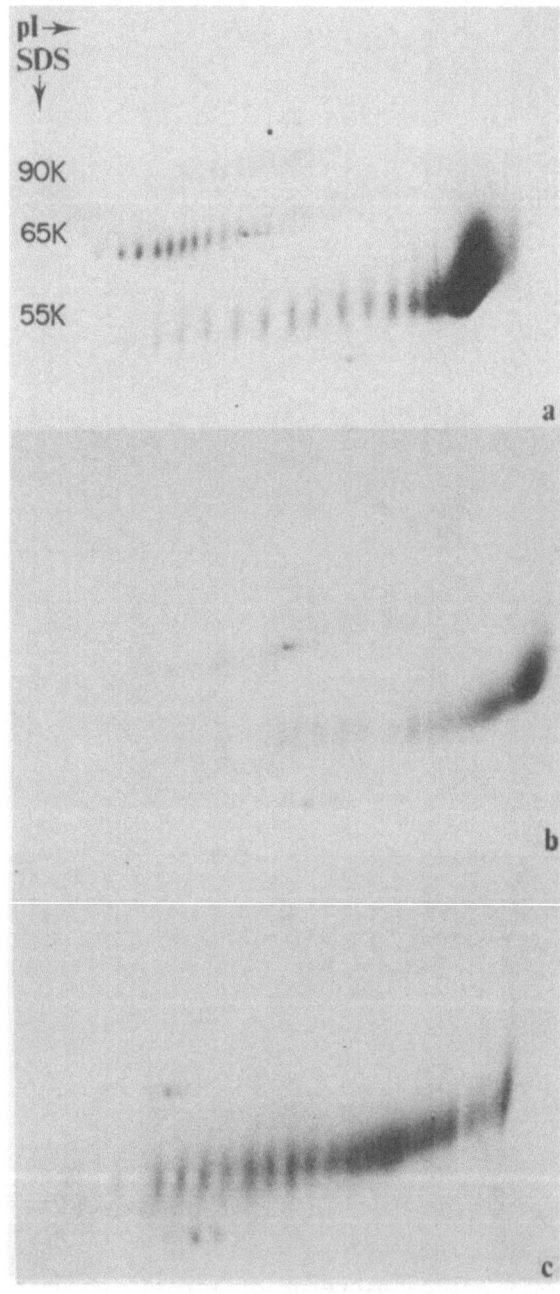

Figure 3. Hydrolysis action patterns of acrosin on particulate ZP.
Figure 3a: 0 time hydrolysis; 3b: 3 h hydrolysis; and
3c: 24 h hydrolysis.

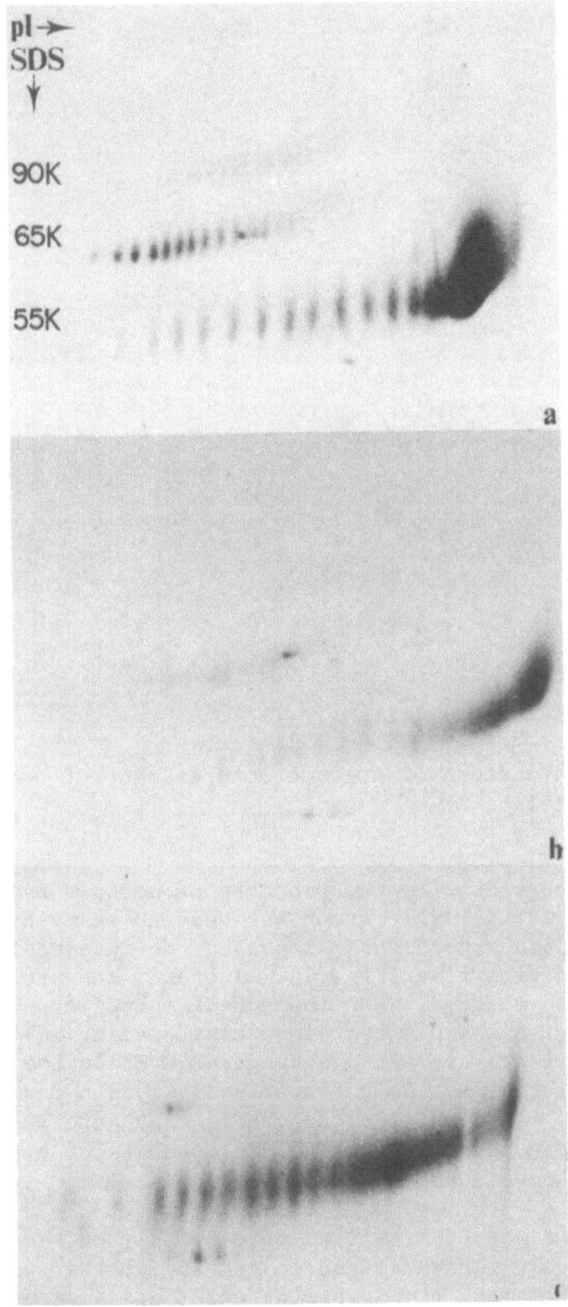

Figure 4. Hydrolysis action patterns of trypsin on particulate ZP.
Figure 4a: 0 time hydrolysis; 4b: 3 h hydrolysis; and
4c: 24 h hydrolysis.

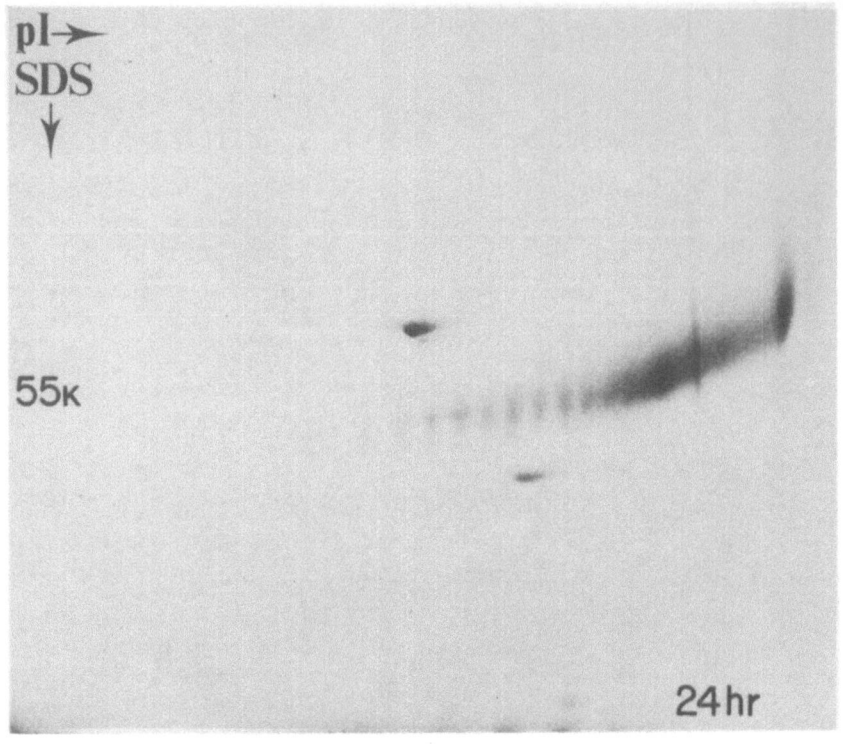

Figure 5. Action pattern of a crude acid extract of boar sperm on
          particulate ZP.

     To further characterize the products of acrosin hydrolysis on
heat solubilized ZP, gel filtration was used to assay hydrolysis time
points. With Sephadex G-75 chromatography, no intermediate sized
peaks were formed with time (Fig. 7). Instead, the excluded volume
peak gradually was reduced. When these peak tubes were electrophor-
esed on SDS-PAGE, the sample contained many smaller molecular weight
species that apparantly remained aggregated even in the presence of
1% SDS on gel filtration. As a comparison, trypsin caused a similar
Sephadex pattern, but upon electrophoresis many more intermediate and
smaller pieces were seen, indicating the products of hydrolysis were
quite different between acrosin and trypsin.

     Since the 90K and 65K families were hydrolyzed specifically,
peptide mapping gels were utilized to examine whether portions of the
individual family member glycoproteins were more susceptible than
others. With electrophoretically isolated zona families, and either
preincubating the isoelectric focusing gel (first dimension for this
peptide mapping system) in acrosin or adding acrosin to the top of
the SDS-PAGE gel (second dimension), the same resultant patterns were

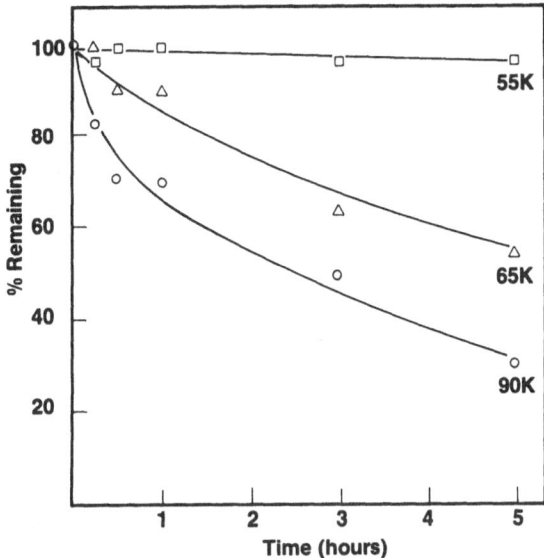

Figure 6. The amount of individual ZP family members remaining after
hydrolysis times measured by integration of the deglycosylated ZP hydrolysis patterns on SDS-PAGE.

obtained (Fig. 8). With the 55K family, no hydrolysis was obtained,
as had been seen in 30 min with all of the above methods. With the
65K family, there was the production of a number of high molecular
weight bands 2-4K dalton smaller than the 65K family; no small macromolecular species were visualized in the 15% acrylamide SDS-PAGE.
With the 90K family, a thirty minute hydrolysis formed products somewhat smaller and in greater number than with the 65K family. Also,
with both the 65K and 90K families, the macromolecules with more
neutral pI's were hydrolyzed first.

With Xenopus laevis sperm, the procedures for extraction and
assay of the sperm lysin activities have proved somewhat reticent to
examination and quite different than the straightforward purification
schemes used for mammalian sperm enzymes. The sperm can be fractured
with cold shock, artificially stimulated to an acrosome reaction with
the calcium ionophone A23187 and broken with hypotonic solutions or
with mild base extraction. The extracts of X. laevis sperm contain at
least three activities on the basis of peptide substrate analysis
(Table I). These activities, chymotrypsin-like, elastase-like and
kallikrein-like, are found in hypotonic shock extracts of sperm.
When the calcium ionophore is used, an agent known to cause the
acrosome reaction in several species, the specific activity of only
the chymotrypsin-like activity increases significantly. Sperm

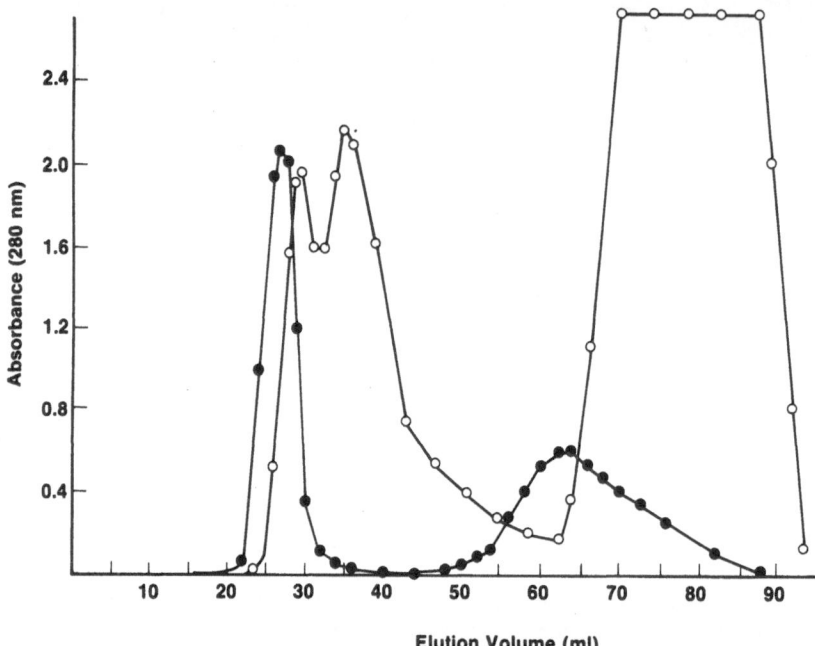

Figure 7. Sephadex G-75 column chromatography patterns of acrosin
         hydrolysis on heat solubilized ZP are displayed under
         reducing and non-reducing conditions. The closed circles
         represent a 24 h hydrolysis reaction developed in 0.1 M
         Tris-HCl, pH 8.0, and 1% SDS added. The open circles
         represent the same hydrolysis time point after boiling 2
         min in 5% β-mercaptoethanol, displayed under similar run-
         ning conditions.

extracts appear to hydrolyze specific portions of the egg vitelline
envelope, the VE, when assigned by SDS-PAGE. This appears to be
consistent with and directly analogous to the boar enzyme system
described above. Sperm extracts readily hydrolyzed the electrophoretically isolated 118 K component of VE. Current investigation is
underway to correlate the VEase activity with the chymotrypsin-like
activity by purifying these activities and determining if peptide
substrates or specific inhibitors of the chymotrypsin-like enzyme
inhibit the hydrolysis of VE and prevent fertilization.

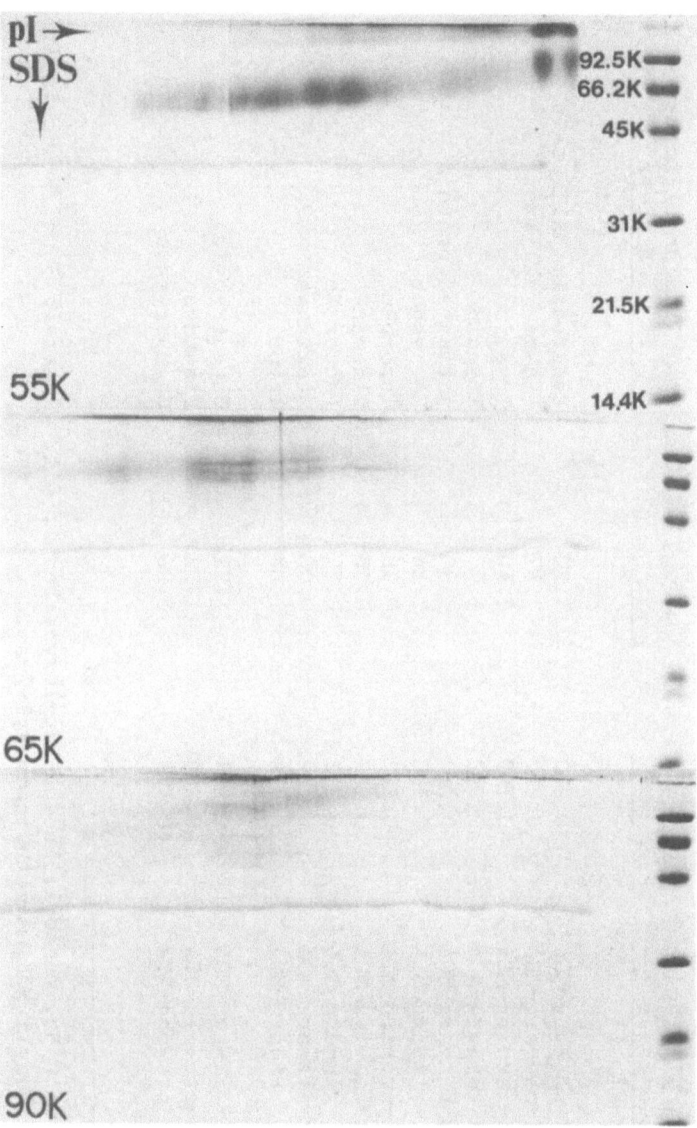

Figure 8. Peptide mapping gels of the action of acrosin on isolated ZP family members.

Table 1. MCA Peptide Activities in <u>Xenopus laevis</u> Sperm Extracts.

4-methylcoumaryl-7-amide peptide substrate P5 P4 P3 P2 P1	mM	mU/mg	Described specificity for substrate
Pro-phe-arg-MCA[a]	0.212	4.22	Pancreatic kallikrein
Z-phe-arg-MCA	0.212	2.87	Plasma kallikrein
Boc-phe-ser-arg-MCA	0.189	2.40	Trypsin
Boc-leu-thr-arg-MCA	0.212	1.77	Trypsin
Boc-leu-ser-thr-arg-MCA	0.212	1.39	Activated protein C
Boc-leu-gly-arg-MCA	0.200	1.34	Horseshoe crab clotting enzyme
Boc-val-pro-arg-MCA	0.212	0.96	Thrombin
Boc-val-leu-lys-MCA	0.212	0.58	Plasmin
Boc-ile-glu-glu-arg-MCA	0.212	0.21	Factor Xa
Glt-gly-arg-MCA	0.212	0.08	Urokinase
Bz-arg-MCA	0.212	0.08	Trypsin
Succ-leu-leu-val-tyr-MCA	0.164	1.71	Chymotrypsin
Succ-ala-pro-ala-MCA	0.243	1.32	Elastase
Succ-ala-ala-pro-phe-MCA	0.189	0.06	Chymotrypsin

[a]Abbreviations: Bz, benzoyl; Boc, t-butyloxycarbonyl; Glt, glutaryl; Succ, N-succinyl; Z, benzyloxycarbonyl; MCA, 4-methylcoumaryl-7-amide.

CONCLUSIONS

There are four main conclusions regarding the effect of boar acrosin on the porcine ZP.

1) Acrosin was found to bind to zona through its active site and to release acetone soluble pieces from radioiodinated zona. The enzyme was found to hydrolyze heat solubilized ZP to small pieces not retained in the electrophoresis gel. So the controversy over whether acrosin aids in zona penetration or not can now be addressed with the information that porcine zona is indeed a substrate for acrosin, and that acrosin specifically binds ZP.

2) Acrosin was found to sequentially hydrolyze the ZP component families in whatever solution state the zona was in. The 90K and 65K families were hydrolyzed first, followed by the 55K family when the ZP was heat solubilized, but not when the ZP was intact. These hydrolyzed components are not the major constituents of the zona, indicating that limited proteolysis of the ZP proceeds without total solubilization of the ZP.

3) Equimolar concentrations of trypsin and acrosin have different hydrolysis kinetics with respect to the zona. While acrosin does share some of trypsin's physicochemical properties, it is sufficiently different that to call it trypsin-like is counterproductive to the full understanding of acrosin's role in fertilization.

4) The ZP is an elastic and resistant structure that remains aggregated even when extensively hydrolyzed. The components are all available to the solution, yet the particulate form is not totally available to either acrosin or trypsin hydrolysis. When heat solubilized ZP is hydrolyzed to component pieces unretained on 2D-PAGE, the resultant products still remain aggregated and are very large as judged by gel filtration.

ACKNOWLEDGEMENTS

I would like to thank Ms. Eda Lim and Ms. Donna Goldhawk for their skillful technical assistance, and Mr. Nate Wardrip for helpful discussions in the course of this work. I would also like to thank Professor J.L. Hedrick for reviewing this manuscript. This work was supported in part by USPHS grant HD-15256.

REFERENCES

1. Ames, G.F.L. and Nikaido, K. 1976. Two dimensional gel electro-
    phoresis of membrane proteins. Biochem. 15:616-623.
2. Austin, C.R. and Bishop, M.W.H. 1958. Role of the rodent acrosome

and perforatorium in fertilization. Proc. R. Soc. London Ser. B. 149:241-248.

3. Bedford, J.M. 1974. Mechanisms involved in penetration of spermatozoa through the vestments of the mammalian egg, in: Physiology and Genetics of Reproduction, Part B (E.M. Coutinho and F. Fuchs, eds.), Plenum Press, New York, pp. 55-68.

4. Bedford, J.M. and Cross, N.L. 1978. Normal penetration of rabbit spermatozoa through a trypsin- and acrosin-resistant zona pellucida. J. Reprod. Fert. 54:385-392.

5. Brown, C.R. 1983. Purification of mouse sperm acrosin, its activation from proacrosin and effect on homologous egg investments. J. Reprod. Fert. 69:289-295.

6. Brown, C.R. and Harrison, R.A.P. 1978. The activation of proacro- sin in spermatozoa from ram, bull, boar. Biochem. Biophys. Acta 526:202-217.

7. Chase, T. and Shaw, E. 1970. Titration of trypsin, phasmin, and thrombin with p-nitrophenyl-p'-quanidinobenzoate HCl. Methods Enzyme. 19:20-27.

8. Cleveland, D.W., Fisher, S.G., Kirschner, M.W., and Laemmli, U.K. 1977. Peptide mapping by limited proteolysis in sodium dodecyl sulfate and analysis by gel electrophoresis. J. Biol. Chem. 252:1102-1106.

9. Dunbar, B.S., Wardrip, N.J., and Hedrick, J.L. 1980. Isolation, physicochemical properties and macromolecular composition of zona pellucida from porcine oocytes. Biochem. 19:356-365.

10. Dudkiewicz, A.G. and Garrison, G.A. 1982. Substrate preference of boar acrosin in zona pellucida lysis. J. Cell Biol. 95:165a.

11. Frazer, L.R. 1982. p-Aminobenzamidine, and acrosin inhibitor, inhibits mouse sperm penetration of the zona pellucida but not the acrosome reaction. J. Reprod. Fert. 65:185-194.

12. Gould, K.G. 1973. Application of in vitro fertilization. Fed. Proc. 32:2069-2074.

13. Green, D.P.L. and Purves, R.D. 1984. Mechanical hypothesis of sperm penetration. Biophys. J. 45:659-662.

14. Greenberg, L.J. 1962. Fluorometric measurement of alkaline phosphatase and aminopeptidase activities in the order of 10^{-14} mole. Biochem. Biophys. Res. Comm. 9:430-435.

15. Hartmann, J.F. and Hutchinson, C.F. 1974. Nature of the prepenetration contact interactions between gametes in vitro. J. Reprod. Fert. 36:49-57.

16. Hartree, E.F. 1977. Spermatozoa, eggs and proteinases. Biochem. Soc. Trans. 5:375-394.

17. Hedrick, J.L. and Wardrip, N.J. 1981. Microheterogeneity in the glycoproteins of the zona pellucida is due to the carbohydrate moeity. J. Cell Biol. 91:177a.

18. Hedrick, J.L. and Wardrip, N.J. 1982. Topographical Radiolabeling of zona pellucida glycoproteins. J. Cell Biol. 95:162a.

19. Hibbard, J. 1928. Contribution a l'étude de l'oogenèse de la fécondation, et de l'histogenèse chez Discoglossus pictus. Otth. Arch. Biol. 38:251-326.

20. Howlett, G.J., Yeh, E., and Schachman, H.K. 1978. Protein-ligand binding studies with a table-top, air-driven high-speed centrifuge. Arch. Biochem. Biophys. 190:809-819.

21. Huneau, D., Harrison, R.A.P., and Flechon, J.E. 1984. Ultrastructural localization of proacrosin and acrosin in ram spermatozoa. Gamete Res. 9:425-440.

22. Karp, D.R., Atkinson, J.P., and Shreffer, D.C. 1982. Genetic variation in glycosylation of the fourth component of murine complement. Association with hemolytic activity. J. Biol. Chem. 257:7330-7335.

23. Laemmli, U.K. 1970. Cleavage of structural proteins during assembly of bacteriophage T4. Nature 227:680-685.

24. Meizel, S. 1978. The mammalian sperm acrosome reaction: A biochemical approach, in: Development in Mammals, Volume 3 (M.H. Johnson, ed.), North-Holland, Amsterdam, pp 1-62.

25. Meizel, S. and Mukerji, S.K. 1976. Biochemical studies of proacrosin and acrosin from hamster cauda epididymal spermatozoa. Biol. Reprod. 14:444-450.

26. McLean, D. and Rowlands, I.W. 1942. Role of hyaluronidase in fertilization. Nature 150:627-628.

27. Müller-Esterl, W. and Fritz, H. 1980. Interactions of boar acrosin with detergents. Hoppe-Seyler's Z. Physiol. Chem. 361:1673-1682.

28. Müller-Esterl, W., Kupfer, S., and Fritz, H. 1980. Purification and properties of boar acrosin. Hoppe-Seyler's Z. Physiol. Chem. 361:1811-1821.

29. Müller-Esterl, W., Wendt, V., Leidl, W., Dann, O., Shaw, E, Wagner, G., and Fritz, H. 1983. Intra-acrosomal inhibition of boar acrosin by synthetic proteinase inhibitors. J. Repro. Fert. 67:13-18.

30. O'Farrell, P.H. 1975. High resolution two-dimensional electro-phoresis of proteins. J. Biol. Chem. 250:4007-4021.

31. Peterson, G.L. 1977. A simplification of the protein assay method of Lowry et al. whch is more generally applicable. Anal. Biochem. 83:346-356.

32. Salacinski, P.R.P., McLean, L., Sykes, J.E.G., Clement-Jones, V.V., and Lowry, P.J. 1981. Iodination of proteins, glycoproteins and peptides using a solid-phase oxidizing agent, 1,3,4,6-tetrachloro-3α, 6α-diphenyl glycouril (Iodogen). Anal. Biochem. 117:136-146.

33. Saling, P.M. 1981. Involvement of trypsin-like activity in binding of mouse spermatozoa to zonae pellucidae. Proc. Natl. Acad. Sci. USA. 78:6231-6235.

34. Schill, W.B., Feifel, M., Fritz, H., and Hammerstein, J. 1981. Inhibitors of acrosomal proteinase as antifertility agents. A problem of acrosomal membrane permeability. Int. J. Androl. 4:25-38.

35. Segel, I.H. 1975. Enzyme kinetics. Behavior and analysis of rapid equilibrium and steady state enzyme systems. Wiley-Interscience, John Wiley and Sons, New York.

36. Stambaugh, R. and Buckley, J. 1968. Zona pellucida dissolution enzymes of the rabbit sperm head. Science 161:585-586.
37. Stambaugh, R., Brackett, B.G., and Mastroianni, L. 1969. Inhibition of in vitro fertilization of rabbit ova by trypsin inhibitors. Biol. Reprod. 1:223-227.
38. Srivastava, P.N., Adams, C.E., and Hartree, E.F. 1965. Enzymatic action of acrosomal preparations on the rabbit ovum in vitro. J. Reprod. Fertil. 10:61-67.
39. Tyler, A. 1939. Extraction of an egg membrane lysin from sperm of the giant keyhole limpet (Megathura crenulata). Proc. Natl. Acad. Sci. USA, 25:317-323.
40. Urch, U.A., Nishihara, T., and Hedrick, J.L. 1979. The use of radioiodination protein substrates for the assay of trypsin and the hatching enzyme from the amphibian Xenopus laevis. Anal. Biochem. 100:325-356.
41. Waldschmidt, M., Hoffman, B., and Karg, H. 1966. Unterschungen uber die tryptische enzymaktivitat in geschlectssekreten von bullen. Zychthygiene 1:15.
42. Wardrip, N.J. and Hedrick, J.L. 1980. The macromolecular composition of the porcine zona pellucida. Fed. Proc. 39:2081.
43. Wardrip, N.J., Goldhawk, D.E., and Hedrick, J.L. 1984. Deglycosylation studies of pig zona pellucida. Third International Congress on Cell Biology. S. Seno and Y. Okada, eds., Academic Press, p. 401.
44. Wintrebert, P. 1933. La fonction enzymatique de l'acrosome spermien du Discoglosse. C.R. Soc. Biol. 112:1636-1640.
45. Wolf, D.P., Nishihara, T., West, D.M., Wyrick, R.E., and Hedrick, J.L. 1976. Isolation, physicochemical properties, and the macromolecular composition of the vitelline and fertilization envelopes from Xenopus laevis eggs. Biochem. 15:3671-3678.
46. Yamane, J. 1930. The proteolytic action of mammalian spermatozoa and its bearing upon the second maturation division of ova. Cytologia 1:394-403.
47. Yamane, J. 1935. Kausal-analytishe studien uber die befruchtung des kaninchenesis II. Die isoliering der auf das eizytoplasma auflosend wirkenden substanzen aus den spermatozoen. Cytologia 6:474-483.
48. Yanagimachi, R. 1981. Mechanisms of fertilization in mammals, in: Fertilization and Embryonic Development in vitro (Biggers, J.D. and L. Mastrioanni, eds), Plenum Press, New York, pp 81-182 .
49. Zaneveld, L.J.D., Robertson, R.T., Kessler, M., Srivastava, P.N., and Williams, W.L. 1970. Inhibition of fertilization in vivo by mammalian trypsin inhibitors. Fedn. Proc. 29:644.
50. Zaneveld, L.J.D., Robertson, R.T., Kessler, M., and Williams, W.L. 1971. Inhibition of fertilization in vivo by pancreatic and seminal plasma trypsin inhibitors. J. Reprod. Fert. 25:387-392.
51. Zimmerman, M., Ashe, B., Yurewicz E.C., and Patel, G. 1977. sensitive assays for trypsin, elastase and chymotrypsin using new fluorogenic substrate. Anal. Biochem. 78:47-51.

BIOCHEMICAL STUDIES OF THE ENVELOPE TRANSFORMATIONS

IN XENOPUS LAEVIS EGGS

George L. Gerton

Division of Reproductive Biology
Department of Obstetrics and Gynecology
University of Pennsylvania School of Medicine
Philadelphia, PA 19104-6080

SUMMARY

The envelopes that enclose the eggs of the amphibian Xenopus laevis were isolated and examined for biochemical correlates of the ultrastructural and sperm penetrability differences among the coelomic egg envelope (CE), the vitelline envelope (VE), and the fertilization envelope (FE). By sodium dodecyl sulfate-polyacrylamide gel electrophoresis (SDS-PAGE), the 43,000 molecular weight glycoproteins of CEs were found to be converted to components with molecular weights of 41,000 in VEs; also, a protein with a molecular weight of 57,000 was added to the envelope during the CE-to-VE conversion. The molecular weights of two components decreased during the VE-to-FE conversion, from 69,000 and 64,000 in the VE to 66,000 and 61,000 in the FE. Components from the cortical granules and the innermost jelly coat were also added to the newly-formed FE. As detected by iodination with lactoperoxidase or IODOGEN, both the CE-to-VE and the VE-to-FE conversions caused conformational changes in envelope glycoproteins. Peptide mapping demonstrated that the 43,000 molecular weight components of CE were precursors to the 41,000 molecular weight components of VE and the 69,000 and 64,000 molecular weight components of VE were precursors to the 66,000 and 61,000 molecular weight components of FE. The CE-to-VE conversion presumably occurs in the first portion of the oviduct. Experiments probing the VE-to-FE conversion demonstrated the need for an intact jelly coat for the molecular weight changes to occur. Sperm were not required for the envelope alteration; the SDS-PAGE pattern of envelopes from jellied eggs activated with the Ca^{++}-ionophore A23187 were indistinguishable

from the FE. These studies show that there are molecular correlates
of the morphological and biological differences among the envelopes.
The CE-to-VE and the VE-to-FE conversions follow a similar pattern:
in both cases, material is added to the envelope and there are
changes in the molecular weights of some of the components.

INTRODUCTION

     In many species, the involvement of the extracellular coverings
or coatings of the egg (collectively termed egg integuments) has been
implicated in several aspects of reproduction. The hypothesized
roles include specific receptors for sperm, participants in the
capacitation and activation of eggs and sperm (including the acrosome
reaction), sites for the blocks to polyspermic fertilization, and
coverings for the protection and sequestration of the embryo until it
hatches (2,4,8,9,13,19,31,32,36,37). In anuran amphibians such as
Xenopus laevis, abnormal and lethal development occurs if more than
one sperm fuses with the egg. As a defense, blocks to polyspermy
occur at the egg plasma membrane and in the egg integuments (15,38).
Changes in the biological, morphological, and biochemical characteristics of the Xenopus egg envelope occur as a function of egg maturation and fertilization. It is the purpose of this paper to review
the current state of knowledge concerning these envelopes and their
alterations during development.

     The eggs of anurans are enclosed by two types of extracellular
integuments, the jelly coats and the egg envelope. These extracellular components are composed of both carbohydrate and protein. The
VE^1 and the inner jelly coat, J_1, from unfertilized eggs of Xenopus
laevis contain protein and carbohydrate in the ratio of 85:15 and
37:60, respectively (38,39). In addition, J_1 also contains small
amounts of sulfate, presumably bound to carbohydrate (20).

     The Xenopus egg envelope is produced during oogenesis. Prior to
the passage of the egg from the body cavity to the oviduct, the envelope is termed the CE. After passage of the egg through the first
region of the oviduct, the CE is converted, presumably by oviductal
factors, to the VE. As it travels down the oviduct, the jelly coats
are secreted around the egg. After oviposition and fertilization of
the egg, the envelope is converted to the FE. This last modification
is caused by cortical granule factors released from the egg. Thus,

-----

[1]Abbreviations used are: CE, coelomic egg envelope; CGL, cortical
granule lectin; FE, fertilization envelope; F-layer, fertilization
layer; J_1, innermost jelly coat; SDS-PAGE, sodium dodecyl sulfate-
polyacrylamide gel electro-phoresis; VE, vitelline envelope; VE*,
vitelline envelope component of the fertilization envelope.

```
CE  ------------------->     VE  --------------------------> FE
    oviductal factors            cortical granule factors
```

These two transformations result in changes in biological and morphological characteristics of the envelopes. Furthermore, there are differences in the molecular composition that are correlated to the biological and morphological features of the egg envelopes. These minor changes in the composition presumably cause major conformational changes in the component molecules, thereby altering the morphology and biological properties of the envelopes.

BIOLOGICAL FEATURES

To obtain large quantities of eggs, Xenopus females are primed by an injection of pregnant mare serum gonadotropin. After three to five days, the animals are given a subsequent injection of human chorionic gonadotropin. Six, eight, and ten hours later, eggs are gently squeezed from the females. Batches of eggs may be fertilized with testicular cell suspensions. Coelomic eggs are surgically removed from the body cavities of females with ligated oviducts (17).

The eggs can be manipulated in several ways. Jelly coats can be removed using reducing agents such as 2-mercaptoethanol (39). Envelopes from eggs can be isolated by lysing the eggs through a 19 gauge syringe needle and collecting the envelopes on nylon screens of the appropriate size (37). Cortical granule exudate can be prepared by activating the dejellied eggs with the calcium ionophore A23187; the exudate then diffuses through the envelope and is collected in the supernatant solution from around the activated eggs (14).

Like other anuran eggs, sperm cannot penetrate the envelopes surrounding fertilized Xenopus eggs. As mentioned above, unfertilized, oviposited eggs with intact jelly coats can be fertilized by the addition of sperm. If the eggs are first treated to remove the jelly coats, washed, and then exposed to sperm, essentially no fertilization takes place. If the solubilized jelly coat substances are then replaced, fertilization does occur, indicating that some component(s) of the jelly has a positive effect on the fertilizing sperm (35). Surprisingly, eggs recovered from the body cavity cannot be fertilized, even in the presence of exogenously added jelly. Several researchers, including Elinson (7) and Katagiri (21), have demonstrated that prior chemical or enzymatic treatment of coelomic anuran eggs can permit fertilization, suggesting a role of the envelope in preventing sperm penetration.

Isolated envelopes demonstrate the same sperm penetration characteristics as do envelopes in situ around the egg. Sperm appear to penetrate isolated VE via either the inner or outer surfaces. Neither isolated CE nor isolated FE can be penetrated by sperm (16, 17).

MORPHOLOGICAL APPROACH

As described above, the three envelope types have distinctly different characteristics with regards to sperm penetration. When the envelopes were examined by scanning and transmission electron microscopy, it was found that each envelope type could also be distinguished on the basis of morphology.

The coelomic egg envelope ultrastructure was examined by Grey et al. (17). The CE is about 1 μm thick and is composed of filaments arranged in thick bundles. The exterior surface of this envelope has an irregular contour which is permeated by large gaps in the fibrillar structure. The inner surface of the CE rests against the tips of the egg microvilli.

Envelopes undergo the conversion to the VE type of morphology after passage through the first centimeter of the oviduct. Like the CE, the inner surface of the VE rests against the tips of egg microvilli. However, the filaments that make up the envelope are now dispersed, giving this envelope a smooth exterior; there are no gaps in the VE as are found in the CE. In addition, after passage through the oviduct, the outermost aspect of the VE is coated with the first jelly coat, termed J_1.

After fertilization or activation with ionophore A23187, the envelope undergoes the second conversion to become the fertilization envelope. The FE contains the same fibrillar structure seen in the VE. However, major changes are seen at this time. First, the envelope is seen to lift away from the egg surface, creating a large perivitelline space between the envelope and the egg plasma membrane (15). Second, an electron-dense fertilization- or F-layer forms at the interface of the envelope and J_1.

MOLECULAR APPROACH

Toward understanding the molecular basis for these biological and morphological differences between CE, VE, and FE, envelopes were isolated and characterized by SDS-PAGE or two-dimensional gel electrophoresis. The electrophoretically separated envelope components were examined in terms of Coomassie blue staining for protein, dansyl hydrazine staining for carbohydrate, labeling with ^{125}I, peptide mapping and amino-terminal amino acids.

CE-TO-VE CONVERSION

When CE and VE were analyzed by SDS-PAGE using slab gels containing 7.5% polyacrylamide, two major differences were found (11). One of the major components of CE was converted from a molecular weight of 43,000 to 41,000. There was also the addition to VE of a new component with a molecular weight of 57,000 (Fig. 1). On two-dimensional gels, the components in the 41,000 to 44,000 molecular weight range of CE encompassed all of the spots found in the corresponding molecular weight region of VE; in addition, there were several spots that were unique to CE (Fig. 2A,B).

If the envelopes were isolated and then iodinated by either the lactoperoxidase method or the IODOGEN method and then analyzed by SDS-PAGE, all of the major Coomassie blue staining proteins of CE were found to be labeled (30). In contrast, the 41K components (i.e., those components with molecular weights of 41,000) of VE could not be labeled unless the envelopes had been previously dissociated with guanidine-HCl or sodium dodecyl sulfate (30, Fig. 3). Further examination by two-dimensional gel electrophoresis (11) demonstrated that the iodinated components of this molecular weight range were those that were unique to CE (Fig. 3B).

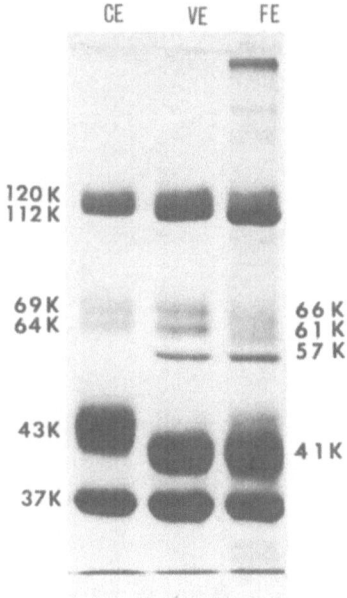

Figure 1. Analysis of envelopes by SDS-PAGE. Molecular weights of envelope components are indicated on the left and right sides of the figure. Data from Gerton and Hedrick (11,12).

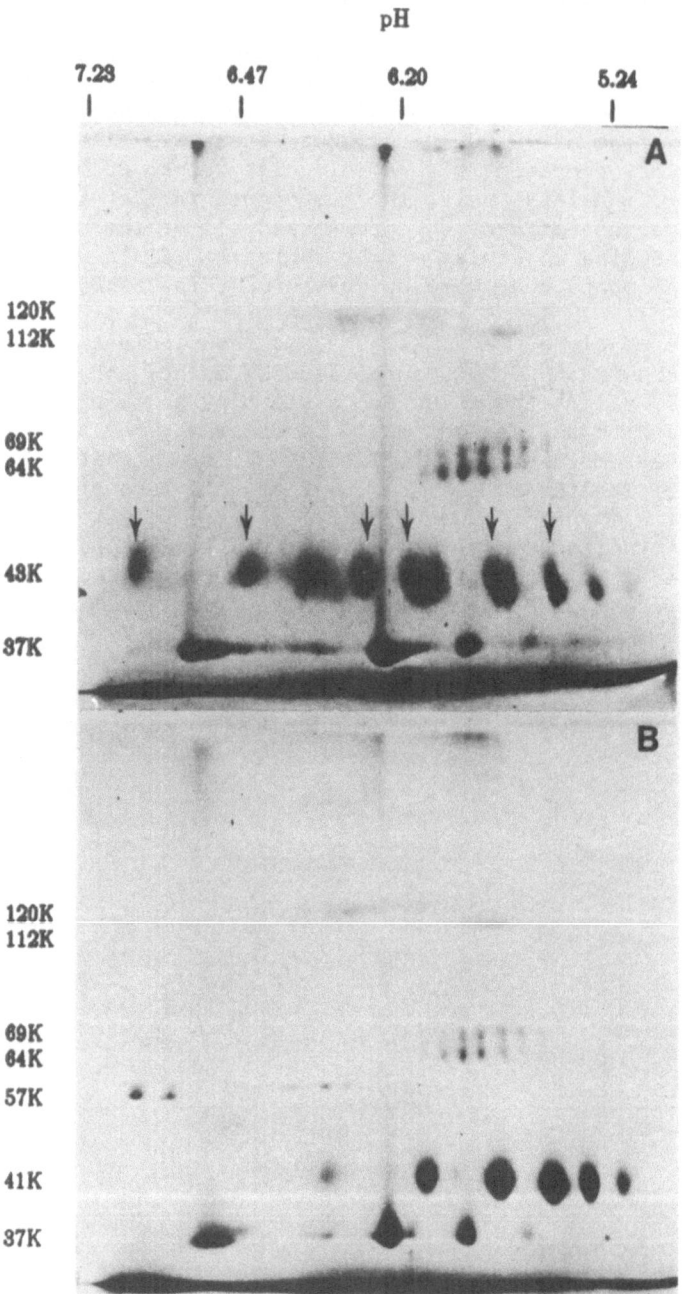

Figure 2. Analysis of envelopes by two-dimensional gel electro-
        phoresis.   A, CE; B, VE.   Arrows in A indicate CE-
        specific components with molecular weights of 43,000.
        Data from Gerton and Hedrick (11,12).

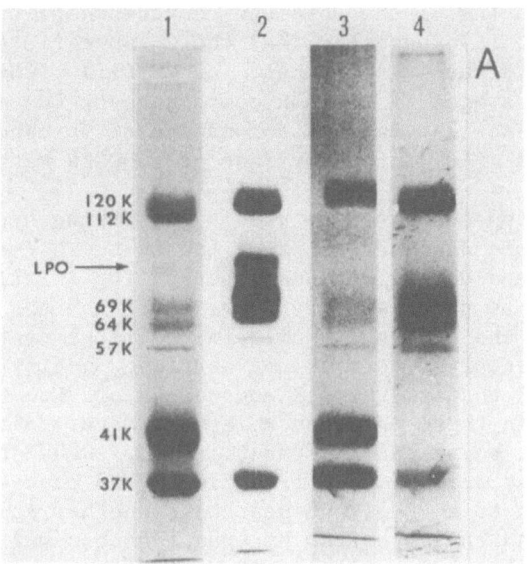

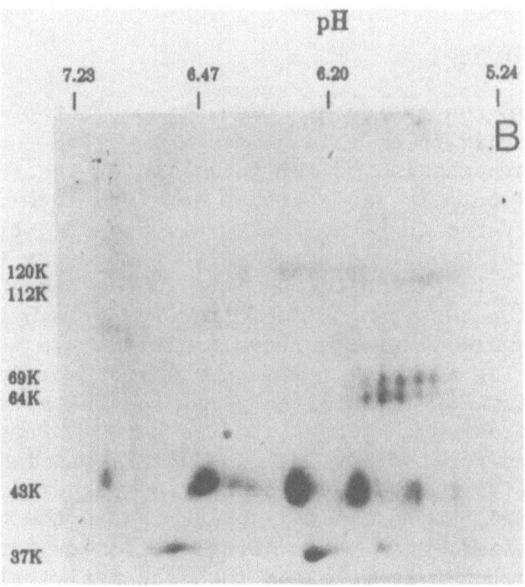

Figure 3. Electrophoretic analysis of envelopes iodinated in the intact state. A. 1: Commassie-blue-stained SDS-PAGE gel of lactoperoxidase (LPO) labeled VE; 2: autoradiogram of 1; 3: Coomassie blue-stained gel of IODOGEN labeled VE; 4: autoradiogram of 3. B. Autoradiogram of two-dimensional gel of CE seen in Fig. 2A. Data in panel B from Gerton and Hedrick (11).

Peptide mapping of the envelope components by the method of
Cleveland et al. (5) suggested that the CE-specific 43K components
are precursors to the 41K components of VE (11). These results
suggest that cleavages of the 43K components of CE cause conformational shifts so that the 41K components of VE cannot be
iodinated by the mild lactoperoxidase or IODOGEN methods.

The following model can be used as a working hypothesis for
future experiments designed to study the CE-to-VE conversion. The
envelope is formed during oogenesis but in its initial state, the
envelope cannot be penetrated by sperm until the egg that encloses
has passed down the oviduct. While in the first centimeter of the
oviduct, some factor(s) acts on the envelope, modifying the 43K
components of CE to form the 41K components of VE. Conformational changes in these envelope molecules occur, altering the
envelope so that sperm can now penetrate it. Grey et al. (17)
demonstrated that acid or base treatment of CE causes the morphological conversion to a VE-like appearance; whether sperm can
penetrate such an envelope from Xenopus eggs has yet to be tested.
The function of the 57K component of VE and its site of synthesis
have also not been examined.

VE-TO-FE CONVERSION

The VE-to-FE conversion has been the subject of several investigations. The results of this research have led to the cortical
granule lectin hypothesis for the formation of the F-layer and the
coincident block to polyspermy at the envelope level. Wyrick et
al. (38) demonstrated that the F-layer is produced by the precipitation of cortical granule contents with components found in jelly
coat layer J_1. This precipitation is a Ca^{++}-dependent lectin-
ligand interaction that can be inhibited by certain galactosides.
The cortical granule lectin has been purified by a one-step
procedure utilizing affinity chromatography of cortical granule
exudate on a column of immobilized jelly coat material (Nishihara
et al., in preparation). Recent studies have indicated that the
jelly coat ligand for the cortical granule lectin is a sulfated
polysaccharide (1). Greve and Hedrick (14) used immunocytochem-
istry to show that the lectin is sequestered in the cortical granules of unfertilized eggs. Upon fertilization or artificial activation of the egg, CGL is released into the perivitelline space.
Some CGL diffuses through the VE and precipitates with its ligand
in J_1 to form the F-layer (Fig. 4). Most of the unprecipitated CGL
remains trapped in the perivitlline space because macromolecules
cannot penetrate the F-layer (28). Unfertilized eggs incubated in
cortical granule exudate or purified CGL exhibit an F-layer and
cannot be fertilized, indicating that the interaction of CGL and
its jelly coat receptor is a block to polyspermic fertilization
(38, Nishihara et al., in preparation).

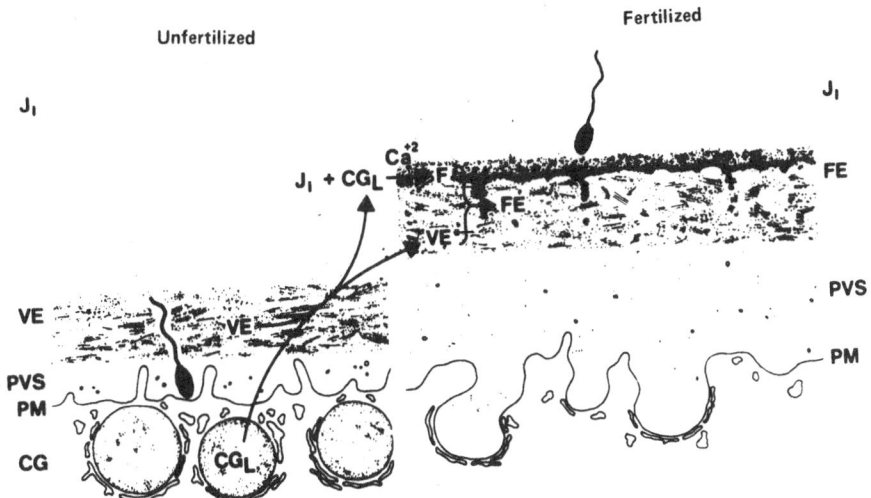

Figure 4. Diagrammatic representation of the structural changes in
an egg from <u>Xenopus</u> <u>laevis</u> at fertilization and the lectin
hypothesis for F-layer formation as suggested by Wyrick et
al. (38). CG, cortical granule; PM, plasma membrane; PVS,
perivitelline space. From Greve and Hedrick (14).

For comparative purposes, the envelopes have been manipulated
in several ways. The VE^* of FE was prepared by treating FE in a
manner that disrupted the $CGL-J_1$ interaction but left the envelope
intact. The F-layer was solubilized by suspending FE in 0.5 M
galactose, 1.0 mM $CaCl_2$, 10 mM Tris-HCl, pH 7.8 for 2h (30). The
envelopes were then pelleted and the supernatant solution (solubilized F-layer components) removed. The envelopes (designated VE^*)
were then rinsed in fresh galactose buffer, followed by several
rinses in ice-cold water to remove the galactose and salts. The
envelopes had different "melting temperatures" and chemical solubi-
lities; in general, VEs were more soluble than FEs or VE^*s (23,34,
37). When solubilized F-layer components were reconstituted with
VE^*s, an F-layer formed on only one surface, presumably what was
originally the outer surface. If the same experiment was performed
with VEs, "F-layers" were formed on both envelope surfaces, demonstrating an asymmetry of VE^* that is not present in VE (29).

Iodination of intact, isolated envelopes also demonstrated the presence of F-layer components on the outer surface of the FE. In studies examining the inner and outer surface labeling profiles for VE and FE, it was concluded that the F-layer components probably masked to some extent the VE* components that were exposed to the radioiodinating reagents previous to fertilization (30).

There are also differences between the SDS-PAGE polypeptide profiles of VE and VE*. Wolf et al. (37) reported that one component derived from the VE decreases in molecular weight during the conversion to the FE, perhaps by proteolysis. We have subsequently shown that there are two molecular weight changes associated with the VE to FE conversion (12, Fig. 1). These are difficult to detect and are best seen by slab gel electrophoresis using 7.5% polyacrylamide gels. The two components decrease from molecular weights of 69,000 and 64,000 to molecular weights of 66,000 and 61,000. (These molecular weights obtained for the envelope components are slightly different than those reported by Wolf et al. (37) and may reflect differences in the electrophoretic conditions used for analysis.)

Envelopes from jellied eggs activated with the ionophore A23187 also exhibited the molecular weight changes associated with fertilization (12). This demonstrated that sperm were not required for the molecular weight changes and that the conversion factor was probably coming from the cortical granules that released their contents upon ionophore stimulation. Unfortunately, if the eggs were dejellied prior to the ionophore treatment, the molecular weights remained the same. This result could mean that the responsible factor was denatured or inhibited by the dejellying procedure, that the substrate was altered, and/or that the jelly coats in some way acted to assist in these changes, perhaps by acting as a diffusion barrier to keep the active factor sufficiently concentrated.

Two-dimensional gel electrophoresis demonstrated that the 69K and 64K components of VEs and the 66K and 61K components of VE*s had heterogeneous isoelectric points (12, Fig. 2B,5). Most of the spots of the 69K component of VE were paired with spots of the 64K group having identical isoelectric points (pI region 5.7 to 6.1). A similar situtation existed for the 66K and 61K components of VE*. Also, as demonstrated by two-dimensional gel electrophoresis, the changes in molecular weights are coincident with shifts to slightly more basic isoelectric points of the component spots involved in the VE-to-VE* conversion.

The paired shifts in the molecular weights and the isoelectric points of component spots of the 69K and 64K components suggested that these two families of components were related in their primary sequences. Peptide mapping by the method of Cleveland et al. (5)

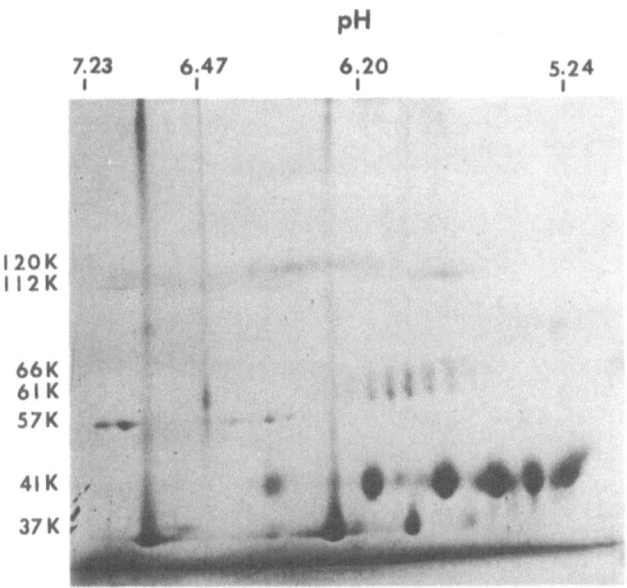

Figure 5. Analysis of FE by two-dimensional electrophoresis. The 66K and 61K components are shifted to slighted more basic isoelectric points in relation to the 69K and 64K components of VE (see Fig. 2b). Data from Gerton and Hedrick (12).

demonstrated this to be the case. In fact, after the peptide maps were stained for carbohydrate with dansyl hydrazine (6) it was concluded that the major difference between the 69K and 64K components was the presence of an extra carbohydrate side-chain in the 69K components (Fig. 6). Conversion of VE-to-VE* resulted in the modification of the same peptide in both the 69K and 64K components (12).

In order to try to understand whether the conversion of VE to VE* was proteolytic or carbohydrolytic in nature, the 69 and 64K components of VE and the 66K and 61K components of VE* were isolated by preparative electrophoresis (10) and analyzed for amino-terminal amino acids (33). In none of the cases could N-terminal amino acids be detected. Nor were any amino-terminal amino acids found when whole envelopes were analyzed. Mild acid hydrolysis was used to cleave the proteins at the rarely occurring aspartyl-prolyl peptide bonds (24). As expected, N-terminal amino acid analysis detected proline as the N-termini of these selectively hydrolyzed proteins, demonstrating that sufficient quantities of material were being used for analysis. Hence it appears that the amino terminal

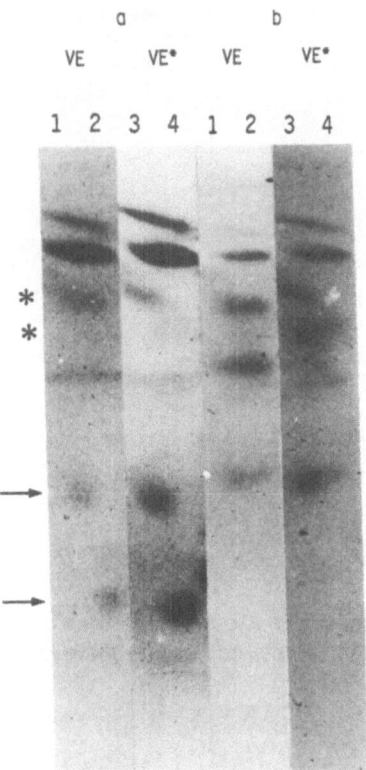

Figure 6. Peptide mapping of the 69K and 64K VE components and the
66K and 61K VE* components. a. Commassie blue-staining
for protein. b. Dansyl hydrazine-staining for
carbohydrate. 1: 69K VE component; 2: 64K VE component;
3: 66K VE* component; 4: 61K VE* component. Asterisks
indicate the glycopeptides involved in the VE-to-VE*
conversion. Arrows indicate the differences between the
69K and 64K components of VE and also indicate the
differences between the 66K and 61K components of VE*.
See Gerton and Hedrick (12) for details.

residues of the envelope components are blocked. It is also interesting to note that the Xenopus CGL also appears to have blocked
N-termini (Birr, Kato, Gerton, and Hedrick, unpublished observations).

It can be argued by analogy with other systems that a cortical
granule protease might function in the Xenopus VE-to-VE* conversion. Based on inhibition studies and changes in the composition
of the zona pellucida of mammalian eggs, a protease has been implicated but not demonstrated in the hamster and mouse zona reaction

(2,19). In sea urchin eggs, the existence of cortical granule proteases has been convincingly demonstrated and these proteases are believed to act in modifying the vitelline envelope by cleaving off sperm receptors and releasing the envelope from its attachment to the egg microvilli (3,32, and the paper in this volume by E.J. Carroll).

The molecular weight changes involved in the VE-to-VE* conversion are correlated with the changes in the observed envelope solubility properties. In agreement with the SDS-PAGE patterns that were observed for envelopes from eggs activated in the presence of jelly, Wolf (34) found that artificial activation of jellied eggs produces envelopes that are indistinguishable from FE in terms of their susceptibility to trypsin and 2-mercaptoethanol dissolution. As found by electrophoresis (12), Wolf (34) also found that envelopes from artificially activated, dejellied eggs are more like vitelline envelopes.

Attempts to demonstrate the VE-to-VE* conversion factor in vitro have so far failed. When isolated VEs were treated with cortical granule exudate, the 69K and 64K components were not cleaved to lower molecular weights (12). When dejellied, unfertilized eggs were activated with A23187, cortical granules underwent exocytosis but the molecular weight conversions of the 69K and 64K components did not occur; if the jelly was left intact, the VE-to-VE* conversion did take place. The molecular weight changes were not inhibited by 1 mM phenylmethylsulfonyl fluoride, an inhibitor of serine proteases.

CONCLUDING COMMENTS

The processing of these envelopes must be carefully controlled. Most likely there are very specific proteases that are required for the CE-to-VE conversion and the VE-to-VE* conversion. These factors may be responsible for the initial gain and subsequent loss of the sperm penetrability of the envelope. Limited proteolysis is required in each case; proteases capable of indiscriminate cleavages would be difficult to control and might cause more modifications of the envelopes than are necessary.

For these reasons, the tasks of elucidating the CE-to-VE and VE-to-VE* conversion factors are difficult ones. First, the enzyme(s) must be found that cause some assayable biological, morphological, or biochemical change in the envelopes. Second, the action(s) of these factor(s) must then be correlated with the associated changes in sperm penetrability, envelope morphology, and physicochemical properties of the envelope glycoproteins.

The two papers by Drs. Katagiri and Miceli also address the question of CE-to-VE conversion factors. It is my belief that the integration of their studies (21, 22, 25, 26, 27) with the work summarized here will lead to a more complete understanding of how these integuments function in fertilization.

ACKNOWLEDGEMENTS

Most of the experiments reported here were carried out by associates of Dr. Jerry L. Hedrick, Department of Biochemistry and Biophysics, University of California at Davis. Thanks are due to the many collaborators who preceded me in his laboratory and to those with whom I had the opportunity to work. I also want to thank Nate Wardrip for help with the photography. This research was funded in part by NIH grant HD-04906 to JLH and NIH predoctoral training grants S-T32-GM 07377-3-0291 and 1-T32-HD 07171-01-0041 to GLG. I also acknowledge the support of Dr. Clarke F. Millette, Harvard Medical School (NIH HD-15269); I was a postdoctoral research associate in his laboratory while writing this review.

REFERENCES

1. Birr, C.A. 1979. Immunoelectophoretic studies of the jelly coat ligand for the cortical granule lectin of Xenopus laevis eggs. Ph.D. Thesis, University of California, Davis.
2. Bleil, J.D., and Wassarman, P.M. 1980. Mammalian sperm-egg interaction: Identification of a glycoprotein in mouse egg zonae pellucidae possessing receptor activity for sperm. Cell 20:873-882.
3. Carroll, E.J., Jr., and Epel, D. 1975. Isolation and biological activity of the proteases released by sea urchin eggs following fertilization. Develop. Biol. 44:22-32.
4. Carroll, E.J., Jr., and Hedrick, J.L. 1974. Hatching in the toad Xenopus laevis: Morphological events and evidence for a hatching enzyme. Develop. Biol. 38:1-13.
5. Cleveland, D.W., Fischer, S.G., Kirschner, M.W., and Laemmli, U.K. 1977. Peptide mapping by limited proteolysis in sodium dodecyl sulfate and analysis by gel electrophoresis. J. Biol. Chem. 252:1102-1106.
6. Eckhardt, A.E., Hayes, C.E., and Goldstein, I.J. 1976. A sensitive method for the detection of glycoproteins in polyacrylamide gels. Anal. Biochem. 73:192-197.
7. Elinson, R.P. 1973. Fertilization of frog body cavity eggs enhanced by treatments affecting the vitelline coat. J. Exp. Zool. 183: 291-302.
8. Florman, H., and Storey, B. 1982. Mouse gamete interactions: The zona pellucida is the site of the acrosome reaction leading to fertilization in vitro. Develop. Biol. 91:121-130.

9. Foerder, C.A., and Shapiro, B.M. 1977. Release of ovoperoxidase from sea urchin eggs hardens the fertilization membrane with tyrosine crosslinks. Proc. Natl. Acad. Sci., USA 74:4214–4218.

10. Gerton, G.L., Wardrip, N.J., and Hedrick, J.L. 1982. A gel eluter for the recovery of proteins separated by polyacrylamide gel electrophoresis. Anal. Biochem. 126:116–121.

11. Gerton, G.L., and Hedrick, J.L. 1986. The coelomic egg envelope to vitelline envelope conversion in eggs of Xenopus laevis. J. Cell. Biochem. in press.

12. Gerton, G.L., and Hedrick, J.L. 1986. The vitelline envelope to fertilization envelope conversion in eggs of Xenopus laevis. Develop. Biol. in press.

13. Glabe, C.G., and Vacquier, V.D. 1978. Egg surface glycoprotein receptor for sea urchin sperm binding. Proc. Natl. Acad. Sci., USA 75:881–885.

14. Greve, L.C., and Hedrick, J.L. 1978. An immunocytochemical localization of the cortical granule lectin in fertilized and unfertilized eggs of Xenopus laevis. Gamete Res. 1:13–18.

15. Grey, R.D., Wolf, D.P., and Hedrick, J.L. 1974. Formation and structure of the fertilization envelope in Xenopus laevis. Develop. Biol. 36:44–61.

16. Grey, R.D., Working, P.K., and Hedrick, J.L. 1976. Evidence that the fertilization envelope blocks sperm entry in eggs of Xenopus laevis: Interaction of sperm with isolated envelopes. Develop. Biol. 54:52–60.

17. Grey, R.D., Working, P.K., and Hedrick, J.L. 1977. Alteration of structure and penetrability of the vitelline envelope after passage of eggs from coelom to oviduct in Xenopus laevis. J. Exp. Zool. 201:73–83.

18. Grey, R.D., Bastiani, M.J., Webb, D.J., and Schertel, E.R. 1982. An electrical block is required to prevent polyspermy in eggs fertilized by natural mating of Xenopus laevis. Develop. Biol. 89:475–484.

19. Gwatkin, R.B.L., and Williams, D.T. 1977. Receptor activity of the hamster and mouse solubilized zona pellucida before and after the zona reaction. J. Reprod. Fert. 49:55–59.

20. Hedrick, J.L., Smith, A.J., Yurewicz, E.C., Oliphant, G., and Wolf, D.P. 1974. The incorporation and fate of [^{35}S]-sulfate in the jelly coat of Xenopus laevis eggs. Biol. Repro. 11:534–542.

21. Katagiri, C. 1974. A high frequency of fertilization in premature and mature coelomic toad eggs after enzymic removal of vitelline membrane. J. Embryol. Exp. Morphol. 31:573–587.

22. Katagiri, C., Iwao, Y., and Yoshizaki, N. 1982. Participation of oviducal pars recta secretions in inducing the acrosome reaction and release of vitelline coat lysin in fertilizing toad sperm. Develop. Biol. 94:1–10.

23. Kutryk, M.A., and Hedrick, J.L. 1983. Melting of glycoprotein egg envelopes as a probe of supramolecular conformational differences. Fed. Proc. 42:1996.

24. Landon, M. 1977. Cleavage at aspartyl-prolyl peptide bonds. Methods Enzymol. 47:145-149.

25. Miceli, D.C., Fernandez, S.N., Raisman, J.S., and Barbieri, F.D. 1978a. A trypsin-like oviducal proteinase involved in Bufo arenarum fertilization. J. Embryol. Exp. Morphol. 48:79-91.

26. Miceli, D.C., Fernandez, S.N., and del Pino, E.J. 1978b. An oviducal enzyme isolated by affinity chromatography which acts upon the vitelline envelope of Bufo arenarum oocytes. Biochim. Biophys. Acta. 526:289-292.

27. Miceli, D.C., and Fernandez, S.N. 1982. Properties of an oviducal protein involved in amphibian oocyte fertilization. J. Exp. Zool. 221:357-364.

28. Nishihara, T., and Hedrick, J.L. 1977a. A molecular mechanism for envelope elevation at fertilization. Fed. Proc. 36:811.

29. Nishihara, T., and Hedrick, J.L. 1977b. Reconstruction of the fertilization envelope from its component parts. J. Cell Biol. 75:172a.

30. Nishihara, T., Gerton, G.L., and Hedrick, J.L. 1983. Radioiodination studies of the envelopes from Xenopus laevis eggs. J. Cell. Biochem. 22:235-244.

31. SeGall, G.K., and Lennarz, W.J. 1979. Chemical characterization of the component of the jelly coat from sea urchin eggs responsible for the induction of the acrosome reaction. Develop. Biol. 71:33-48.

32. Vacquier, V.D., Tegner, M.J., and Epel, D. 1973. Protease released from sea urchin eggs at fertilization alters the vitelline layer and aids in preventing polyspermy. Exp. Cell Res. 80:111-119.

33. Weiner, A.M., Platt, T., and Weber, K. 1972. Amino-terminal sequence analysis of proteins purified on a nanomole scale by gel electrophoresis. J. Biol. Chem. 247:3242-3251.

34. Wolf, D.P. 1974. On the contents of the cortical granules from Xenopus laevis eggs. Develop. Biol. 38:14-29.

35. Wolf, D.P., and Hedrick, J.L. 1971a. A molecular approach to fertilization. II. Viability and artificial fertilization of Xenopus laevis gametes. Develop. Biol. 25:348-359.

36. Wolf, D.P., and Hedrick, J.L. 1971b. A molecular approach to fertilization. III. Development of a bioassay for sperm capacitation. Develop. Biol. 25:360-376.

37. Wolf, D.P., Nishihara, T., West, D.M., and Hedrick, J.L. 1976. Isolation, physicochemical properties, and macromolecular composition of the vitelline and fertilization envelopes from Xenopus laevis eggs. Biochemistry 15:3671-3677.

38. Wyrick, R.E., Nishihara, T., and Hedrick, J.L. 1974. Agglutination of jelly coat and cortical granule components and the block to polyspermy in the amphibian Xenopus laevis. Proc. Natl. Acad. Sci., USA 71:2067-2071.

39. Yurewicz, E.C., Oliphant, G., and Hedrick, J.L. 1975. The macromolecular composition of *Xenopus laevis* egg jelly coat. Biochemistry 14:3101-3107.

THE ROLE OF OVIDUCAL SECRETIONS IN MEDIATING GAMETE FUSION

IN THE TOAD, BUFO BUFO JAPONICUS

Chiaki Katagiri

Zoological Institute
Faculty of Science
Hokkaido University
Sapporo 060, Japan

SUMMARY

A fertilizing sperm of the anuran amphibians has to pass through the jelly envelopes and the vitelline envelope (VE) before making a successful fusion with the egg plasma membrane. Of these the jelly envelopes, secreted by the long pars convoluta (PC) portion of the oviduct, have long been known to be indispensable for the sperm entrance in the egg. The most recent experiments employing dejellied uterine eggs of the toad, Bufo bufo japonicus, revealed that the jelly plays its role in fertilization by its unique capacity of retaining divalent cations (Ca^{2+} and/or Mg^{2+}) which are essential for a fertilizing sperm. There are other lines of evidence which implicate that the secretions of the uppermost portion of oviduct, p. recta (PR), render the VE penetrable by sperm. We show that the secretory granules (PRG) isolated from PR of ovulating Bufo females by centrifugation in Percoll possess such biological activities as an increase of fertilizability of coelomic eggs and the induction of both the acrosome reaction and a release of the VE lysin from sperm. In addition, the activities of the PRG are inhibited by trypsin inhibitors, and this trypsin-like activity is dependent on Ca^{2+}. These results, combined with the previous immunohistochemical demonstration of the deposition of PR-substance(s) in the VE, lead us to propose that a fertilizing toad sperm is acrosome-reacted in response to the PRG substance deposited in the VE and finds a way of traversing the VE by the released lysin, both of which may be dependent on Ca^{2+} supplied by jelly envelopes, the product of PC.

151

INTRODUCTION

The oviduct, the major female reproductive tract of amphibians, attracts attention because it provides the substances that evidently mediate a succussful sperm entry into eggs. One of these substances is the jelly which is deposited in several layers during passage of eggs down this long tract to the uterus. These jelly layers have been regarded as a prerequisite for a fertilizing sperm, since artificially dejellied uterine eggs are unfertilizable but become fertilizable when readded with appropriate jelly preparations (2,10). Despite many experimental studies employing the dejellied uterine egg bioassay system (4,12), however, none of the previous studies succeeded in defining the molecular entities that are essential for the biological activity of egg-jelly.

Besides these questions, recent studies implicate that the amphibian oviduct contributes to gamete fusion by offering another role than supplying jelly. This new role of the oviduct has been attributed to the pars recta, the uppermost extremely short portion of a long reproductive tract. Thus, there is morphological (6,21) and experimental (3,5,13) evidence that after passage through this portion eggs are fertilizable probably due to the increased penetrability of the vitelline envelope (VE) by sperm. Although a trypsin-like enzyme has been isolated from the pars recta (18,19), there is controversy as to the mode of action of pars recta derived substances in mediating gamete fusion (14, see also D.C. Miceli, this volume).

In this paper, I would like to describe our most recent experiments (8) in which the biologically active substance in the egg-jelly has been defined in the toad. Then I will review our recent experiments with the isolated granules from the oviducal pars recta, together with a discussion about possible functions of these oviducal secretions.

MATERIALS AND METHODS

Procurement of Gametes

Mature toads, Bufo bufo japonicus, were purchased from dealers during the hibernation period, and stored at 4°C until use. Ovulation was induced by subcutaneous injection of pituitaries or human chorionic gonadotropin. Twelve to thirteen h after the injection at 18°C, eggs were found mostly in the coelom, occasionally a few passing through the oviduct. Uterine eggs were obtained 18-24 h after the injection. Coelomic eggs were placed in De Boer's solution (DB: 110 mM NaCl, 1.3 mM KCl, 1.3 mM $CaCl_2$, pH 7.3 with $NaHCO_3$ or 5 mM Tris-HCl) until 18-24 h after the hormone injection, to allow attainment of second meiotic metaphase. Unless otherwise stated, sperm suspension was made by macerating testes in 0.05 DB.

For the dejellied-egg bioassay, uterine eggs were dejellied using 0.5-1% NaCN in Na-free DB, and were washed thoroughly with DB. The eggs were placed in 2 ml test solutions to be described below, to which 0.05 ml sperm suspension was added dropwise. Thirty min after insemination, test solutions were replaced by 0.5 DB or Ca-free 0.5 DB to suppress sperm fertilizing capacity. Dejellied eggs inseminated in 0.05 DB served as a "negative control".

Egg-jelly Materials

Diffusible jelly material (DF) was prepared according to the method described previously (12). To obtain the UV-irradiated jelly (UVJ), uterine eggs were placed in deionized water to allow hydration of the jelly coat. The jelly solubilized by ultraviolet irradiation was lyophilized, dialyzed against deionized water, and the dialyzate (UVJD) was concentrated by evaporation at 40°C. The highly concentrated UVJD was freed from a precipitate by a brief centrifugation, and the supernatant was heated at 400°C for one h and 600°C for 16 h. The resulting ash was dissolved in 0.01 N HCl, and diluted to the original volume (adjusted to pH 7.3 with NaOH) for assaying its biological activity.

Oviducal Pars Recta Materials

The oviducal pars recta, the uppermost 0.05 portion of oviduct extending for about 3 cm from the ostial opening, was removed from the ovulating females 10-13 h after the hormone injection. To prepare the crude extract of this oviducal portion, the tissues were frozen at -20°C, thawed, and homogenized in Steinberg's solution or DB at a concentration of 4 tissues/ml. The 25,000g supernatant was used as the pars recta extract (PRE).

To obtain secretory granules from the pars recta (PRG), the freshly isolated tissues were homogenized in the medium consisting of 0.25 M sucrose, 1 mM EDTA, and 10 mM Tris-HCl, pH 8.0 (SET), and filtered through several layers of gauze. The filtrate was centrifuged at 600g for 10 min to sediment tissue debris, and the supernatant was centrifuged at 7,000g for 15 min. The precipitate was suspended in 3 ml SET, and centrifuged in 67% Percoll in 0.25 M sucrose at 36,000g for 60 min. The granules collected at the density of 1.126 - 1.141 g/ml were resuspended in sucrose-Tris-HCl solution, and pelleted by centrifugation at 100,000g for 90 min.

Enzyme Activities

The VE lysin was prepared from 0.5% Triton X-100-treated sperm, and partially purified according to the method described previously (9). The lytic activity on the VE was assayed by placing 10-20 coelomic eggs in 0.5 ml test solutions, either with or without pretreatment with PRE or PRG, and incubation in a moist chamber at 18°C

for periodic binocular microscopic observations for the swelling or
lysis of the VE. Enzyme activities were determined in vitro by
using N,N-dimethylated casein, BAEE, or TAME as substrates, as
described previously (9).

Immunohistochemistry

Rabbits were immunized by subscapular injections of the PRE or
PRG emulsified in Freund's complete adjuvant. Antisera used for
immunofluorescence studies were rendered specific to the PRE or PRG
by absorption with both glutaraldehyde-precipitated liver and serum
from mature females. The crude IgG fraction was obtained by precipitation with 33% ammonium sulfate. For immunohistochemical
studies, oviducal tissues and eggs were fixed in Carnoy's fixative,
and 5 μm thick sections were reacted with the above-mentioned
absorbed antisera, followed by incubation with fluorescein
isothiocyanate-conjugated goat IgG against rabbit IgG (14).

Chemical Analyses

UVJD was fractionated by gel-filtration on Sephadex G-25 or
G-10 (8). Each fraction was determined both for absorption at
260 nm and for contents of neutral sugars by anthrone method. The
fertilization-supporting activity of each fraction was also determined by the dejellied-egg bioassay method described above. The
determination of inorganic cations were carried out by using an
electron microprobe X-ray analyzer (Shimazu EMX-SM) and an atomic
absorption photometer (Hitachi Model 170-50).

RESULTS AND DISCUSSION

Egg-Jelly Factors that Support Fertilization of Dejellied Eggs

We have repeatedly shown (12) that a high frequency of fertilization is obtained when dejellied uterine eggs are inseminated in
the presence of UVJ or UVJD. It has also been demonstrated that
this fertilization-supporting activity of jelly preparations is
small enough to be dialyzable, stable after heating at 100°C for 15
min (12), resistant to extensive proteolysis by pronase (10), and
functions less species-specifically (11,12). Gel-filtration of the
dialyzable egg-jelly components suggested that the biologically
active fractions have a molecular weight of less than 500 daltons,
are basic in nature and include some reducing sugars (12).

In the present study, UVJD was concentrated 10-fold by evaporation, and dejellied eggs were inseminated in the presence of UVJD
at various concentrations. The rate of fertilization was highest
at the concentration equivalent to that in the original jelly. A
surprising finding was that the fertilization-supporting activity

of UVJD was retained even after heating at 600°C for 16 h. The
active preparation thus contained, in contrast to native UVJD (12),
neither sugars, ninhydrin-positive substances, nor any substances
with absorption at 260 nm. Determinations of inorganic salts in the
active preparations (Table 1) revealed that, compared with uterine
egg-jelly, it contained relatively high levels of Ca^{2+} and Mg^{2+},
variable levels of K^+, and a much lower level of Na^+.

The above finding prompted an inquiry of the role of inorganic
salts in fertilization of dejellied eggs. A reconstituted salt solution (RSS) was prepared according to the composition of UVJD described in Table 1: 3.4 mM NaCl, 1.6 mM KCl, 0.33 mM $CaCl_2$, and 0.48
mM $MgCl_2$. The dejellied-egg bioassay summarized in Table 2 clearly
revealed that RSS is equivalent to UVJD in supporting fertilization.
Further analyses were made to determine what kinds of salts contained
in RSS are effective in this biological activity. The results showed
that (a) at 5.5 mM which is close to the total cationic concentration
of UVJD (cf., Table 1), $CaCl_2$ and $MgCl_2$ were highly active, KCl was
somewhat active and NaCl was not active in supporting fertilization;
(b) $CaCl_2$ tends to give a higher activity than $MgCl_2$, but its activity could be replaced by the latter (Fig. 1); (c) insemination of
dejellied eggs in the presence of various amounts of EDTA and $CaCl_2$
or $MgCl_2$ gave rise to successful fertilization when divalent cations
exceeded the molarity of EDTA; (d) dejellied eggs were fertilized in
0.05 DB supplemented with 2-5 mM $CaCl_2$. These results strongly indicate that Ca^{2+} and/or Mg^{2+} are involved in the fertilization-supporting activity of UVJD.

Table 1. Ionic composition of biologically active egg-jelly
          preparations in comparison with 0.05 De Boer's solution

Sample	Concentration in mM of			
	Na^+	K^+	Ca^{2+}	Mg^{2+}
Dialysate from UV-irradiated jelly (UVJD)	3.4	1.6	0.33	0.48
Ash from UVJD	2.7	0.2	0.13	0.087
Uterine egg-jelly[a]	95.0≠ 3.0	21.0 ± 6.0	8.5 ± 2.8	6.0 ± 2.2
0.05 De Boer	5.5	0.065	0.065	

[a] Based on ten separate determinations, and expressed as mean
± SD.

Table 2. Fertilization of dejellied uterine eggs inseminated in the
presence of jelly preparation or salt solutions

	Insemination medium			
Experiment	UVJD[a]	0.05 DB[b]	UVJD-ash[c]	RSS[d]
1	45.1 (51)[e]	2.2 (45)	44.9 (49)	--
2	60.0 (55)	0 (56)	39.6 (58)	70.9 (55)
3	63.6 (55)	8.7 (46)	--	63.8 (55)

[a] Dialysate from UV-irradiated jelly.
[b] 0.05 De Boer's solution.
[c] Ash obtained by heating UVJD at 600°C for 16 h.
[d] Reconstituted salt solution according to the ionic composition of
UVJD.
[e] Numbers in parentheses indicate the number of eggs used.

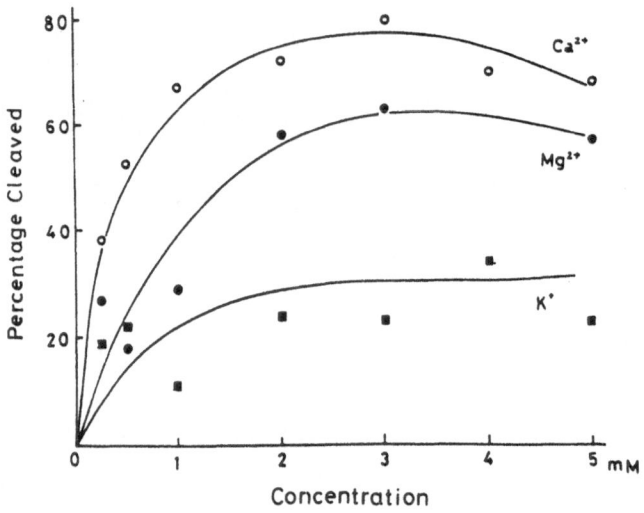

Figure 1. Fertilization-supporting activity of Ca^{2+}, Mg^{2+}, and K^+ as
a function of their concentrations. The total concentra-
tion was adjusted to 5.8 mM by addition of NaCl.

UVJD or RSS has to be present at the moment of sperm penetration of the egg, as shown by the experiments summarized in Table 3. In these experiments, sperm were pretreated with UVJD or RSS for 20 min, centrifuged then washed and resuspended in 0.05 DB for insemination of dejellied eggs. Sperm pretreated with the biologically active preparations did not fertilize dejellied eggs in 0.05 DB, although they possessed a high fertilizing capacity in UVJD or RSS. In this respect, the term "sperm capacitation", often used in describing the role of amphibian egg-jelly (12,20) by analogy with the concept developed in mammals (7), should now be abandoned.

Egg-Jelly as a Source of Divalent Cations Necessary for Fertilization

The finding of Ca^{2+} and/or Mg^{2+} as an essential factor for supporting fertilization of dejellied eggs demands an examination of ionic conditions in the egg-jelly at the time of fertilization. Uterine eggs were immersed in deionized water for various periods up to 5 min, taken out of water, and were rolled briefly on a filter paper to remove excess water. Insemination of these eggs with a small amount of sperm suspension (<0.05 ml) and placing them in a moist chamber revealed that the eggs became fertilizable after contact with water for at least 2 min (Fig. 2). This necessity of

Table 3. Necessity for successful fertilization of the presence of UVJD[a] or RSS[b] at the time of insemination[c]

Pretreatment of sperm	Inseminated in	Number of eggs used	Percentage cleaved
UVJD	UVJD	49	69.4
	0.05 De Boer	43	0
RSS	RSS	44	70.5
	0.05 De Boer	41	0
0.05 De Boer (control)	UVJD	39	89.7
	RSS	41	82.9
	0.05 De Boer	46	2.2

[a] Dialysate from UV-irradiated jelly·
[b] Reconstituted salt solution (see text for explanation)·
[c] Dejellied eggs were inseminated in the presence or absence of UVJD or RSS, by sperm which had been pretreated for 20 min with UVJD or RSS.

the contact of eggs with water may be explained in terms of the
requirement of a low salinity for the motility of anuran sperm. In
fact, the sperm of <u>Bufo bufo</u> were motile in media whose total
cation concentration is less than 20–25 mM. In another series of
experiments, uterine eggs were immersed in deionized water for
various periods, and the concentration of inorganic salts in the
jelly accompanying their hydration was determined (see 8 for
details of these experiments). The results proved that salts,
particularly Na^+, contained in the egg-jelly rapidly diffused away
into water, so that in 2–3 min the salinity within the jelly was
low enough to assure the motility of sperm.

A more important finding, however, may be the differential
loss of cations from the egg-jelly upon its hydration, as evidenced
by the following experiment. UVJ was dialyzed against deionized
water, and the concentrations of inorganic cations remaining in the
retentate were determined at various times. It is remarkable in
Figure 3 that in contrast to Na^+ and K^+, Ca^{2+} and Mg^{2+} were not
readily diffusable in water. Even dialysis against 10 mM EDTA was
not effective in removing these divalent cations from jelly. This
differential affinity of ions to egg-jelly was observed also in
Sephadex G-10 gel-filtration of UVJD: a single peak of
fertilization-supporting activity contained Ca^{2+}, Mg^{2+}, and K^+ in
association with sugars and A_{260nm} absorbing material, but Na^+ and
Cl^- were eluted in more retarded fractions.

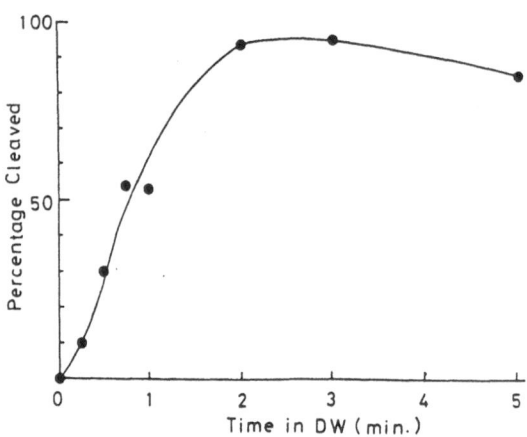

Figure 2. Fertilizability of uterine eggs after immersion in
deionized water for short periods.

The unique nature of egg-jelly in terms of hydration and pre-
ferential retention of divalent cations is intriguing in view of
the ionic requirements of the anuran sperm, which are as follows.
First, anuran sperm are not motile in physiological salt solutions,
but require much lower concentrations for motility. The jelly
layers surrounding uterine eggs immediately after oviposition
contain cations at approximately 130 mM (c.f., Table 1), which is
apparently too high to permit the sperm motility. Most of these
ions are lost during the first 2-3 min immersion of eggs in pond
water or 0.05 DB, so that the sperm motility is assured. Second,
however, a simple dilution of ordinary physiological saline results
in the divalent cation concentration far below the level that is
necessary for the sperm incorporation into eggs (e.g., 0.065 mM
Ca^{2+} in 0.05 DB). In this context, the successful fertilization of
anurans is dependent on a delicate balance of a decrease in the
total salinity while maintaining minimal divalent cation levels.
The ability of anuran egg-jelly to retain divalent cations as
elucidated in our study is thus regarded to be of special adaptive
value in the strategy of fertilization.

Role of the Oviducal Pars Recta in Fertilization

A number of studies showed that the coelomic eggs of anurans
are not fertilizable when inseminated in the biologically active
jelly preparations, even though the eggs are physiologically
mature. The success of fertilizing them after the coelomic

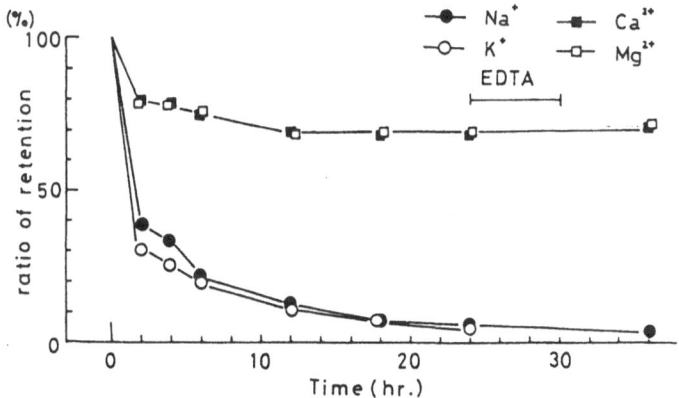

Figure 3. The rate of retention of cations in UV-irradiated jelly
(UVJ) as a function of the time of dialysis against
deionized water. At 24-30 h after the beginning of
dialysis, dialysis was against 10 mM EDTA-4NH$_4$$^+$. Initial
concentration for each cation: 10.1 mM Na$^+$; 2.3 mM K$^+$;
1.1 mM Ca^{2+}; 0.48 mM Mg^{2+}.

envelope (CE) has been removed or weakened enzymatically or mechanically (5,13) postulated a role of oviduct in rendering the CE accessible for sperm penetration. This role of the oviduct other than
secreting the jelly has been ascribed to the secretions of the pars
recta portion, on the basis of: (a) the demonstrated enhancement of
fertilization of Bufo arenarum coelomic eggs by pretreatment with the
pars recta extract (16,19, see also G.L. Gerton, this volume), and
(b) a morphological transformation and increased susceptibility to
sperm attack of the CE of Xenopus (6), R. japonica (21), or B.
arenarum (15) during the passage through the pars recta. More recently, characterization of proteolytic activity has been achieved from
B. arenarum pars recta secretions (17). All these studies thus favor
the view that the pars recta secretions contribute to enhance the egg
fertilizability by a limited hydrolytic processing of the CE (16).
We have also found a striking effect of the pars recta substances in
enhancing the fertilizability of coelomic Bufo bufo eggs (14). In
the following pages, I would like to summarize our observations, with
emphasis on the cellular basis of the pars recta secretions and their
possible function in the fertilization process.

     The TEM of the pars recta revealed that, like in R. japonica
(21), the epithelial wall contains unique secretory cells in addition
to the cilia cells common to the whole length of oviduct. These
secretory cells are apparently distinct from the jelly secreting
cells in the pars convoluta portion by possessing zymogen granule-
like, electron dense granules of 0.4 μm in size (Fig. 4). The
granules were few in number in the hibernating females, but increased
significantly in number in the ovulating females so that the whole
cytoplasm was occupied by them. The granules then decreased after
oviposition. In cross sections, the CE of coelomic eggs comprised
conspicuous filament bundles running both parallel and perpendicular
to the egg surface. In the VE of the eggs after passing through the
pars recta, these bundles were less conspicuous, but instead extremely fine electron dense particles were present among the bundles.

     Immunohistochemical studies employing antibodies against the PRE
revealed that the antigen(s) specifically localized in the pars recta
epithelial cells are deposited in the VE during passage of eggs
through this uppermost portion of oviduct (14). These observations
lead us to propose that the alteration of the CE during passage of
the p. recta includes a deposition of certain molecules to the CE.

     To determine that the biological activities observed previously
in the PRE (14) are ascribable to the contents of secretory granules
from the pars recta epithelial cells, the PRG were isolated by
density gradient centrifugation on Percoll (Fig. 5). Examination of
their biological activities proved that in all respects those found
with the PRE (14) were also associated with the PRG, as follows: (a)
the coelomic eggs pretreated with the PRG for 4 h and inseminated in
the presence of UVJD or RSS were fertilized at a similar level to

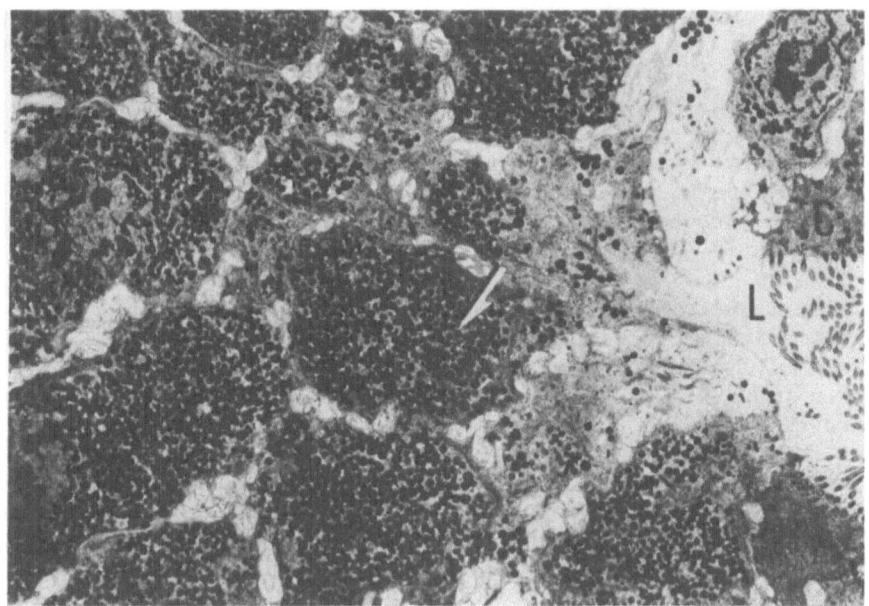

Figure 4. Electron micrograph of epithelial cells in oviducal pars
recta from an ovulating toad, showing abundance of
cells possessing secretory granules (arrow). C, cilia cell;
L, lumen. x 2,400.

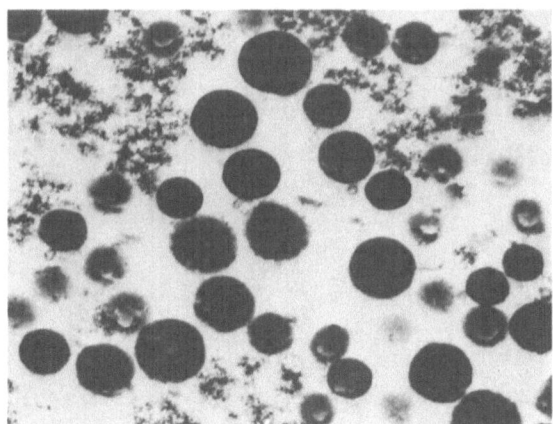

Figure 5. Secretory granules isolated from oviducal pars recta cells
by a Percoll gradient centrifugation. x 26,000.

those pretreated with the PRE, (b) the coelomic eggs pretreated with the PRG for 1 h and then added to sperm (1×10^6 cells/ml) were not fertilized, but their envelopes were lysed in 1-2 h, as in the PRE-treated eggs, (c) sperm were treated with the PRG for 30 min, centrifuged, and the supernatant solution possessed a strong VE lytic activity, similarly to the preparation obtained by incubation with the PRE, (d) the sperm incubated with the PRG for 30 min showed a high incidence of acrosome reaction which was inhibited by soybean trypsin inhibitor (Table 4).

These results emphasize the role of the pars recta secretory granules in inducing the acrosome reaction and a concomitant release of acrosomal lysin from individual sperm, rather than their direct effect on the VE. Several lines of evidence are compatible with this interpretation. First, the sperm incubated with the PRE have lost most of their VE lysin (14). Second, the sperm passing through the jelly envelopes possess intact acrosomes, but those attaching to and passing through the VE are acrosome-reacted probably in response to the PRG substances deposited on the VE (22). Third, the PRG substan-

Table 4. Acrosome reaction inducing activity of oviducal pars recta granules (PRG) and its inhibition by soybean trypsin inhibitor (SBTI)

Treatment[a]	Total number of sperm	Percentage reacted
Experiment 1		
PRG	147	51.7
PRS[b]	138	8.7
Steinberg (Control)	126	11.9
Experiment 2		
PRG	191	47.1
PRG + SBTI[c]	168	7.3
Steinberg (Control)	97	9.3

[a] Incubation for 30 min at 20-22°C.
[b] Fraction containing cilia and membranous organelles on Percoll gradient centrifugation.
[c] 1 mg/ml (final concentration).

ces contain trypsin-like enzyme(s) as evidenced by their hydrolytic
activity on BAEE. This last point is consistent with the observed
inhibitory effect of SBTI on the acrosome reaction as well as on the
fertilization of PRE-treated coelomic eggs or dejellied uterine eggs
(Table 5).

It should be mentioned, however, that the hydrolysis by tryptic
enzyme(s) included in the pars recta secretions is also tenable, as
evidenced by the spectrofluorometric changes of the B. arenarum CE
upon incubation with the partially purified pars recta substance
(16). The conformational changes in the CE induced by the
secretions may increase its susceptibility to the sperm lysin by
exposing the lysin-binding sites.

Apparently the exact role of the pars recta secretions in the
amphibian fertilization process awaits further studies with more
purified preparations. It is not known whether the acrosome
reaction-inducing and the VE hydrolytic activities are ascribable to
the dual function of a single molecule or represent the activities
by different molecules. In functional aspects, it is relevant to
mention that the tryptic activities both from the pars recta
secretions of R. arenarum (17) and our PRG (unpublished) are highly
dependent on Ca^{2+}. In addition, our preliminary observation reveals
that the chymotryptic activity of our VE lysin on the VE is also
dependent on

Table 5. Fertilizability of oviducal pars recta granule
          (PRG)-treated coelomic eggs or dejellied uterine eggs
          inseminated in the presence of reconstituted salt solution
          RSS[a] with or without addition of soybean trypsin inhibitor
          (SBTI)

Eggs	Inseminated in	Number of eggs used	Percentage cleaved
Coelomic eggs	RSS	35	74.3
	RSS + SBTI[b]	32	0
Uterine eggs	RSS	67	85.1
	RSS + SBTI[b]	63	0
	RSS + SBTI[c]	58	10.3

[a] See text for ionic composition.
[b] 1 mg/ml (final concentration).
[c] 0.5 mg/ml (final concentration).

Ca^{2+}, regardless of the pretreatment of the VE with the PRG. Evidences are also available that the induction of acrosome reaction by the PRG requires an extremely small amount of Ca^{2+}. Taken together, the elucidation of the exact role of oviducal pars recta secretions in amphibian fertilization should be done in relation to the analyses of their functional dependence on Ca^{2+}, which is evidentally retained in the jelly as discussed above.

CONCLUDING REMARKS

In normal anuran fertilization, a fertilizing sperm passes through jelly layers and the vitelline envelope before it successfuly fuses with the egg within 3-4 min. During this initial phase of fertilization the passage of sperm through jelly is assured by a rapid loss of cations due to the hydration of the jelly, while Ca^{2+} and/or Mg^{2+} are differentially retained by the jelly molecules. Sperm are acrosome-intact during passage through jelly, but undergo the acrosome reaction when they attach to the VE in response to the pars recta granule-derived substance. This reaction includes a breakdown of the apical and the outer acrosomal membranes, accompanied by a release of the VE lysin (22). Thus the sperm will eventually fuse with the egg plasma membrane by means of the inner acrosomal membrane. Currently several biological activities can be ascribed to the pars recta-derived trypsin-like enzyme(s), e.g., the induction of acrosome reaction, the hydrolytic effect on the CE, and activation of the VE lysin. Whatever their exact molecular basis may be, it is intriguing that these activities are evidently dependent on certain levels of Ca^{2+} that can be preferentially retained in the jelly. Further studies are required to elucidate the molecular basis of the divalent cation-binding nature of the egg-jelly, and to characterize more exactly the pars recta substances and their mode of action in the Ca^{+2} dependent steps preceding the gamete fusion.

ACKNOWLEDGEMENTS

I thank Drs. J. Hosono, K. Ishihara, Y. Iwao, and N. Yoshizaki for their collaboration and stimulating discussion in pursuing the studies reviewed here.

REFERENCES

1. Barbieri, F.D., and Oterino, J.M. 1972. A study of the diffusable factor released by the jelly of the egg of the toad, Bufo arenarum. Dev. Growth Diff. 14:107-117.
2. Barbieri, F.D., and Raisman, J.S. 1969. Non-gamete factors involved in the fertilization of Bufo arenarum oocytes. Embryologia 10:363-372.

3. Cabada, M.O., Mariano, M.I., and Raisman, J.S. 1978. Effect of trypsin inhibitors and concanavalin A on the fertilization of Bufo arenarum coelomic oocytes. J. Exp. Zool. 204:409-416.

4. Elinson, R.P. 1971. Fertilization of partially jellied and jelly-less oocytes of the frog Rana pipiens. J. Exp. Zool. 176:415-428.

5. Elinson, R.P. 1973. Fertilization of frog body cavity eggs enhanced by treatments affecting the vitelline coat. J. Exp. Zool. 183:291-302.

6. Grey, R.D., Working, P.K., and Hedrick, J.L. 1977. Alteration of structure and penetrability of the vitelline envelope after passage of eggs from coelom to oviduct in Xenopus laevis. J. Exp. Zool. 201:73-84.

7. Gwatkin, R.B.L. 1977. "Fertilization Mechanisms in Man and Mammals," Plenum Press, New York.

8. Ishihara, K., Hosono, J., Kanatani, H., and Katagiri, C. 1984. Toad egg-jelly as a source of divalent cations essential for fertilization. Dev. Biol. 105:435-442.

9. Iwao, Y., and Katagiri, C. 1982. Properties of the ·vitelline coat lysin from toad sperm. J. Exp. Zool. 219:87-95.

10. Katagiri, C. 1966. Fertilization of dejellied uterine toad eggs in various experimentsl conditions. Embryologia 9:159-169.

11. Katagiri, C. 1968. Immunological relationship among anuran egg-jellies and their contribution to the analysis of fertilization. SABCO J. 4:33-43.

12. Katagiri, C. 1973. Chemical analysis of toad egg-jelly in relation to its "sperm-capacitating" activity. Dev. Growth Diff. 15:81-92.

13. Katagiri, C. 1974. A high frequency of fertilization in premature and mature coelomic toad eggs after enzymic removal of vitelline membrane. J. Embryol. exp. Morphol. 31:573-581.

14. Katagiri, C., Yoshizaki, N., and Iwao, Y. 1982. Participation of oviducal pars recta secretions in inducing the acrosome reaction and release of vitelline coat lysin in fertilizing toad sperm. Dev. Biol. 94:1-10.

15. Mariano, M.I., DeMartin, M.G., and Pisano, A. 1984. Morphological modifications of oocyte vitelline envelope from Bufo arenarum during different functional states. Dev. Growth Diff. 26:33-42.

16. Miceli, D.C., and Fernandez, S.N. 1980. Effect of oviducal proteinase upon Bufo arenarum vitelline envelope: A fluorescence approach. Dev. Growth Diff. 22:639-643.

17. Miceli, D.C., and Fernandez, S.N. 1982. Properties of an oviducal protein involved in amphibian oocyte fertilization. J. Exp. Zool. 221:257-364.

18. Miceli, D.C., Fernandez, S.N., and Del Pino, D.J. 1978. An oviducal enzyme isolated by affinity chromatography which acts upon the vitelline envelope of Bufo arenarum coelomic oocytes. Biochim. Biophys. Acta 526:289-292.

19. Miceli, D.C., Fernandez, S.N., Raisman, J.S., and Barbieri, F.D.
    1978. A trypsin-like oviducal proteinase involved in Bufo
    arenarum fertilization. J. Embryol. exp. Morphol. 48:79-91.
20. Wolf, D.P., and Hedrick, J.L. 1971. A molecular approach to
    fertiliza-tion III. Development of a bioassay for sperm
    capacitation. Dev. Biol. 25, 360-376.
21. Yoshizaki, N., and Katagiri, C. 1981. Oviducal contribution to
    alteration of the vitelline coat in the frog, Rana japicona.
    Dev. Growth Diff. 23:495-506.
22. Yoshizaki, N. and Katagiri, C. 1982. Acrosome reaction in sperm
    of the toad, Bufo bufo japonicus. Gamete Res. 6:343-352.

OVIDUCAL PARS RECTA AS FACTOR IN FERTILIZATION PROPERTIES

AND HORMONAL REGULATION IN THE TOAD BUFO ARENARUM

Dora C. Miceli

Departamento de Biologia del Desarrollo
Instituto Superior de Investigaciones Biologicas (INSIBIO)
Chacabuco 461
4000-San Miguel de Tucuman, Argentina

SUMMARY

In this communication I have attempted to present an overview of some contributions to the understanding of the oviduct-egg interaction in amphibians.

According to data from other authors, the vitelline envelope of the newly ovulated egg constitutes a barrier for the passage of spermatozoa. Our results demonstrated that only after they have been affected by substances released by the first 1-3 cm of the oviduct (pars recta), is the envelope sensitive to spermlysins and the oocytes fertilizable. This functional change is matched by biological, physicochemical and morphological differences in the vitelline envelope.

The fact that the pars recta activity is affected by the sexual cycle and that in ovariectomized females – devoid of active pars recta – the biological activity can be restored by steroid hormones, strongly suggests that the molecules involved in fertilization are synthesized and secreted during specific steps of the reproductive cycle.

The pars recta-oocyte interaction probably involves more than one type of molecules, considering the observations made on the carbohydrate metabolism of coelomic eggs, which could be altered by the oviducal secretions.

Several explanations for the pars recta mechanism of action have been suggested. One is a direct action on the sperm; pars recta molecules – engulfed in the vitelline envelope – would trigger the acrosome reaction. We propose the unmasking of specific vitelline envelope sites for sperm interaction. Material on the outer surface of the VE can be removed or altered by the enzymatic activity – similar to plasmin and trypsin – detected in the pars recta secretions.

INTRODUCTION

Data derived from observation on amphibians can serve to build a useful model of the mechanism of spermatozoa interaction with the coats of the oocytes. In this model, one of the most interesting egg components is the vitelline envelope (VE) which protects or covers the fusible plasma membrane "until the right moment" according to Epel (6). The sperm must bind specifically to VE and bore through this coat; only then can it fuse with the egg-plasma membrane.

There is a good deal of evidence indicating that the VE possesses a special molecular pattern which determines a structural configuration for the spermatozoa to interact with and penetrate it. This has been demonstrated in Xenopus laevis (8), Bufo arenarum (14,18), Rana japonica (26) and Bufo bufo japonicus (10). The VE structural and molecular requirements for sperm penetration are not definitively established during oogenesis but are acquired when the oocytes travel along the uppermost region of the oviduct, named the pars recta (PR). The changes produced in the VE by the PR secretions are essential for fertilization. Comparison of coelomic oocytes fertility with oocytes previously treated with an extract of pars recta (PRE) proved that oocyte fertilizability increases in Bufo arenarum (1,16) and Bufo bufo japonicus (10). Our results suggest that the oocytes, during their transit down the PR, still uncovered by jelly coats, are bathed with PR secretion fluid containing enzymes (19) which through a mild proteolysis modify the VE which becomes permeable to sperm. This change in the VE from an impenetrable to penetrable condition (8,16) is probably mediated by an increase of the sensitivity of the VE to sperm lysins (1,16,27).

In Xenopus laevis, when the envelope of coelomic oocytes (CE) is converted to the envelope of oviposited eggs (VE) in the oviduct, a new glycoprotein component is added (MW: 57,000) and another is altered (MW of 43,000 reduced to 41,000) (25). Somewhat analogous results reported that a substance containing carbohydrate is supplied to the vitelline coat by the oviducal tissue in R. japonica and B. bufo (10,26). From this and other results obtained in these species, they conclude that the fertilizing sperm undergoes the acrosome reaction in response to the PR-derived substances when it attaches to the VE.

In addition to its role in fertilization, the PR of B. arenarum
through its effect on carbohydrate metabolic pathways, seems to be
involved in the establishment of cytoplasmatic biochemical conditions
required for embryonic development (11). The full grown oocytes of
this species exhibited two alternative metabolic pathways of carbohydrates: one similar to that of the adult tissue, the other characteristic of early stages of development (12). The transition from
the first to the second type of metabolism is essential for development (13). Such transformation would depend, to some extent, upon
the secretion of the PR and would involve the establishment of
embryonic metabolism.

In this presentation, the functional aspecta of PR in B.
arenarum as well as the hormonal regulation of its function will be
considered, since all known PR biological activities show variations
in the different stages of the sexual cycle.

INVOLVEMENT OF PARS RECTA IN FERTILIZATION

Striking differences are observed between jellyless oocytes
removed from the pars recta and body cavity oocytes, inseminated at
high sperm concentration (1.7 x 10^7 cells/ml). While a fertilization
rate of about 80% was obtained from oocytes taken from pars recta,
coelomic oocytes could not be fertilized under the same conditions.
Likewise, "in vitro" treatment of coelomic oocytes with an extract of
PR tissue or with PR secretion fluid for a relatively short period
(20 min) significantly increased their fertilizability. When the
oocytes are inseminated with sperm concentration normally used for
oviposited eggs (3 x 10^6 cell/ml) in the presence of egg-water, a
little higher rate of fertilization with respect to high sperm concentration without egg-water is observed.

Only PR-treated oocytes showed the typical penetration path,
clearly indicated by a trail of pigment in the cytoplasm. These
results reinforce the evidence that isolated VE from Xenopus leavis
coelomic oocytes are refractory to sperm penetration under conditions
that facilitate penetration of VE isolated from oviposited eggs (8).
Both results further support the view that the participation of the
PR in fertilization is through a change in VE sperm penetrability.

It is generally accepted that in amphibians, sperm penetration
through VE is mediated by chemical attack on the envelope by enzymes
released as a result of the acrosome breakdown (9,15,20). The
envelope of Bufo arenarum coelomic oocytes is not attacked by sperm
lysins.

However, if the same oocytes are pretreated with PRE, the VE
becomes sensitive to sperm lysins. It was consistently observed that
after a 3-min period of incubation in the presence of a preparation

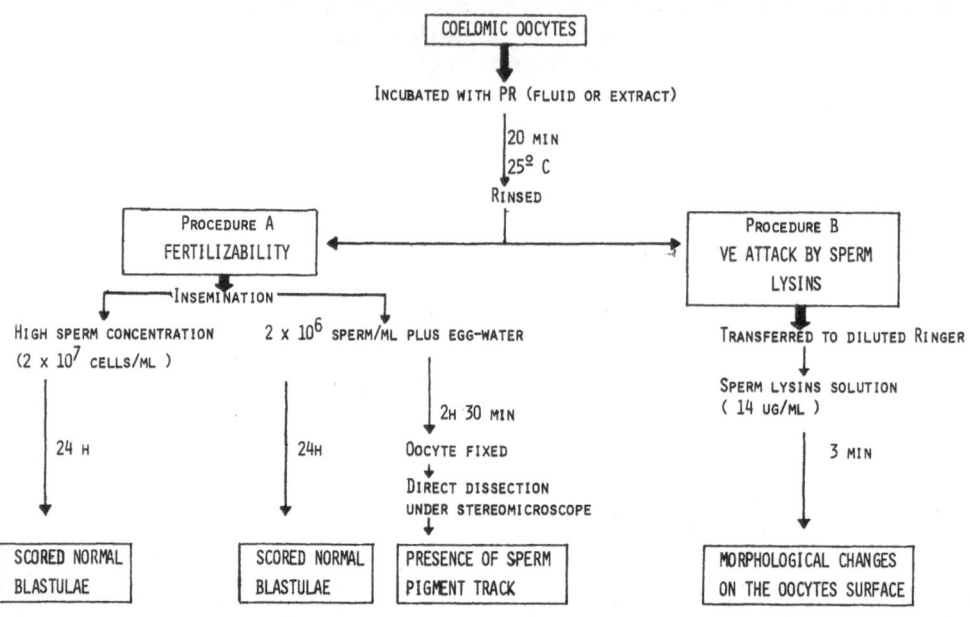

Figure 1.

of lysins, all PR-treated coelomic oocytes becomes less turgid and depressions and wrinkles appear on the egg surfaces. Later on, the eggs collapsed and the VE appeared swollen, with evidence of partial dissolution. These are the typical signs of sperm attack on the ovi-posited egg envelopes (4,5,20). We find significant correlation among PR function as fertilization factor and its effect on the VE: when VE from coelomic oocytes become sensitive to sperm lysins attack and permeable to spermatozoa, the oocytes turn fertilizable (16).

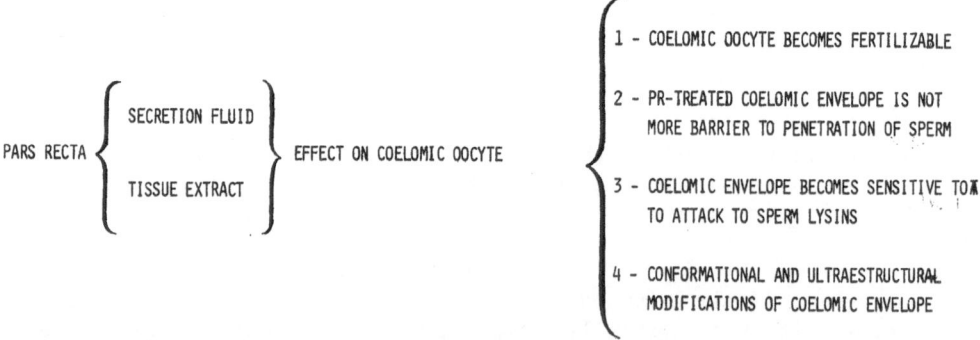

Figure 2.

Table 1.   Fertilization of Pars Recta Oocytes

	Percentage Cleaved	VE Attack by Sperm Lysin
Coelomic oocytes	0.8 ± 0.8^B (5)	No
Pars Recta oocytes	84.8 ± 5.3^B (5)	Lysis

Fertilization of Pars Recta - Treated Coelomic Oocytes

Treatment	Percentage Cleaved	VE Attack by Sperm Lysins
Ringer solution	0.0^{A-B} (7)	No
Pars Recta secretion fluid		
(40 μg of protein/ml)	73.1 ± 5.6^A (7)	Lysis
Pars Recta extract	88.3 ± 3.5^A (7)	Lysis
(40 μg of protein/ml)	66 ± 15.1^B (3)	Lysis

[A]Eggs were inseminated in the presence of egg-water (final sperm concentration was 3×10^6 cells/ml).

[B]Eggs were inseminated with a high sperm concentration (1.7×10^7 cells/ml).

Results are expressed as the mean and standard error from experiments carried out with different animals. ( ): number of experiments.

STRUCTURAL CHANGE OF THE COELOMIC OOCYTE VITELLINE ENVELOPE BY OVIDUCAL SECRETION

Fluorescence probes. The enhanced susceptibility of PR-treated coelomic oocytes envelopes (PR-CE) to sperm lysin attack and their change from an impenetrable to a penetrable condition, is probably due to coelomic oocyte envelope (CE) modifications produced by the PR secretion. Spectrofluorometric methods used to study these

changes can detect small conformational disturbances in free and
membrane-bound protein measured by changes in tryptophan or tyrosine
fluorescence (3,23).

PR-CE showed a marked enhancement of the intrinsic fluorescence
when compared with CE. This increase in the fluorescence efficiency
of tryptophan could be ascribed to the removal of quenching molecules
from the neighborhood of tryptophan residues.

Binding experiments on isolated CE and PR-CE with fluorescent
8-anilino-1-naphtalene sulfonate (ANS) also showed differences between both envelopes. ANS binding assay is sensitive to small localized changes in the membrane proteins which cannot be detected by
electron microscopy and light scattering. Both kinds of envelope, CE
and PR-CE, show a characteristic saturation curve when ANS concentration is increased, but the fluorescence of ANS bound to PR-CE is
twice as much as that of CE. When envelopes are solubilized by heating and the structural organization is disjoined, the differences
between CE and PR-CE disappear.

The increase in the fluorescence of PR-CE, compared to CE, could
be due either to an increase in the quantum yield of ANS-bound to PR-
CE or alternatively to an increment in the number of molecules bound
to the envelope. These alternatives were tested and the results
clearly show that the CE increase the total number of ANS binding
sites from 190 to 270 nmoles/mg of protein by the PR-treatment. The
observed increase in the number of sites for ANS could originate from
rearrangements of the proteins as a consequence of the sublytic
attack produced by the proteolytic enzymes existing in PR secretion
fluid. Such rearrangements may include reduction of the lateral
interaction between envelope proteins, change in their tertiary or
quaternary structures, and/or dissociation of some superficial peptides (18).

VE ULTRASTRUCTURAL STUDY

The structural alteration suffered by the vitelline envelope
during the passage of the oocyte through the PR in Bufo arenarum
seems coincident with that reported for Xenopus laevis (8) and Rana
japonica (26).

A cross section of a coelomic oocyte stained with Ruthenium red,
shows a CE, resembling a grill-like network, formed by fibrils
arranged into fascicles approximately 45 nm thick running parallel to
the egg surface. The network shows randomly distributed varyingly
sized holes.

The structural reorganization of the CE that occurs in the oviduct appears to modify the primary structural configuration established during oogenesis. The distinguishing features of PR-CE are 1) The fibril bundles are dispersed and the network presents a relatively smaller and more uniform arrangement of the holes, providing a more homogenous appearance and 2) a well-defined perivitelline space resembling that of a uterine egg is seen. The morphological similarities between PR-CE and VE are matched by a functional one: both, PR-treated coelomic oocyte and oviposited egg, can be penetrated by sperm. Nevertheless, the structure of the envelopes in not identical. In VE the bundles are less tightly organized, resulting in a looser network where holes can no longer be distinguished. This configuration is not observed even after two hours of PR-treatment of coelomic oocytes, which suggests that some other process might take place during oocyte transit through the entire oviduct (14).

PARTICIPATION OF PARS RECTA IN INDUCING ACROSOME REACTION

Results obtained from experiments with Bufo bufo japonicus led to a re-evaluation of the role of pars recta in fertilization. It was proposed that some PR secretion components (mainly glycoproteins) retained by the vitelline envelope, induce the acrosome reaction (10,27). This assumption seems attractive as it would explain the origin and nature of the factor that triggers the acrosome reaction, an aspect of the fertilization process which remains unclear.

Nevertheless, some observations on Bufo arenarum cannot be explained solely by this mechanism of action. Structural changes illustrate the direct effect of PR on the VE (14,18). Moreover, the fact that the PR-CE is sensitive to sperm lysin solution (without sperm), support the idea that in this case, the PR effect is on the CE and not on the sperm through induction of acrosome breakdown (16). Taking into account that the PR effect in triggering the acrosome reaction has already been demonstrated on Bufo bufo japonicus, this mechanism was analyzed for other species: Bufo arenarum and Leptodactylus chaquensis. The advantage of Leptodactylus chaquensis is that its sperm acrosome are easily visualized under the light microscope and its acrosome breakdown may be followed along sequen ces clearly determined (22). To facilitate comparison of the acrosome stage of sperm populations, an Acrosomic Index (AI) has been devised (2). This AI ranges from 1 (population with 100% of acrosome in stage 1, i.e. unreacted) to 5 (100% of acrosome in stage 5, i.e. completely reacted). After incubating spermatozoa for four hours in 10% Ringer, the AI was near 4, indicating that the process of acrosome breakdown is well advanced in the sperm population (Table 2). It could be argued that acrosome breakdown was occurring as the fastest possible velocity due to media tonicity, and no

Table 2.   Influence of <u>Bufo</u> <u>arenarum</u> PRE on the acrosome breakdown of
           <u>Leptodactylus</u> <u>chaquensis</u> sperm.   Incubation in 10% Ringer.

Period of Incubation (Minutes)	Incubation Media		
	10% Ringer	PRE	Heated PRE
0	1.4 ± 0.4	1.4 ± 0.6	1.4 ± 0.4
20	1.5 ± 0.3	1.5 ± 0.3	1.5 ± 0.2
30	1.7 ± 0.1	1.9 ± 0.3	2.0 ± 0.6
40	1.6 ± 0.7	1.9 ± 0.2	1.9 ± 0.6
60	2.1 ± 0.5	1.9 ± 0.6	2.0 ± 0.6
120	2.9 ± 0.5	2.4 ± 0.7	2.9 ± 0.9
180	3.5 ± 0.5	3.5 ± 0.1	3.7 ± 0.4

        2×10^6 <u>L</u>. <u>ch</u>. spermatozoa were incubated in 200 μl of 10%
Ringer, containing <u>B</u>. <u>a</u>. Pre or heated <u>B</u>. <u>a</u>. PRE (200 μg of PRE
protein).   Incubations were carried out at 22°C for different
periods.   After incubations, aliquots were fixed in 3% glutaraldehyde
prepared in phosphate 0.1 M (pH 7.4) buffer, and the acrosomal stage
of at least 200 spermatozoa was scored.   Results are expressed as AI
mean + SD (n = 4).

acceleration could be noticed by addition of PRE.   However, the same
experiments carried out with standard Ringer, in which the acrosome
breakdown occurs very slowly, indicate that the addition of active
or inactive PRE neither induces nor accelerates acrosome breakdown.
It is worthwhile to point out that when <u>L</u>. <u>chaquensis</u> spermatozoa
were incubated with PRE from <u>B</u>. <u>arenarum</u>, the results were similar to
those obtained with homologous PR (Miceli et al., unpublished
results).   The <u>L</u>. <u>chaquensis</u> sperm can penetrate both <u>B</u>. <u>arenarum</u> and
<u>L</u>. <u>chaquensis</u> oocytes (21).

        Evidence indicates that in amphibians, the acrosome reaction
occurs when the fertilizing spermatozoa is close to the VE (22,27).
A spermatozoon undergoing the acrosome reaction far from the VE would
lose its fertilizing capacity (21).

Table 3. Influence of <u>Bufo arenarum</u> PRE on <u>B. a.</u> sperm fertilizing capacity.

Sperm Incubation Media	Fertilized Oocytes (%)	p
100% Ringer	88	
	50	
	80	
PRE in 100% Ringer	86	0.35
	55	
	90	
10% Ringer	63	
	60	
	63	
PRE in 10% Ringer	58	0.20
	76	
	58	

Results (summarized in Table 3) indicate that no difference in the fertilizing capacity can be ascribed to the presence of the PRE in the incubation media. A decrease in fertility is observed when spermatozoa are incubated in 10% Ringer. However, this effect is likely to depend on the media tonicity rather than on the PRE effect since both control and PRE-treated sperm present the same degree of fertility declination.

<u>Bufo arenarum</u> spermatozoa were incubated for five hours at 22°C in the conditions described in Tables 1 and 2. After incubation, sperm suspensions were centrifugated, resuspended in one ml of 10% Ringer and used to inseminate strings of about 300 oocytes each. Results are expressed as percentage of oocytes fertilized. The data, analyzed by the Mann-Whitney test, indicate that the differences between control and treated sperm were not significantly different.

Other differences exist between <u>Bufo arenarum</u> and <u>Bufo bufo japonicus</u> as regards the "in vitro" treatment conditions to realize the PR effect. In order to render <u>Bufo arenarum</u> CE sensitive to sperm lysin and the coelomic oocytes fertilizable, a relatively low protein concentration of PRE in the incubation media (0.2 PR/ml) and a 20-minute treatment are enough. Under such conditions, fertilized oocytes reach gastrulae stage and their cells show normal diploid cariotype. <u>Bufo</u> <u>bufo</u> <u>japonicus</u> oocytes are incubated in a medium containing at least 4 PR/ml for 4 h.

The incubation of <u>Leptodactylus</u> <u>chaquensis</u> spermatozoa with <u>Bufo</u> <u>arenarum</u> PRE was specially relevant since <u>Bufo</u> <u>arenarum</u> PRE was found to be very active in conditioning coelomic oocytes. Regarding this point it is important to consider that cross fertilization between <u>L.</u> <u>chaquensis</u> spermatozoa and <u>Bufo</u> <u>arenarum</u> oocytes is easily carried out under experimental conditions, and polyspermy is the rule. If it were present, any acrosome reaction-inducing agent in the PR secretion entrapped in the VE of the <u>Bufo</u> <u>arenarum</u> oocyte, would also be present in the PRE; this molecule should also induce the acrosome breakdown "in vitro" as was reported for <u>Bufo</u> <u>bufo</u> <u>japonicus</u> (10). However, the results of the experiments incubating <u>Bufo</u> <u>arenarum</u> or <u>L.</u> <u>chaquensis</u> spermatozoa with PRE from <u>Bufo</u> <u>arenarum</u> did not suggest the presence of any substance that could trigger the acrosome reaction (Miceli, unpublished results).

METABOLIC SWITCH IN COELOMIC OOCYTES BY PR EFFECT

Fully grown ovarian oocytes from <u>Bufo</u> <u>arenarum</u> exhibit two different and consecutive metabolic routes for glucose related to different periods of the sexual cycle. During the hibernation period, carbohydrates are metabolized mainly through the Embden-Meyerhof route followed by a typical Kreb's cycle, as in adult tissue (Adult Metabolism). During the breeding season, the oocytes acquire the characteristic behavior of early stages of development in which glucose is metabolized via the penthose phosphate cycle with a relatively low participation of the glycolytic pathway (12). The tricarboxylic acid cycle, operating as the glutamic-aspartic route, provides an important amount of precursors for nucleotide biosynthesis (Embryonic Metabolism) (24). The relative participation of the Kreb's cycle might be regulated by hypophyseal activity through changes in the mitochondrial permeability to certain intermediates of metabolism (12). The transition of adult to embryonic metabolism is essential for development: segmentation is only attained when the oocytes acquire embryonic metabolism (13). Thus, the oviposited eggs always show an embryonic metabolism, although, coelomic oocytes with adult metabolism can be obtained by hypophysis induction of ovulation in animals captured during the hibernation period. The results show that citrate and succinate are oxidized in coelomic oocyte mitochondria at a faster rate than fumarate and pyruvate. The oviductal oocyte mitochondria oxidize fumarate and pyruvate at the fastest rates, while citrate is oxidized at a considerably slower rate (11). These results can be considered as a modification of the mitochondrial permeability to citrate which reduces its oxidation significantly (12).

The relative activity of the pentose phosphate cycle and glycolysis was estimated in coelomic PR-treated and oviducal oocytes by
means of the Shunt Index, i.e., the relative yields of $^{14}CO_2$ produced
from C-1 or C-6 labeled glucose. The average Shunt Index value for
coelomic oocytes was 0.87 while for oviducal oocytes it was 0.26.
Similar data have been reported for adult tissue (about 1) and early
stages of development (about 0.2) respectively (24). When coelomic
oocytes are pre-incubated one hour in PRE or PR secretion from
hypophysis-stimulated animals, the index value drops significantly
to the value of oviposited eggs, suggesting that this portion of the
oviduct plays an important role in the metabolic transition Bufo
arenarum oocytes undergo (11).

The metabolic change from adult to embryonic metabolism is also
produced "in vitro" by hyphohysis extract (12), FHS and serum of
hypophysis-stimulated females (Budeguer de Atenor, personal
communication).

It is likely that the components which produce the metabolic
switch could be different from that implicated in fertilization:
neither FSH nor serum are effective in inducing coelomic oocyte
fertilizability. Thus, development appears to be indirectly regulated by the PR secretion. The metabolic switch at the level of pars
recta might be interpreted as a double assurance mechanism for insur-
ing the establishment of metabolic maturation before fertilization.

COMPARATIVE BIOLOGICAL AND BIOCHEMICAL PROPERTIES OF PARS RECTA
SECRETION FLUID AND PROTEIN(S) PURIFIED FROM PARS RECTA EXTRACT

In order to imitate natural conditions of the oviduct in "in
vitro" experiments, it is more convenient to use PR secretions
instead of PRE tissue extracts because the presence of tissue
proteins and other substances in PRE can interfere with and even blur
the real effect of the biologically active substances under study.
In B. arenarum, this situation was overcome in part by studying
properties of the PR factor, using secretion fluid and partially
purified active protein from PRE.

An active protein preparation was obtained by filtration chromatography through Sephadex gels (16). This preparation could be preserved by freeze drying for several weeks. It was also purified by
affinity chromatography on concanavalin A bound to Sepharose 4-B.
This procedure yielded highly active protein(s). Unfortunately,
this preparation rapidly lost activity and could not be preserved,
either frozen nor lyophylized (17).

Isolation by affinity chromatography using concanavalin A indicates that this factor is a glycoprotein with multiple α-D-manno-
pyransoyl or α-D-glucopyranosyl end groups or internal 2-0-linked

α-D-mannopyranosyl residues. However, the reaction of lectin with oligosaccharides is highly complex, depending not only on the types of monosaccharides present but also on their sequence and on the nature of the glycosidic linkages involved. Thus, although the PR factor-lectin interaction provides some information about this oviducal protein, further characterization of its carbohydrate content is necessary for a better knowledge of its function.

We postulate that the enhancement of the CE sensitivity to sperm lysins is probably due to modification of the CE by a proteolytic enzyme secreted by PR epithelium since the biological activity of PR secretion fluid and PRE is inhibited by trypsin inhibitors. Both the enzyme(s) present in the secretion fluid and those purified from a PRE were identified by their esterase activity toward α-N-benzoyl-L-arginine ethyl ester-HCl(BAEE) and p-tosyl-L-arginine methylester HCl (TAME), and amidase activity toward α-N-benzoyl-DL-arginine-p-nitroanilide HCl (BAPNA). Some properties of PR proteolytic enzymes resemble the properties of pancreatic trypsin and plasmin: it is activated by calcium ion, it hydrolyzes TAME faster than BAEE and it is inhibited by N-α-P-tosyl-L-lysine chloromethyl ketone HCl (TLCK) and soybean trypsin inhibitor (SBTI) but not by L-1-tosyl-amide-2-phenylethylchloro-methyl-ketone (TPCK). On the other hand, it differs from trypsin in that it has higher molecular weight (MW:47,000 app), is stable at alkaline pH and is not inhibited by lima bean trypsin inhibitor (LBTI).

It is important to point out that PR secretion is characterized by a low proteolytic activity upon synthetic substrates. The protein content of the secretion fluid obtained from females captured at different seasons ranged from 0.69 to 1.2 mg of protein/ml. Enzymatic activity of the secretion expressed as I.U. TAME/ml, ranged from 0.100 to 0.200 units/ml (one unit is equal to the hydrolysis of one micromole of substrate per minute).

Even though purified PRE has the same biological effect as the secretion fluid, some differences are detected as regard the viability of the fertilized oocytes pretreated with the purified extract or secretion fluid. Thus, direct incubation of coelomic oocytes with purified PRE (1.4 units of TAME/ml) led to dissolution of the vitelline envelope and severe egg collapse. On the other hand, secretion fluid was not found to produce this effect at a similar concentration.

Coelomic oocytes treated either with trypsin (1.4 units of TAME/ml) or purified PRE at a concentration which partly solubilized the CE demonstrated similar behavior. Both are also fertilizable but the development is generally arrested before grastula stage. Even at lower concentration of purified PRE (0.2-0.3 units of TAME/ml), the proportion of embryos that reach gastrula stage is significantly lower than that of embryos treated with the secretion fluid at

similar concentration that reach the gastrula stage. These observations suggest the involvement of some additional molecules in the
secretion. This molecule, separated during purification, could be
responsible for the protection of certain oocyte structures such as
the vitelline envelope itself and the plasma membrane, the integrity
of which is fundamental for normal embryogenesis (19).

Even though the PR factor is active toward synthetic substrates
for proteolytic enzymes and partially solubilizes the CE, it was not
possible to estimate its activity using isolated CE as substrate.
This might be attributed to the fact that it hydrolyzes internal
links leading to conformational alterations without the release of
free peptides.

HORMONAL REGULATION OF BIOLOGICAL ACTIVITY

An additional point observed in early experiments on B. arenarum
was that the PR effect is clearly hypophysis-dependent. However,
hyphophysis extracts have been found to be ineffective in inducing PR
activity on ovariectomized toads, suggesting that its action is
indirect, through an initial stimulation of ovarian follicle cells.
Indeed, the biological activity of PR is closely correlated with the
ovarian cycle. Maximum PR activity is detected during the breeding
season and does not disappear immediately after spawning, but a slow
reduction ensues. The maximum percentage (about 75%) of animals
lacking PR activity is usually reached several months after the
breeding season. However, lack of PR activity is seen only in
spawned toads. Even six months after the breeding period, PR
activity does not disappear in unspawned toads.

Ovariectomy steadily decreases PR activity at a rate depending
on the season the animals are captured and when ovariectomy takes
place. The data from about 300 animals analyzed by variance and
multiple range test prove that the season during which the animals
are captured and ovariectomized, as well as the number of days
elapsed after castration, introduce modifications in the decreasing
PR activity by ovariectomy. The higher percentage of animals with
inactive PR was obtained among those captured and ovariectomized
during autumn-winter, 25 days after castration. The blood serum
estradiol level decreased sharply below the control values (4.16 -
1.89 ng/ml) after castration: within eleven days to 2-3 pg/ml and
after 18 days to 0.86-0.31 pg/ml.

Evidence that PR activity disappears at a slower rate in normally spawned toads than in ovariectomized ones suggest that the endogenous levels of estrogens might be an important factor in PR activity regulation. Data prove that a single dose of estradiol 17-β,
diethylstilbestrol, estrone and dihydrotestosterone is effective in
inducing PR activity in ovariectomized toads. Among the estrogens

tested, estradiol 17-β appears to be the most potent inducer of PR
activity: a dose of 150 μg/ 150 g of body weight is capable of
inducing PR activity 60 minutes after its administration. Progesterone does not induce activity in ovariectomized animals. However, in
toads captured during hibernation periods, with a low value of blood
serum estradiol and inactive PR, progesterone functions as an inducer. Probably both hormones are functionally correlated at the
molecular level in terms of their receptors and the presence of endogenous estradiol is a prerequisite for progesterone action (7).

There is considerable evidence that the primary action of estradiol 17-β is to accelerate genetic expression in the target organ.
The inhibition of an estradiol effect on PR by actinomycin D and
cycloheximide suggests that stimulation of RNA synthesis is involved
in the observed biological response of estradiol. The use of double-
isotope labelling techniques for studying "de novo" synthesis of PR
proteins (pre-estradiol labelling with[3] H and post-estradiol labeling
with[14] C) indicated that one of the effects of estradiol is to pro-
mote an increment of protein synthesis in general in the PR tissue.
This occurs between 1 and 4 h after estradiol administration to
ovariectomized mature toads. Induction of PR protein synthesis can
also be obtained "in vitro" by incubating the first segment of oviduct from castrated females in an appropriate medium containing
estradiol-17-β . However, early attempts to detect an increase in the
incorporation of label into the electrophoretic band corresponding to
the PR protein with biological activity were inconclusive.

An interesting, but still poorly-known, phenomenon is that
experiments carried out "in vivo" showed the presence of a labeled
acid precipitable protein in the secretion fluid. Whether one or
more of such proteins are involved in the aforementioned biological
effects is not yet known. In agreement with these results, secretion
of labeled proteins can also be obtained in the system of "in vitro"
organ culture. In control cultures not induced by hormones, labeled
proteins were also detected, but in lower cocentration. The important fact is that the culture medium of estradiol-treated PR, as distinguished from control medium, shows biological activity characteristic of secretion fluid because this system provides the opportunity
for quantitative comparisons of secreted compounds in terms of their
ability to induce a biological response (unpublished results).

CONCLUSIONS

As regards the mechanisms of action of PR secretions on the
sperm-vitelline envelope interaction, during their transit down the
pars recta, the oocytes, still uncovered by jelly coats are bathed
with the secretion fluid containing enzymes, which modify the vitelline envelope. This modification leads to a decrease of superficial
protein interaction and a loosening of fibril bundles of the VE. It

may be postulated that due to this ultrastructural transformation, the "uncovering" of binding sites for the male gamete, or of sites with greater affinity for sperm lysins which assist the penetration of the spermatozoa, is produced.

Concerning the effect on metabolic behavior, the results indicate that the dividing capacity of the oocytes is attained only when, through biochemical modifications, the oocytes acquire the metabolic behavior characterizing embryonic cells.

It is postulated that the metabolic changes observed in the oocyte constitute a fundamental aspect of cytoplasmic maturation. In this case, the PR would constitute an assurance mechanism for the establishment of a suitable metabolism before fertilization.

ACKNOWLEDGEMENTS

I am grateful to Dr. Marcelo Cabada for many suggestions and stimulating discussions, and to Dr. Francisco Barbieri for critically evaluating this manuscript. Appreciation is expressed to Mrs. Maria Juarez de Doz Costa for her assistance in the preparation of this manuscript. This work was supported in part by grants from the Consejo Nacional de Investigaciones Cientificas y Tecnologicas (Argentina) (CONICET).

REFERENCES

1. Cabada, M.O., Mariano, M.I., and Raisman, J.S. 1978. Effect of trypsin inhibitors and concanavalin A on the fertilization of Bufo arenarum coelomic oocytes. J. Exp. Zool. 204:409-416.
2. Cabada, M.O., Bloj, B., Peralta de Ortiz, L., Valz-Gianinet, J.N., and Diaz-Fontdevila, M. 1984. Effect of phosphatidyl-choline on acrosome breakdown and fertilizing capacity of amphibian spermatozoa. Develop. Growth Differ. 126:515-523.
3. Dufourcq, J., and Faucon, J.F. 1977. Intrinsic fluorescence study lipid protein interaction in membrane model-binding of melitin, an amphipatic peptide, to pholpholipid vesicles. Biochim. Biophys. Acta. 467:1-11.
4. Elinson, R.P. 1971. Sperm lytic activity and its relation to fertilization in the frog Rana pipiens. J. Exp. Zool. 177:207-218.
5. Elinson, R.P. 1974. A comparative examination of amphibian sperm proteolytic activity. Biol. Reprod. 11:406-412.
6. Epel, D. 1980. Fertilization. Endeavour, New series 4:26-31.
7. Fernandez, S.N., Mansilla, C., Miceli, D.C. 1984. Hormonal regulation of an oviducal protein involved in Bufo arenarum fertilization. Comparative Biochim. Physiol. 78A:147-152.

8. Grey, R.D., Working, P.K., and Hedrick, J.L. 1977. Alteration of structure and penetrability of the vitelline envelope after passage of eggs from coelom to oviduct in Xenopus laevis. J. Exp. Zool. 219:87-95.

9. Iwao, Y., and Katagiri, Ch. 1982. Properties of the vitelline coat lysin from toad sperm. J. Exp. Zool. 219:87-95.

10. Katagiri, Ch., Iwao, Y., and Yoshizaki, N. 1982. Participation of oviducal pars recta secretions in inducing the acrosome reaction and release of vitelline coat lysin in fertilizing toad sperm. Develop. Biol. 94:1-10.

11. Legname, A.H., Salomon de Legname, H., Sanchez, S.S., Sanchez Riera, A.N., and Fernandez, S.N. 1972. Metabolic changes in Bufo arenarum oocytes induced by oviducal secretions. Develop. Biol. 29:283-292.

12. Legname, A.H., Salomon de Legname, H., Miceli, D.C., Sanchez, S.S. and Sanchez Riera, A.N. 1976. Endocrine control of amphibian oocyte metabolism. Acta Embryol. Exp. 1:37-49.

13. Legname, A.H., and Buhler, M.I. 1978. Metabolic behavior and cleavage capacity in the amphibian egg. J. Embryol. Exp. Morph. 47:161-168.

14. Mariano, M.I., Gomez de Martin, M.I., and Pisano, A. 1984. Morphological modifications of oocyte vitelline envelope from Bufo arenarum during different functional states. Develop. Growth Differ. 26:33-42.

15. Miceli, D.C., Del Pino, E.J., Barbieri, F.D., Mariano, M.I., and Raisman, J.S. 1977. The vitelline envelope-to-fertilization envelope transformation in the toad Bufo arenarum. Develop. Biol. 59:101-110.

16. Miceli, D.C., Fernandez, S.N., Raisman, J.S., and Barbieri, D.F. 1978a. A trypsin-like oviducal proteinase involved in Bufo arenarum fertilization. J. Embryol. Exp. Morph. 48:79-91.

17. Miceli, D.C., Fernandez, S.N., and Del Pino, E. 1978b. An oviducal enzyme isolated by affinity chromatography which acts upon the vitelline envelope of Bufo arenarum coelomic oocytes. Biochim. Biophys. Acta. 526:289-292.

18. Miceli, D.C., Fernandez, S.N., and Morero, R. 1980. Effect of oviducal proteinase upon Bufo arenarum vitelline envelope. A fluorescence approach. Develop. Growth Differ. 22:639-643.

19. Miceli, D.C., and Fernandez, S.N. 1982. Properties of an oviducal protein involved in amphibian oocyte fertilization. J. Exp. Zool. 221:357-364.

20. Raisman, J.S., and Barbieri, F.D. 1969. Lytic effects of sperm suspensions on the vitelline membrane of Bufo arenarum oocytes. Acta Embryol. Exp. 1:17-26.

21. Raisman, J.S., and Cabada, M.O. 1977. Acrosome reaction and proteolytic activity in the spermatozoa of an anuran amphibian, Leptodactylus chaquensis. Develop. Growth Differ. 19:227-232.

22. Raisman, J.S., Wasserman de Cunio, R., Cabada, M.O., Del Pino, E.J., and Mariano, M.I. 1980. Acrosome breakdown in Leptodact-ylus chaquensis (Amphibia anura) spermatozoa. Develop. Growth Differ. 22:289-297.

23. Rubalcava, B., Martinez de Munoz, D., and Gitler, C. 1969. Inter-action of fluorescent probes with membranes. 1) Effect of ions on erytrocyte membranes. Biochemistry, 8:2742-2747.

24. Salomon de Legname, H., Sanchez Riera, A.N., and Sanchez, S.S. 1975. Source of precursors for nucleotide biosynthesis in Bufo arenarum segmenting eggs. Acta Embryol. Exp. 123-136.

25. Schmell, E.D., Gulyas, B.J. and Hedrick, J.L. 1983. Egg surface changes during fertilization and the molecular mechanisms of the block to polyspermy, in: "Mechanism and control of animal fertilization," John F. Hartmann, Academic Press, New York, pp. 398.

26. Yoshizaki, N., and Katagiri, Ch. 1981. Oviducal contribution to alteration of the vitelline coat in the frog, Rana japonica. An electron microscopic study. Develop. Growth Differ. 23:495-506.

27. Yoshizaki, N., and Katagiri, Ch. 1982. Acrosome reaction in sperm of the toad, Bufo bufo japonicus. Gamete Res. 6:343-352.

THE CORTICAL ENDOPLASMIC RETICULUM AND ITS POSSIBLE ROLE IN

ACTIVATION OF <u>DISCOGLOSSUS</u> <u>PICTIS</u> (Anura) EGGS

Chiara Campanella, Riccardo Talevi, Umberto Atripaldi, and Lucia Quaglia

Dipartimento Di Biologia Evolutiva E Comparata
University of Naples
Naples, Italy

SUMMARY

The role of endoplasmic reticulum was investigated in the egg of <u>Discoglossus pictus</u>; recent findings suggest that this organelle is the source of Ca^{2+} sequestration and release at activation. In the egg of <u>Discoglossus</u> the dimple is the only site where sperm-egg fusion occurs. Microvilli containing microfilament bundles penetrate into the dimple cytoplasm and thus define a cortical layer containing cortical granules, tubular cisternae 35 nm thick and vacuoles. In the underlying cytoplasm are clusters of small cisternae and mitochondria. In the region of the egg cortex outside of the dimple, clusters of cisternae, some vacuoles and a heterogeneous population of small granules are found. In eggs activated by pricking contraction starts from the site of pricking and travels to the antipode. Sections of eggs, fixed 20 to 60 s following fertilization or pricking, show that the tubular cisternae have disappeared and the clusters of cisternae have opened to give rise to longer cisternae arranged in chains. These chains pile up below the vacuoles which are now flat and long because of multiple fusion. A network of cisternae is thus formed whose constituents lean against the microfilament bundles and within 5 min from activation approach the plasma membrane. The flattened vacuoles and the cortical granules are exocytozed. Some cisternae fuse with the microvillar membrane thus participating in microvilli elongation. The microfilament rootlets grandually rearrange and become shorter. In the cortex outside the dimple the cisternae clusters do not open, whereas exocytosis of granules occurs. In eggs treated with A23187, vacuoles and cortical

granules undergo exocytosis; however a cisternal network is not
formed.

The possible stimuli that cause the cisternae network to form at
activation only in the dimple and in a matter of seconds are discussed. Our observations, namely exocytosis, formation of cortical
wave of contraction and changes in the cytoskeleton organization at
activation are consistent with an increase of free Ca^{2+}. Vacuoles
and cisternae appear to be good candidates for the sequestration and
release of Ca^{2+} as well an important source for the increase in total
plasma membrane at activation.

INTRODUCTION

Upon fertilization of fish and sea urchin eggs, Ca^{2+} is released
from intracellular storage sites, and the resulting rise in the free
Ca^{2+} concentration leads to cortical granule (CG) exocytosis and the
activation of metabolism (17,19,38). In the fish, Oryzias, a wave
of Ca^{2+} release and, following by about 15 sec, of CG exocytosis,
spreads from the site of sperm entry to the opposite pole of the egg
(19,33). The search for the intracellular site of Ca^{2+} storage and
the mechanism of Ca^{2+} release at activation are intriguing lines of
research generated by these findings (17,37,41). In the present
paper we will describe some recent data related to these questions
and describe our recent studies on the egg of the frog, Discoglossus
pictus.

A particularly attractive subject for study in this regard has
been the egg of Xenopus laevis. The ultrastructure of its cortex
suggests that the site of Ca^{2+} storage may be a specialized network
of the smooth endoplasmic reticulum (SER). A shell of smooth
cisternae surrounds each CG (20). These shells are interconnected
and are part of a network of SER which includes clusters of cisternae
situated at regular intervals in the subcortical layer, as well as
cisternae lying between the CG, in close contact with the plasma
membrane (5,18). Junctions between these cisternae and the plasma
membrane are similar to those between the sarcoplasmic reticulum and
the plasma membrane of muscle cells (18).

The following lines of evidence strongly suggest that the cortical SER network is the cellular compartment which stores and releases
Ca^{2+} at fertilization (and is probably involved in the propagation of
CG exocytosis). First, during progesterone-stimulated oocyte maturation, the oocyte becomes able to respond to a pricking stimulus at
the same time that the cortical SER becomes fully organized (7,11).
Second, at fertilization the organization of the network changes and
then the network gradually disappears (5). Finally, the results of
staining eggs with osmium and pyroantimonate to visualize Ca^{2+}
suggest that Ca^{2+} release originates at these cortical cisternae.

There are no calcium antimonate deposits in unfertilized eggs, but in eggs activated by Ca^{2+} ionophore in Ca^{2+} -free Ringer's containing EGTA, deposits are first seen on the plasma membrane and the adjacent smooth cisternae, and later are found over organelles more centrally located (1,2).

In sea urchin eggs, calcium antimonate deposits are found on the cytomembranes of all egg organelles (8). However, according to Poenie and co-workers (32), as a consequence of fertilization these deposits are lost from tubular sacs located in the egg cortex but not from sacs and other vacuoles located in deeper cytoplasm. Therefore also in the sea urchin, tubular sacs located in the egg cortex appear to be good candidates for storage and release of Ca^{2+} at fertilization.

There is as yet no direct evidence that the free Ca^{2+} concentration increases in the cortex of Xenopus eggs, but it seems likely that a wave of Ca^{2+} release starts at the point of sperm penetration and then spreads around the egg, leading to the wave of CG exocytoses. Recently Busa and Nuccitelli (personal communication), using a Ca^{2+} -selective intracellular microelectrode, found that the subcortical free Ca^{2+} level increases from 0.4 to 1.2 μM about 1 min after fertilization. The increase is transient; it begins in the animal hemisphere and traverses the egg as a wave moving at 10 $\mu m/sec$. This Ca^{2+} pulse is probably the subcortical extension of a cortical Ca^{2+} wave. Alternatively, a change in membrane occurring when the membrane depolarizes at activation (10,13,21,24,25,28,35) could be transmitted to the subjacent cisternae through the junctions they make with the membrane and stimulate them to release Ca^{2+}. This model has obvious parallels to the release of Ca^{2+} from the sarcoplasmic reticulum upon depolarization of the plasma membrane and transverse tubule system (2,18,19), and is supported by the presence of the plasma membrane-cisternae junctions (18), and by the above mentioned data obtained with pyroantimonate fixation. This hypothesis, however, does not easily explain the fact that CG exocytosis spreads as a wave from the point of sperm entrance (20,44). The factors which cause the propagation of CG exocytosis may be better understood when electrophysiological studies are coupled to an accurate analysis of CG exocytosis.

In contrast to the Xenopus egg, the unfertilized eggs of other amphibians have randomly distributed vesicles in the cortex and they lack the cortical vesicular network. In the present study we describe the presence of numerous, well-organized cisternae in eggs of Discoglossus pictus. Interestingly, this pattern of cisterna organization is first observed shortly after fertilization, and is seen only at the site where the sperm fuses with the egg: the "animal dimple" (4,22). These conditions may aid in further elucidating the role of the SER in egg activation, including its function as a Ca^{2+} sequestering and releasing organelle.

MATERIALS AND METHODS

Adult <u>Discoglossus pictus</u> (the painted frog) were collected near
Palermo, Italy, during the period February to May. To induce ovulation, females were injected in the dorsal lymph sac with 100 I.U. of
human chorionic gonadotropin (Pregnyl, Organon Oss, Holland) in
amphibian Ringer's solution, containing (in mM):NaCl, 111; $CaCl_2$,
1.3; KCl, 2.0; and $NaHCO_3$, 2.0. Eggs were surgically removed from
the uterus 18 h later. Sperm in seminal fluid was obtained by
pricking a male's seminal vesicles 24-48 h after a similar injection
of hCG. To insure the optimal percentage of fertilized eggs, this
sperm suspension was spread over "dry" eggs.

To artificially activate eggs, the regions of jelly designated
J2, J3, and the "animal plug" (14) were first removed by immersing
the eggs in a solution of 5 mM dithiothreitol (DTT), 0.1 M NaCl, and
5 mM Tris, pH 8.0. The eggs were then activated in full strength
Ringer's by pricking with a tungsten needle, or by exposure to a
solution of the Ca^{2+} ionophore, A23187, made by dilution from a stock
solution of a 2 mM in absolute ethanol. (See Results for the concentrations used.) At intervals of 30 s to 20 min after insemination or
artificial activation, eggs were fixed for electron microscopy as
previously described (6).

RESULTS

Unfertilized Eggs

The animal hemisphere of the <u>Discoglossus</u> egg is indented by an
asymmetrical jelly component, the "animal plug". The center of this
indentation is further invaginated and forms the cup-shaped dimple.
The dimple is about 150 μm in diameter and filled with a colloidal
content (Fig. 1, and inset a). In the dimple cytoplasm are finger-
shaped microvilli containing microfilament bundles implanted at
regular intervals. The bundles penetrate into the cytoplasm for
about 12 μm; contractile protein localization in the dimple has been
described elsewhere (6). The thickness of the peripheral layer (zone
A), where there are CGs, which are about 0.5 μm in diameter, tubular
cisternae, glycogen, and vacuoles, and where there are few mitochondria, is defined by the presence of the microfilament bundles, and
therefore is about 12 μm deep (Figs. 1 and 2a). The depth of this
layer and also the concentration of the vacuoles are maximal at the
center of the dimple, where the sperm fuses with the egg, and they
gradually decrease toward the perimeter of the dimple (4,6).

The tubular cisternae of zone A have electron-dense contents,
and are mostly present just beneath the plasma membrane. They
average about 35 nm in thickness. Vacuoles are about 0.2 to 0.3 μm
in diameter, contain sparse material, and are characterized by the

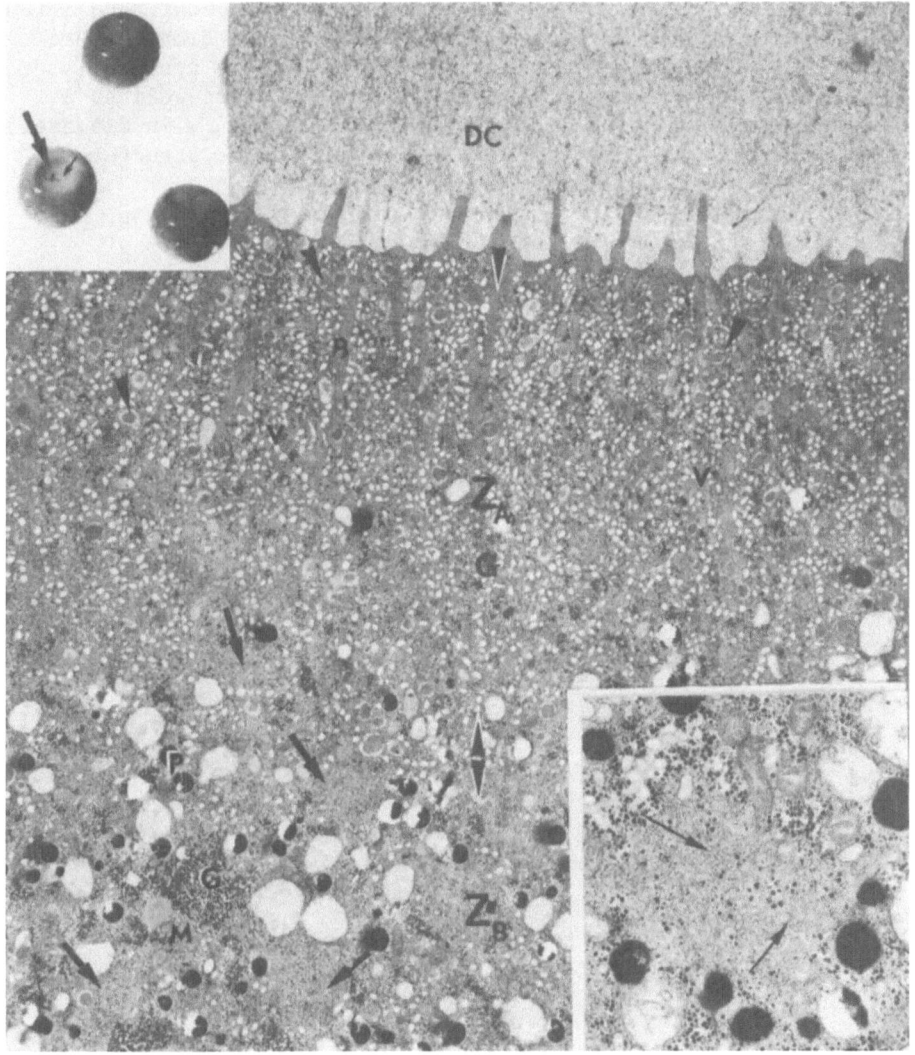

Figure 1. Dimple of unfertilized egg. At the dimple surface are
          finger-shaped microvilli containing microfilament bundles.
          In zone A (Z_A), one can observe the bundle rootlets (B),
          cortical granules (arrowheads), vacuoles (V), and glycogen.
          In zone B (Z_B) are clusters of cisternae (large arrows),
          pigment granules (P), mitochondria (M), and large islets of
          glycogen (G). DC = dimple content x 5000. Inset a: the
          large arrow points to the indentation at the animal half
          and the small arrow points to the dimple x 10. Inset b:
          clusters of small cisternae (arrows) x 14,000.

presence of small electron-dense invaginations (Fig. 2a). In zone B
(Fig. 1), several clusters of small cisternae are present, and a few
CGs. In the region of the egg cortex outside of the dimple where
microvilli and microfilament bundles are short, there are similar
clusters of cisternae, vacuoles with invaginations, as well as a
heterogenous population of small granules (about 0.3 μm in diameter).
These granules differ from the CGs in size and in the structure of
their contents (compare Figs. 2a and 2b).

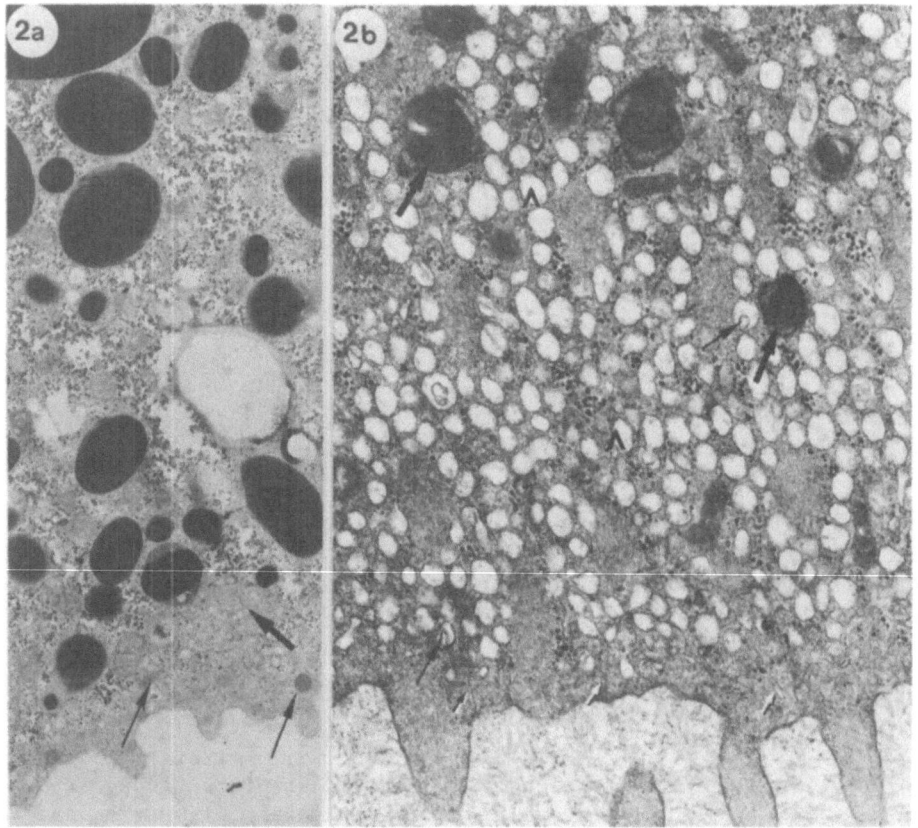

Figure 2. Unfertilized egg. 2a: Dimple. Beneath the plasma membrane
          are tubular cisternae about 35 nm in diameter (small
          arrows). Cortical granules (large arrows) and vacuoles (V)
          containing sparse material and invaginations (arrows), and
          located among microfilament bundles x 26,000. 2b: Vegetal
          hemisphere. Small granules (about 0.3 μm) (arrows) with
          variable content and clusters of cisternae (large arrow)
          are in the egg cortex x 7,000.

Fertilized Eggs

In eggs fixed 30 s to 1 min after sperm addition, conspicuous changes are evident in the central portion of the dimple. In zone B the cisternae have lengthened, and are arranged in chains rather than clusters. The chains of cisternae are parallel to the egg surface, but bend to extend into zone A, where they are perpendicularly oriented with respect to the plasma membrane (Fig. 3). Some chains bend laterally and anastomose with adjacent chains of cisternae (Fig. 4a). The cisterna chains pile up below the vacuoles, which are now flatter and longer than those in the unfertilized egg, so that an apparent continuity is formed between these two kinds of vacuolar structures (Fig. 3). Furthermore, the cisternal chains lean against the microfilament bundles in zone A (Fig. 4a). The CGs and flattened vacuoles are also closely associated with the microfilament bundles (Fig. 4a). Another feature of interest observed at this stage is the absence of the 35 nm thick cisternae.

After insemination, several types of exocytosis are seen. Some of the flattened vacuoles appear to have fused with the plasma membrane by 1 min after fertilization (Fig. 4a). At later stages (1 to 3 min after insemination), when the microfilament bundles have partly disaggregated, flattened vacuoles and/or cisterna chains are seen in the microvilli (Fig. 5). The cisternae appear to fuse with the microvillar membrane, so that they become inserted by exocytosis into the plasma membrane and cause the microvilli to elongate (Fig. 5). Other cisternal chains terminate in close contact with the plasma membrane in the spaces between the microvilli, but we were unable to detect fusion between the plasma membrane and these cisternae. Furthermore, at this stage the number of flattened vacuoles and CGs has decreased. In fact, cases of CG exocytosis are observed: where the CG membrane becomes inserted into the plasma membrane, flat vesicles form which may be hybrid vesicles containing portions of both the CG membrane and the dimple membrane; they are also seen in the dimple contents (Fig. 5).

Later (3 to 5 min after insemination), longer microvilli are observed, and the microfilament rootlets lose their regular orientation in the peripheral cytoplasm. Interestingly, the cisternal clusters in the peripheral cytoplasmic layer outside the dimple do not change following activation. However, about 5 min following insemination, most granules disappear, showing that exocytosis does also occur in these regions of the egg (data not shown).

By about 20 min after fertilization, the microfilament rootlets become shorter, few cisternae are visible in the dimple, and vacuoles are arranged in a layer beneath the plasma membrane (Fig. 6). During this period two other changes also occur: the dimple regresses in the first 6 min after insemination, and the larger concavity gradually rounds out (by 20 min after insemination) as the animal plug

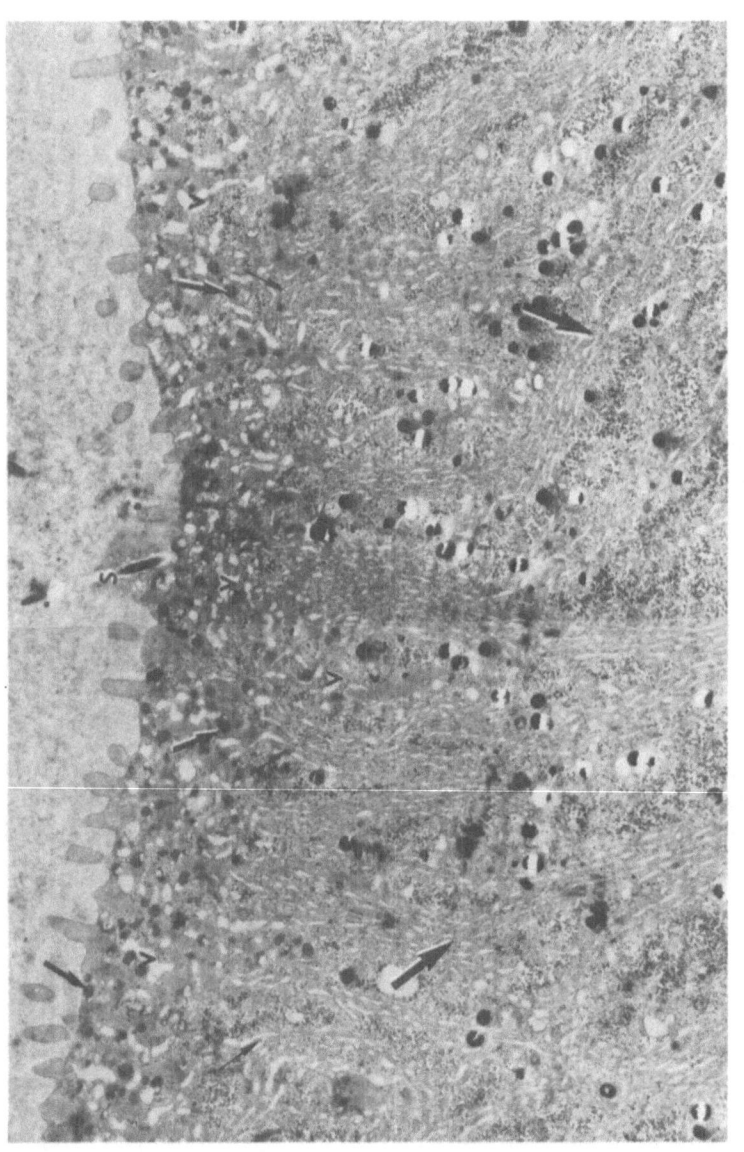

Figure 3. Dimple, 1 min following insemination. Cisternae clusters of zone B have transformed into chains of longer cisternae (large arrows) and are perpendicularly oriented with respect to the plasma membrane. Zone A vacuoles (V) are now flatter and longer than in the unfertilized egg. The small arrows indicate regions of apparent continuity between the two components of this network. Arrows = cortical granules; S = spermatozoon x 6,000.

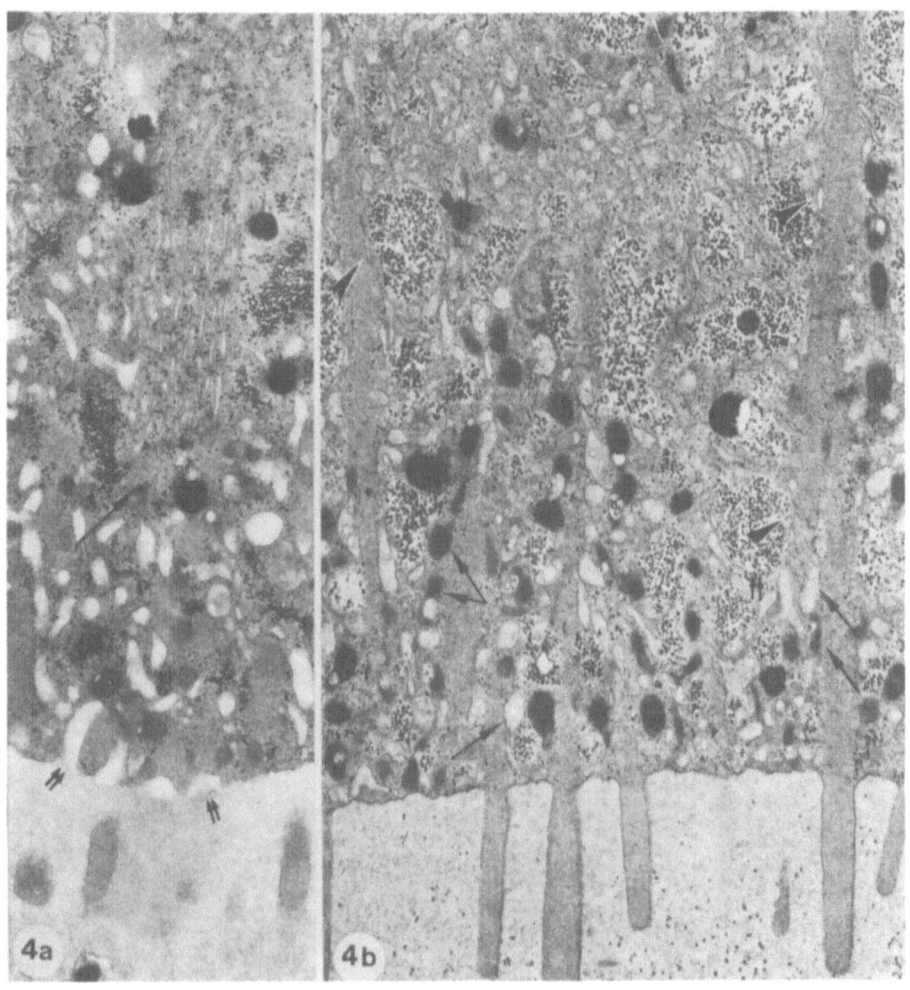

Figure 4. Dimple of fertilized egg. 4a: One min following insemina-
          tion. Chains of cisternae bend laterally and anastomose
          with adjacent chains of cisternae (double small arrows).
          Arrowheads = sites of cisternae and bundle associations.
          Arrows = association between either cortical granules or
          flattened vacuoles and microfilament bundles x 10,000. 4b:
          About 2 min following insemination, flattened vacuoles are
          undergoing exocytosis (double small arrows). Arrow = region
          of continuity between flattened vacuoles and cisternae x
          10,000.

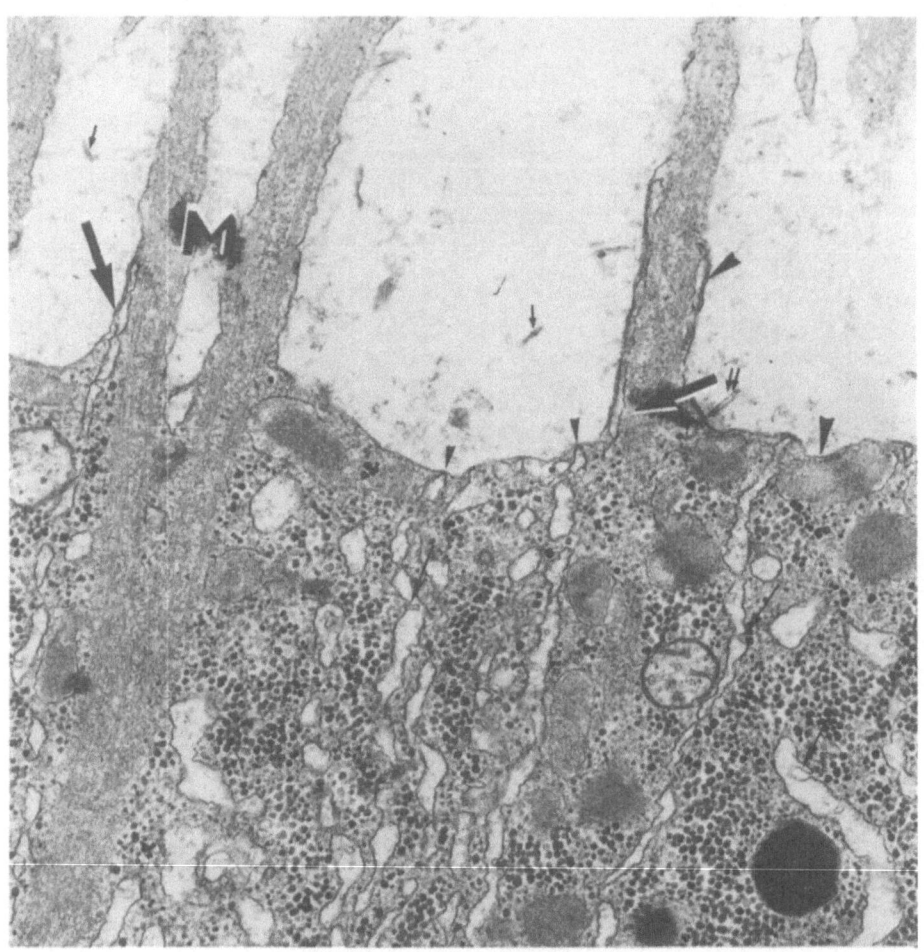

Figure 5. Dimple 3 min following insemination. In the microvilli (M)
        microfilament bundles are slightly disaggregated. Cis-
        ternal chains have entered the microvilli (large arrows).
        Large arrowheads indicate probable sites of prospective
        exocytosis of cortical granules and cisternae. Hybrid flat
        vesicles – deriving from partial fusion of these organ-
        elles' membranes with the plasma membrane are observed next
        to the plasma membrane (double small arrow) or in the
        extracellular dimple content (small arrow). Small arrow-
        heads = cisternae contacting the plasma membrane. Arrows =
        invaginations in the cisternae lumen x 38,000.

of jelly dissolves (43). As a result of these changes, the egg loses
its asymmetry (Fig. 6, inset) and the dimple surface is now similar
to that of the rest of the egg (6).

Pricked Eggs

To provide a more precise timing of the unusual changes occurring in the cisternal clusters and vacuoles immediately following
activation, we have examined DTT-treated eggs pricked in the dimple.
The first changes occur in the cisternal clusters in zone B. Indeed,
within 20 s of pricking, the clusters are less compact and give rise
to whorls of longer cisternae, most of which run obliquely to the
plasma membrane. The vacuoles lose their spherical shape and become
markedly elongated (Fig. 7), because they have fused together. They
may also incorporate the invaginations which were present in their
lumen before activation (Fig. 7, inset). Furthermore, similar

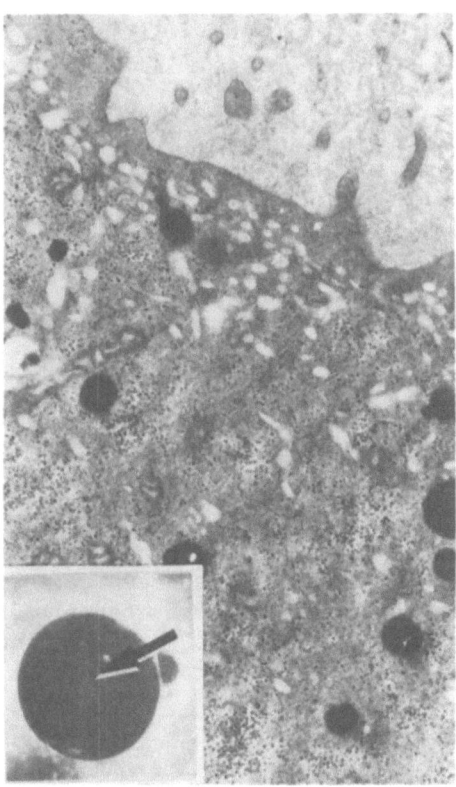

Figure 6. Dimple 20 min after insemination. The cortex is filled
          with vesicles; at this time both indentation and dimple
          (arrow) have regressed (inset) x 14,000. Inset x 10.

changes may also occur in the cisternae. These ultrastructural changes, and those observed in pricked eggs at later times, are similar to those described for fertilized eggs.

One interesting difference observed when these partially dejellied eggs were pricked was a wave of cortical contraction, beginning 3-4 min after pricking and spreading over the egg, starting from the site of the stimulus. This wave has not been detected in fertilized eggs, probably because the presence of the plug makes it difficult to observe changes in the dimple.

A23187 Treated Eggs

Eggs treated with DTT and then immersed in $5 \mu M$ ionophore do not show any signs of dimple regression. At $10-12 \mu M$ ionophore, the dimple regresses in about 20-30 min (the exact time depends on the egg clutch). At $35-50 \mu M$, the dimple starts regressing about 5 min after treatment and there is some variation in the pigmentation around the dimple itself. Such a higher concentration of ionophore is required to obtain a reasonably rapid activation probably because of the presence of the jelly layer J1, the vitelline envelope, and in particular, the dimple contents (see Fig. 1), which could slow down the diffusion of A23187.

In eggs exposed to $35-50 \mu M$ ionophore, vacuoles and CGs undergo exocytosis, even in Ca^{2+}-free Ringer's containing 10 mM EGTA. The reaction of the egg to A23187 is not usually uniform: there are regions of "unreacted" dimple peripheral cytoplasm next to regions where exocytosis is already underway. Again, this may be the result of an unequal diffusion of ionophore and the fact that the response to this drug is not propagated (9).

The time of the first response to ionophore varied with the clutch of eggs; the earliest response found was at 1 min after treatment. Once the response began, most vacuoles and CGs disappeared within the next 5 min (data not shown). Importantly, in only 1 egg of 25 did we observe a network of SER cisternae similar to that found in fertilized eggs. In the remaining eggs neither the transformation nor the re-orientation of the cisternal clusters and vacuoles were detected. Apparently the response of the egg to activation by ionophore is not equivalent to that induced by fertilization or by pricking, as ionophore does not lead to the formation of a cisternal network, at least in our experimental conditions.

DISCUSSION

The most intriguing features of the activation response of Discoglossus eggs are the changes in the morphology and organization of the organelles in the dimple, and the rapidity of these changes.

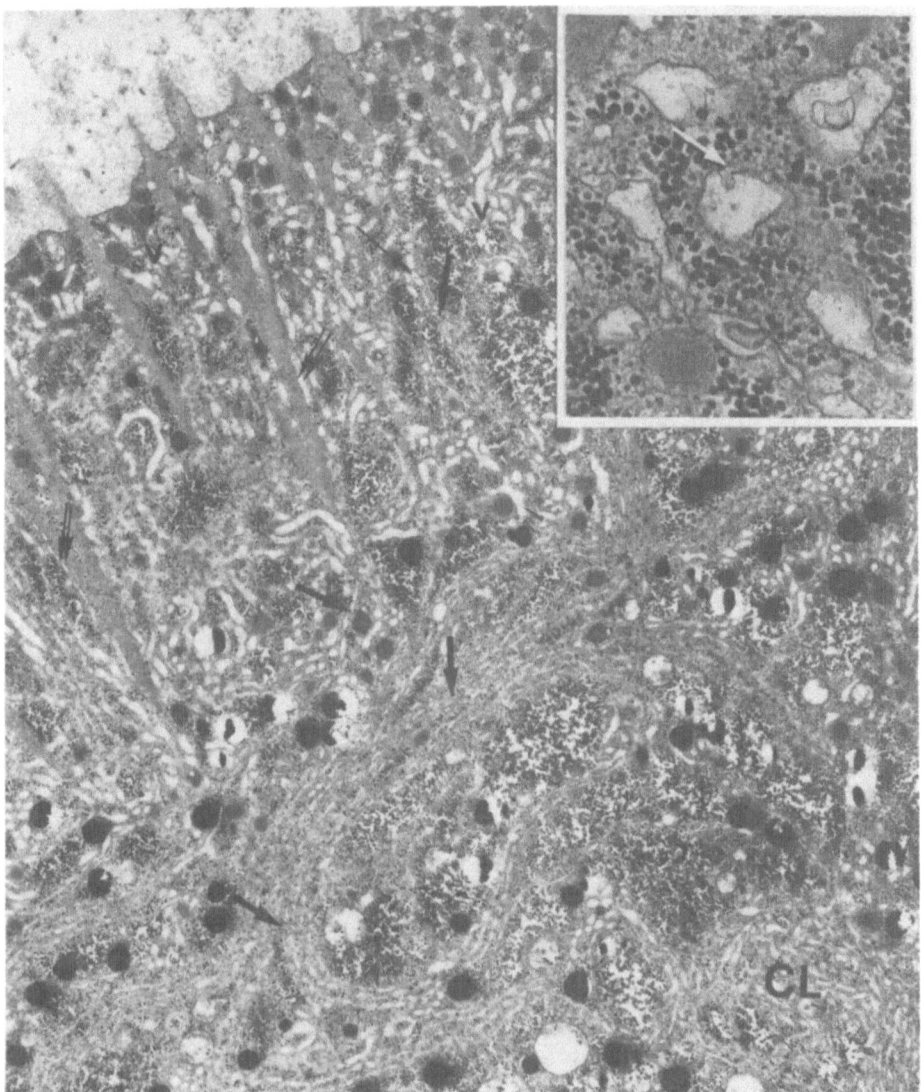

Figure 7. Dimple about 30 sec after pricking. In this early stage of
          cisternal network formation, zone B clusters (CL) have lost
          their compactness and are continuous with whorls of cis-
          ternae (arrows). Cisternae chains are already in the pro-
          cess of invading zone A, and pile up below the vacuoles
          (small arrow). Vacuoles (V) have elongated, probably
          because they have fused together and have incorporated some
          of the invaginations (inset, arrow) present in their lumen.
          Double small arrows = sites where either cisternae or flat-
          tened vacuoles lean against the microfilament bundles x
          7,500. Inset, x 25,000.

Vacuoles and cisternae of the clusters are both competent to respond
at the same time to the sperm by transforming into the peripheral
network of cisternae. The transformation of the vacuoles consists of
their flattening and elongation due to the fusion of vacuoles, and to
the incorporation of the invaginations present in the lumen. Similar
changes may occur in the cisternae of the clusters during their
transformation into arrays of longer cisternae. The elaboration of
the peripheral network creates a transient communication to bridge
the superficial and the deeper regions of the egg periphery.

It is interesting to compare the formation of this membranous
network with the origin of the analogous cortical complex in Xenopus
oocytes. The Discoglossus cisternae and vacuoles probably form when
the annulate lamellae (AL) disaggregate, as is true for Rana and
Xenopus fully grown oocytes following progesterone treatment (26,7).
Indeed, in the fully grown Discoglossus oocyte, the AL are located at
the future site of the dimple (the "germinative area"). At the same
time that the AL disappear following germinal vesicle breakdown, the
vacuoles which are later found in oviposited eggs gradually accumu-
late in the peripheral cytoplasm of the forming dimple (14). The
observation that AL in Discoglossus may give rise to both cisternae
and vacuoles is in agreement with the situation in Xenopus, where AL
disaggregation results in the formation of cisterna clusters as well
as the SER shells which surround the CGs (7). However, while in
Xenopus the cisternae and shells are already interconnected in
oviposited eggs, in Discoglossus the two sets of organelles become
continuous only at the time of egg activation. They then regain the
association they probably had in the AL of the ovarian oocyte.

There is much indirect evidence suggesting that the free Ca^{2+}
concentration increases in amphibian eggs at fertilization and this
has recently been confirmed by measurements in Xenopus oocytes using
intracellular Ca^{2+}-selective microelectrodes (Busa and Nuccitelli,
personal communication). Three lines of evidence suggest that the
free Ca^{2+} increases at activation of Discoglossus eggs. First, there
is exocytosis of CGs cisternae and vacuoles. Second, there is a cortical contraction wave after activation, which is most likely due to
a Ca^{2+} induced contraction of cortical acto-myosin (16,36,39). Actin
and myosin are present in the Discoglossus egg cortex (6). Finally,
the microfilament bundles gradually disaggregate. They become less
compact, less organized, and are shortened, probably due to severing.
These effects may be the result of Ca^{2+} regulation of actin-
associated proteins (42,29). It is interesting to note, however,
that in the Discoglossus egg a wave of contraction spreads over the
entire egg from the site of pricking, and exocytosis occurs in both
animal and vegetal halves after activation, but only in the dimple is
there a transformation of the cisternae and vacuoles. These observations suggest that a transient increase in free Ca^{2+} may propagate
from the dimple to the rest of the egg following fertilization, and

that this ionic change alone is not sufficient to cause the transformation of vacuoles and clusters.

What stimulus could reorganize the organelles located in a peripheral region more than $12 \mu m$ deep in a matter of seconds? One good
candidate is a flow of current through the egg after activation (30).
Transcellular currents have been detected at several stages of egg
development. In the fully grown, germinal vesicle stage Xenopus
oocyte, a Ca^{2+}-mediated Cl^- current enters the oocyte at the animal
pole (34). The current is lost during meiotic maturation, but a Cl-
efflux reappears at fertilization (27,30). In Rana eggs the Cl^-
channels, which may be more concentrated in the animal hemisphere,
can be opened by an increase in free Ca^{2+} (Cross, 1981). Sperm enter
the Rana egg throughout the animal hemisphere, but in Discoglossus,
the site of sperm entry is restricted to the dimple, which is an
expression of the exaggerated polarity of this egg (6). If the Cl^-
channels are more concentrated in the dimple than in the rest of the
egg, then at activation a current by Cl^- efflux through the dimple
membrane would traverse the egg along its polar axis, and this
current could reorient the cisternae and vacuoles towards the plasma
membrane.

Alternatively, the driving force for reorganizing the dimple
cytoplasm might be provided by contractile proteins. It is only in
the dimple that there are both the striking rearrangement of
membrane-bound organelles and the extensive arrays of microfilament
bundles. A role for myosin and for a spectrin-like protein in the
movement of vesicles along core rootlets in intestinal epithelial
cells has been suggested by Pearl et al. (31) and by Hirowaka et al.
(23).

The next obvious question is the role of the constituents of
such a network, and of the network itself. From our data it is evident that both transformed vacuoles and cisternae constitute,
together with CG membranes, an important source for an increase in
total plasma membrane at activation as they are inserted in the egg
plasma membrane, and in particular, participate in microvilli elongation. Furthermore, as most vacuoles and cisternae undergo exocytosis, it is highly probable that they may provide the egg surface or
the perivitelline space with substances related to fertilization or
to later stages of development, as observed in, for example, Acipenser (15) and Limulus (3).

In ionophore-treated eggs, CGs and vacuoles undergo exocytosis,
but neither the transformation nor the re-orientation of the cisternal clusters and vacuoles were detected. The fact that exocytosis
may occur also when the network fails to form (as in ionophore
activation) allows us to distinguish between further possible roles
of the network itself and of its constituents. Although other
possibilities exist, it may be that vacuoles and cisternal clusters

transform into the network only as a response to stimuli such as fertilization or pricking, in which a propagating reaction spreads from the site of activation. If so, the network may be instrumental in the propagation of the reaction itself. It cannot be excluded, however, that A23187 treatment might be exerting effects in addition to the release of Ca^{2+} from intracellular stores; these additional effects might inhibit the transformation of cisternal clusters and vacuoles into a network.

Even if the organization in a cisternal network is not needed for the exocytotic process, its constituents may well be the source of Ca^{2+} storage and release at activation in view of the above-mentioned analogy to the cortical SER of Xenopus, which, however, in contrast to the Discoglossus organelles, is not exocytozed at activation. Organelles other than the peripheral vacuoles could be totally or partially responsible for Ca^{2+} sequestation and release at activation (for example, mitochondria or the 35 nm thick cisternae observed in the unactivated but not in the fertilized egg cortex). However, at least in the case of mitochondria, this possibility seems less likely because they are predominantly located in zone B, far from the dimple plasma membrane, and also because in sea urchin egg homogenates little of the Ca^{2+} sequestering activity is in the mitochondria (40).

In contrast, the hypothesis that the network constituents are the source of Ca^{2+} release is strengthened by the observation that, as a result of fertilization or of activation by pricking, the network cisternae are in close contact, directly or through interconnections, with the microfilament bundles. This association, which closely resembles that found between sarcoplasmic reticulum cisternae and microfilaments, suggests a calcium release by the cisternae that could account for most of the observed changes in the cytoskeletal organization, and CG exocytosis.

ACKNOWLEDGEMENTS

This paper is dedicated to Professor Mario Galgano on the occasion of his seventy-fifth birthday. We thank Dr. Nicholas L. Cross and Dr. Brian Dale for suggestions on the manuscript. Part of this work was carried out in the Center for Electron Microscopic Studies (Centro per gli Studi di Microscopia Elettronica) of the University of Naples. Supported by a 40 percent M.P.I. grant to Dr. Gianfranco Ghiara.

REFERENCES

1. Andreuccetti, P., Burrini, A.G., Denis-Donini, S., and Campanella, C. 1980. Ca^{2+} localization in smooth endoplasmic reti-

culum in Xenopus eggs. Abstract of the XLV International
Embryological Conference, Patras, Greece.

2. Andreuccetti, P., Denis-Donini, S., Burrine, A.G., and
   Campanella, C. 1984. Calcium ultrastructural localization in
   Xenopus laevis eggs following activation by pricking or by
   Calcium ionophore A23187. J. Exp. Zool. 229:295-308.

3. Bannon, G.A., and Brown, G.G. 1980. Vesicle involvement in the
   egg cortical reaction of the horseshoe crab, Limulus polyphemus L. Dev. Biol. 76:418-427.

4. Campanella, C. 1975. The site of spermatozoon entrance in the
   unfertilized egg of Discoglossus pictus (Anura): An electron
   microscope study. Biol. Reprod. 12:439-447.

5. Campanella, C., and Andreuccetti, P. 1977. Ultrastructural
   observa-tions on cortical endoplasmic reticulum and on residual cortical granules in the egg of Xenopus laevis. Dev.
   Biol. 56:1-10.

6. Campanella, C., and Gabbiani, G. 1980. Cytoskeletal and contractile proteins in coelomic oocytes, unfertilized and fertilized
   eggs of Discoglossus pictus (Anura). Gam. Res. 3:99-114.

7. Campanella, C., Andreuccetti, P., Taddei, C., and Talevi, R.
   1984. The modification of cortical endoplasmic reticulum
   during in vitro maturation of Xenopus laevis oocytes and its
   involvement in cortical granules exocytosis. J. Exp. Zool.
   229:283-293.

8. Cardasis, C.A., Schuel, H., and Herman, L. 1978. Ultrastructural
   localization of calcium in unfertilized sea urchin eggs. Cell
   Sci. 31:301-315.

9. Charbonneau, M., and Picheral, B. 1983. Early events in anuran
   amphibian fertilization: an ultrastructural study of changes
   occurring in the course of monospermic fertilization and artificial activation. Dev. Growth and Differ. 25:23-37.

10. Charbonneau, M., Moreau, M., Picheral, B., Vilain, J.P., and
    Guerrier, P. 1983. Fertilization of amphibian eggs: a compar-
    ison of electrical responses between Anurans and Urodeles.
    Dev. Biol. 98:304-318.

11. Charbonneau, M., and Grey, R.D. 1984. The onset of activation
    responsiveness during maturation coincides with the
    formation of the cortical endoplasmic reticulum in oocytes of
    Xenopus laevis. Dev. Biol. 102:90-97.

12. Cross, N.L. 1981. Initiation of activation potential by an
    increase in intracellular calcium in eggs of the frog Rana
    pipiens. Dev. Biol. 85:380-384.

13. Cross, N.L., and Elinson, R.P. 1980. A fast block to polyspermy
    in frogs mediated by changes in the membrane potential. Dev.
    Biol. 75:187-198.

14. Denis-Donini, S., and Campanella, C. 1977. Ultrastructural and
    lectin binding changes during the formation of the animal
    dimple in oocytes of Discoglossus pictus (Anura). Dev. Biol.
    61:140-152.

15. Dettlaff, T.A. 1962. Cortical changes in acipenserid eggs during

fertilization and artifical activation. J. Embryol. Exp. Morph. 10:1-26.

16. Elinson, R.P. 1975. Site of sperm entry and a cortical contraction associated with egg activation in the frog Rana pipiens. Dev. Biol. 47:257-268.

17. Epel, D. 1982. The physiology and chemistry of calcium during the fertilization of eggs, in: "Calcium and Cell Function" Vol. II, pp. 355-383, W.Y. Cheung, ed., Academic Press, New York, London.

18. Gardiner, D.M., and Grey, R.D. 1983. Membrane junctions in Xenopus eggs: their distribution suggests a role in calcium regulation. J. Cell Biol. 96:1159-1163.

19. Gilkey, J., Jaffe, L.F., Ridgway, E.B., and Reynolds, G.T. 1978. A free calcium wave traverse the activating egg of the medaka, Oryzias latipes. J. Cell Biol. 76:448-466.

20. Grey, R.D., Wolf, D.P., and Hedrick, J.L. 1974. Formation and structure of the fertilization envelope in Xenopus laevis. Dev. Biol. 36:44-61.

21. Grey, R.D., Bastiani, M.J., Webb, D.J., and Shertel, E.R. 1982. An electrical block is required to prevent polyspermy in eggs fertilized by natural mating of Xenopus laevis. Dev. Biol. 89:475-484.

22. Hibbard, H. 1928. Contribution a l'etude de l'ovogenese chez Discoglossus pictus. Otth. Arch. Biol. 38:251-326.

23. Hirokawa, N., Cheng, R.E., and Willard, M. 1983. Location of a protein of the fodrin-spectrin-TW 260/260 family in the mouse intestinal brush border. Cell 32:953-965.

24. Ito, S. 1972. Effects of media of different ionic composition on the activation potential of anuran egg cell. Dev. Growth Diff. 14:217-227.

25. Iwao, Y., Ito, S., and Katagiri, C. 1981. Electrical properties of toad oocytes during maturation and activation. Dev. Growth and Diff. 23:89-100.

26. Kessel, R.G., and Subtelny, S. 1981. Alteration of annulate lamellae in the "in vitro" progesterone-treated full-grown Rana pipiens, oocytes. J. Exp. Zool. 217:119-135.

27. Kline, R. and Nuccitelli, R. 1983. Activation current in the frog egg. J. Cell Biol. 97:25a.

28. Maeno, T. 1959. Electrical characteristics and activation potential of Bufo eggs. J. Gen. Physiol. 43:139-157.

29. Mooseker, M.S. 1983. Actin binding proteins of the brush border. Cell 35:11-13.

30. Nuccitelli, R. 1983. Transcellular ion currents: signals and effectors of cell polarity, in: "Modern Cell Biology", 2:451-481, Alan R. Liss, N.Y.

31. Pearl, M., Fishkind, D., Mooseker, M., Keene, D., and Keller, T. III. 1984. Studies on the spectrin-like proteins from the intestinal brush border, TW260/240, and characterization of its interaction with the cytoskeleton and actin. J. Cell Biol. 98:66-78.

32. Poenie, M., Patton, C., and Epel, D. 1983. Ultrastructural
    visualization of calcium movements associated with
    fertilization of S. purpuratus eggs. J. Cell Biol. 97:181a.
33. Ridgway, E.B., Gilkey, J.C., and Jaffe, L.I. 1977. Free calcium
    increases explosively in activating medaka eggs. Proc. Natl.
    Acad. Sci. USA 74:623-627.
34. Robinson, K.R. 1979. Electrical currents through full-grown and
    maturing Xenopus oocytes. Proc. Natl. Acad. Sci. USA
    76:837-841.
35. Schlichter, L.C., and Elinson, R.P. 1981. Electrical responses of
    immature and mature Rana pipens oocytes to sperm and other
    activating stimuli. Dev. Biol. 83:33-41.
36. Schroeder, T.E., and Strickland, D.L. 1974. Ionophore A23187,
    calcium and contractility in frog eggs. Exp. Cell Res.
    83:139-142.
37. Shapiro, B., and Eddy, E.M. 1980. When sperm meet egg: biochem-
    ical mechanisms of gamete interaction. International Review of
    Cytology, Bourne and Danielli, eds., Academic Press, N.Y.,
    66:257-295.
38. Steinhardt, R.A., Zucker, R., and Shatten, G. 1977. Intracellular
    calcium release at fertilization in the sea urchin egg. Dev.
    Biol. 25:232-247.
39. Stewart-Savage, J., and Grey, R.D. 1982. The temporal and spatial
    relationship between cortical contraction, sperm trail forma-
    tion, and pronuclear migration in fertilized Xenopus eggs.
    Wilhelm Roux's Arch. Dev. Biol. 191:241-245.
40. Suprynowicz, F.A., Poenie, M. and Mazia, D. 1983. Calcium
    sequestering system of the sea urchin embryo. J. Cell Biol.
    97:30a.
41. Vacquier, V.D. 1981. Dynamic changes of the egg cortex. Dev.
    Biol. 84:1-26.
42. Weeds, A. 1982. Actin-binding protein-regulators of cell archi-
    tecture and motility. Nature 296:811-816.
43. Wintrebert, P. 1933. La mecanique du developpment chez
    Discoglossus pictus Otth, de l'ovogenese a la segmentation.
    Arch. Zool. Exp. Genet. 75:501-539.
44. Wolf, D.P. 1974. The cortical response in Xenopus laevis. Dev.
    Biol. 40:102-115.

URODELE EGG JELLY AND FERTILIZATION

Patrick Jego,[1] Hubert Lerivray,[1] Amand Chesnel,[1] and
Michael Charbonneau[2]

Laboratoire de Biologie de la Reproduction[1]
Laboratoire de Cytologie Experimentale[2]
Groupe de Recherches en Biologie Cellulaire et Reprod
Laboratoier Associe au C.N.R.S.
Campus de Beaulieu
Universite de RENNES I
35042 RENNES cedex - France

SUMMARY

Fertilization of urodele amphibians is physiologically polyspermic. These amphibians lack sperm entry blocking mechanisms at
the egg surface, such as a cortical reaction or a membrane depolarization. Although, egg jelly is necessary for sperm capacitation, a
late block to sperm entry does occur about 30 min after fertilization
at a precise interface between jelly layers. The jelly is secreted
by oviductal cells. In order to investigate its role in fertilization, we studied some biochemical properties of the oviductal
secretions of eight species. 1) In double diffusion experiments on
agarose plates, some components secreted by the anterior and the
middle parts of the oviduct interacted together and formed precipitin
lines. This reaction might be responsible for the formation of the
dense zone that delimits the capsular chamber. 2) A hemagglutinating
activity was found in the anterior or in the posterior part of the
oviduct depending on the species. A 18K or 26K lectin was purified
respectively from the oviduct of Ambystoma mexicanum and Pleurodeles
waltl. In both species, the site where the late block to sperm entry
is operative was spatially related to the location of the lectin in
the jelly. However, sperm in contact with the purified lectins did
not undergo any visible morphological change.

INTRODUCTION

       Amphibian eggs have been a prime material for fertilization
studies for a very long time. Bataillon (3) was the first to high
highlight the importance of egg jellies in amphibian fertilization.
This was confirmed by many authors (see review by Metz, 26) carrying
out many anatomical and biochemical studies on this material (see 21
for references). The principal objective for these authors was to
find a correlation between the presence of a particular structure or
substance and fertilization. Sperm-jelly interactions initiate sperm
transformation allowing its entry into the egg and then block the
entry of additional sperm. Anuran egg jellies have been the most
studied in this way. Sperm incubated in jellies are able to
fertilize dejellied eggs (1,23,38,39,41,44). The jelly component
which is responsible for this sperm transformation has been isolated
(2,24,28). In the earliest block to sperm entry the jellies are not
concerned since it depends upon egg membrane potential changes, the
fertilization potential (7,13). In the late block to sperm entry,
however, the inner jelly is implicated since the fertilization
envelope is the result of interactions between this jelly, the egg
cortical granule components and the vitelline envelope (11,12) Wyrick
et al. (46) proposed that the cortical granule lectin (whose
secretion starts just after the first sperm entry) is principally
responsible for this late block to sperm entry.

       Urodeles could be seen as a less suitable material for gamete
interactions studies since sperm penetration appears to be a more
complex phenomena than in the anurans. Sperm incubated in jelly are
not able to fertilize dejellied eggs, and polyspermy is the rule.
However, a late block to sperm entry ocurrs in urodeles (25,31).
This late block to sperm entry can be observed in unfertilized eggs.
As Salthe (36) pointed out, there is no anatomical differences in
urodele egg jellies of fertilized and unfertilized eggs. Cortical
granules are absent (43). Moreover, the main difference between
anurans and urodeles appears to be the lack of signals from the eggs
at fertilization in urodeles. Therefore, all gamete interactions,
and particularly those which develop progressively after laying and
are responsible for the late block to sperm entry, seem to be jelly
dependent. In such a situation, urodele egg jelly becomes an interesting material because it soley contains the components which interact with sperm. Under physiological conditions, fertilization occurs
a very few seconds after the laying down of jellies around the egg.
It is very important to know the properties of freshly secreted egg
jelly especially if reactions occur which progressively modify them.
Therefore, the search for egg jelly component interactions has been
done directly on oviduct secretory products. From such studies, we
observed the absence of an early block to sperm entry in urodeles.
Among the jelly components, a lectin has been well characterized in
some species. Involvement of the lectin in the late block to sperm

entry would support the idea of a functional role for this molecule
in gamete interactions.

MATERIALS AND METHODS

    1. Animals; preparation of the oviductal extracts. Pleurodeles
waltl and Ambystoma mexicanum were reared in the laboratory.
Salamandra salamandra, Triturus helveticus, Triturus marmoratus, Bufo
bufo, Rana esculenta and Alytes obstetricans were collected near
Rennes. A few experiments used Hynobius nebulosus a gift from M. R.
Thorn, Luxembourg or Notophtalmus viridescens, purchased from Xenopus
Ltd., Redhill, England.

    In the oviduct of the urodeles, it is easy to distinguish three
parts: the anterior part (AP) which secretes the inner jelly coat
layer, J_1 and the posterior part (PP), which secretes the external
jelly coat layer, J_3, J_4. They are separated by a more milky-
white and opaque part, the middle part (MP) which secretes J_2 (Fig.
1). The oviduct of Alytes obstetricians is similar, but in the
anurans (Bufo and Rana), such morphological differences cannot be
seen.

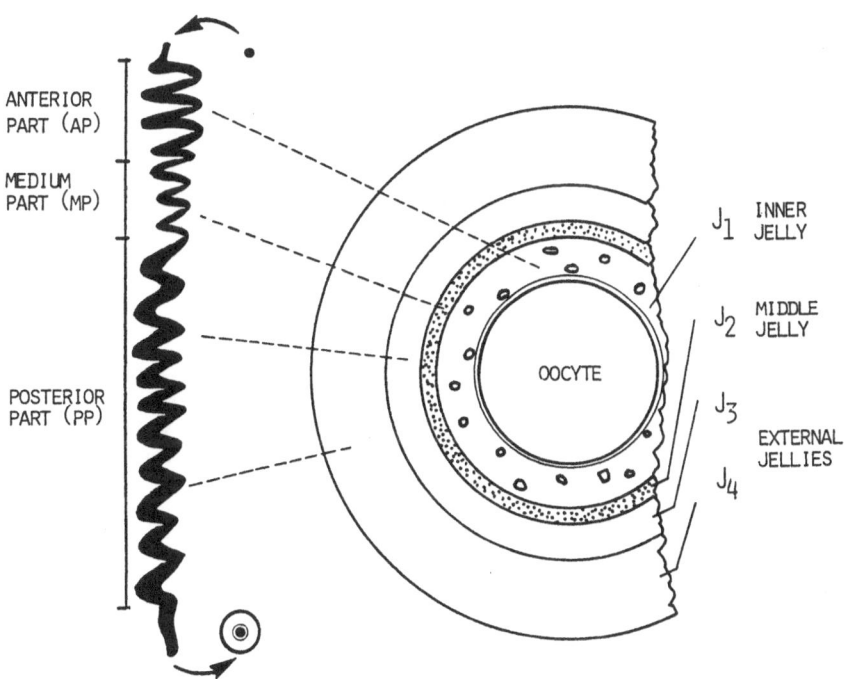

Figure 1. Simplified diagrammatic representation of Pleurodeles
           oviduct and egg jelly.

After dissection, the oviduct parts were homogenized (Potter apparatus) in a Tris-HCl buffer (0.1M, pH 7 or pH 8.2) and centrifuged (20,000 g for 1 h). The supernatant solution constitutes the oviductal extracts.

2. Separation of the egg jellies. Egg jellies of Pleurodeles, Ambystoma, Salamandra, Triturus and Hynobius were collected after spontaneous laying or stimulated laying (human chorionic gonadotrophin injection). Just after laying, it is very difficult to separate the different egg jelly layers because of their viscosity and adhesiveness. Whole freshly laid jellies or grossly separated fractions were tested to confirm the presence of well characterized oviduct secretory component. Several hours after laying, the hydration of the jellies allows for more precise separation. For chemical analysis, Pleurodeles jellies were separated using a method described before (19). In this method UV light (λ = 254nm; 30 min) and mechanical separation (dissecting forceps, filtrations, centrifugation) yielded four cleanly separated jelly coat layers which correspond to J_1, J_2, J_3 and J_4.

3. Chemical analysis. Proteins, amino acids, neutral sugars, hexoseamines, uronic acids, sialic acids and sulphate were determined using standard methods (see 18,19 for details).

4. Measurement of glycosyltransferase activities. The glycosyltransferase activities were measured in the 15,000 g supernatant solution of homogenized parts of Pleurodeles oviduct in distilled water containing Triton X-100 (0.5%) (termed crude extracts) or in the microsomal fraction prepared from these oviductal parts. This fraction was prepared by successive centrifugations in 1 mM Tris-HCl buffer, pH 7 containing 0.25 M sucrose; the last pellet (100,000 g for 1 h) constituted the microsome fraction (33).

Saccharide acceptors in the different parts of the oviduct or in the egg jelly coat layers were tested by measuring the enhancement of saccharide incorporation in the glycoprotein fraction after addition of adult oviduct parts or egg jelly coat extracts in a medium containing glycosyltransferases and radioactive saccharide donors. Fucosyltransferase and galactosyltransferase extracts were prepared from the β -estradiol stimulated oviduct of immature Pleurodeles (20). Guanosine diphospho-L-[$U^{14}C$] fucose was used for fucose acceptors measurements and uridine D-[$U^{14}C$]-diphosphogalactose for galactose acceptors. Adult oviduct parts were heat denatured (100°C for 10 min) before incubation to denature endogenous glycosyltransferases; egg jelly coats do not contain glycosyltransferases (see 20 for details).

5. Agarose diffusion. We used the classical method for agarose diffusion (30). Oviduct or jelly extracts (2-8 mg protein/ml) were placed in the wells of an agarose gel (1% in 0.1 M Tris-HCl buffer,

pH 7 or 8.2). Diffusion was performed at 37°C for 3 days in a humid
chamber. Agarose plates were then dessicated and precipitin lines
were stained with amido-schwartz.

6. Hemagglutination experiments. Rabbit red cells were trypsin
and glutaraldehyde treated according to the method described by Nowak
et al. (27).

Hemagglutinations were performed in U shaped microtiter plates.
In each well was successively added: 25 μl of oviduct or jelly
extract (or of purified lectin solution) with successive dilutions in
T. S. buffer (10mM Tris-HCl, 150mM NaCl; pH7.6); 25 μl of T.S.
containing 8mM CaCl$_2$ and sometimes inhibitory saccharides (with
adequate dilutions); 25 μl of T.S. containing 1% bovine serum
albumin; 25 μl of the treated rabbit red cells (2.5% v/v in T.S.
buffer).

7. Lectin isolation. The Pleurodeles and Ambystoma lectins
were isolated using an affinity chromatography method with epichlor-
hydrin treated starch according to Osawa et al. (29).

For Pleurodeles, monosaccharide addition during starch resin
preparation did not improve the efficiency. Elution was performed
with 1M glucose in T.C.S. (Tris-HCl 50 mM, pH 7.5, containing 150 mM
NaCl and 5 mM CaCl$_2$.

For Ambystoma, addition of xylose during starch resin preparation enhanced significantly the efficiency. Elution was performed
with 1 M xylose in T.C.S.

8. Inseminations. Eggs of Pleurodeles, Ambystoma and Triturus
helveticus were collected a few minutes after laying onto
dry plastic Petri dishes for insemination experiments. Sperm was
often applied without dilution by wiping the slightly minced Wolf's
duct over the egg jelly. When necessary, it was diluted in 1/10 OR2
solution (42).

In addition to microscopic observation of spermatozoa, the
criteria for sperm penetration were as follows:

- Fertilization craters [when sperm penetrated inside the oocyte, a
  large depression (45 μm diameter) could be seen at the surface of
  the oocyte at the penetration site (32); these craters remain
  visible for at least 3 h];
- Reorganizations of the pigmented area, emergence of the second
  polar body, fertilization bodies (32);
- Cleavage (about 6 h after insemination) and segmentation.

9. Electrical measurements. Records were obtained with two
intracellular microelectrodes filled with 3M KCl. The membrane

potential was measured differentially between one intracellular elec-
trode and the 1/10 OR2 medium (42) with a WPIM 707 preamplifier.
The method used for voltage-clamping was similar to that described by
Hagiwara et al. (14). For details see Charbonneau et al. (6).

RESULTS

A. ORGANIZATION OF URODELLE EGG JELLY

     Urodele egg jelly has been described in detail by Salthe (36),
Humphries (15), Vilter (40) and Boisseau et al. (5). Oocytes are
first surrounded by a thin envelope secreted by the ovary, the
vitelline envelope. As the oocyte moves through the oviduct, the
cells of this organ secrete jelly which is successively laid down
around the vitelline envelope. Four jelly coat layers can be easily
distinguished with the inverted microscope in all urodeles. The
inner jelly coat (J_1) (Fig. 1) is often very fluid. It contains
large granules just after egg laying; these granules are not
surrounded by a membrane and dissolve quickly as J_1 is diluted (i.e.
hydrated under natural conditions). The middle jelly layer (J_2) is
dense and light refringent. The outer jelly layers which vary in
number depending upon the species (36), are the thickest. J_3 appears
to be made of fibrous elements lying parallel to the egg surface. In
the outermost layers, fibers are usually orientated perpendicular to
the egg surface. It is difficult to distinguish with the inverted
microscope the different outermost layers that have already been
described following histochemical observations (usually named J_4 and
J_5; 15,16). These layers are referred to as J_4 in this paper.

     Between J_1 and J_2, a very dense zone appears progressively after
these layers have been laid down around the Pleurodeles egg (5,31).
Such a dense zone appears between the J_1 and J_2 jelly layers of all
urodele eggs we have observed including Ambystoma (Fig. 2).

B. CHEMICAL COMPOSITION OF PLEURODELES EGG JELLY

     Egg jellies contain essentially proteins and sugars (in a rela-
tive proportion of about 30:70). J_1 and J_3 contain more serine and
threonine than the two other layers (Table 1). The only neutral
sugars found in detectable amounts are galactose and fucose. Acid
hydrolysis liberated fucose before galactose. The fucose:galactose
ratio increased from J_1 to J_4 (0.35 to 0.93). This increase correlated well with the fucosyltransferase : galactosyltransferase
activity ratio (Table 2). J_1 components are able to incorporate
fucose if they were added in an exogenous fucosyoltransferase solution; this was not the case for J_2, J_3 or J_4 (Table 2). Glucosamine
and galactosamine were liberated in about a 1:1 ratio after 4N HCl
hydrolysis of J_2 and J_4. J_1 contained more glucosamine and J_3 more

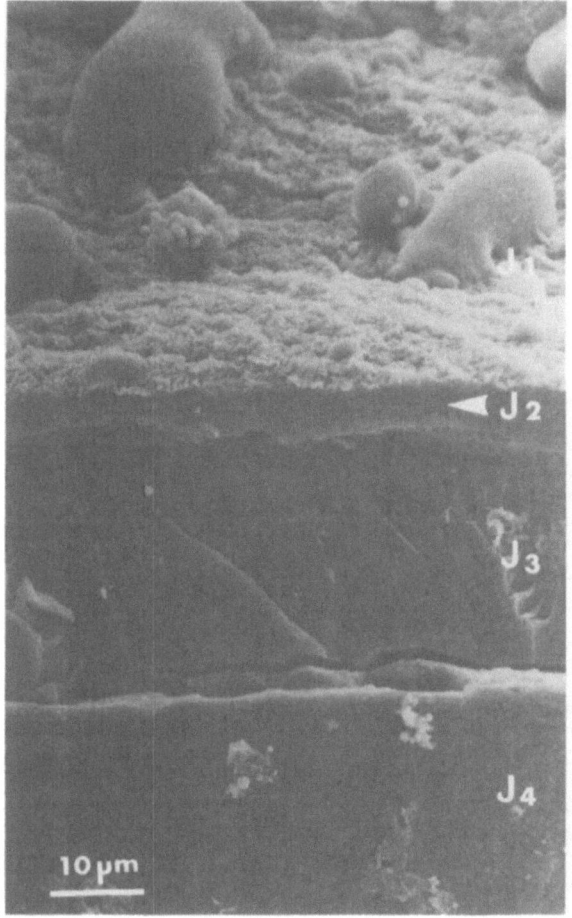

Figure 2. Scanning electron micrographs of Ambystoma egg jellies.
At the limit between J_1 and J_2 villosities appear
progressively; they are probably the result of the L_1-l_1
reaction.

galactosamine. Egg jelly coats were rather poor in sialic acids; J_1
contained slightly more than other layers. Uronic acids have not
been detected in any significant amounts. J_1, and to a lesser extent
J_3, contained sulphate groups in significant amounts.

SDS-PAGE revealed a great heterogeneity of each jelly coat
layer: five to eleven bands were observed in each case (not shown).
J_2 and J_4 contain the largest components (equal or greater than 10^6

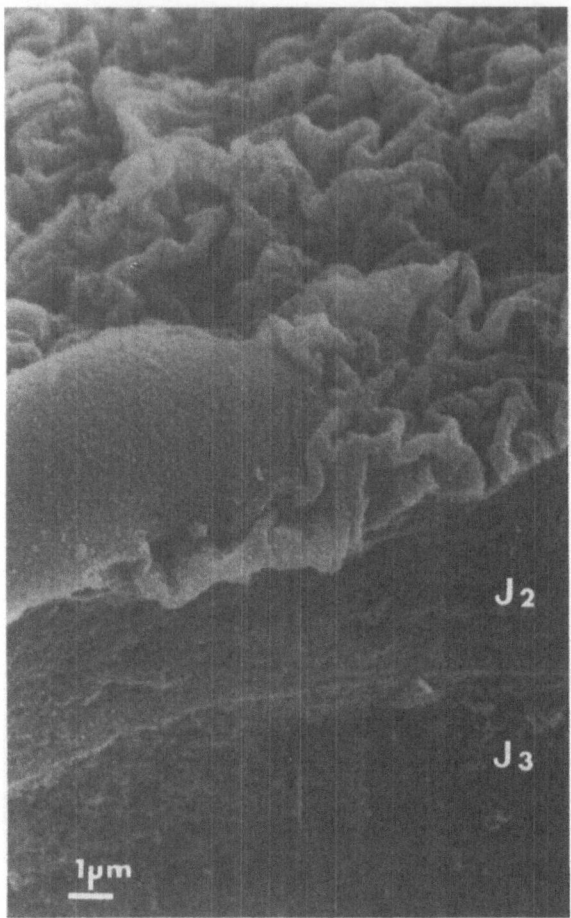

Figure 2. continued.

daltons even after SDS and β–mercaptoethanol treatment). The main components of J_1 and J_3 had a molecular weights of about 20K to 50K (19).

C. PRECIPITIN PROPERTIES OF URODELE EGG JELLY COMPONENTS

   When the products of secretion from the anterior and middle part of the urodele oviduct were put side to side on a glass plate, precipitation occured spontaneously between them. When these products were put in the wells of an agarose gel (30) precipitin lines appeared (Fig. 3). With Pleurodeles products, which were the most studied, two different groups of precipitin lines were distinguished. The first group of precipitin lines were not calcium dependent because EDTA or EGTA had no effect (Fig. 4). These reactions are termed L_1-l_1. The second group of precipitin lines

Table 1. Chemical composition of <u>Pleurodeles</u> egg jelly coat layers.

	J_1	J_2	J_3	J_4
Total amino acids	100	100	100	100
(Ser + Thr)	(39.6)	(20.0)	(46.5)	(26.1)
Glucosamine	35.7	11.2	14.6	17.2
Galactosamine	12.5	12.8	31.1	17.5
Galactose	49.5	20.3	38.7	22.9
Fucose	17.5	12.8	24.3	21.4
Sialic acids	3.4	1.4	1.9	1.5
Uronic acids	0 - 1.6	0 - 0.6	0 - 1.2	0 - 0.8
Sulphate	25.9	0 - 1.4	14.4	0 - 4.2

Results are expressed as the mole % of each compound relative to the total amino acids.

disappeared in the presence of EDTA or EGTA, these lines were continuous with precipitin lines obtained between anterior oviduct products and starch (Fig. 4). These calcium dependent reactions are termed L_2-l_2.

     1) Calcium independent precipitin reactions (L_1-l_1). The exist-
ence of such a reaction has been demonstrated with different species of urodeles. All of the species examined gave continuous precipitin lines between anterior and medium part of the oviduct, and this res-
ponse was not species specific. All species tested crossreacted in precipitin-line formation (Fig. 5). The products responsible for these reactions were secreted by anterior oviduct cells (L_1) and middle oviduct cells (l_1) (Table 3). The L_1-l_1 precipitin lines appeared between pH 4.0 and pH 8.5. High ionic strength did not inhibit the reaction. L_1 products were thermolabile; they were destroyed by a 1 min incubation at 100°C; l_1 products were very thermostable (they were not destroyed by a 20 min incubation at 100°C). Pronase (1 mg/ml) or trypsin (10 mg/ml) treatments inhibit the L_1-l_1 reaction; these proteolytic enzymes dissolved L_1-l_1 precipitin lines on agarose. Dithiothreitol or β-mercaptoethanol (100 mM) inhibited the L_1-l_1 reaction; however they did not dissolve L_1-l_1 precipitin lines in agarose even when they were applied for several days. We have not found any potent inhibitory saccharide.

Table 2. Glycosyl transferases and sugar acceptors in _Pleurodeles_ oviduct parts and egg jellies.

	AP	J_1	MP	J_2	PP_1	J_3	PP_2	J_4	-(5)
Fucose: galactose ratio	0.22	0.35	0.73	0.63	0.52	0.63	0.88	0.93	
Fucosyltransferase:	0.96(1)	–	2.47(1)	–	3.57(1)	–	2.51(1)	–	
Galactosyltransferase ratio	0.73(2)	–	2.75(2)	–	2.86(2)	–	2.42(2)	–	
Fucose acceptor (3)	450	462	107	104	164	217	ND	102	100
Galactose acceptor (4)	227	139	108	113	143	172	ND	118	100

(1) and (2): Glycosyltransferases have been measured in crude extracts (1) or in mocrosomal fractions (2).

(3) and (4): Enhancement of incorporated cpm in glycoprotein fraction in an incubation medium containing fucosyltransferase, Guanosine diphospho-L-[$U^{14}C$] fucose or Uridine diphospho-D-[$U^{14}C$] galactose after addition of heat denatured oviduct extracts or egg jellies.

Results are expressed as percent of endogenous of the incubation medium(5).

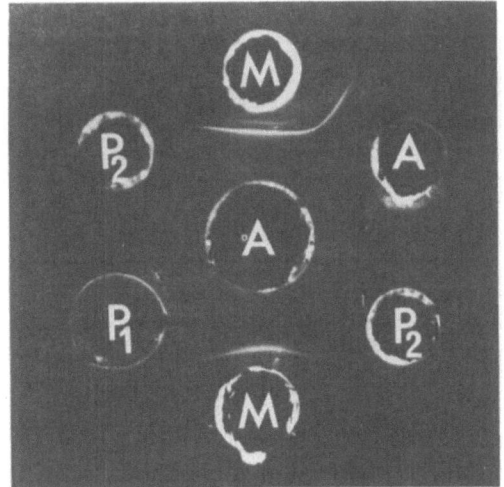

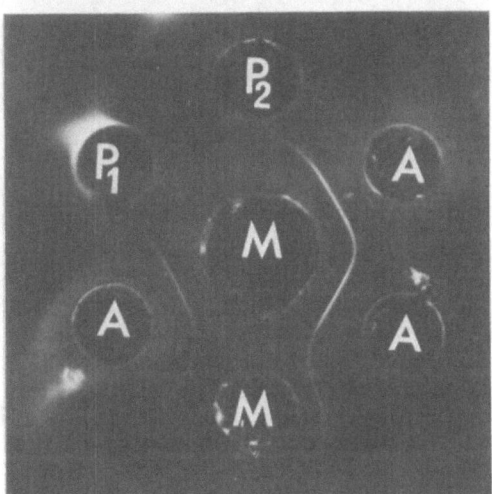

Figure 3. Precipitin lines on agarose between secretory products of
anterior part (A) and medium part (M) of the oviduct. P_1
and P_2 are respectively the first third and the two other
thirds of posterior oviduct. (Pleurodeles on the left,
Ambystoma on the right).

Such a reaction was not found with anuran oviduct secretory
products using Alytes obstetricans.

When jellies were taken from eggs about one hour after laying,
it was impossible to obtain precipitin lines on agarose gels although
the different jellies can be separated at this time. When they were
taken within a few minutes after laying (or after squeezing the
female), homogenates of the jellies (it was not possible to separate

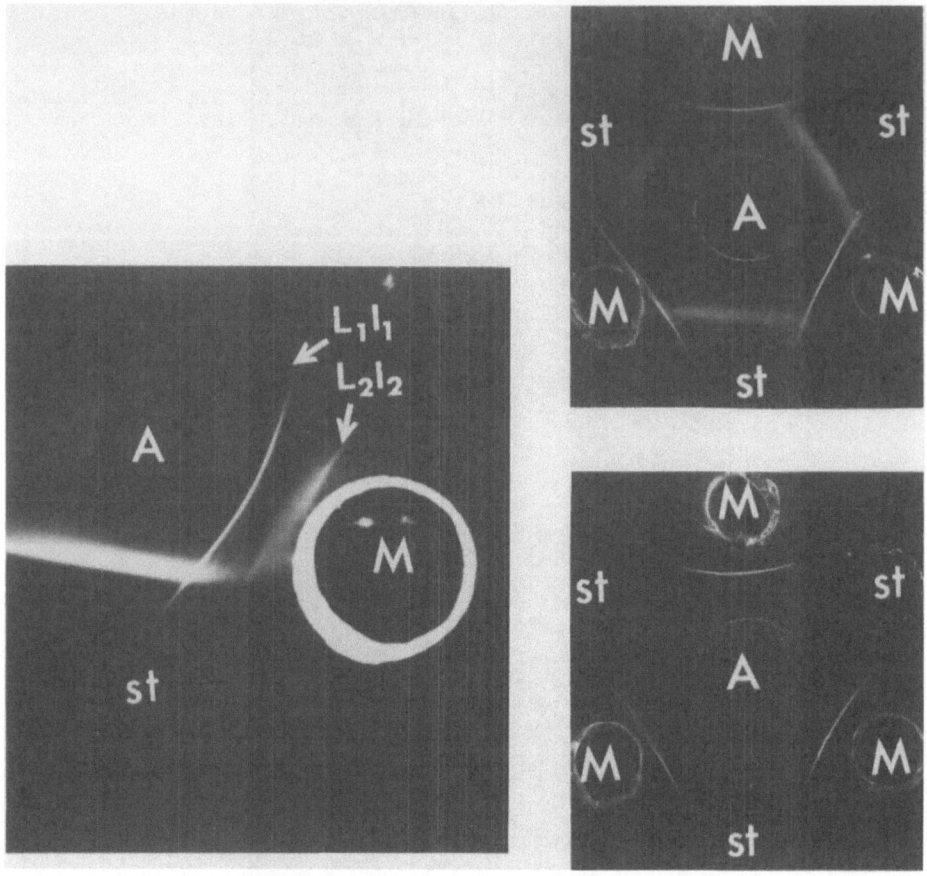

Figure 4. Action of EGTA (10 mM) upon precipitin lines observed on
agarose gel between the secretory products from anterior
part of <u>Pleurodeles</u> oviduct (A) and middle part (M) or
starch (ST). On the left micrograph, a continuity between
L_2-l_2 and A-starch precipitin lines can be observed.

them properly at this time) were able to develop precipitin lines
against oviduct products (Fig. 6).

2) Calcium dependent reactions (L_2-l_2). The agarose precipitin
lines which resulted from this reaction could be dissolved by some
saccharides but not by others. This suggested a lectin-ligand reaction. The presence of such a molecule in several urodele jellies was
demonstrated using hemagglutination and the ability to precipitate
with polysaccharides. The lectins were isolated by affinity chromatography in two species (<u>Pleurodeles</u>, <u>Ambystoma</u>).

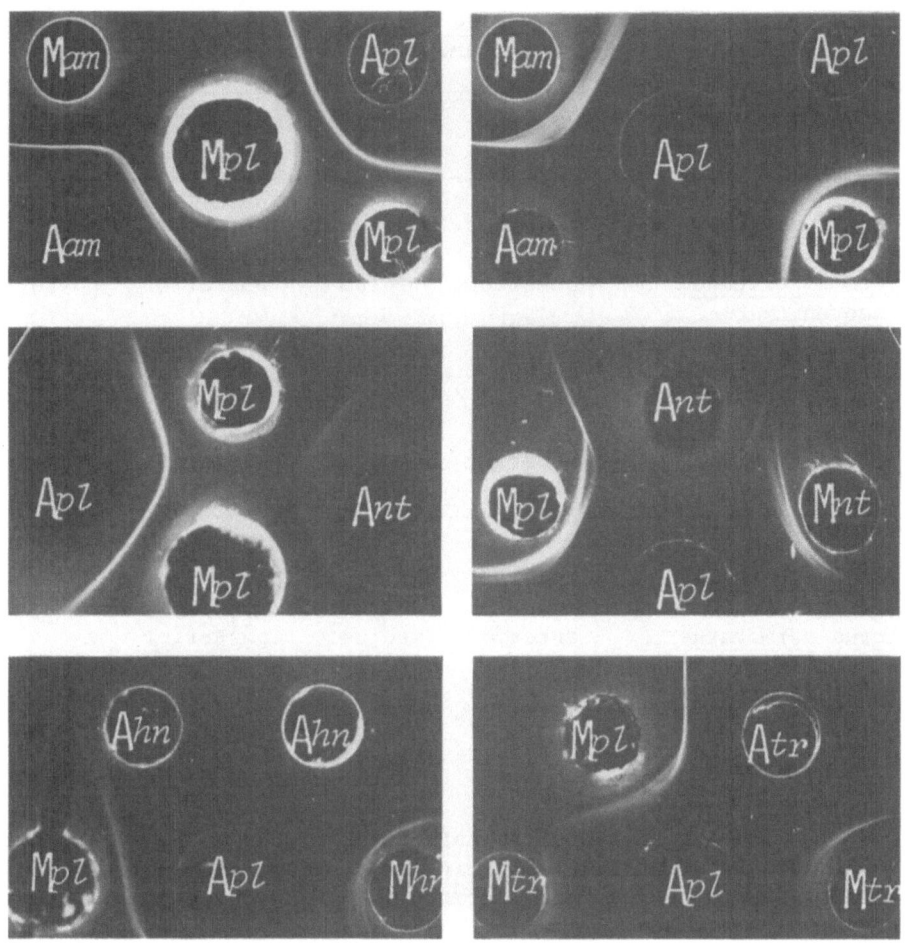

Figure 5. Cross species L_1-l_1 precipitin lines. Continuous
          precipitin lines appear between anterior and medium part
          of the urodele oviduct.

          pl : Pleurodeles waltl; am : Ambystoma mexicanum;
          nt : Notophtalmus viridescens; hn : Hynobius nebulosus;
          tr : Triturus helveticus.

     All urodele species studied exhibited an oviduct hemagglutina-
tion activity. However, this activity was not always present in the
same part of the oviduct (Table 3). Two groups were distinguished.
Pleurodeles, Hynobius and Salamandra had an hemagglutination activity
in the anterior part of the oviduct. Ambystoma, Triturus helveticus,
alpestris and marmoratus had an hemagglutination activity in the
posterior part of the oviduct.

Table 3. Localization in <u>Urodeles</u> oviduct of L_1, l_1, hemagglutination activity and <u>lectin</u>.

	L_1	l_1	Hemagglutination activity	lectin
<u>Hynobius</u> <u>nebulosus</u>	anterior	medium	anterior[+]	anterior[+]
<u>Ambystoma</u> <u>mexicanum</u> (neotenic)	anterior	medium	posterior	posterior
<u>Amybstoma</u> <u>mexicanum</u> (metamorphosed)	ND	ND	posterior	posterior
<u>Pleurodeles</u> <u>waltl</u>	anterior	medium	anterior	anterior
<u>Triturus</u> <u>helveticus</u>	anterior	medium	posterior	
<u>Triturus</u> <u>alpestris</u>	anterior	medium	posterior	
<u>Triturus</u> <u>marmoratus</u>	anterior	medium	posterior	
<u>Notophtalmus</u> <u>viridescens</u>	anterior	medium	ND	
<u>Salamandra</u> <u>salamandra</u>	anterior	medium	anterior	

L_1 and l_1 compounds of the different <u>Urodeles</u> species were recognized by their activity to form precipitin lines on agarose plates which were continuous with the well characterized <u>Pleurodeles</u> L_1-l_1 precipitin lines.

ND = not determined.

[+] The hemagglutination activity and lectin were tested on egg jelly from <u>Hynobius</u> (only inner jelly layers were positive).

Specific saccharide inhibition was observed only in three species: <u>Pleurodeles</u>, <u>Ambystoma</u> and <u>Hynobius</u>. In other species, we have not found a clearly demonstratable specific inhibition with the saccharides used. The three lectins which have been identified recognize D-glucose derivatives (Table 4). D-galactose was not a potent inhibitory saccharide. Differences appeared in saccharide specificity (Table 4). <u>Pleurodeles</u> lectin recognized D-glucose better than other lectins. Precipitin lines on agarose gel were obtained with <u>Pleurodeles</u> lectin against starch or glycogen (Fig. 4)

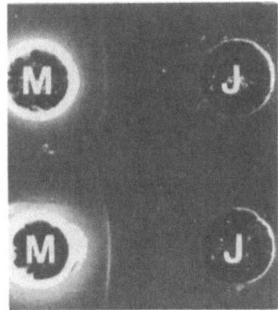

Figure 6. Precipitin lines on agarose plate between homogenates of
fresh laid jelly (J) and medium part of oviduct (M) in
Pleurodeles.

but not when the other two lectins were used. Interestingly with
Hynobius, which was the most ancient species studied, the lectin
specificity was for N-acetyl-D-glucosamine whereas the two other lectins had a specificity for D-glucosamine.

The Pleurodeles and Ambystoma lectins were isolated on epichlorhydrine cross linked starch. Purification increased the specific
activity (Table 5). With Pleurodeles, the lectin was the most important product of secretion (about 20 percent of total anterior oviduct
proteins) and so the increase of specific activity after isolation
was less than with Ambystoma (Table 5).

On SDS-PAGE, the Pleurodeles lectin appeared as a 26 K band; the
Ambystoma lectin had a molecular weight of about 18 K (Fig. 7).

Saccharide specificity of purified lectins was very similar to
that of oviduct extracts (Table 4). The two lectins required calcium
for activity. They were heat-sensitive; the activity was destroyed
after 5 min incubation at 60°C and pH 7.5. Dithiothreitol and
β-mercaptoethanol inhibited the lectin activity; they dissolved
precipitin lines using starch and glycogen and the Pleurodeles
lectin. Proteolytic enzymes (pronase 1 mg/ml or trypsin 10 mg/ml)
inhibited the lectin activity; however, they did not dissolve starch
or glycogen precipitin lines formed with the Pleurodeles lectin.

When the Pleurodeles inner jelly coat layer J_1 or the Ambystoma
outer jelly coat layer J_4 was separated from other jelly layers (on
eggs about one h after laying, without any chemical or UV treatment),
they exhibited hemagglutinating properties; the other jelly coat
layers did not. Pleurodeles inner jelly coat layer precipitated
soluble starch. The precipitate was dissolved by EGTA or D-
glucosamine (Fig. 8). Therefore, the lectin was present in the jelly
and retained its saccharide-precipitating properties up to several
hours after egg laying.

Table 4. Saccharide inhibition of the hemagglutination activity of oviduct crude extracts and purified lectins.

	AP Pleuro-deles	PP Ambystoma	PP metamorp. Ambystoma	AP Hynobius	Lectin Pleuro-deles	Lectin Ambystoma
D-xylose	0.32	0.04	0.04	0.08	0.32	0.04
D-arabinose	2.56	0.64	0.32	2.56	2.56	0.64
D-ribose	5.12	5.12	5.12	5.12	2.56	10.24
D-glucose	0.16	0.08	0.64	0.16	0.16	0.08
D-mannose	1.18	2.56	2.56	0.32	1.28	5.12
D-galactose	5.12	1.28	1.28	2.56	2.56	1.87
D-fructose	0.64	1.40	2.56	0.64	0.64	1.28
L-fucose	2.56	1.28	2.56	0.32	1.28	1.87
L-rhamnose	>>10	>>10	>>10	>>10	>>10	>>10
D-glucosamine	0.08	0.04	0.16	10.24	0.04	0.04
D-galactosamine	>>10	>>10	>>10	>>10	>>10	
NAc-D-glucosamine	>>10	10.24	5.12	0.64	>>10	>>10
NAc-D-galacto-samine	>>10	>>10	>>10	>>10	>>10	>>10
maltose	0.08	0.16	0.16	0.08	0.04	0.16
D-trehalose	0.32	0.16	0.32	0.32	0.32	0.32
melibiose	0.08	0.16	0.16	0.64	0.08	0.16
lactose	10.24	0.64	0.64	1.28	5.12	1.28
saccharose	0.32	0.64	0.64	0.32	0.16	1.28
raffinose	0.04	0.32	0.32	0.64	0.08	1.28
melezitose	0.64	1.28	1.28	0.32	0.64	2.56

Results are expressed as the concentration of saccharide (mM) that inhibits hemagglutination by 50%.

AP: anterior part of the oviduct.
PP: posterior part of the oviduct.

Table 5. Lectin purification from Pleurodeles and Ambystoma oviducts by one step on epichlorhydrine cross linked starch.

	Pleurodeles oviduct (AP) crude extracts	Ambystoma oviduct (PP) crude extracts	Pleurodeles purified lectin	Ambystoma purified lectin
Total protein mg	33.6	87	7.5	1.2
Total activity	7.700	10.270	13.500	9.600
Specific activity (titer^{-1} divided by mg of protein per ml)	230	132	1.800	8.000
Purification (fold)			7.9	60.6
Recovery %	100	100	175	93

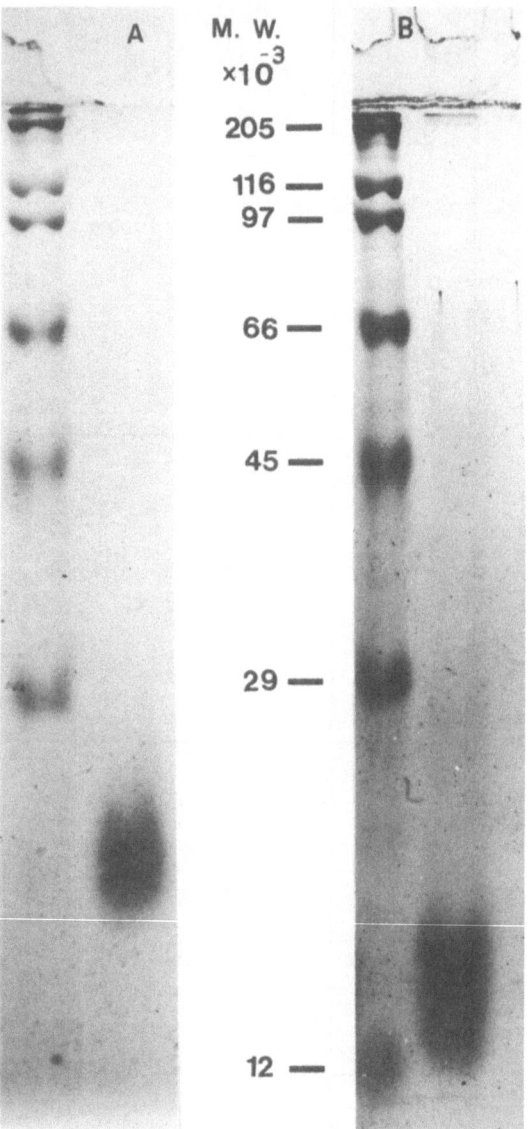

Figure 7. Polyacrylamide gel electrophoresis in SDS of purified
Pleurodeles oviduct lectin (A) or Ambystoma oviduct
lectin (B).

The molecular weight markers used were : myosin (205,000),
β-galactosidase (116,000), phosphorylase B (97,400), bovine
albumin, (66,000), egg albumin (45,000) and carbonic
anhydrase (29,000). Cytochrome C was sometimes used
12,400).

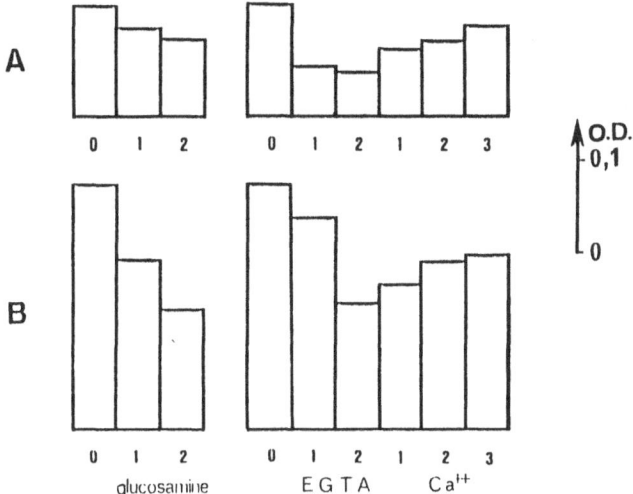

Figure 8. Presence of a lectin in <u>Pleurodeles</u> inner jelly coat layer
J_1. J_1 was collected mechanically from swollen egg jellies
without the use of UV or chemicals.

A - The optical density of J_1 alone (500 nm) decreases in
the presence of D-glucosamine (0 : 0 mM = control ; 1 :
4 mM ; 2 : 8 mM) or EGTA (0 : control ; 1 : 10 mM ; 2 :
20 mM). If calcium was then added (1 : 10 mM ; 2 :
20 mM ; 3 : 50 mM) the optical density increases
again.

B - Same experiments as in A but in the presence of soluble
starch (0.5 mg / ml in 0.1M Tris-HCl, pH 7). Optical
measurements were done after 2 h at 40°C.

D. URODELES ARE POLYSPERMIC; THERE IS NO FAST BLOCK TO SPERM ENTRY

      If the surface of fertilized eggs (<u>Pleurodeles</u> and <u>Ambystoma</u>)
are examined after natural mating, 3 to 15 sperm entry sites can be
identified with the stereomicroscope. Such eggs divide normally. If
sperm are deposited on the outer jelly layer of freshly laid or
expressed eggs, sperm are seen 2 to 3 min later in the inner jelly.
Sperm entry sites can be detected 3 to 4 min later. The number of
sperm entry sites was found to be proportional to the concentration
of the sperm at the time of insemination. It was sometimes possible
to induce more than 60 sperm to enter the egg. In order to improve
the similarity between our experimental conditions and the natural
ones, we artificially inseminated with concentrations of sperm that
resulted in the appearance of about 10 to 20 sperm entry sites.

Urodele eggs seem to be particularly susceptible to electrode insertion. This problem did not appear to be due to the techniques we used since maturing oocytes (prophase I or II) were found to display very stable membrane potentials (respectively around −70 and −30 mV). However, we succeeded in overcoming the problem by performing many careful recordings. Before fertilization, the membrane potential and resistance of mature eggs of Pleurodeles were −17.9 mV ± 4.1 SD (N = 79) and 4.1 MΩ ± 1.8 SD (N = 29), whereas similar measurements in Ambystoma gave values of −25.5 mV ± 9.4 (n = 22) and 2.5 mΩ ± 1.1 (n = 22). These data were compiled only from recordings lasting more than 30 min and when the electrical parameters were stable for at least 5-10 min.

Despite the problems encountered during electrode insertion, we carried out around 300 experiments with Pleurodeles and about 50 with Ambystoma during which eggs remained impaled for more than 15 min after insemination. This was long enough to verify the success of fertilization. In no case could we observe an abrupt change in the membrane potential in response to insemination (Fig. 9). A very slow hyperpolarization was in some cases recorded after fertilization or artificial activation. This was associated with an increase in membrane resistance indicating a change in membrane conductance. However, in all other cases there was no change at all in membrane potential and resistance.

In voltage-clamp experiments, the membrane potential of mature eggs of Pleurodeles was held constant at values ranging from −40 mV to +40 mV from before insemination until 10-15 min after insemination. This had no effect on fertilization (Table 6). With respect to control eggs present in the recording chamber, the clamped eggs did not display any delay in the events occurring between insemination and fertilization or any significant difference in the number of sperm entry sites. Fertilization of anuran eggs was prevented when the membrane was held at positive potentials using the same voltage clamp technique.

E. A LATE BLOCK TO SPERM ENTRY IN URODELES

When eggs that have been laid for some time were used for insemination experiments, the number of sperm entry sites decreased. After a delay (which is different from species to species) additional penetration was not observed. Microscopic observations indicated that progress of the sperm towards the ooctye was always stopped at the same level in the jelly. Initial experiments were done on Pleurodeles eggs with B. Picheral and suggested that J_2 was the barrier. It has been suggested that hydration modifies the properties of jelly (Picheral, 1977). More detailed observations clearly indicated that sperm were able to cross J_2 but were stopped at the limit between J_1 and J_2 (Fig. 10). At this level a dense zone progressively appeared

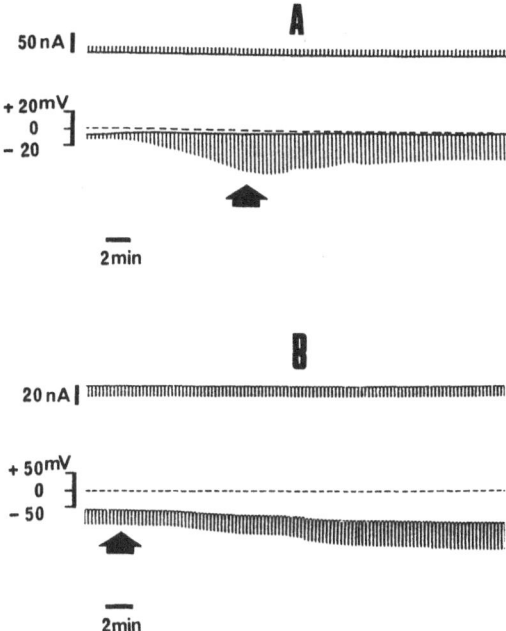

Figure 9. Electrical records from the egg during fertilization
(A : Pleurodeles ; B : Ambystoma).

The arrows indicate the time of insemination. There was no
significant change indicating the presence of a fertilization potential. The changes in membrane resistance were
not significant (6).

as the time after laying increased. It was evident that this dense
zone was the place where the late block to sperm entry developed in
Pleurodeles.

With Ambystoma, the late block to sperm entry did not occur at
the dense zone level. In jelly layers of Ambystoma eggs, sperm were
stopped at the limit between J_3 and J_4 (Fig. 10). This property was
not dependent on the sperm properties since Pleurodeles sperm were
stopped at the J_3-J_4 limit in Ambystoma and Ambystoma sperm were
stopped at the J_1-J_2 limit in Pleurodeles. This late block to sperm
entry was observed in hydrated egg jelly but also in egg jelly
collected in a dry Petri dish. This was more evident with Ambystoma
eggs than with those of Pleurodeles. With Ambystoma eggs, having
been kept in dry conditions 30 to 120 min after laying, sperm moved
easily into the outermost jelly coat layer (as quickly as in fresh
jelly) but they were all stopped at the J_3-J_4 limit.

Table 6. Effect of maintaining the membrane potential by voltage clamp upon fertilization of Pleurodeles eggs.

Clamped values (mV)	Fertilized eggs / eggs inseminated[+]	Sperm entry sites	Cleaved eggs / Fertilized eggs
+ 40	6 / 6	multiple	5 / 6
+ 30	8 / 8	multiple	3 / 8
+ 20	6 / 6	multiple	3 / 6
+ 10	6 / 6	multiple	3 / 6
0	6 / 6	multiple	4 / 6
- 10	5 / 5	multiple	3 / 5
- 20	6 / 6	multiple	4 / 6
- 30	6 / 6	multiple	5 / 6
- 40	5 / 5	multiple	3 / 6

The potential was maintained up to the appearance of the sperm entry sites (6 to 15 min. after insemination).

+ Eggs tested were in second metaphase stage before insemination.

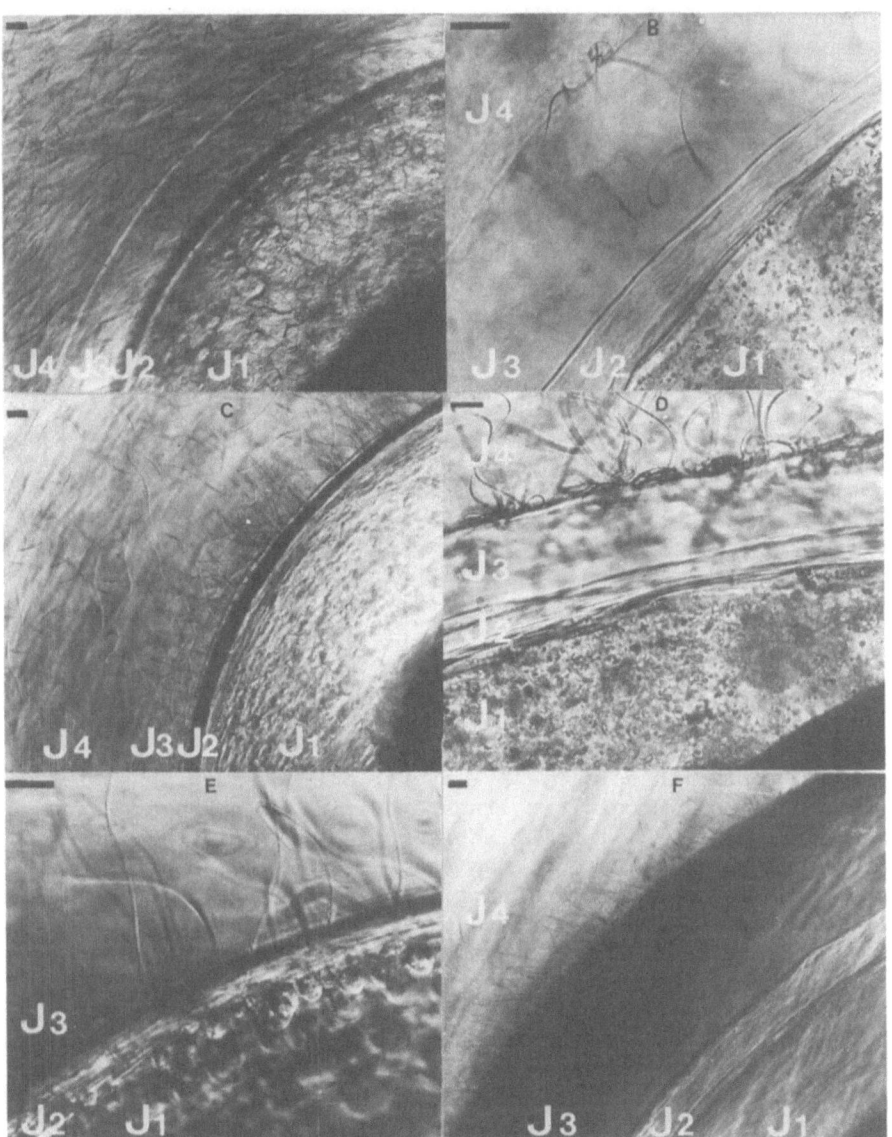

Figure 10. Sperm in the urodele egg jelly coat layers. A, C, E :
Pleurodeles; B, D : Ambystoma egg and sperm; F : Ambystoma
egg and Pleurodeles sperm.   Sperm were able to penetrate
all jelly coat layers of freshly laid eggs (A,B). About
1 h (Pleurodeles) or 20 min (Ambystoma) after laying,
sperm were stopped at the J_1-J_2 limit in Pleurodeles (C
and E) or J_3-J_4 limit in Ambystoma (D). Pleurodeles sperm
were also stopped at the J_3-J_4 limit in Ambystoma egg
jelly (F).

Xylose applied to Ambystoma eggs (50 mM) delayed the establish-ment of the block to sperm entry but did not suppress it. With Ambystoma eggs, 120 min after laying in water containing 50 mM xylose a few sperm reached the inner jelly coat layer J_1 in contrast to what happened in water alone.

ADDITIONAL RECENT RESULTS

We recently found a lectin in the posterior part of the oviduct of Notoplitalmus viridesiens. The most potent inhibitory saccharide was heparin.

DISCUSSION

Most of the results on chemical analysis of Pleurodeles egg jellies do not throw much light on their possible interacting proper-ties. However, the great amount and heterogenous distribution of carbohydrates in the jelly coat layers suggest that they could play a physiological function. Fucose, which is in a terminal position in glycoprotein structure, could be of special importance. Moreover, the fucosyltransferase which is necessary for the fucose incorpora-tion into glycoproteins, has a particular distribution in the oviduct cells and is β-estradiol dependent (20).

The reactions between jelly layers J_1 and J_2 of all the urodeles studied (L_1-l_1), which were first demonstrated on oviduct secretion products and confirmed using jelly layers, are not due to simple ionic interactions. Precipitin lines appearing on agarose plates indicate that, as in the case of antigen-antibody reactions or lectin-carbohydrates reactions, the reactants must have particular concentrations to optimally interact. The participation of a lectin in this reaction remains possible even if J_1 products (including L_1 products) have no hemagglutination activity in, for example, Amby-stoma or Triturus. The existence of a very complicated inhibitory oligosaccharide is possible as is the case for some plant lectins (10,45). A precipitation reaction was not observed with the oviduct secretory products of anurans. However, there are similarities with CGL (cortical granule lectin) –J_1-VE (vitelline envelope) inter-actions demonstrated in Xenopus laevis (46). In both cases, a sul-phated glycoprotein is implicated from J_1 in anurans (4) and from J_2 in urodeles (22). In both cases, the reaction products become progressively insoluble with mercaptans.

In anurans, these reactions are responsible for the formation of the fertilization envelope; in urodeles, a lectin-ligand reaction could be responsible for the dense zone which appeared between J_1 and J_2 (Fig. 2). The idea of a homology between perivitelline space of anurans and capsular chamber of urodeles was advanced by Salthe (36) after comparative histochemical observations. Anuran egg orientation

is possible only after fetilization, when a fluid space appears
around the egg plasma membrane as a consequence of vitelline envelope
elevation and fertilization envelope formation. Urodele egg orientation occurs independent of fertilization as it takes place inside the
inner jelly layer. Rotation of the urodele egg in the perivitelline
fluid can occur, but it is much slower and therefore, it is less
likely to be physiologically important (36).

In anurans and urodeles the fluidity of the perivitelline space
or inner jelly layer J_1 increases as a result of water entry into
these regions. This is possible because of the appearance of a
compartment in which proteins and other molecules are confined,
thereby generating an osmotic pressure differential which causes an
influx of water. We believe that the anuran vitelline envelope alone
or urodele J_2 alone are not able to constitute such an osmotic
barrier. The osmotic barrier could be formed bya J_1-J_2 reaction in
urodeles and CGL-J_1-VE reaction in anurans.

A significant feature of the lectins characterized in urodele
egg jelly is their specificity. They are all inhibited by D-glucose
derivatives and are different from anuran lectins which recognize D-
galactose derivatives better (34,35,46). The Ambystoma and Pleurodeles lectins recognize D-glucosamine better than N-acetylated
glucosamine. This is an unusual result even with plant lectins.
However, a human platelet lectin has a similar preference for non
acetylated hexosamines (9). The uniqueness of the Hynobius lectin
specificity could be related to the fact that it is the most primitive species we have studied.

In many groups, including anurans, the earliest mechanism acting
as a block to polyspermy is a rapid depolarization of the egg plasma
membrane (7,13). This fertilization potential dependent early block
to polyspermy was first demonstrated in sea urchin eggs (17). Our
results clearly indicate that such a membrane depolarization does not
occur in the two urodeles studied (Pleurodeles and Ambystoma). We
have also demonstrated that even if a modification of the membrane
potential is imposed (voltage clamp experiments), sperm entry is not
blocked; fertilization is possible even when the membrane potential
is held at +40 mV. So in these species, the lack of an early block
to sperm entry is not a simple consequence of the absence of a
fertilization potential but rather a consequence of the properties of
the egg membrane which does not contain the voltage dependent system
able to block the sperm entry in other organisms.

It is surprising to observe that the late block to sperm entry
does not appear in the same jelly coat layer in the urodeles jellies.
Sperm entry block seems to appear always at the limit between the
jelly layer which contains the lectin and the neighboring jelly
layer: J_3-J_4 for Ambystoma and J_1-J_2 for Pleurodeles. We have not
yet succeeded in demonstrating a direct interaction between the puri-

fied lectins and the sperm membrane. The sperm barrier may result from an interaction of the lectin and carbohydrates of neighboring jelly layers. However, experimental evidence is lacking. Hydration of egg jellies has been advanced as a possible cause of the late block (8,25,31). It is our opinion that hydration slows down the sperm in the outer jelly layer and so decreases the number of sperm reaching the inner layer. But it seems difficult to imagine that hydration alone establishes such a localized barrier (sometimes between J_1-J_2, sometimes between J_3-J_4).

In urodeles, the late block does not prevent polyspermy, but an increase in the number of sperm entering in the egg results in abnormal development and death of embryos (25). So, in urodeles, the late block seems to be more important in ensuring embryo development rather than having a role in fertilization.

Differences between urodeles and anurans exist in the organization of egg jelly coats and fertilization. However, some of these differences could be more a consequence of variation in the sequence of events than in the nature of the mechanisms involved. Comparable mechanisms may exist in anurans and urodeles: the formation of a fluid chamber around eggs and embryos and perhaps a lectin dependent protection of them. The essential difference could be the appearance in anurans of signals from the egg or transformations of egg integuments which arise as a result of fertilization.

ACKNOWLEDGEMENTS

We are grateful to Dr. D. Webb, Dr. R. Dohmen and Dr. B. Jego for criticisms on our laborious translation of the manuscript. We also thank Mrs. M. R. Allo, A. Bernard, G. Bonnec and M. M. J. Le Lannic, D. Blanchet, and P. Souchet for their technical assistance.

REFERENCES

1. Barbieri, F.D., Raisman, J.S. 1969. Non gamete factors involved in the fertilization of Bufo arenarum ovocytes. Embryologica 10:363-372.
2. Barbieri, F.D., Oterino, J.M. 1972. A study of the diffusible factor released by the jelly of the egg of the toad, Bufo arenarum. Develop. Growth Different. 14:107-117.
3. Bataillon, E. 1919. Analyse de l'activation par la technique des oeufs nus et la polyspermie expérimentale chez les Batraciens. Ann. Sci. Nat. Zool. 10:1-38.
4. Birr, C. and Hedrick, J.L. 1979. Immunological identification of the jelly coat ligand for the Xenopus laevis cortical granule lectin. Fed. Prod. 38:1252.

5. Boisseau, C., Jego, P., Joly, J., and Picheral, B. 1974. Organization et caracterisation histochimique des gangues ovulaires
   sécrétées par l'oviducte de Pleurodeles waltlii Michah.
   (Amphibien, Urodèle, Salamandridé). C. R. Soc. Biol. (Paris)
   168:1102-1107.
6. Charbonneau, M., Moreau, M., Picheral, B., Vilain, J.P., and
   Guerrier, P. 1983. Fertilization of amphibian eggs: a comparison of electrical responses between anurans and urodeles.
   Dev. Biol. 98:304-318.
7. Cross, N.L. and Elinson, R.P. 1980. A fast block to polyspermy in
   frogs mediated by changes in the membrane potential. Dev.
   Biol. 75:187-198.
8. Del Pino, E.M. 1973. Interactions between gametes and environment
   in the toad Xenopus laevis (Daudin) and their relationship to
   fertilization. J. Exp. Zool. 185:121-132.
9. Gartner, T.K. 1984. The endogenous platelet lectin and platelet
   aggregation, in: "Endogenous lectins," I.N.S.E.R.M. -C.N.R.S.
   Congress, Aussois, France, 18-24 mars.
10. Goldstein, I.J., and Hayes, C.E. 1978. The lectins:
    carbohydrate-binding proteins of plants and animals, in:
    "Advances in carbohydrate chemistry and biochemistry," Tipson,
    R.S. and Horton, D., eds., Vol. 35:127-340, Acad. Press.
11. Grey, R.D., Wolf, D.P., and Hedrick, J.L. 1974. Formation and
    structure of the fertilization envelope in Xenopus laevis.
    Dev. Biol. 36:44-61.
12. Grey, R.D., Working, P.K. and Hedrick, J.L. 1976. Evidence that
    the fertilization envelope blocks sperm entry in eggs of
    Xenopus laevis: interaction of sperm with isolated envelopes.
    Dev. Biol. 54:52-60.
13. Grey, R.D., Bastiani, M.J., Webb, D.J. and Schertel, E.R. 1982.
    An electrical block is required to prevent polyspermy in eggs
    fertilized by natural mating of Xenopus laevis. Dev. Biol.
    89:475-484.
14. Hagiwara, S., Ozawa, S., and Sand, O. 1975. Voltage clamp analy-
    sis of two inward current mechanism in the egg cell membrane
    of starfish. J. Gen. Physiol. 65:617-644.
15. Humphries, A.A. 1966. Observations on the deposition, structure
    and cytochemistry of the jelly envelopes of the egg of the
    newt, Triturus viridescens. Dev. Biol. 13:214-230.
16. Humphries, A.A.J., and Hughes, W.N. 1959. A study of the polysac-
    charide histochemistry of the oviduct of the newt, Triturus
    viridescens. Biol. Bull. 116:446-451.
17. Jaffe, L.A. 1976. Fast block to polyspermy in sea urchin eggs is
    electrically mediated. Nature (London) 261:68-71.
18. Jego, P. 1974. Composition en glucides des differents segments de
    l'oviducte et des gangues ovulaires chez Pleurodeles waltlii
    Michah. (Amphibien, Urodèle). Comp. Bioch. Physiol. 47:435-
    446.

19. Jego, P. 1976. Analyse des protides des gangues ovulaires de
    Pleurodeles waltlii Michah. (Amphibien, Urodèle). Ann. Biol.
    Anim. Bioch. Biophys. 16:13-24.

20. Jego, P. 1977. Action differentielle de l'oestradiol sur les
    activités de la fucosyltransférase et de la galactosyltrans-
    férase de l'oviducte du triton pleurodèle. Gen. Comp.
    Endocr., 31:475-481.

21. Jego, P., Joly, J., and Boisseau, C. 1980. Les gangues ovulaires
    des Amphibiens (protéines sécrétées par l'oviducte) et leurs
    rôles dans la fécondation. Reprod. Nutr. Dévelop. 20(2):557-
    567.

22. Jego, P. Chasnel, A., Lerivray, H., and Le Tallec, H. 1983.
    Caractéristiques des réactions de précipitations entre les
    produits de sécrétion de l'oviducte du Pleurodèle: identifi-
    cation d'une lectine. Reprod. Nutr. Dévelop. 23(3):537-552.

23. Katagiri, C. 1966. Fertilization of dejellied uterine toad eggs
    in various experimental conditions. Embryologica 9:159-169.

24. Katagiri, C. 1973. Chemical analysis of toad egg jelly in rela-
    tion to its "sperm capacitating" activity. Develop. Growth
    Different. 15:81-92.

25. MacLaughlin, E.W., and Humphries, A.A. Jr. 1978. The jelly envel-
    opes and fertilization of eggs of the newt, Notophthalmus
    viridescens. J. Morphol. 158:73-90.

26. Metz, C.B. 1967. Gamete surface components and their role in
    fertilization, in: "Fertilization: Comparative Morphology,
    Biochemistry and Immunology," C.B. Metz and A. Monroy, Eds.,
    Acad. Press, New York, Vol. 1:163-236.

27. Nowak, T.P., Haywood, P.L., and Barondes, S.H. 1976. Develop-
    mentally regulated lectin in embryonic chick muscle and a
    myogenic cell line. Biochem. Biophys. Res. Commun. 69:621-
    627.

28. Oliphant, G., and Hedrick, J.L. 1971. Isolation and physico-
    chemical characterization of a sperm capacitation factor from
    the jelly coat of Xenopus laevis eggs. Fed. Proc. 30:1280.

29. Osawa, T., Terao, T., Matsumoto, I., and Imbe, K. 1973. Affinity
    chromatography of carbohydrate-binding proteins, in: "Actes
    du colloque international du CNRS sur la methodologie de la
    structure et du metabolisme des glycoconjuges," CNRS ed., p.
    725-731.

30. Ouchterlony, O. 1949. Antigen-antibody reactions. Acta Pathol.
    Microbiol. Scand. 26:507-515.

31. Picheral, B. 1977a. La fécondation chez le triton Pleurodèle. I.
    La traversée des enveloppes de l'oeuf par les spermatozoïdes.
    J. Ultrastructure Res. 60:106-120.

32. Picheral, B. 1977b. La fécondation chez le triton Pleurodèle.
    II. La pénétration des spermatozoïdes et la réaction locale de
    l'oeuf. J. Ultrastructure Res. 60:181-202.

33. Richard, M., and Louisot, P. 1974. Glycoprotein biosynthesis in
    splenic cells. IV. Subcellular localization of glycosyltrans-
    ferases. Biochime 56:1381-1385.

34. Roberson, M.M., and Barondes, S.M. 1982. Lectin from embryos and oocytes of Xenopus laevis: purification and properties. J. Biol. Chem. 257:7520-7524.
35. Sakakibara, F., Takayanagi, G., Kawauchi, H., Watanabe, K., and Hakomori, S. 1976. An anti-a-like lectin of Rana catesbiana eggs showing unusual reactivity. Biochem. Biophys Acta 444:386-395.
36. Salthe, S. 1963. The egg capsules in the amphibia. J. Morph. 113:161-171
37. Schmell, E.D., Gulyas, B.J., and Hedrick, J.L. 1983. Egg surface changes during fertilization and the molecular mechanism of the block to polyspermy, in: "Mechanism and control of animal fertilization," F.J. Hartmann, Ed., Acad. Press p. 365-413.
38. Shivers, C.A., and James, J.M. 1970. Capacitation of frog sperm. Nature, 227:183-184.
39. Subtelny, S., and Bradt, C. 1961. Transplantations of blastula nuclei into activated eggs from the body cavity and from the uterus of Rana pipiens. II. Development of the recipient body cavity eggs. Develop. Biol. 3:96-114.
40. Vilter, V. 1967. Histochimie des gaines ovulaires chez le triton aspestre. C. R. Soc. Biol. 161:63-67.
41. Vorps, M.M., and Elinson, R.P. 1979. Extraction of oviducal materials important in the fertilization of Rana pipiens eggs. Biol. Reprod. 20 (suppl. 1):91A
42. Wallace, R.A., Jared, D.W., and Sega, M.W. 1973. Protein incorporation by isolated amphibian oocytes. III. Optimum incubation conditions. J. Exp. Zool. 184:321-334.
43. Wartenberg, M., and Schmidt, W. 1961. Elektronenmicroskopische Untersuchungen der Strukturellen Veranderungen im Rindenbereich des Amphibianeies im Ovar und nach des Befruchtung. Zeit. Zellforsch. 54:118-146.
44. Wolf, D.P. and Hedrick, J.L. 1971. A molecular approach to fertilization. III. Development of a bioassay for sperm capacitation. Develop. Biol. 25:360-376.
45. Wright, C.S. 1980. Crystallographic elucidation of the saccharide binding mode in wheat germ agglutinin and its biological significance. J. Mol. Biol. 141(3):267-291.
46. Wyrick, R.E., Nishihara, T., and Hedrick, J.L. 1974. Agglutination of jelly coat and cortical granule components and the block to polyspermy in the amphibian Xenopus laevis. Proc. Nat. Acad. Sci. USA 71:2067-2071.

INDUCTION OF THE ACROSOMAL REACTION IN SPERM FROM

THE WHITE STURGEON, ACIPENSER TRANSMONTANUS

Gary N. Cherr[1] and Wallis H. Clark, Jr.[2]

Department of Obstetrics and Gynecology[1]
and the Bodega Marine Laboratory[2]
University of California, Davis, CA 95616

SUMMARY

Sperm from the white sturgeon, Acipenser transmontanus possess an acrosome while the eggs possess numerous micropyles. The sperm undergo an acrosome reaction that includes exocytosis and process formation. This acrosomal reaction can be induced in freshwater with ionophore A23187, high pH, high Ca^{++}, or egg water. In addition, this event is dependent on extracellular Ca^{++} and Mg^{++}. Upon immersion of eggs in freshwater, a water-soluble jelly layer hydrates and forms a highly adhesive coating surrounding the eggs. Jelly release and hydration appears dependent on both a trypsin-like protease and Ca^{++} and Mg^{++} ions. Isolated egg jelly (110,000 dalton glycoprotein) does not possess acrosome reaction inducing activity. The acrosome reaction inducing activity in egg water resides in a water-soluble 66,000 dalton glycoprotein. Egg water appears species specific in its ability to elicit a response. Prior to freshwater exposure, the egg envelope possesses a water-insoluble glycoprotein with a molecular weight of 70,000 daltons. When isolated from polyacrylamide gels, the 70,000 dalton component induces acrosome reactions in a species specific manner. Following freshwater exposure, the egg envelope possesses the water-soluble 66,000 dalton glycoprotein in addition to the 70,000 dalton component. The 70,000-66,000 dalton conversion can be blocked by incubating eggs in freshwater containing inhibitors of trypsin-like proteases. Both the 66,000 dalton and 70,000 dalton glycoproteins originate and reside in layer three of the egg envelope.

INTRODUCTION

The acrosome reaction that occurs in many animal sperm involves exocytosis of the acrosomal contents and, in many higher invertebrate species, includes the formation of an acrosomal process or filament. In most vertebrates however, the acrosome reaction involves only exocytosis; an acrosomal process is not formed. This event is usually triggered by components of the egg and is considered necessary for the sperm to traverse the extracellular coats which surround many animal eggs.

Typically, species which possess sperm that lack an acrosome, accordingly, possess eggs that lack investments or have vestments perforated by channels (micropyles). For example, most fish sperm do not possess an acrosome since they are provided unobstructed passage to the egg's plasma membrane by the micropyle, a funnel-shaped perforation in the investment coats that surround fish eggs.

In this paper, we will discuss investigations on the gametes from a primitive vertebrate, the white sturgeon (Acipenser transmontanus). Early reports indicated that sturgeon sperm possessed acrosomes and exhibited acrosomal processes while the eggs possessed not one, but numerous micropyles (see 9 for a review). The presence of acrosomes in the sperm and micropyles in the eggs of sturgeon seems to contradict many ideas on sperm and acrosomal evolution (1,2). In this paper we describe the fine structure of the white sturgeon sperm and its acrosomal reaction, the physiological parameters of this event and the egg envelope component that induces the reaction.

MATERIALS AND METHODS

Obtaining sturgeon is quite different from that of most experimental animals. These large, anadromous fish (>30 Kg) are collected in San Francisco Bay, California prior to their winter spawning migration up the Sacramento River. Sturgeon are caught with hook and line, transferred to a boat in which they are assessed (surgically) for gonadal maturity. If the animals are suitable for spawning, they are acclimatized to fresh water and induced to ovulate or spermiate with carp pituitary extract. Males can be repeatedly induced to spermiate throughout an 8 mo holding period (5). Ovulated eggs are collected from the coelomic cavity of the female (4). The preparation of gametes for electron microscopy has been previously described (3,5).

Egg water is collected by incubating eggs in freshwater (water with a known ionic composition) for 30 min (5). The water is then collected and centrifuged at 30,000 g for 30 min and the supernatant solution is used in subsequent experiments. It should be noted that sturgeon eggs do not activate when they are exposed to freshwater;

thus the presence of egg components in the surrounding medium is
independent of egg activation and the cortical reaction (4,6).

Sturgeon egg jelly can be solubilized by treating eggs that
possess hydrated jelly (eggs that have been exposed to freshwater)
with 4 M urea in a Tris-HCl buffer at pH 7.8 (4). This soluble
jelly is then centrifuged to remove particulates and the supernatant
is dialyzed against freshwater in an ultrafiltration cell with a
1000 molecular weight cut-off (MWCO) filter (6).

Induction of the acrosomal reaction

A. transmontanus sperm are only motile upon dilution in freshwater. This motility lasts for approximately 4 min. In order to
score for acrosome reactions, sperm are diluted into the appropriate
medium (See Table 1) and fixed following a 2 min incubation. Acrosomal process formation is scored using phase contrast microscopy
with a 40 x objective lens (5).

Gel electrophoresis

Sodium dodecyl sulfate polyacrylamide gel electrophoresis (SDS-
PAGE) was carried out as previously described (5), the gels being
silver stained for both protein and carbohydrate. Protein was
isolated from gels by scanning slab gels at 280 nm, cutting out the
appropriate band(s) and electroeluting and electrodialyzing the
protein in dialysis tubing as previously described (6).

RESULTS AND DISCUSSION

Morphology

The morphology of sturgeon sperm had been reported in early
light microscopic and electron microscopic studies of whole-mounted
cells (8,9); however, the intracellular fine structure and the
morphology of the acrosome reaction has only been recently described
(5). The sperm of A. transmontanus is an elongate cell that possesses a $10\mu m$ long head with a conspicuous acrosome, a midpiece, and a
flagellum (Fig. 1). The sperm's acrosome is a scalloped cap on the
apical end of the cell and consists of a bell-shaped acrosomal
vesicle and a subacrosomal region (Fig. 2). The subacrosome
consists of two regions: one immediately posterior to the acrosomal
vesicle that contains 6 nm diameter filaments and a more lateral
region that apears granular. The filamentous material in the
subacrosome extends posteriorly, traversing the nucleus in three
membrane-lined, helically oriented channels (Figs. 1,2). These
channels are lined by nuclear membrane and terminate at the midpiece
(5).

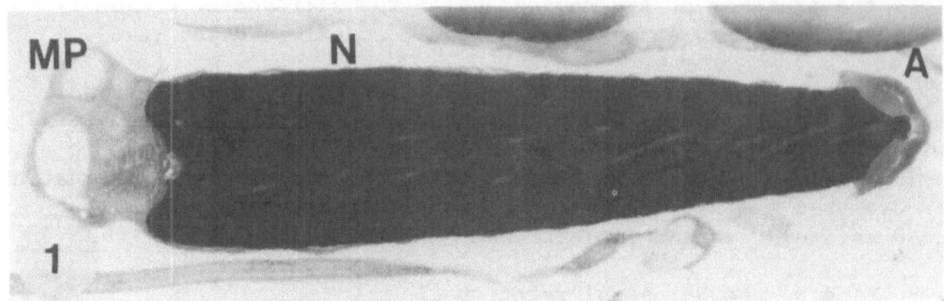

Figure 1. A transmission electron micrograph of an unreacted A.
          transmontanus sperm. A, acrosome; N, nucleus; MP, mid-
          piece; C, channels. X18,500 (5).

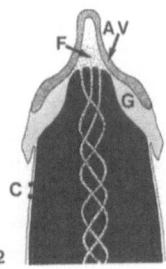

Figure 2. Schematic drawing of the acrosomal region. AV, acrosomal
          vesicle; F, filamentous material in subacrosome and chan-
          nels; G, granular material in lateral margins of subacro-
          some; C, channels in nucleus (5).

     The sturgeon sperm acrosome reaction can be induced with the
inophore A23187, high Ca++, high pH, or egg water, and results in
acrosomal exocytosis and process formation (5). The acrosomal
process is 10-12 μm in length and extends anteriorly from the
subacrosome; however, the rigidity of this process is rapidly lost.
The A. transmontanus sperm before and after the acrosome reaction is
shown in Figs. 3A and 3B. The acrosomal process appears to form
from the filamentous material in the subacrosome and channels. In
addition, actin can be isolated from a preparation of sperm heads
(nucleus and intact acrosome) and is a major component of whole
sperm (5).

Physiology of the acrosomal reaction

     Some of the physiological parameters of an acrosome reaction
occurring in freshwater have been examined (5). The effects of

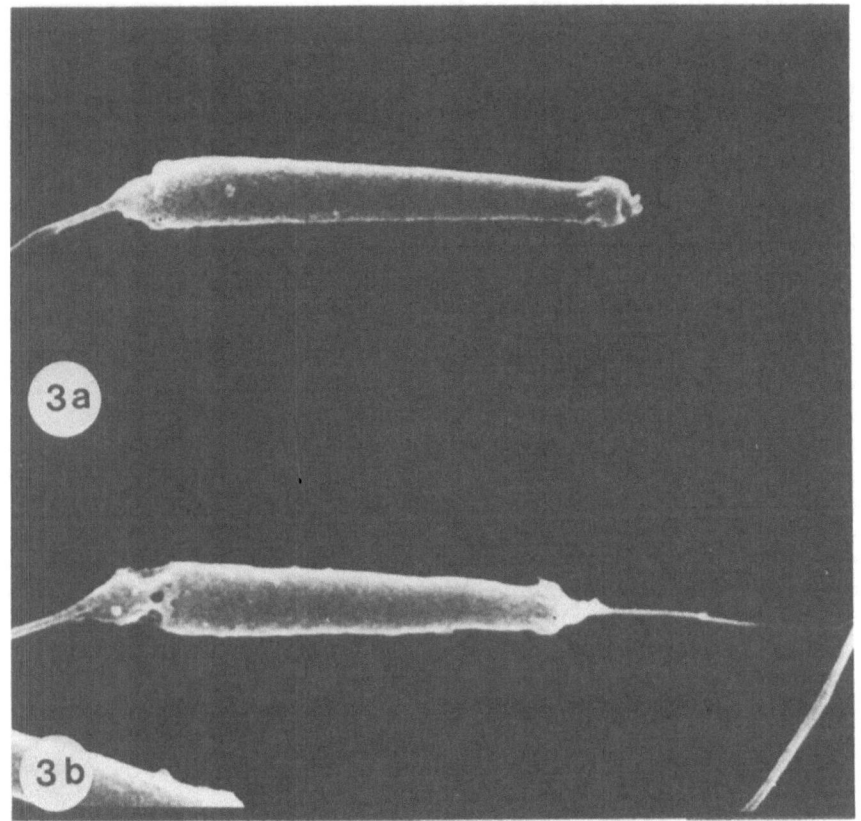

Figure 3a. Scanning electron micrograph of A. transmontanus sperm
           prior to induction of the acrosomal reaction. X10,000.

Figure 3b. A. transmontanus sperm following induction of the acro-
           somal reaction. Note the acrosomal proces. X10,000.

various ion specific media, with and without inophore A23187, on
sperm are presented (Table 1). When sperm are diluted into artificially prepared freshwater (1.0 mM NaCl; 0.4 mM KCl; 0.5 mM $CaCl_2$;
0.05 mM $MgSO_4$; pH 7.6), very few sperm undergo the acrosome reaction. However, when A23187 is present, a high percentage of the
sperm form acrosomal processes. An absence of Ca^{++} and Mg^{++} ions
from the extracellular medium results in an inhibition of inophore-
induced reactions. Deleting only Ca^{++} or Mg^{++} from the medium
results in a partial decrease in the percentage of inophore-induced
reactions. In addition, increasing the Ca^{++} concentration of the
extracellular medium (up to 30 mM) induces acrosome reactions without A23187. However, increasing the extracellular concentration of
Mg^{++} (up to 30 mM) does not elicit a response. This data indicates

that Mg^{++} as well as Ca^{++} may be involved in the sturgeon sperm acrosome reaction; however, this point needs clarification.

It was also found that simply increasing the pH of the extra-cellular medium with either NaOH or NH_4OH induced acrosome reactions (Table 1). However, by decreasing the pH of the medium with HCl, the effect of the inophore was blocked. This data is consistent with data from other sperm that utilize actin in acrosomal process formation (10). In addition, this supports the morphological and molecular data from A. transmontanus sperm, indicating a role for actin in acrosomal process formation (5).

In general, the above data appears similar to data from other sperm in that the acrosome reaction in A. transmontanus sperm involves Ca^{++}. However, these data suggests that Mg^{++} may act synergistically with Ca^{++} to elicit the acrosome reaction. In other sperm, Mg^{++} appears to inhibit the acrosome reaction (7). This is of great interest and needs further research attention.

While a great deal of attention has been paid to the physiolog-ical controls of the acrosome reaction, relatively less information is available regarding "natural" inducers of this event. Before discussing data on the egg components responsible for eliciting the sturgeon sperm acrosome reaction, it is necessary to orient one to the structure of the white sturgeon egg and its investments.

Table 1. Percentage of acrosome reactions generated in A. transmontanus sperm by acrosome reaction inducers[1] (5).

Treatment	% acrosome reaction
Freshwater pH 7.6	6 ± 2.30
Freshwater + A23187	74 ± 13.22
Ca^{++} - Mg^{++} - free freshwater + A23187	6 ± 4.36
Ca^{++} - free freshwater + A23187	19 ± 1.73
Mg^{++} - free freshwater + A23187	19 ± 6.99
High Ca^{++} - freshwater (30 mM)	48 ± 2.90
High Mg^{++} - freshwater (30 mM)	6 ± 1.99
Freshwater pH 5.5 + A23187	9 ± 3.98
Freshwater pH 9.0	65 ± 13.32
Egg water	47 ± 2.51

[1] Percentages are expressed as means (n = 3) ± S.D. The concentra-tion of A23187 was 200 $\mu g/ml$.

Egg and investment morphology

The mature white sturgeon egg is 4 mm in diameter, is heavily pigmented, and is surrounded by a multilayered envelope (four layers) that is 50 μm in diameter at the time of spawning (3). Following immersion of the eggs in freshwater, a water-insoluble jelly layer hydrates to form a highly adhesive coat that surrounds the egg and anchors it to most substrates (4). A schematic drawing of the egg before and after freshwater exposure is presented in Figure 4. The morphology of the egg envelope before and after freshwater exposure (30 min) is shown in Figures 5A, 5B, 6A, and 6B. Prior to freshwater exposure, the jelly resides in the ducts and on the surface of the electron dense layer 3 (L3) (Fig. 5). Following freshwater exposure, the ducts of L3 expand and jelly is released and hydrates. Concomitantly, L3 loses electron density (4).

It has been determined that jelly release and hydration is dependent on the activity of a trypsin-like protease since specific inhibitors of trypsin activity (soybean trypsin inhibitor, SBTI; benzamidine hydrochloride, N-α-p-tosyl-L-lysine chloromethyl ketone, TLCK) block this event (Table 2) (4). In addition, these inhibitors block the loss of electron density of L3. Inhibitors of chymotrypsin activity (2-nitro-4-carboxyphenyl-N,-N-diphenylcarbamate, NCDC; 1-tosylamide-2-phenylethyl-chlormethyl-ketone, TPCK) do not block jelly release and hydration or the loss of electron density in L3. In addition to the above, an absence of Ca^{++} and Mg^{++} from the media inhibits these events.

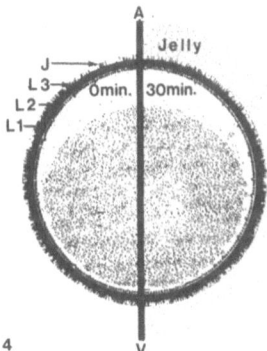

Figure 4. Schematic drawing of white sturgeon egg before (0 min) and after (30 min) freshwater exposure. Note the four layers of the egg envelope (jelly, L3,L2,L1). A, animal pole; V, vegetal pole.

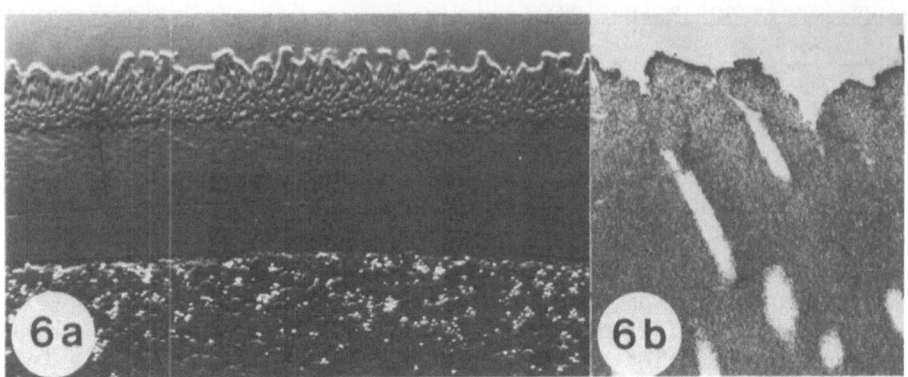

Figure 5a. Thick plastic section of egg envelope prior to freshwater exposure. J, jelly. X300 (4).

Figure 5b. Transmission electron micrograph of outer edge of the egg
envelope. Note the presence of jelly (J) on the surface
and in the ducts of the electron dense L3. X3,500 (4).

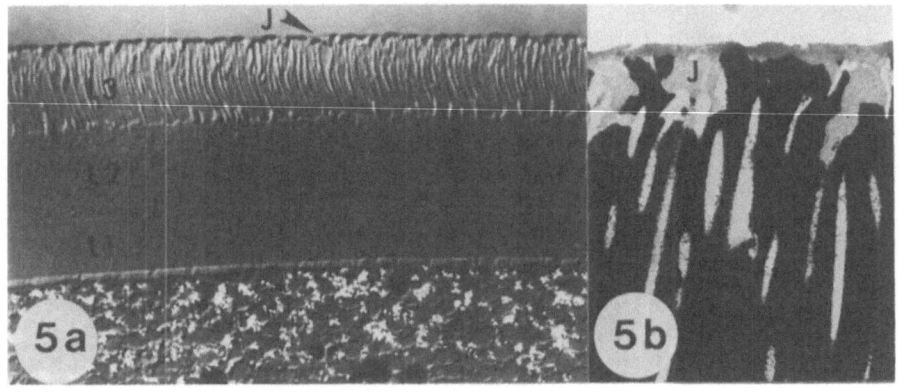

Figure 6a. Thick plastic section of egg envelope after exposure to
freshwater. Note the expansion of the ducts of L3. X300
(4).

Figure 6b. Transmission electron micrograph of the egg envelope
following exposure to freshwater. Jelly has been
released and has hydrated while L3 has lost electron
density. X4,000 (4).

Table 2. Jelly release and hydration.

Treatment	Jelly release
Freshwater (pH 7.6)	++
Ca^{++} - Mg^{++} - Free Freshwater (2 mM EDTA)	--
Freshwater + SBTI (0.01%)	--
Freshwater + Benzamidine (5 mM)	--
Freshwater + Aprotonin (0.01%)	--
Freshwater + TLCK (10 μm)	--
Freshwater + TPCK (up to 40 μm)	++
Freshwater + NCDC (up to 40 μm)	++

In order to contact the oolemma, sperm must traverse the egg envelope via micropyles. The micropyles (3-15) are located in a 100-200 μm region at the animal pole. These micropyles taper such that the inner opening (at the oolemma) is the diameter of a single sperm. L3 lines the outer 2/3 of the micropylar canal. The jelly only extends a short distance into the micropyle (3).

It is assumed that sperm must traverse the jelly layer, enter a micropyle that is lined with L3, undergo the acrosome reaction, and fuse with the egg in order for successful fertilization to occur. In many invertebrate species, a jelly component is purported to be the inducer of the acrosome reaction. Sturgeon egg jelly was solubilized with urea, dialyzed against freshwater, and assayed for acrosome reaction inducing activity. The dialyzed jelly consists of a 110,000 dalton (KD) glycoprotein that is 89% protein and 11% carbohydrate, of which 4% is sialic acid (4). Jelly is the only layer of the egg envelope that contains sialic acid residues. When dialyzed jelly was incubated with sperm at increasing protein concentrations, acrosome reactions were not elicited even at 100 μg/ml of jelly protein (Fig. 7) (6).

Sturgeon egg water has been reported to elicit acrosome reactions in sperm from at least two species of sturgeon (5,6,8). The effect of increasing egg water protein concentrations on A. transmontanus sperm is shown in Figure 8. In addition, Great Lake sturgeon (Acipenser fulvescens) sperm were incubated in white sturgeon egg water in order to determine specificity.

As shown in Figure 8, A. transmontanus egg water does not elicit acrosome reactions in A. fulvescens sperm even at maximum protein concentrations. It should be noted that A. fulvescens sperm will form acrosomal processes in the presence of A23187 or high pH freshwater.

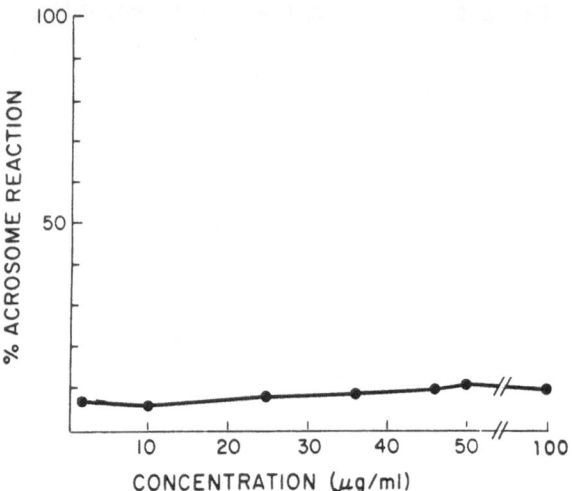

Figure 7. A. transmontanus sperm incubated at increasing protein
         concentrations of dialyzed egg jelly do not undergo the
         acrosomal reaction.

     Electrophoretically, egg water consists of three major pro-
teins at 66KD, 23KD, and 20KD (Fig. 9). Only the 66KD band stains
positively for carbohydrate, indicating it is a glycoprotein. Egg
water contains 89.2% protein and 10.8% carbohydrate. It does not
contain sialic acid residues or a 110KD glycoprotein, both of which
are components of jelly. When egg water is fractionated on a
Sepharose CL6B column, the acrosome reaction inducing activity
corresponds to the fractions that contain the 66KD glycoprotein. In
addition, the 66KD glycoprotein was isolated from the gels (see
Materials and Methods) and incubated with sperm. Acrosome reactions are elicited at comparable levels to that of egg water (6).

     In order to determine the origin of the 66KD inducer from egg
water, envelope preparations from eggs prior to (0 min) and following (30 min) freshwater exposure were analyzed (6).

Origin of the inducer

     SDS-PAGE of egg envelopes prior to freshwater exposure demon-
strates that the 66KD glycoprotein is not present; however, a 70KD
glycoprotein is present and is the single most prominant band (Fig.
10). Following freshwater exposure the 66KD glycoprotein is apparent in addition to the 70KD glycoprotein (Fig. 10) (6).

     The 70KD band was cut from the gels of 0 min egg envelopes and
electroeluted and electrodialyzed. When incubated with A. transmontanus and A. fulvescens sperm at increasing protein concentra-

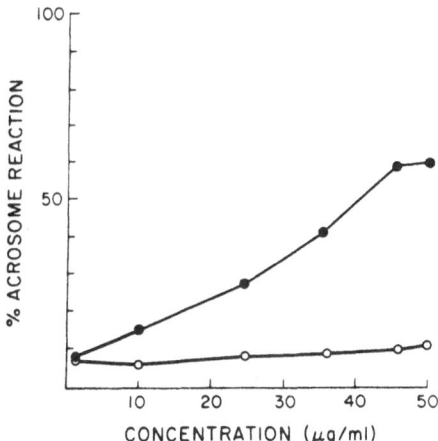

Figure 8. Effect of increasing concentrations of egg water protein
on <u>A</u>. <u>transmontanus</u> (0) and <u>A</u>. <u>fulvescens</u> (0) sperm (6).

tions, a response was only elicited in <u>A</u>. <u>transmontanus</u> sperm,
indicating that the 70KD glycoprotein is species specific in its
ability to elicit a response (Fig. 11). The 70KD glycoprotein is
water insoluble and was only solubilized in these experiments by
the denaturing conditions of SDS-PAGE (6).

It was determined that the 66KD inducer is proteolytically
derived from the 70KD glycoprotein in the 0 min egg envelope (6).
By incubating eggs in freshwater containing benzamidine, TLCK,
SBTI, or aprotinin, the appearance of the 66KD inducer in both egg
water and the egg envelope can be blocked. TPCK and NCDC do not
inhibit its appearance. The enzyme responsible for the 70KD-66KD
conversion may also be the same protease that is responsible for
jelly release. The enzyme(s) would appear to reside in the egg
envelope since the eggs have not been activated. While the 70KD
glycoprotein is capable of inducing acrosome reactions when it is
denatured (solubilized), it may not be capable of eliciting a response when it is in an insoluble form in the egg envelope. Apparently, the 70KD-66KD conversion enables a soluble form of the
inducer (the 66KD glycoprotein) to diffuse into the surrounding
medium and elicit acrosomal activation in a species specific
manner.

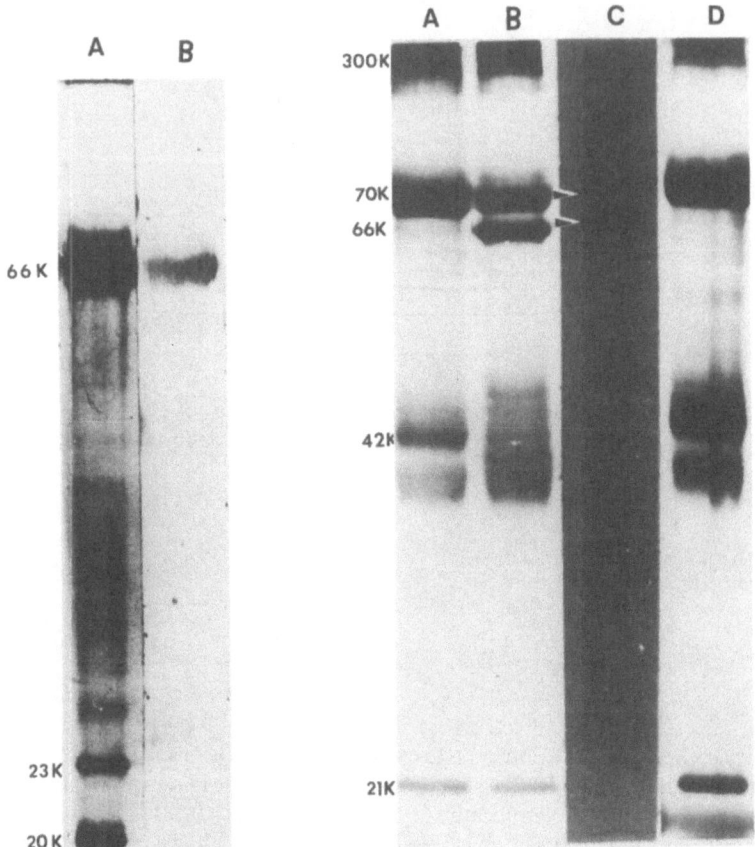

Figure 9. SDS–PAGE (10%) of 1.5 μg protein of egg water components silver stained for both protein (A) and carbohydrate (B). Note only the 66KD band stains positively for carbohydrate (6).

Figure 10. SDS–PAGE (10%) of egg envelope components. Prior to freshwater exposure, the 70KD band is most prominent (A). After freshwater exposure, the 66KD band is present (B). Both the 66KD and 70KD components silver stain positively for carbohydrate (C). Egg envelopes incubated in freshwater containing benzamidine, SBTI, Aprotonin, or TLCK lack the 66KD glycoprotein (D). Lanes A–C contain 1.75 μg protein; Lane D contains 2.5 μg protein (6).

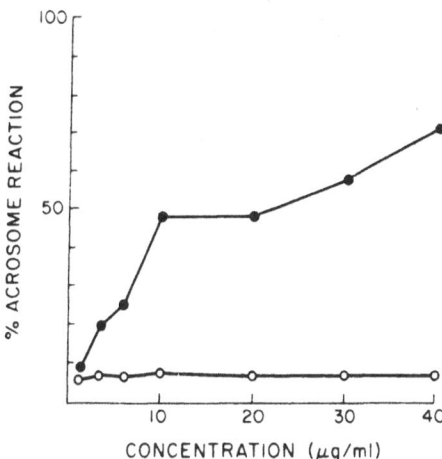

Figure 11. Effect of increasing concentrations of the 70KD glyco-
          protein on A. transmontanus (0) and A. fulvescens (0)
          sperm (6).

Localization of the inducer

     In order to localize the 66KD and 70KD glycoproteins in the
egg envelope, one can take advantage of their inherent marker
(carbohydrate). When egg envelopes were sectioned and stained for
carbohydrate with periodic-acid schiff, only L3 (and the jelly)
stain positively (Fig. 12) (6). L3, a galactose-containing layer,
stains positively for carbohydrate before and after fresh-water
exposure, and after urea extraction (Cherr, unpublished observation). Thus, it appears that the 66KD and 70KD glycoproteins
reside in L3 of the egg envelope.

     Fortuitously, L3 lines the micropylar canal and is situated
such that a sperm is 10-12 μ m away from the oolema when it is in
the L3 portion of the micropylar canal. This is the same length as
the sperm's acrosomal process. In addition, concentrations of the
66KD water-soluble inducer would probably be greatest within the L3
portion of the canal. Finally, the loss of rigidity of the acrosomal process could dictate the site of occurrence of the acrosome
reaction in successful sperm. If the acrosome reaction occurred
outside of a micropyle, the sperm would probably be unsuccessful.
Therefore, a requirement for successful sperm to undergo an acrosome reaction in a micropyle could explain the presence of numerous
micropyles. For example, unsuccessful, reacted sperm entering a
micropyle would exclude unreacted, potentially successful sperm
from gaining access to the micropyle. In this situation, numerous
micropyles would enable successful fertilization.

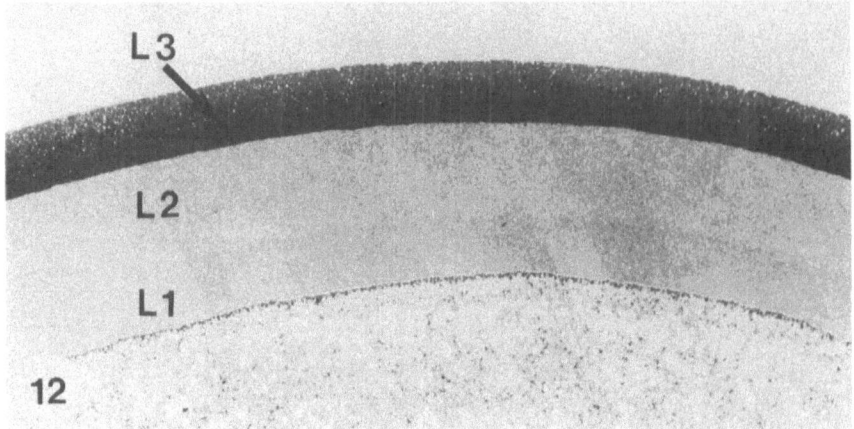

Figure 12. Periodic acid Schiff staining of an egg envelope follow-
          ing freshwater exposure. Only L3 stains positively.
          X300 (6).

CONCLUSIONS

At the present time it is unclear how sturgeon gametes fit
into schemes of gamete evolution and function. While acrosomes in
sperm and micropyles in eggs seem contradictory, data suggest that
the acrosomal reaction in sturgeon sperm may function as a species
specific block to fertilization as it does in many other species.
Since sturgeon are broadcast spawners, this could be of considerable importance. While it is apparent that the acrosome of sturgeon sperm is not involved in vestment penetration, the role of the
acrosome reaction and the acrosomal process in sperm-egg fusion is
unknown and deserves further attention.

Now that a baseline of knowledge has been obtained in this
system, we hope future studies may be revealing in the areas discussed in this paper.

ACKNOWLEDGEMENTS

The authors wish to thank Fred J. Griffin for his illustrations. Supported by USDC 81-ABH-001101, NMFS contract 80-ABC00017,
and NOAA, National Sea Grant College Program, Dept. Commerce
#NA80AA-D-00120. Through the California Sea Grant Program and
California State Resources Agency, Proj. RA45.

REFERENCES

1. Baccetti, B. and Afzelius B.A. 1976. The biology of the sperm cell. Monogr. Dev. Biol. 10:1-254.
2. Baccetti, B. 1979. The evolution of the acrosomal complex, in: "The Spermatozoon," D.W. Fawcett and J.M. Bedford, eds., Urban and Schwarzenber, Baltimore, pp. 305-329.
3. Cherr, G.N. and Clark, W.J., Jr. 1982. Fine structure of the envelope and micropyles in the eggs of the white sturgeon, Acipenser transmontanus. Richardson. Dev. Growth Differ. 24(4):341-352.
4. Cherr, G.N. and Clark, W.H., Jr. 1984a. Jelly release in the eggs of the white sturgeon, Acipenser transmontanus: An enzymatically mediated event. J. Exp. Zool. 230:145-149.
5. Cherr, G.N. and Clark, W.H., Jr. 1984b. An acrosome reaction in sperm from the white sturgeon, Acipenser transmontanus. J. Exp. Zool. 232:129-139.
6. Cherr, G.N. and Clark, W.H., Jr. 1984c. An egg envelope component induces the acrosome reaction in sturgeon sperm. J. Exp. Zool. 234:75-85.
7. Collins, F. and Epel, P. 1977. The role of calcium ions in the acrosome reaction of sea urchin sperm: Regulation of exocytosis. Exp. Cell Res. 106:211-222.
8. Detlaf, T.A. and Ginzburg, A.S. 1963. Acrosome reaction in sturgeons and the role of calcium ions in the union of gametes. Doklady An SSSR. 153(6):1461-1464.
9. Ginzburg, A.S. 1968. Fertilization in fishes and the problem of polyspermy, in: "Israel Program for Scientific Translations, Ltd.," Tel Aviv, 1972, T.A. Detlaf, ed., pp. 1-175.
10. Tilney, L.G., Kiehart, D.P., Sardet, C., and Tilney, M. 1978. Polymerization of actin IV: Role of Ca^{++} and H^+ in the assembly of actin and in membrane fusion in the acrosomal reaction of echinoderm sperm. J. Cell Biol. 77:536-550.

SPERM GLYCOSIDASE AS A PLAUSIBLE MEDIATOR OF SPERM BINDING TO THE

VITELLINE ENVELOPE IN ASCIDIANS

Motonori Hoshi

Department of Biology
Nagoya University
Chikusa, Nagoya, Japan[*]

SUMMARY

In the Ascidian, <u>Ciona intestinalis</u>, sperm α-L-fucosidase is concluded to be a recognition protein for the sperm receptor in the vitelline envelope. The spermatozoa bind to the vitelline envelope probably by forming rather stable substrate (α-L-fucoside)-enzyme complex at many spots. Experimental evidence indicates a good correspondence in the properties between receptor binding activity and α-L-fucosidase activity such as the effects of saccharides, of pH and of monoclonal antibodies raised against sperm α-L-fucosidase, and locus in the sperm. A general mechanism involving an enzyme as a recognition protein for the cell to cell binding is also discussed in relation to the molecular evolution of the recognition proteins.

INTRODUCTION

Fertilizing spermatozoa bind to the vitelline envelope as an obligatory step for fertilization in various animals. The binding is species-specific and achieved by the association between sperm receptors at the outer surface of vitelline envelope and recognition proteins on the surface of sperm head (for a review see 14). No sperm receptors have been purified till now except ZP-3, a glycoprotein isolated from the zona pellucida of mouse eggs (1). The sperm receptors have been claimed to have saccharide chains as the ligands, although their structures have not yet been elucidated. Terminal L-fucose residue is reported to be an indispensable part of

[*]Present Address: Department of Biology, Tokyo Institute of
                     Technology, Meguro, Tokyo, Japan

the ligand in the Ascidian, Ciona intestinalis (2,3,15,18), in horseshoe crab (20), and in some mammals (12). Participation of D-
N-acetylglucosamine moiety in the binding is implied in the mouse
(23, see also B. Shur, this volume) and in the Ascidian, Phallusia
mammillata (5).

     Sea urchin bindin is the first recognition protein in the sper-
matozoa that has been isolated as a single protein entity (24). In
sea urchins, only acrosome-reacted spermatozoa bind to the vitelline
coat with bindin, a lectin-like protein exposed as a result of the
acrosome reaction (4,24). The bound spermatozoa then start boring
through the vitelline coat with the aid of lysins (7). In contrast,
only unreacted spermatozoa bind to the vitelline coat in the mouse
(23) and in Ciona (2). In these animals, therefore, the recognition
molecule for sperm receptor has to exist at the sperm surface prior
to the acrosome reaction.

     We found the activities of several glycosidases in the acrosome-
unreacted spermatozoa of various marine invertebrates including the
Ascidians in the course of searching for lysins (Hoshi, unpublished
data). Most of them do not seem to relate directly to sperm penetration and they have optimum pHs in the acidic region. If a lectin at
the surface of acrosome-unreacted spermatozoa binds to the vitelline
envelope, the binding will be pretty much static because there may be
no agents that modify or remove the binding complex and because the
vitelline coat is merely a coat of glycoproteins apart from egg
plasma membrane. What happens actually is that the binding is a
temporary event followed by dynamic changes in the gametes. Thus we
thought that, in the Ascidians, sperm glycosidase might be a better
candidate for a recognition protein than the lectin-like proteins
such as bindin.

     We have examined such possibilities by using Ciona intestinalis,
in which sperm binding could be analysed independent of the events
leading to the acrosome reaction and sperm penetration by using gly-
cerinated eggs (2,17). This paper summarizes the experimental
evidence for our hypothesis that α-L-fucosidase is a recognition
protein for the sperm receptor in Ciona intestinalis (6,8-11,13).

EXPERIMENTAL

     Gametes   Gametes of Ciona intestinalis from the Gulf of Naples
and the Pacific Coast of Japan were collected as described previously
(17). Spermatozoa of Phallusia mammillata were also used for enzyme
assays.

     Enzyme Assay   The activities of various glycosidases were
fluorometrically assayed at 25°C by using 4-methylumbelliferyl glycosides (4-MU glycosides) as the substrates (8,10). The intact sperma-

tozoa of Ciona and of Phallusia hydrolyzed various 4-MU glycosides in
acidic sea water. p-Nitrophenyl (p-NP) derivatives were also favorable substrates for sperm glycosidases. The activity of α-L-fucosi-
dase was much higher than any others in Ciona where L-fucose is suggested to be a ligand for sperm binding (2,3,15,18), while β-
D-N-acetylglucosaminidase was the strongest glycosidase in Phallusia
where D-N-acetylglucosamine is implied to be essential (5). The
optimal pH of Ciona α-L-fucosidase was 3.9 and the activity in normal
sea water was only 2% of the maximum. No appreciable loss of the
activity was found after one month storage of the spermatozoa at
-20°C. The enzyme was not extracted at all by washing spermatozoa
three times with sea water, but was quantitatively solubilized from
the frozen sperm by sonication or by extraction with Triton X-
100.

Purification and Characterization of α-L-Fucosidase in Ciona
Spermatozoa α-L-Fucosidase was extracted from the frozen spermatozoa by sonication. The enzyme was 5400-fold purified by a procedure including ammonium sulphate precipitation, gel filtration and
affinity chromatography using Sepharose-ϵ-aminocaproylfucosylamine
(8). The affinity column adsorbed the enzyme appreciably at pH 8.2,
but less than at 3.9. No contamination was detected in the purified
preparation by SDS-PAGE (silver staining), HPLC using TSK-Gel
G3000SW, or various enzyme assays. The molecular weight was 105K by
SDS-PAGE and 112K by HPLC. The enzyme activity was competitively
inhibited by 4-MU-β-L-fucoside as well as p-NP-β-L-fucoside; K_i for
4-MU-β-L-fucoside was comparable to K_m for 4-MU-α-L-fucoside. L-
Fucose, one of the products, was a poor inhibitor of the enzyme. No
other saccharides so far tested inhibited the enzyme activity. The
kinetic parameters for purified enzyme and for the activity in situ
were similar.

Binding Assay   Sperm binding to the vitelline envelope of
glycerol-treated eggs was assayed by the published method (18) with
slight modifications (10). In some experiments, the number of
spermatozoa along the optical cross-section of the vitelline envelope
was directly scored (11). Numbers obtained by the two methods were
reasonably similar.

The number of spermatozoa bound to the vitelline envelope was
variable using different gamete preparations, but the inhibition of
binding by saccharides was very reproducible (10). Aryl-L-fucosides
(both 4-MU and p-NP-L-fucosides, and both α and β anomers) remarkably
inhibited the sperm binding at 2.5 mM or less. In some experiments,
the binding was considerably inhibited by p-NP-L-fucosides at a
concentration as low as 125 μM. L-Fucose was also inhibitory, but it
was effective only at the concentration above 25 mM. No significant
inhibition of binding has been observed thus far with any other free
sugars or p-NP-glycosides including D-fucose and D-fucosides. Some

of the monoclonal antibodies against sperm α-L-fucosidase inhibited
the binding as will be shown in the following section.

The spermatozoa bound to the vitelline envelope in normal sea
water mostly detached within several minutes of decreasing the pH of
the mixture to 5 or 4 at 20°C, but no significant detachment was
observed at 0°C (10). These results are illustrated schematically in
Fig. 1.

Monoclonal Antibodies Against α-L-Fucosidase   Mouse monoclonal
antibodies have been raised against sperm α-L-fucosidase. Five monoclonal antibodies (1C7, 2B5, 2G2, 6E4 and 7F12) inhibit the sperm
binding, when the spermatozoa have been preincubated with one of them
before mixing with eggs. Antibody 2G2 agglutinate spermatozoa, which

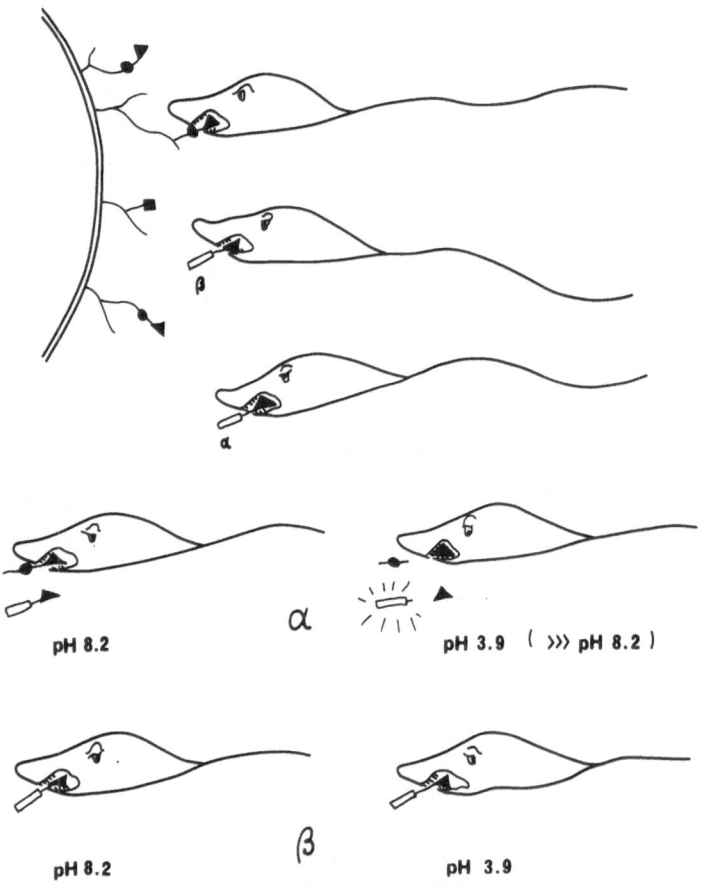

Figure 1. Cartoon showing the sperm binding to the vitelline envelope
          and the effects of aryl fucosides and of decreasing pH.

decreased the number of freely swimming spermatozoa and eventually
blocked the binding. Antibodies 1C7 and 7F12 inhibited the activity
of sperm α-L-fucosidase (11). Furthermore it has been shown by
indirect immunofluorescence that the enzyme is located in sperm head,
mostly at the tip and probably at the surface, as expected from our
hypothesis (12).

DISCUSSION

    If sperm α-L-fucosidase is really the recognition protein for
the sperm receptor having terminal L-fucose residue as (a part of)
ligand, naturally there must be a good correspondence in the properties between binding activity and enzyme (hydrolysing) activity.
Table 1 summarizes the corresponding properties of each activity
based upon the data presented in this paper. Collectively, these
data along with those published previously (2,3,15,17,18) lead us to
believe that α-L-fucosidase at the surface of sperm tip binds to the
sperm receptor in the vitelline envelope by forming an enzyme-
substrate complex. In the normal sea water, the complex may be maintained for a while, because the physicochemical conditions do not
favor hydrolysis. The spermatozoa may firmly bind to the vitelline
envelope by forming many complexes as shown in Fig. 2.

    Several problems remain to be solved. We are not definitely
sure that the enzyme is really an ectoenzyme, even though the antibodies bind to the spermatozoa tip. This question will be answered
simply by examining whether the spermatozoa bind specifically to the
beads conjugated with the monoclonal antibody, either 1C7 or 7F12.
Similar experiments using Sepharose-ε-aminocaproylfucosylamine did
not give us clear results. This may mean that the spacer was too
short and/or a certain length of saccharide chain is necessary for

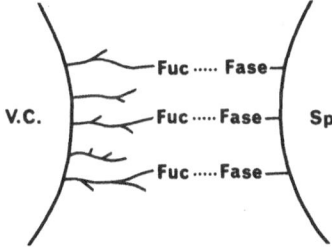

Figure 2. Diagrammatic illustration of sperm binding to the vitelline
          envelope based on the hypothesis described in the text.
          Fuc, terminal α-L-fucose residue of the sperm receptor;
          Fase, α-L-fucosidase at the sperm surface; V.C., vitelline
          envelope; Sp, spermatozoon. (Reproduced from 6).

Table 1. Is α-L-Fucosidase a Mediator of Sperm Binding to the Vitelline Envelope?

	Binding Activity***	α-L-Fucosidase Activity
Ligand/Substrate	Terminal L-Fucoside	Terminal α-L-Fucoside
Locus in Sperm	Surface of the Tip of Head	Surface of the Tip of Head
Normal Sea Water	Stable Binding	Binding to Fucosylamine
Acidic Sea Water	Temperature Dependent Detachment of Bound Sperm	Optimum pH at 3.9 Binding to Fucosylamine
Aryl* α-L-Fucoside	Inhibitor (~ mM)	Synthetic Substrate
Aryl* β-L-Fucoside	Inhibitor (~ mM)	Competitive Inhibitor
Aryl* β-D-Fucoside	No effect	No effect
L-Fucose	Inhibitor (>25 mM)	Product (Poor Inhibitor)
D-Fucose	No effect	No effect
Monoclonal Antibody** (1C7, 7F12)	Inhibitor	Inhibitor

*   P-Nitrophenyl or 4-Methylumbelliferyl.
** Monoclonal antibody against sperm α-L-Fucosidase.
*** If living eggs are used, some, not many, of the bound sperm undergo the acrosome reaction.

stabilizing the binding. The latter seems consistent with the fact
that fucosides are much more inhibitory to sperm binding than free
fucose.

We do not have any data concerning the possible hydrolysis of
L-fucose when the bound spermatozoa detach from the vitelline
envelope. This will be technically difficult to answer, because we
have not yet succeeded in obtaining vitelline envelopes labelled
adequately with radioactive fucose and because the sperm contain some
free fucose.

The more serious problem here is that we do not know exactly
what kinds of structural changes have happened after glycerol treatment of eggs. We know at least the vitelline envelope loses its
ability to induce the acrosome reaction after glycerol treatment,
which is actually a great advantage for analysis of sperm binding
mechanisms (17). Although the glycerol model apparently retains all
the properties of sperm receptor (17), this question must be evaluated carefully before applying the conclusions obtained from the
model system to the living gametes.

A similar mechanism involving a cell surface glycosidase, α-
mannosidase, has been reported in mammalian fibroblasts (16) suggesting that glycosidase is a recognition molecule for cell binding in
both gametes and somatic cells. Besides glycosidases, glycosyltransferase has been suggested for sperm binding to the vitelline envelope, zona pellucida (23, also see B. Shur, this volume). Thus both
of the two known categories of complementary carbohydrate-binding
proteins, namely lectins and carbohydrate metabolizing enzymes are
proposed as recognition proteins for sperm receptors (Table 2). It
is certainly too early to generate a unifying principle from these
three cases. However, it should be noticed that a lectin-type
molecule is used by the spermatozoon binding to the vitelline
envelope only after the acrosome reaction, whereas an enzyme, either
anabolic or catabolic, is used by the spermatozoon that binds before
the acrosome reaction.

Not only the enzymes relating carbohydrate metabolism, a
trypsin-like protease in the sperm plasma membrane is also claimed to
be necessary for sperm binding in the mouse (21). These reports,
including the present paper, imply a mechanism for cell binding; an
enzyme mediates cell binding by expressing the binding activity in
preference to the catalytic one under certain circumstances. This
could give an important clue for understanding the molecular evolution of recognition molecules such as various lectins, which mediate
cell to cell binding or adhesion, immunoglobulins and other components in self-defense mechanisms.

Table 2. Proposed Mechanisms of Sperm Binding to the Vitelline Envelope.

Animal	Acrosome Reaction	Recognition Molecule	Ligand	References
Sea Urchin*	Reacted Sperm	Bindin	Fucose Sulfate? (Glycosaminoglycan-like Structure?)	4,19,20,26
Sea Squirt**	Unreacted Sperm	α-L-Fucosidase	L-Fucose (Glycoprotein)	3,8,9,18
Mouse	Unreacted Sperm	Galactosyltransferase	β-D-N-Acetylgluco-samine (glyco-protein ZP-3)	1,23

* Strongylocentrotus purpuratus
** Ciona intestinalis

ACKNOWLEDGEMENTS

    The author is grateful to Drs. R. DeSantis and M. R. Pinto, Zoo-
logical Station at Naples, for their generous supply of the gametes.
He is also indebted to Ms. I. Kubo for the diagrams. Supported in
part by the grants from the Ministry of Education, Science and
Culture, Japan (57480018, 59540456) and Yamada Science Foundation,
and by the Naito Foundation for Inter-Institute Researches.

REFERENCES

1. Bleil, J.D., and Wassarman, P.M. 1980. Structure and function of
   the zona pellucida: Identification and characterization of
   proteins of the mouse oocyte's zona pellucida. Develop. Biol.
   76:185-202.
2. DeSantis, R., Jammunno, G., and Rosati, F. 1980. A study of the
   chorion and the follicle cells in relation to the sperm-egg
   interaction in the ascidian, Ciona intestinalis. Develop.
   Biol. 74:490-499.
3. DeSantis, R., Pinto, M.R., Cotelli, F., Rosati, F., Monroy, A.,
   and D'Allesio, G. 1983. A fucosyl glycoprotein component with
   sperm receptor and sperm-activating activities from the
   vitelline coat of Ciona intestinalis eggs. Exp. Cell Res.
   148:508-513.
4. Glabe, C.G., Grabel, L.B., Vacquier, V.D., and Rosen, S.D. 1982.
   Carbohydrate specificity of sea urchin sperm bindin: A cell
   surface lectin mediating sperm-egg adhesion. J. Cell Biol.
   94:123-128.
5. Honneger, T.G. 1982. Effect of fertilization and localized
   binding of lectins in the ascidian Phallusia mammillata. Exp.
   Cell Res. 138:446-450.
6. Hoshi, M. 1984. Roles of sperm glycosidases and proteases in the
   ascidian fertilization. Advances in invertebrate reproduction
   3, W. Engels et al., ed., Elsevier Sci. Publ., Amsterdam,
   pp. 27-40.
7. Hoshi, M. 1985. Lysins, Biology of Fertilization, C. B. Metz and
   A. Monroy, eds., Academic Press, New York, pp. 431-462.
8. Hoshi, M. 1985. Purification and characterization of
   α-L-fucosidase from the spermatozoa fo Ciona intestinalis,
   (submitted).
9. Hoshi, M., DeSantis, R., Pinto, M.R., Cotelli, F., and Rosati,
   F. 1983. Is sperm α-L-fucosidase responsible for sperm-egg
   binding in Ciona intestinalis, The Sperm Cell, J. Andre, ed.,
   Martinus Nijhoff Publishers, The Hague, pp. 107-110.
10. Hoshi, M., De Santis, R., Pinto, M.R., Cotelli, F., and Rosati,
    F. 1985. Sperm glycosidases as mediators of sperm-egg binding
    in the ascidians. Zool. Sci. 2:65-69.
11. Hoshi, M., and Iwata, M. 1985. Monoclonal antibodies raised
    against sperm α-L-fucosidase block sperm-egg binding and

inhibit the activity of the purified enzyme (in preparation).

12. Huang, T.T.F., Jr., Ohzu, E., and Yanagimachi, R. 1982. Evidence suggesting that L-fucose is part of recognition signal for sperm-zona pellucida attachment in mammals. Gamete Res. 5:355-361 (1982).

13. Iwata, M., and Hoshi, M. 1985. Distribution of α-L-fucosidase in the spermatozoa of Ciona intestinalis (in preparation).

14. Monroy, A., and Rosati, F. 1983. A comparative analysis of sperm-egg interaction. Gamete Res. 7:85-102.

15. Pinto, M.R., DeSantis, M., D'Alessio, G., and Rosati, F. 1981. Studies on fertilization in the ascidians. Fucosyl sites on vitelline coat of Ciona intestinalis. Exp. Cell Res. 132:289-295.

16. Rauvala, H., and Hakomori, S. 1981. Studies on cell adhesion and recognition III. The occurrence of α-mannosidase at the fibroblast cell surface, and its possible role in cell recognition. J. Cell Biol. 88:149-159.

17. Rosati, F., and DeSantis, R. 1978. Studies on fertilization in the ascidians I. Self-sterility and specific recognition between gametes of Ciona intestinalis. Exp. Cell Res. 112:111-119.

18. Rosati, F., and DeSantis, R. 1980. Role of the surface carbo-hydrates in sperm-egg interaction in Ciona intestinalis. Nature 283:762-764.

19. Rossignol, D.P., Earles, B.J., Decker, G.L., and Lennarz, W.J. 1984. Characterization of the sperm receptor of the surface of eggs of Strongylocentrotus purpuratus. Develop. Biol. 104:308-321 .

20. Ruttenberg-Barnum, S., and Brown, G.G. 1983. Effect of lectins and sugars on primary sperm attachment in the horseshoe crab, Limulus polyphemus L.

21. Saling, P.M. 1981. Involvement of trypsin-like activity in binding of mouse spermatozoa of zonae pellucidae. Proc. Natl. Acad. Sci. USA 78:6231-6235.

22. Saling, P.M., and Storey, B.T. 1979. Mouse gamete interactions during fertilization in vitro. Chlortetracycline as a fluorescent probe for the mouse sperm acrosome reaction. J. Cell Biol. 83:544-555.

23. Shur, B.D., and Hall, N.G. 1982. A role for mouse sperm surface galactosyltransferase in sperm binding to the egg zona pellucida. J. Cell Biol. 95:574-579.

24. Vacquier, V.D., and Moy, G.W. 1977. Isolation of Bindin: The protein responsible for adhesion of sperm to sea urchin eggs. Proc. Natl. Acad. Sci. USA 74:2456-2460.

STRUCTURE, ASSEMBLY AND FUNCTION OF THE SURFACE ENVELOPE

(FERTILIZATION ENVELOPE) FROM EGGS OF THE SEA URCHIN,

STRONGYLOCENTROTUS PURPURATUS

Edward J. Carroll, Jr.[1], Mildred Acevedo-Duncan,
Robin Wilby Justice, and Lyric Santiago

Department of Biology
University of California
Riverside, CA 92521

SUMMARY

The sea urchin fertilization envelope (FE) is formed following
initial sperm-egg interaction from the egg surface vitelline envelope
(VE) and the paracrystalline protein fraction (PCF), derived from
cortical granules. Although mature FEs are physicochemically hard-
ened postinsemination, a major protein fraction consisting of seven
major polypeptides was extracted from Strongylocentrotus purpuratus
FEs and the major, separated components were immunologically cross-
reactive with the principal polypeptides in PCF and isolated cortical
granules. Antibodies prepared against extracted, core FEs were
immunologically crossreactive with isolated VEs, but not with PCF,
suggesting that only VE components are covalently crosslinked. Based
on protease inhibitor experiments, our model of FE development is
that a benzamidine-sensitive, cortical granule protease cleaves a 200
kD VE polypeptide during initial envelope elevation to set up the
morphological change in FE papillae which occurs later. Divalent
cations precipitate the PCF and form metal proteinate bridges between
the VE and PCF. Based on peroxidase inhibitor experiments, we suggest that the cortical granule peroxidase crosslinks VE polypeptides,
beginning at 2-3 min postinsemination, to restrict the permeability
of the VE so that normal envelope thickening occurs. A 305 kD VE
polypeptide was isolated and appears to be important in sperm-egg
interaction based on inhibition of sperm binding and fertilization by
antibodies against the purified polypeptide.

INTRODUCTION

The sea urchin fertilization envelope (FE) is an extracellular
structure formed from the surface vitelline envelope (VE) and a
selected set of cortical granule polypeptides collectively known as
the paracrystalline protein fraction. The transmission electron
micrographs of unfertilized and fertilized Strongylocentrotus purpuratus egg cortices presented in Figure 1 illustrate that the VE is
quite thin and is closely associated with the egg plasma membrane
when compared to the FE. The VE is also the substrate upon which
sperm binding (Fig. 2) and its modulation take place during the VE to
FE transition. In addition, the definitive, fully formed FE also
serves a protective role during early embryogenesis until hatching.
In this review, we will consider three principal questions:

What is the ultrastructure and polypeptide composition of the
fully developed FE?

What are the ultrastructural and biochemical stages in FE development?

What is the molecular nature of the VE sperm receptor and how is
its activity modulated during the VE to FE transition?

ULTRASTRUCTURE AND POLYPEPTIDE COMPOSITION OF THE FE

The ultrastructure of the fully formed FE of S. purpuratus is
shown in Figure 3A, B. By 30 min postinsemination at 17°C, the
envelope papillae are quite pointed, have differentially thickened
sides and a "floor" which often appears to have more than one layer.
Large quantities of these fully formed envelopes are easily prepared
with a high degree of purity [at least 94%, (9)]. Such isolated
preparations show the same center-to-center spacing of envelope
papillae (0.35 ± 0.06 μm) as has been reported for FEs in situ (49,
50). To test the hypothesis that the fully formed FE is totally
crosslinked and insoluble, we quantitated extraction of purified
envelopes. Using multiple extractions under "harsh" conditions (6.0
M urea − 1.5 M 2−mercaptoethanol at pH 10.1 and 100°C for 10 min),
followed by quantitative amino acid analysis after acid hydrolysis,
we found that 71% of the amino acid content of the fully formed FE
was extracted. The envelopes are not solubilized by this procedure,
but scanning and transmission electron microscopy of extracted,
washed envelopes clearly showed that extraction dramatically altered
envelope ultrastructural features (Fig. 4). A variety of media were
tested for the ability to extract protein from the fully formed envelope; the effects of temperature and pH on extraction efficiency were
also examined. Buffers which contained 2−mercaptoethanol were the
most efficient with an inflection in the rate of extraction vs. pH
curve at pH = 9.8, which is close to the pK_a for the sulfhydryl group
of 2−mercaptoethanol. A "T_m" of approximately 60°C for extraction of

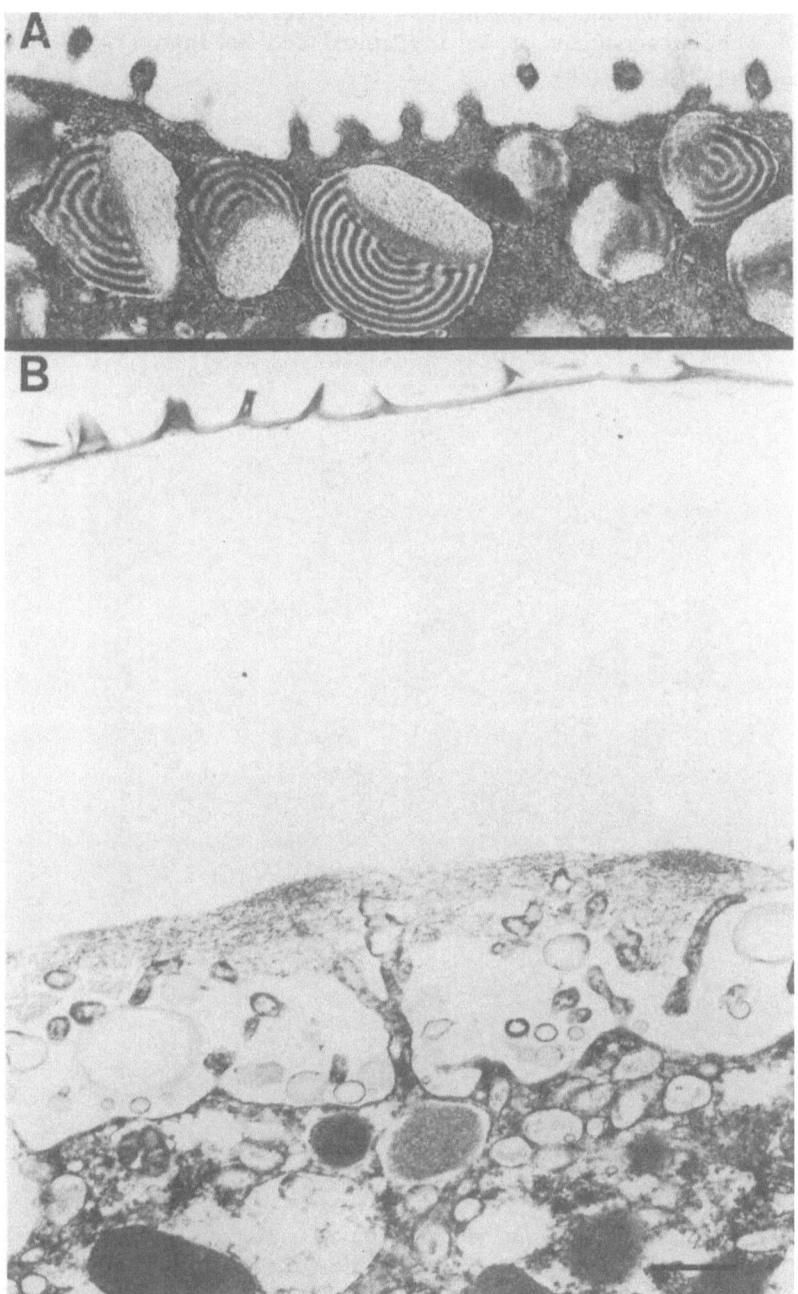

Figure 1. Transmission electron micrographs of the cortex of unfertilized (A) and fertilized (B) Strongylocentrotus purpuratus
eggs. Panel A is taken from Carroll et al. (10), panel.

B is from Carroll and Lackey (unpublished observation).
Fixation and preparations for microscopy were according to
the procedures of Gould-Somero and Holland (24). Bar
equals $0.5\,\mu\mathrm{m}$.

Figure 2. Scanning electron micrograph of sperm binding to an S.
purpuratus egg inseminated in 40 μg/mL soybean trypsin inhi
bitor and diluted into Ca^{2+}-and Mg^{2+}-free seawater at 30
sec postinsemination. The vitelline-like envelope [termed
VE* by Carroll and Endress (11)] which elevates under these
conditions retains intact sperm receptors (1). Bar equals
$1.0\,\mu\mathrm{m}$.

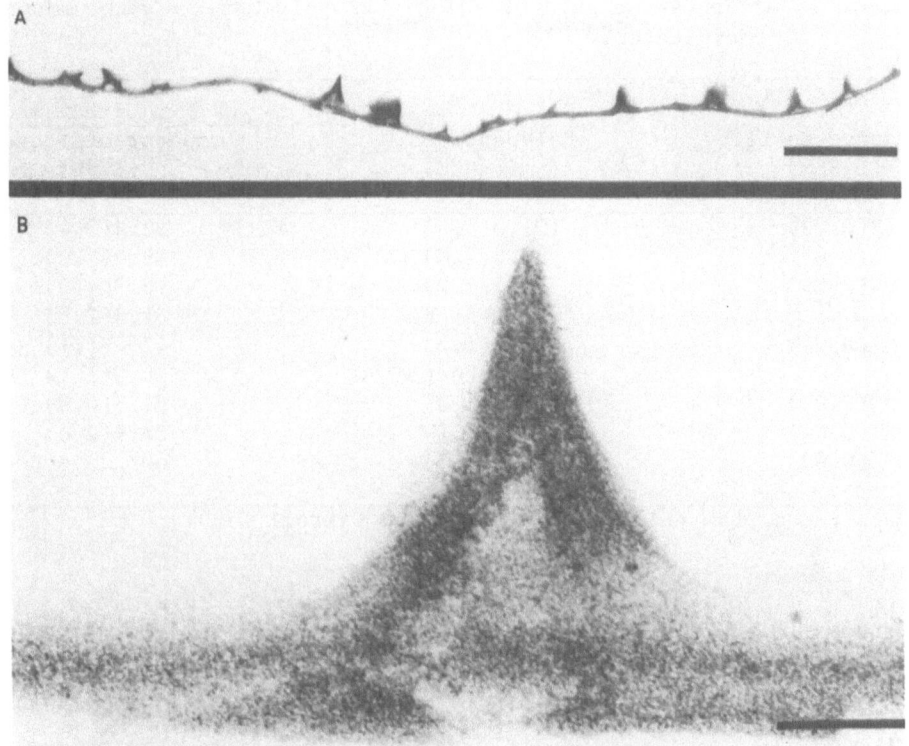

Figure 3. Transmission electron micrograph of a fully formed S.
purpuratus FE. An insemination mixture was fixed and
prepared for microscopy according to the procedures of
Gould-Somero and Holland (24) at 30 min postinsemination
(Justice and Carroll, unpublished observation). In panel
A, the bar equals 1.0 μm; in panel B, the bar equals 0.1
μm.

fully formed FEs in an ethanolamine buffer at pH 10.0 was determined
with a plateau in the extraction rate between 80°C and 100°C (9).

The macromolecular characteristics of the extractable protein
fraction prepared from fully formed FEs were determined directly
using sodium dodecyl sulfate polyacrylamide gel electrophoresis. In
addition, antisera against envelope extracts were prepared in rabbits
and also used to determine the size and number of polypeptides which
were present. As shown in Table 1, seven major Coomassie blue staining polypeptides and eight immunoprecipitates (detected by two-
dimensional immunoelectrophoresis) were observed; molecular weights
ranged from 16,800 to 101,000. The same number and proportion of
Coomassie blue staining polypeptides were obtained regardless of the

Table 1. Molecular weights of soluble FE polypeptides and immuno-
         precipitates.[a]

| Relative Mobility (± SEM) | Molecular Weight x 10^{-3} | |
	Polypeptides (± SEM)	Immunoprecipitates (± SD)
0.188(0.03)	101.0(4.5)	98.8(5.4)
0.278(0.03)	80.0(2.0)	89.6(2.0)
0.309(0.03)	75.0(2.2)	75.9(4.0)
	not detected	64.3(2.5)
0.484(0.04)	46.8(1.7)	47.4(1.1)
	not detected	36.5(2.1)
0.620(0.04)	32.8(1.3)	31.8(0.9)
0.687(0.04)	27.4(1.1)	27.9(2.0)
0.877(0.03)	16.8(0.3)	not detected

[a] Data taken from Villacorta-Moeller and Carroll (51).

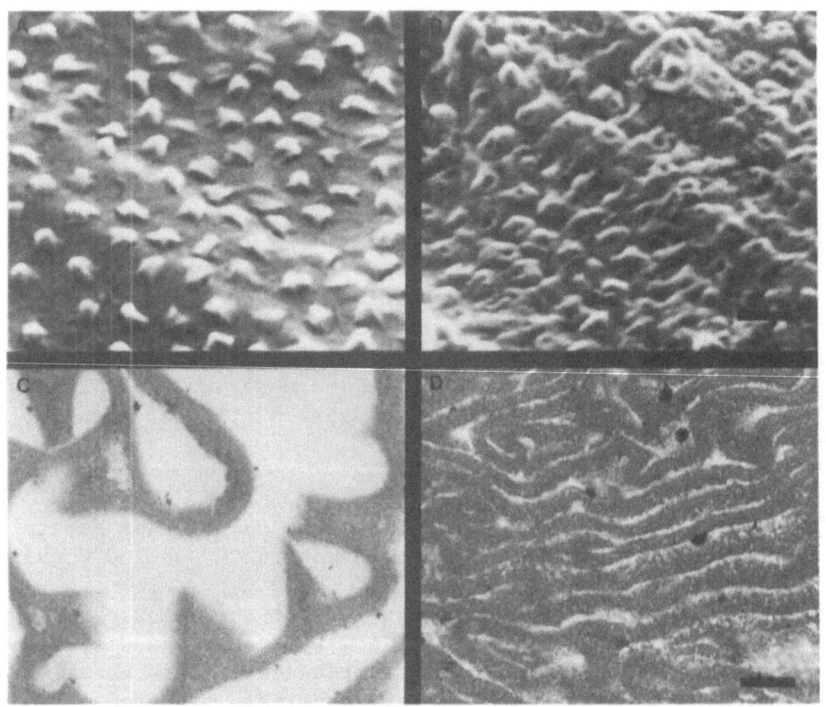

Figure 4. Electron micrographs of fully developed S. purpuratus FEs
          before and after extraction. This data is taken from
          Carroll and Baginski (9). Scanning electron micrographs of
          control (A) and extracted (B) envelopes; bar equals 0.5 μm.
          Transmission electron micrographs of control (C) and
          extracted (D) envelopes; bar equals 0.1 μm.

medium used for extraction. Periodic acid-Schiff base staining of gels showed that the 101,000, 75,000, 48,600 and possibly the 32,800 molecular weight polypeptides were glycoproteins (9). We concluded from these studies that a major fraction of the fully formed FE structure is extractable and consists of a small number of polypeptides and glycoproteins of moderate size (51).

To determine the subcellular origin of the polypeptides extracted from the fully formed FE, antisera prepared against envelope extracts were tested for crossreactivity [using immunoelectrophoresis) against extracts of VE preparations, isolated cortical granules and purified paracrystalline protein fraction. Extracts of VEs (isolated according to the methods of Glabe and Vacquier (22)] and extracts of unfertilized egg surfaces known to contain VE proteins (10,18) did not form immunoprecipitates. Extracts of isolated cortical granules and polypeptides of the paracrystalline protein fraction which were purified from cortical granule exudate showed complex patterns of immunoprecipitates (Table 2), but a major finding was that a 193,000 molecular weight cortical granule antigen was not found in the paracrystalline protein fraction. These results suggested that proteolytic processing of at least one cortical granule antigen occurs during FE development and that not all of the cortical granule polypeptides are incorporated into the envelope by di- and trityrosine crosslinks with VE proteins.

Since a significant portion of FE structure is not solubilized by any of the extraction media examined thus far, we wished to determine the qualitative composition of the insoluble "core" envelope. Our approach was to prepare antisera against exhaustively extracted core envelopes and utilize an enzyme-linked immunosorbent assay to determine the crossreactivity of core envelope antisera with various VE preparations, the paracrystalline protein fraction and FE polypeptides (Fig. 5). Hyalin, the major calcium-insoluble polypeptide isolated from the hyaline layer (45) was included in the analysis as a control since hyalin is not believed to be a component of the FE , but does crossreact immunologically with the paracrystalline protein fraction (Justice and Carroll, unpublished observation). Antiserum against the core envelope crossreacted with both VE preparations and an egg surface extract known to contain VE antigens. Crossreactivity of antiserum against the core envelope was also detected using soluble FE polypeptides whereas no reaction was observed with either the paracrystalline protein fraction or hyalin as antigens. These results show that the crosslinked core envelope contains VE antigens and antigens which are related to the soluble envelope polypeptides.

Table 2. Molecular weights of cortical granule and paracrystalline
protein fraction antigens which crossreact with soluble
FE polypeptide antiserum.[a]

Antigen Molecular Weight ($\pm$ SD) x 10^{-3}

Cortical Granule Fraction[b]	Paracrystalline Protein Fraction[c]	FE Polypeptide[d]
193 (7.5)	not detected	not detected
not detected	not detected	98.8(5.4)
not detected	not detected	89.6(2.0)
85.0(2.5)	75.5($--$)[e]	75.9(4.0)
not detected	61.2(1.6)	64.3(2.5)
49.0(0.5)	53.0(4.0)	47.4(1.1)
not detected	41.5(1.5)	not detected
36.8(0.8)	37.0($--$)[e]	36.5(2.1)
31.7(0.2)	31.0($--$)[e]	31.8(0.9)
27.0(0.3)	not detected	27.9(2.0)
20.0(1.0)	19.8(1.8)	not detected

[a] Data taken from Villacorta-Moeller and Carroll (51).
[b] One preparation was analyzed in duplicate.
[c] Two preparations were used in duplicate determinations.
[d] Data from Table 1.
[e] Observed in one of two replicates.

Since a clear immunological relationship has been shown between
the soluble envelope polypeptides, the paracrystalline protein fraction and extracts of isolated cortical granules (51), it was somewhat
surprising that crossreactivity was detected between core envelope
antiserum and soluble envelope polypeptides, but not with isolated
paracrystalline protein fraction. One possible explanation of these
results is that the soluble envelope polypeptides represent an
extremely small subset of immunological determinants present in the
paracrystalline protein fraction. We directly examined the immunological relatedness between the paracrystalline protein fraction and
the soluble FE polypeptides using enzyme-linked immunosorbent assay
(Table 3). Antiserum against the paracrystalline protein fraction
reacted with the soluble FE polypeptides at a titer of 1.5×10^4 and
antiserum against envelope polypeptides reacted with the paracrystalline protein fraction at a titer of 1.5×10^3.

These results suggest that sufficient immunological crossreactivity exists between the paracrystalline protein fraction and the
soluble envelope polypeptides to detect antibodies specific to the
paracrystalline protein fraction in antiserum against the core envelope. An alternative hypothesis to explain the lack of crossreactivity between antiserum against the core envelope and paracrystalline protein fraction is that the soluble envelope fraction may

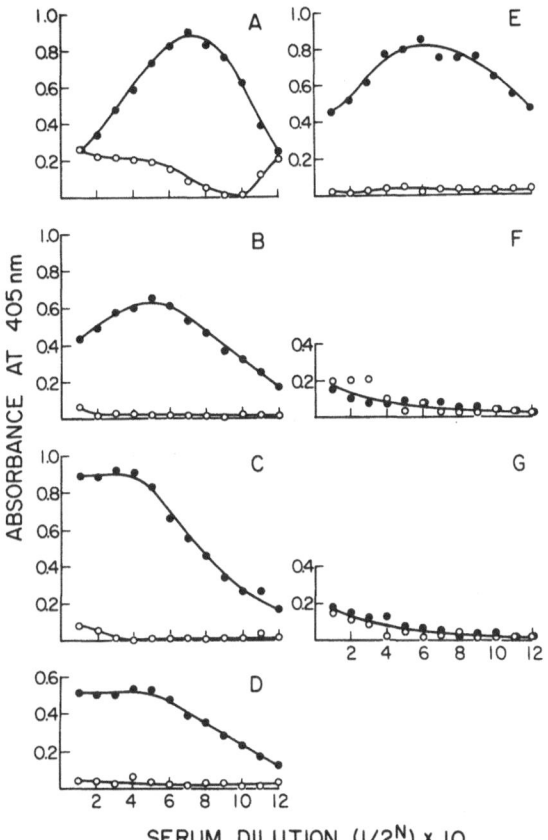

Figure 5. Crossreactivity of antiserum against extracted S. purpuratus FEs and various antigens by an enzyme-linked immunosorbent assay (42). A) extracted FEs, 22 nmol of glycine
equivalents/well; ● – immune and o – preimmune sera; B) VE*
[prepared according to Carroll and Endress (11)], 79 nmol
of glycine equivalents/well; ● – immune serum and o –
coating buffer control. C) VE [prepared according to
Glabe and Vacquier (22)], 10 μg of protein/well. ● –
immune serum and o – coating buffer control. D) dithiothreitol extract of unfertilized egg surfaces [prepared
according to Villacorta-Moeller and Carroll (51)], 6.8 μg
of protein/ well. ● – immune serum and o – coating buffer
control. E) extracted FE polypeptides, 2 μg of protein/
well; ● – immune serum and o – coating buffer control. F)
paracrystalline protein fraction [prepared according to
Baginski et al. (5) as modified by Justice and Carroll
(29)], 38 μg of protein/well. ● – immune and o – pre-
(continued)

Fig. 5 Cont. immune sera. G) hyalin prepared according to McBlaine
           and Carroll (36) as modified by Gray et al. (25), 41 μg of
           protein/well. ● – immune and o – preimmune sera. The pre-
           immune sera and coating buffer groups presented here are
           equivalent controls as purified immune and preimmune
           immunoglobulin fractions showed similar ratios of reactiv-
           ity as a function of concentration (42).

contain previously undetected levels of VE antigens which are now
detectable with the increased sensitivity provided by enzyme-linked
immunosorbent assay. The latter hypothesis was tested using three
different types of VE preparations as antigens and antiserum against
soluble FE polypeptides. As shown in Table 4, all three types of VE
preparations crossreacted with antiserum prepared against soluble FE
polypeptides.

Our current understanding of the molecular nature of the fully
developed FE in S. purpuratus is that the envelope contains cortical
granule material (paracrystalline protein fraction) which is not held
in the envelope structure by di- and trityrosine crosslinks, but
rather by Ca^{2+} and Mg^{2+}-salt linkages and non-covalent protein-
protein interactions. The components which are extractable from the
envelope as the soluble envelope fraction are principally derived
from cortical granules via the paracrystalline protein fraction; some
VE antigens are also present in the soluble fraction and are probably
present at low prevalence. The insoluble, core FE does not appear to
contain any paracrystalline protein fraction antigens, but rather
consists entirely of VE antigens. Peroxidase-catalyzed crosslinking
of the envelope presumably occurs among some of the VE antigens.
Work is currently in progress in our laboratory to determine which VE
antigens are extracted from the fully formed FE by reducing and
denaturing solvents and thus which antigens are crosslinked during
development.

ULTRASTRUCTURAL AND BIOCHEMICAL STAGES IN FE DEVELOPMENT

As noted, the fully formed FE is derived from the VE as a temp-
late onto which paracrystalline protein fraction antigens derived
from cortical granules bind. Some VE polypeptides in the initially
formed FE are subsequently crosslinked by cortical granule peroxidase
activity (20). We have examined the role(s) of calcium and magnesium
ions in the initial assembly process and have studied the function(s)
of identified cortical granule enzymes in FE development. Our
approach was to fertilize eggs in normal seawater, then remove Ca^{2+}
and Mg^{2+} by dilution and chelation; the eggs were prepared for transmission electron microscopy and the altered FEs which developed were
isolated and characterized by sodium dodecyl sulfate polyacrylamide
gel electrophoresis. In separate groups of experiments, inhibitors

Table 3. Immunological crossreactivity between the paracrystalline protein fraction and soluble FE polypeptides.[a]

Antigen	Antiserum	Max. Absorbance at 405 nm[b]	Titer[c]
soluble envelope polypeptides	soluble envelope polypeptides	0.950	2.0×10^4
paracrystalline protein fraction	paracrystalline protein fraction	0.800	1.5×10^4
soluble envelope polypeptides	paracrystalline protein fraction	0.450	1.7×10^4
paracrystalline protein fraction	soluble envelope polypeptides	0.200	1.5×10^3

[a] Enzyme-linked immunosorbent assay was used (52); all antigens were loaded at 1.0μ g/well (42).
[b] The net maximum absorbance was calculated as the difference between the immune and preimmune serum signals.
[c] The titer was the inverse of the serum concentration at which one-half maximum absorbance was observed.

Table 4. Immunological crossreactivity between antiserum against the soluble FE polypeptides and VE preparations.[a]

Antigen	Maximum Absorbance at 405 nm	Titer[b]
soluble envelope polypeptides	0.580	2.7×10^5
VE*	0.460	2.7×10^5
VE	0.427	1.2×10^4
egg surface extract	0.341	1.7×10^3

[a] Enzyme-linked immunosorbent assay was used (52); all antigens were loaded at 1.0μg/well (42).
[b] The titer is reported as the inverse of the serum dilution at which one-half the maximum absorbance was observed.

of cortical granule protease (benzamidine), peroxidase (3-amino-1,2,4-triazole) and β-1,3-glucanohydrolase (DL-threitol) activities were added at various times postinsemination. Portions of the egg suspensions were prepared for transmission electron microscopy and FEs were isolated for analysis of solubility under reducing and denaturing conditions. The polypeptide spectra of isolated envelopes were also determined by sodium dodecyl sulfate polyacrylamide gel electrophoresis.

To determine the effect(s) of Ca^{2+} and Mg^{2+} on the development of envelope structure, egg suspensions were inseminated in normal seawater and diluted into either normal seawater as a control or a Ca^{2+} and Mg^{2+}-free artificial seawater containing ethyleneglycol-bis [β-aminoethyl ether] N,N'-tetracetic acid. As shown in Figure 6, the envelope which forms under these conditions of reduced divalent cations is very thin [approximately 11 nm, (11)] and rather like the

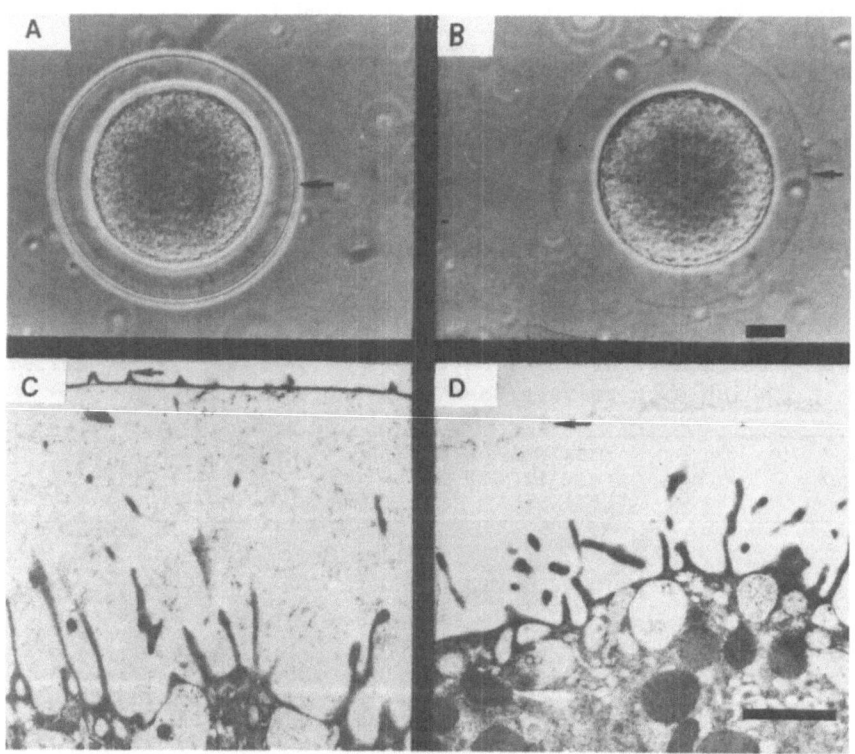

Figure 6. Effect of Ca^{2+} and Mg^{2+} deletion on FE structure. <u>S. pur-puratus</u> egg suspensions were inseminated in normal seawater and either diluted with 40 volumes of normal seawater (A,C) or 40 volumes of a Ca^{2+}/Mg^{2+}-free artificial seawater (B,D); 0.575 M NaCl-0.01 M KCl - 0.005M ethyleneglycol-bis

[β -aminoethyl ether] N,N'-tetraacetic acid–0.01 M Tris–HCl,
pH 8.0). (A,B) phase contrast micrographs, bar = 20 μm;
(C,D) transmission electron micrograph, bar = 0.1 μm. The
arrows delineate the fully formed FE (A,C) and the altered
envelope (B,D) termed VE* [data taken from Carroll and
Endress (11)].

VE in ultrastructure. In separate experiments, we also showed that
the permeability of the envelope [termed VE* (11)], was equivalent to
that of eggs stripped of the VE as determined from measurements of
secreted protein (11) and cortical granule enzymes (5). Analysis of
the polypeptide spectrum of VE* preparations determined using reducing and denaturing conditions showed eight major polypeptides which
ranged from 30,500 to 270,000 in molecular weight (Table 5). Our
interpretation of the preparation is that it represents what is
normally only a transient stage in FE development. Cortical granule
exocytosis is initiated at one site by the fertilizing sperm, propagated over the egg surface and is accompanied by assembly of the
paracrystalline protein fraction to the elevated envelope; however,
our procedure has uncoupled exocytosis and envelope elevation from FE
assembly. Therefore, the VE* is an elevated VE which has undergone
whatever limited proteolysis is required for disruption of the plasma
membrane–VE connections (12,13).

Table 5. Molecular weights of VE* polypeptides.[a]

Relative Mobility (mean ± SEM)	Molecular Weight x 10^{-3} (mean ± SEM)[b]	
0.119 (0.011)	270.0 (25)	*
0.179 (0.009)	175.0 (8.8)	*
0.285 (0.017)	114.0 (6.8)	
0.291 (0.012)	113.0 (4.7)	*
0.323 (0.010)	100.0 (3.0)	*
0.343 (0.012)	95.0 (3.3)	*
0.437 (0.017)	74.0 (2.9)	*
0.576 (0.012)	54.0 (1.1)	*
0.620 (0.049)	49.0 (3.9)	
0.660 (0.013)	45.0 (0.9)	
0.819 (0.026)	33.5 (1.1)	
0.849 (0.021)	32.0 (0.8)	
0.863 (0.017)	30.5 (0.6)	*

[a] Data of Carroll and Endress (11). Six different preparations were
analyzed by duplicate or triplicate determinations using electrophoresis on 9% separating gels according to the procedures of
Laemmli (32).
[b] Major polypeptides are indicated by an asterisk.

To determine the sequence of biochemical events in FE development, separate egg suspensions were inseminated and inhibitors of the
three classes of known cortical granule enzymes were added at various
times postinsemination. As shown in Figure 7, a typical control egg
fixed at 60 sec postinsemination had a thickened envelope (46 nm
compared to the VE thickness of approximately 10 nm) and the papillar
structures which represent the impressions of the cytoplasmic microvilli in the VE were still rounded. By 30 min postinsemination,
envelope development was considered complete [criterion of Carroll
and Baginski (9)] and the envelope was approximately 45 nm in thickness between the papillar structures (11). Benzamidine (5 mm final
concentration, protease inhibitor) was added at 15 sec postinsemination and the egg suspension was fixed at 30 min postinsemination.
The FEs which formed had a lower electron density compared to the
controls and were greatly thickened (96 nm), but the papille retained
the rounded morphology of the VE. In experiments not shown here, we
found that this benzamidine-induced effect on envelope development.
had a very early window, e.g., if the inhibitor was added at 60 sec
postinsemination the envelopes were essentially normal. The peroxidase inhibitor 3-amino-1,2,4-triazole (1 mM final concentration) when
added to an egg suspension at 60 sec postinsemination followed by
fixation at 30 min postinsemination caused development of an envelope
which had the pointed papillar ultrastructure of a fully developed
envelope (Fig. 7D), but was very thin (33 nm). DL-threitol (a β-
glucanohydrolase inhibitor) when present at 5 mM concentration 60 sec
postinsemination (followed by fixation at 30 min postinsemination)
induced development of an envelope which had normally shaped angular
papillae, but the envelope was of lesser electron density when compared to the control and the perivitelline space was filled with
granular material (Fig. 7E). When the inhibitors were mixed in all
possible combinations of two at a time and added to insemination
mixtures, no apparent synergistic effects were seen and in the case
of benzamidine containing mixtures, the compounds had to be added
between 15 sec and one min postinsemination for the benzamidine
effect to be seen (30).

We wished to correlate the effects of cortical granule enzyme
inhibitors on development of envelope ultrastructure with polypeptide
spectra of the extractable FE fraction. Separate batches of eggs
were inseminated, inhibitors were added at various times postinsemination and FEs were isolated at 30 min postinsemination. As a
control for each treatment group, inhibitors were added at 30 min
postinsemination and FEs were immediately isolated. As shown in
Figure 8, benzamidine addition at 20 sec postinsemination increased
the relative amount of a polypeptide of approximately 200,000 molecular weight which was reduced to the level of the 30 min control if
the inhibitor was added later (even as early as one min postinsemination). In addition, very early benzamidine addition greatly reduced
the amount of the major 90,000 and 75,000 molecular weight polypeptides present at later times of development. Thus, the window for

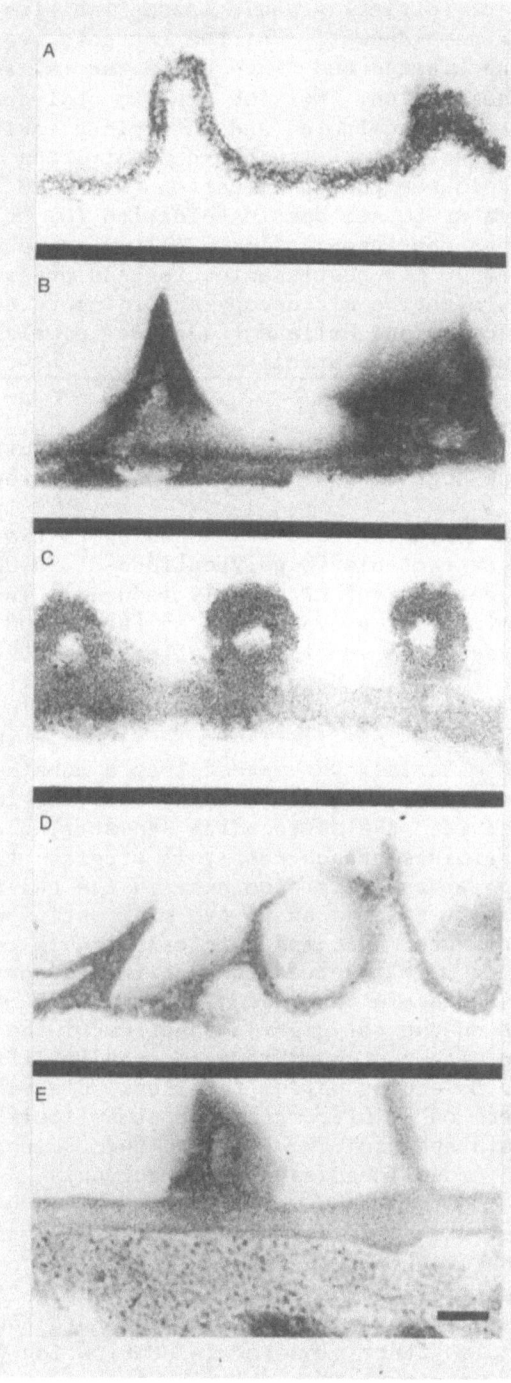

Figure 7. (continued)

Figure 7. Effect of cortical granule enzyme inhibitors on S. purpuratus FE development. For the controls, egg suspensions
were inseminated and fixed at 60 sec (A) and 30 min (B)
postinsemination. For the experimental groups, egg suspensions were inseminated and sufficient inhibitor added to
the suspension to bring the concentration to 5 mM benzamidine at 15 sec postinsemination (C), 1 mM 3-amino-1,2,4-
triazole at 60 sec postinsemination (D) or 5 mM DL-threitol
at 60 sec postinsemination. All experimental groups were
fixed at 30 min postinsemination and prepared for transmission electron microscopy according to the procedures of
Gould-Somero and Holland (24). Bar equals 0.1 μm. Data
from Justice and Carroll (30).

the benzamidine-induced effects on envelope structure and polypeptide
spectra was at the beginning of the cortical reaction.

When 3-amino-1,2,4-triazole was added up to two min postinsemination, the major extractable FE polypeptides at 90,000 and 75,000
molecular weight were absent or greatly reduced. In addition, the
appearance of families of polypeptides in the 30,000 and 17,000 molecular weight ranges was essentially unaffected by the peroxidase
inhibitor. Since FEs which develop with early addition of 3-amino-
1,2,4-triazole are very thin and the gel profile (loaded at equal
envelope concentrations for all time points) shows the presence of a
smaller amount of material, we suggest that a considerable fraction
of cortical granule protein that would normally be incorporated into
the FE is secreted into the surrounding seawater under these
conditions. Experiments are currently in progress to test the pre-
diction that VE polypeptides predominate in the gel pattern when the
peroxidase inhibitor is added up to two min postinsemination and that
the surrounding seawater contains cortical granule components. No
synergistic effects were observed in experiments where benzamidine
and 3-amino-1,2,4-triazole were added together. DL-threitol had no
detectable effect on the polypeptide spectrum of the extractable
envelope fraction with the possible exception of differences in the
30,000 and 17,000 molecular weight families. Since DL-threitol had
significant effects on FE ultrastructure, but little effect on the
extractable protein spectrum, we conclude that DL-threitol may affect
the status of the crosslinked envelope fraction.

In a parallel study, we determined the effect of cortical
granule enzyme inhibitors on the solubility of isolated envelopes
(Table 6). Benzamidine and DL-threitol had no significant effect on
envelope solubility during early development. In contrast, 3-
amino-1,2,4-triazole either alone or in combination with benzamidine
when added up to two min postinsemination induced formation of an FE
which was completely soluble.

Table 6. Effect of cortical granule enzyme inhibitors on the
solubility of S. purpuratus FEs.[a]

Group	Time of Inhibitor Addition (min)	Percent Solubility
5 mM benzamidine	0.33	25
	1.0	25
	4.0	36
	30.0	39
1 mM 3-amino-1,2,4-triazole	0.5	100
	1.0	100
	2.0	100
	5.0	43
	30.0	30
5 mM benzamidine and 1 mM 3-amino-1,2,4-triazole	0.5	100
	1.0	100
	2.0	81
	4.0	35
	8.0	45
	15.0	43
5 mM DL-threitol	0	14
	0.5	15
	1.0	17
	30.0	16

[a] Envelopes were isolated and the solubility was determined according
to the procedures of Carroll and Baginski (9). Extraction of
envelopes was as described in the legend to Figure 8. The average
overall recovery for envelopes isolated at 30 min postinsemination
and extracted as described was $72 \pm 11\%$ (mean $\pm$ SD, n = 5).

Our current working hypothesis of FE development is that cortical granule exocytosis releases a benzamidine-sensitive protease
which cleaves a 200,000 molecular weight VE polypeptide during the
elevation process to "set up" the envelope for the morphological
change in the papillae which occurs later. Divalent cations (especially Ca^{2+} and Mg^{2+}) act to precipitate the paracrystalline protein
fraction and form metal proteinate bridges between the VE and paracrystalline protein fraction. We suggest that the principal role of
the peroxidase is to crosslink some VE polypeptides (beginning at 2-3
min postinsemination) which results in a restriction in permeability
of the envelope so that normal envelope thickening occurs.

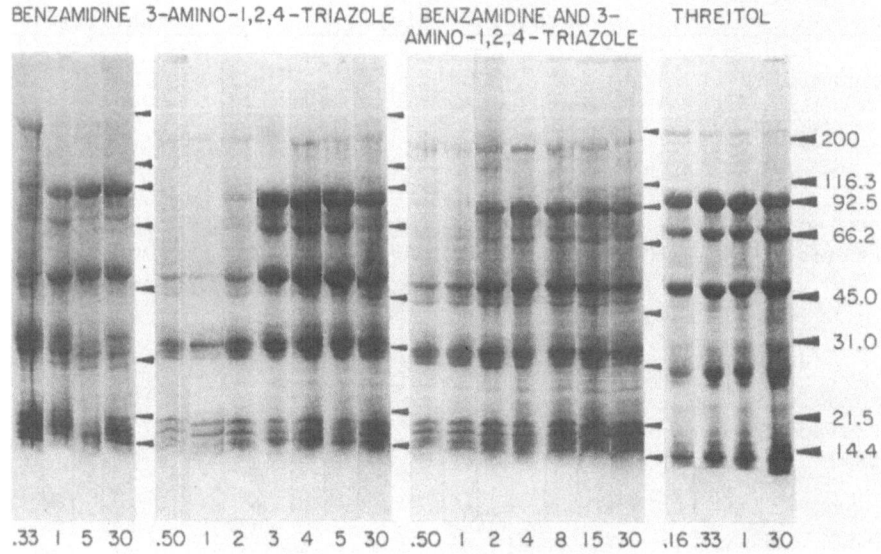

MINUTES POST INSEMINATION

Figure 8. Effect of cortical granule enzyme inhibitors on the polypeptide spectra of S. purpuratus FEs. Egg suspensions were
inseminated and the inhibitors were added (as described in
the legend to Fig. 7) at the indicated times. At 30 min
postinsemination the FEs were isolated as described by
Carroll and Baginski (9) and prepared for electrophoresis
on sodium dodecyl sulfate polyacrylamide gels (4% stacking
gel, 4-18% linear gradient running gel) using the buffer
system described by Laemmli (32). Isolated envelopes were
extracted in 2% sodium dodecyl sulfate-5% 2-mercaptoethanol
-0.0625 M Tris-HCl, pH 6.8 at 100°C for 3 min. The envelope suspension was centrifuged at 20,000xg for 10 min at
20°C. Approximately 300 nmol of envelope glycine equivalents [analyzed as described by Carroll and Baginski (9)]
was loaded in each lane. Separate gels were run for each
experimental group. The arrowheads indicate the position
of the standard proteins which were: 200,000 (myosin),
116,300 (β-galactosidase), 92,500 (phosphorylase B), 66,200
(bovine serum albumin), 45,000 (ovalbumin), 31,000 (carbonic anhydrase), 21,500 (soybean trypsin inhibitor) and
14,400 (lysozyme).

VE SPERM RECEPTOR AND MODULATION OF ITS ACTIVITY DURING THE VE TO FE
TRANSITION

Our approach to the isolation of a VE sperm receptor was to
determine conditions for VE elevation in the absence of receptor
inactivation and FE assembly. Such a modified VE could be isolated
and its constituent polypeptides identified and surveyed for sperm
receptor activity. We have described a procedure for uncoupling
cortical granule exocytosis and VE elevation from FE assembly (11)
which uses insemination in normal seawater followed by dilution into
Ca^{2+} and Mg^{2+}-free artificial seawater prior to the cortical reaction. We combined this technique with the addition of soybean
trypsin inhibitor to insemination mixtures to induce formation of a
VE with the desired properties. A control egg which was inseminated
in normal seawater showed normal FE elevation and complete sperm
detachment (Fig. 9A). In contrast, insemination in normal seawater
supplemented with $40\,\mu g/mL$ soybean trypsin inhibitor resulted in
normal FE elevation, and a lack of sperm detachment (Fig. 9C).
Combination of insemination in $40\,\mu g/mL$ soybean trypsin inhibitor and
dilution into Ca^{2+} and Mg^{2+}-free artifical seawater permitted normal
elevation of a thin envelope which retained many bound sperm (compare
Fig. 9D with 9B). The percentage of eggs with normally elevated
envelopes and sperm spared from detachment was determined as a function of soybean trypsin inhibitor concentration (Table 7). Control
eggs fixed 10 sec postinsemination had 34.9 ± 2.4 sperm/egg attached
to the unelevated VE (in the plane of maximum cell diameter) and all
of the control eggs fixed at 60 sec postinsemination had elevated FEs
with essentially no attached sperm (0.9 ± 0.2 sperm/egg). Increasing
concentrations of soybean trypsin inhibitor gave progressively
greater numbers of attached sperm and a decreasing percentage of eggs
with normally elevated envelopes. In this experiment, $30\,\mu g/mL$ soybean trypsin inhibitor maximized both normal envelope elevation (81%)
and sperm attachment (64% of the 10 sec control value). We concluded
from these results that eggs which are inseminated in a low concentration of soybean trypsin inhibitor and diluted into Ca^{2+}-Mg^{2+}-free
artificial seawater have thin, elevated VE*s with intact sperm receptors (1).

VE*s isolated from eggs inseminated in the absence and presence
of a low concentration of soybean trypsin inhibitor were completely
soluble (0.0625 M Tris-HCl – 2% sodium dodecyl sulfate – 5% 2-
mercaptoethanol – 10% glycerol at 80°C for three min) and yielded an
average amino acid content of $0.88\,\mu mole$ glycine equivalents per mL
of packed eggs used (1) analyzed as described by Carroll and Baginski
(9). To determine if VE*s isolated from eggs inseminated in soybean
trypsin inhibitor had any additional high molecular weight components, solubilized preparations were run on 4% separating gels
[reducing and denaturing conditions, system of Laemmli (32)]. As
shown in Figure 10, envelopes prepared from eggs inseminated in soybean trypsin inhibitor had an additional polypeptide (of 305,000

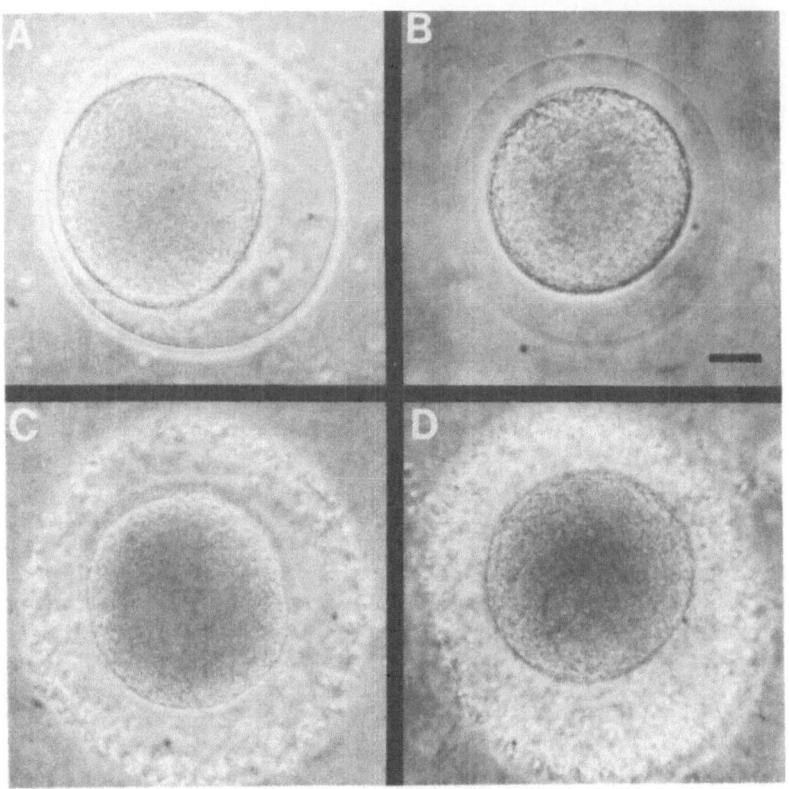

Figure 9. Effects of soybean trypsin inhibitor and Ca^{2+}-Mg^{2+}-free
artificial seawater on envelope elevation and sperm
binding. S. purpuratus egg suspensions were inseminated in
normal seawater (A); inseminated in normal seawater and
diluted into four volumes of Ca^{2+}-Mg^{2+}-free artificial
seawater (0.575 M NaCl-0.01 M KCl-0.005 M ethyleneglycol-
bis [β-aminoethyl ether] N,N'-tetraacetic acid-0.01 M Tris-
HCl, pH 8.0) at approximately 30 second postinsemination
(B); inseminated in seawater containing 40 μg/mL soybean
trypsin inhibitor and diluted into seawater at approximately 30 sec postinsemination (C); inseminated in normal
seawater containing 40 μg/mL soybean trypsin inhibitor and
diluted into Ca^{2+}-Mg^{2+}-free artificial seawater (composition described above) at approximately 30 sec postinsemination (D). Eggs were fixed at two min postinsemination with
1.5% glutaraldehyde in either normal seawater or the
Ca^{2+}-Mg^{2+}-free artificial seawater described above; bar =
20 μm. Data are from Acevedo-Duncan and Carroll (1).

Table 7. Sperm attachment and FE elevation as a function of soybean trypsin inhibitor concentration.[a]

Sperm Attachment or Envelope Elevation[b]	Soybean Trypsin Inhibitor (µg/mL)				
	0		17	23	35	45
	10 sec	60 sec				
Sperm/egg[c]	34.9±2.4	0.9±0.2	9.0±1.6	12.0±1.8	21.8±1.7	35.8±2.7
Envelope elevation	---	100%	100%	100%	81%	7%

[a] Data taken from Acevedo-Duncan and Carroll (1).

[b] Insemination mixtures were fixed either at 10 and 60 sec (0 µg/mL inhibitor) or at 60 sec postinsemination (all treatment groups) with 2.6% formaldehyde-seawater.

[c] The number of sperm/egg (± SEM) was determined from phase contrast microscopic observations focused on the plane of maximum diameter of the egg cell or elevated envelope.

molecular weight based on comparison with the same standards described in the Legend to Figure 8 run on 7% separating gels). FEs prepared from eggs inseminated in the presence and absence of soybean trypsin inhibitor did not show the 305,000 molecular weight polypeptide, suggesting that it is crosslinked by phenolic coupling of tryosyl residues [described by Hall (26) and Foerder and Shapiro (20)] during FE development (1).

To determine if the 305,000 molecular weight polypeptide is a sperm receptor, we prepared polyclonal antibodies in rabbits against electrophoretically purified material and examined the effects of immune sera and purified Fab fragments on sperm binding and fertilization. Antisera were also prepared against an electrophoretically purified 225,000 molecular weight VE* polypeptide as a control since this polypeptide is present in VE*s prepared from eggs inseminated both in the absence and presence of soybean trypsin inhibitor. Antibody titers were determined using enzyme-linked immunosorbent assay with 2 µg of antigen per well. The titers for the 305,000 and 225,000 polypeptides were 3.0×10^4 and 1.0×10^3, respectively (2).

The VE origin of the 305,000 and 225,000 molecular weight polypeptides was confirmed by indirect immunofluorescence. Unfertilized S. purpuratus eggs were incubated in seawater containing either immune or preimmune sera, washed and prepared for light microscopy as described. Prior to observation, the eggs were activated with the

A B

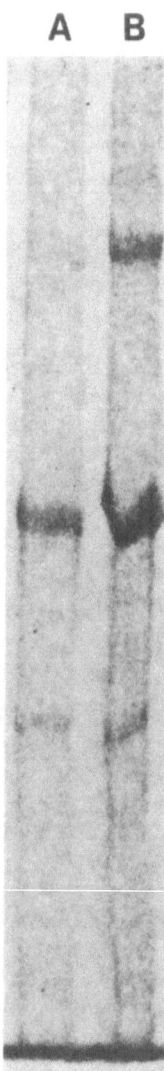

Figure 10. Sodium dodecyl sulfate polyacrylamide gels [4% separating
gel, system of Laemmli (32)] of envelopes from S. purpuratus eggs activated in the absence (A) and presence (B) of
40 μ g/mL soybean trypsin inhibitor. Data of Acevedo-
Duncan and Carroll (1).

Ca^{2+} ionophore A23187. Data for antiserum against the 305,000 molecular weight polypeptide and a polyacrylamide matrix control are shown in Figure 11. The immunofluorescence was specifically localized in the FE and no fluorescence was observed with either anti-acrylamide serum (shown) or any preimmune sera tested (data not shown). The 225,000 molecular weight polypeptide was also localized in the VE using indirect immunofluorescence (data not shown).

If the 305,000 molecular weight VE polypeptide is a sperm receptor, then we would expect antiserum against the polypeptide to block sperm binding and fertilization. To insure that fertilization inhibition experiments were done under sperm-limiting conditions, the minimum sperm concentration required for 95-100% fertilization was determined for the eggs of each female used. We report inhibition of fertilization in these experiments and in each case where fertilization inhibition was observed, sperm were motile, but did not bind to the unfertilized eggs (data not shown). As shown in Figure 12A, inhibition of fertilization by antiserum against the 305,000 molecular weight polypeptide was proportional to serum protein concentration. Antiserum against the 225,000 molecular weight polypeptide and control preimmune sera had no detectable effect on sperm binding and fertilization. Univalent Fab fragments were prepared from immune sera to determine if the sperm binding and fertilization inhibition effects of whole antiserum against the 305,000 molecular weight polypeptide were due to antibody crosslinking of surface antigens and consequent nonspecific masking of sperm receptors. Univalent Fab fragments directed against the 305,000 and 225,000 molecular weight polypeptides had titers (determined as described above) of 7.1×10^2 and 2.0×10^{-2} mL·μg^{-1}, respectively. Sperm binding and fertilization were inhibited by both Fab preparations in a concentration-dependent manner (Fig. 12B). Since antiserum against the 305,000 molecular weight polypeptide was bound to the VE[*] of eggs inseminated in the absence of soybean trypsin inhibitor (2) and no electrophoretically detectable polypeptide was observed in this preparation (refer to Fig. 10), we determined if these envelope polypeptides were immunologically related using enzyme-linked immunosorbent assay. As shown in Table 8, the 305,000 molecular weight polypeptide was recognized by antiserum and purified Fab fragments against the 225,000 molecular weight polypeptide.

We suggest that a 305,000 molecular weight VE polypeptide is the VE sperm receptor which may be cleaved to produce a 225,000 molecular weight polypeptide(s) by sperm receptor hydrolase. Currently, we are attempting fertilizability titrations of acrosome reacted sperm by purified 305,000 molecular weight polypeptide and are determining the precise peptide fragmentation pattern produced by in vitro incubation of purified receptor and sperm receptor hydrolase.

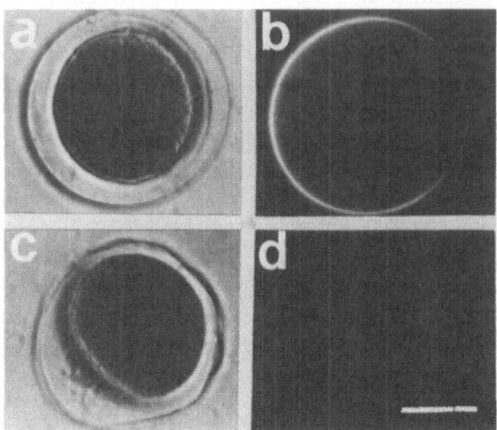

Figure 11. Indirect immunofluorescent localization of the 305,000
molecular weight VE polypeptide. S. purpuratus eggs were
dejellied (at pH 5.8 for two min), washed and resuspended
to 2.5% (v/v). Antiserum against the 305,000 molecular
weight polypeptide or the polyacrylamide matrix was
dialyzed against seawater, absorbed with sperm [see
Acevedo-Duncan and Carroll (2)] and 200 μL was mixed with
100 μL of the above egg suspension and incubated for one
hour at 17°C. The eggs were observed to clump very rapidly (Justice, unpublished). The eggs were washed three
times with approximately 9 mL of normal seawater and
resuspended to 1 mL. Five microliters of rhodamine-
conjugated swine anti-rabbit immunoglobulin (Dakopatts via
Accurate, Westbury, New York) was added and incubation was
continued for 30 min at 17°C. The rhodamine-conjugated
swine anti-rabbit immunoglobulin was preabsorbed with an
excess of glutaraldehyde fixed sea urchin eggs. The egg
suspensions were washed three times with 9 mL of normal
seawater and resuspended to 1 mL. The suspensions were
activated with approximately 25 μM A23187 (dissolved in
dimethyl sulfoxide) and photographed. An egg incubated in
antiserum against the 305,000 molecular weight polypeptide
and photographed using either phase (A) or fluorescence
(B) optics, an egg incubated in antiserum against the
polyacrylamide gel matrix and photographed using either
phase (C) or fluorescence (D) optics. Bar = 40 μm.

Table 8. Immunological crossreactivity of the 305,000 and 225,000
molecular weight VE polypeptides.[a]

Antibody Directed Against	Multivalent[b] x 10^{-3}	Univalent[c] $(g/mL)^{-1}$ x 10^2
305,000 molecular weight polypeptide	4.0	2.9
225,000 molecular weight polypeptide	6.3	1.8

[a] Data of Acevedo-Duncan and Carroll (2). Two micrograms of electrophoretically purified 305,000 molecular weight polypeptide per
well were loaded and the titers were determined by enzyme-linked
immunosorbent assay as described (52).

[b] The titer is reported as the inverse of the serum dilution at which
one-half the maximum absorbance was observed.

[c] The titer is reported as the inverse of the Fab concentration at
which one-half the maximum absorbance was observed.

CONCLUDING REMARKS

Our understanding of FE development has evolved considerably
since early experimental (28,46-48) light (37) and electron microscopic (15) studies which suggested that cortical substances and Ca^{2+}
were important in FE formation. More recent ultrastructural studies
have described the envelope assembly process using the freeze-
fracture/deep etching technique (14) as well as confirming the
earlier work (3,4,16,17). Inoue and Hardy (29) have described the
fully formed FE as a trilaminar structure consisting of a central VE
core 20 nm in thickness, symmetrically flanked by a layer of cortical
granule material 17.5 nm in thickness. Several other studies have
reported a partial dissection of the envelope assembly process (6,7,
20,26,31,44). Our own earlier work, reviewed here, has complemented
these studies and has shown that a major protein fraction of the
fully developed FE is not retained in the structure by di- and trityrosine crosslinks (9) and is principally derived from the cortical
granules (51) with a small contribution from the VE (42). We have
also reported isolation of a VE[*] preparation which represents an FE
intermediate normally present transiently during development (11). A
recent report (39) confirms and extends the earlier description of VE
polypeptide composition published by Glabe and Vacquier (23).

The immunological studies reported here regarding the biochemical
nature of the core FE structure (insoluble in reducing and denaturing
solvents) clearly showed that no detectable cortical granule antigens

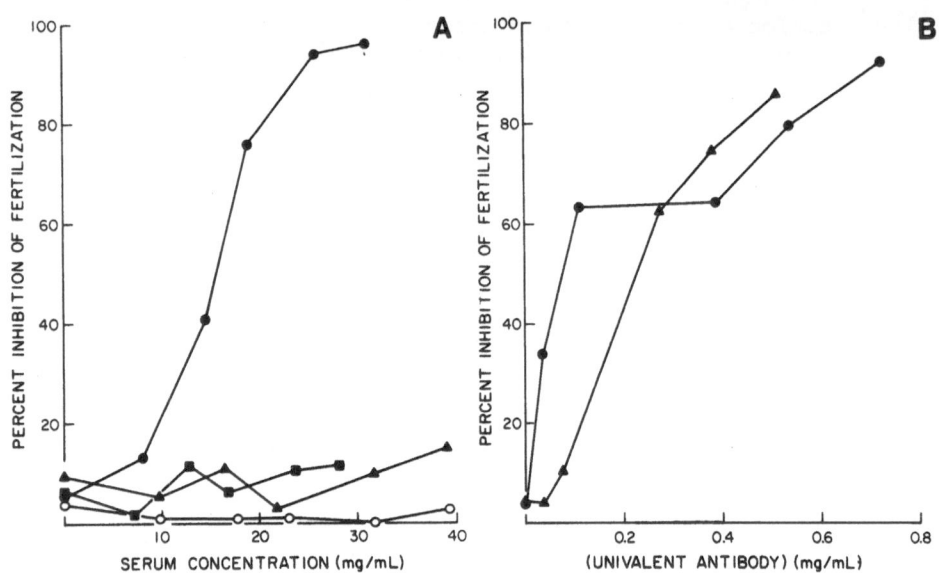

Figure 12. Effect of whole antiserum and purified Fab fragments
against the S. purpuratus 305,000 and 225,000 molecular
weight VE polypeptides on fertilization. Sera were
dialyzed against normal seawater. A sperm agglutinating
factor present in preimmune and immune sera was removed as
described (2). Insemination mixtures containing 50 μL of
2.5% (v/v) egg suspension and 200 μL of serially diluted
serum or Fab fragments prepared as described (2) were
incubated for 90 sec prior to addition of 20 μL of sperm
suspension. The sperm density was adjusted to give
95-100% fertilization in the absence of serum or Fab
fragments. Percent fertilization was scored by counting
one hundred eggs for FEs at two min postinsemination.
Preimmune (o,▲), and immune (●,■) sera against the 305,000
and 225,000 molecular weight VE polypeptides, respectively
(A): purified Fab fragments against the 305,000 (●) and
225,000 (▲) molecular weight VE polypeptides (B).

remain after multiple extraction and that some VE components are
extracted by the procecure. These results suggest that FE hardening
via peroxidase-catalyzed crosslinking occurs between selected VE
polypeptides. Our laboratory is currently determining whether a
selected set of VE polypeptides are crosslinked or a percentage of
all VE polypeptides are crosslinked.

Our experimental analysis of FE development using cortical
granule enzyme inhibitors established that a proteolytic event occurs

beginning at the time the cortical reaction starts and is required for the morphological change in envelope papillae from a rounded form to a pointed, angular form. Gel electrophoretic analysis identified an extractable envelope polypeptide of approximately 200,000 molecular weight that may be the substrate acted upon by the protease. Vitelline delaminase, sperm receptor hydrolase (8,12) or some as yet undescribed protease may be involved. Experiments using a peroxidase inhibitor showed that peroxidase activity was not involved in the morphological transition of the FE papillae, but was involved in normal envelope thickening. A restricted permeability of the normal FE is a classical observation (27,33-35). The relationship between this permeability change and the developmental processes described above is unclear. The precise role of the β-1,3-glucanohydrolase in sea urchin fertilization has been elusive ever since its discovery by Epel et al. (19). Our ultrastructural work shows that DL-threitol, a glucanohydrolase inhibitor causes the formation of an FE with an aberrant electron density and an accumulation of granular material in the perivitelline space. Since DL-threitol had virtually no effect on the polypeptide spectrum of the extractable protein fraction of isolated FEs, we conclude that DL-threitol is either affecting the crosslinked portion of the envelope or the processing of cortical granule material prior to addition to the VE. We are currently attempting to distinguish these hypotheses.

The currently accepted model of sea urchin sperm-egg interaction is that sperm adhesion to the VE occurs via sperm bindin (38). A molecular description of the egg receptor for sperm (presumably a bindin receptor) has been elusive until now. Glabe and Vacquier (23) suggested that the egg receptor for bindin was proteolytically cleaved and released during the cortical reaction. Lennarz and co-workers (40,41,43) have used pronase digests of unfertilized eggs and cell surface complexes to isolate preparations which contain sperm receptor activity as defined by the ability of such preparations to interact with bindin and inhibit fertilization. These workers have not as yet identified a discrete macromolecule with receptor activity and precise reasons for the differences between their findings and ours are unclear. Gache et al. (21) have prepared monoclonal antibodies to isolated VEs. Antibodies produced by several clones block sperm binding and fertilization, but a single target antigen has not been identified. Yoshida and Aketa (53) have identified a 225,000 molecular weight "core polypeptide" which appears to be the active component of the sperm receptor in urea extracts of Anthocidaris crassispina eggs. In contrast to these approaches, we feel that isolation of the VE[*] with intact sperm receptors (prior to extraction of constituent envelope polypeptides) is an important prerequisite to isolation of the VE sperm receptor. It remains to be seen if the 305,000 molecular weight VE polypeptide is also a bindin receptor and experiments are currently in progress in our laboratory to test this hypothesis.

ACKNOWLEDGEMENTS

     This paper is dedicated to the memory of Edward J. Carroll, a
fellow scientist and father to the senior author.

     It is with great pleasure that we recognize the expert technical
assistance of John Kitasako, Marcia Kooda-Cisco, Debora Lackey, Paul
McBlaine and Maria Villacorta-Moeller during various phases of this
work. Alice Lider-Carroll is thanked for her patient word processing. The senior author is grateful to: grants from the National
Science Foundation over the past eight years, to National Institutes
of Health grant BRSG-RR07010-15, grants from the Riverside Division
of the Academic Senate, I. M. Newell Graduate Research Awards and the
support of Dean Irwin W. Sherman through grants from the SCAR fund.

REFERENCES

  1. Acevedo-Duncan, M., and Carroll, E.J., Jr. 1986. Isolation of a
      sea urchin egg vitelline envelope with sperm receptor acti-
      vity. Dev. Biol. in review.
  2. Acevedo-Duncan, M., and Carroll, E.J., Jr. 1986. Immunological
      evidence that a 305 kilodalton vitelline envelope polypeptide
      is a sperm receptor. Dev. Biol. in review.
  3. Afzelius, B.A. 1956. The ultrastructure of the cortical granules
      and their products in the sea urchin egg as studied with the
      electron microscope. Exp. Cell Res. 10:257-285.
  4. Anderson, E. 1968. Oocyte differentiation in the sea urchin,
      Arbacia punctulata, with particular reference to the origin of
      cortical granules and their participation in the cortical
      reaction. J. Cell Biol. 37:514-539.
  5. Baginski, R.M., McBlaine, P.J., and Carroll, E.J., Jr. 1982.
      Novel procedures for collection of sea urchin egg cortical
      granule exudate: Partial characterization and evidence for
      postsecretion processing. Gamete Res. 6:39-52.
  6. Bryan, J. 1970. The isolation of a major structural element of
      the sea urchin fertilization membrane. J. Cell Biol.
      44:635-644.
  7. Bryan, J. 1970. On the reconstitution of the crystalline
      components of the sea urchin fertilization membrane. J. Cell
      Biol. 45:606-614.
  8. Carroll, E.J., Jr. 1976. Cortical granule protease from sea
      urchin eggs. Meth. Enzymol. 45:343-353.
  9. Carroll, E.J., Jr., and Baginski, R.M. 1978. Sea urchin fertili-
      zation envelope: Isolation, extraction and characterization
      of a major protein fraction from Stronglyocentrotus purpuratus
      embryos. Biochemistry 15:2605-2612.
 10. Carroll, E.J., Jr., Byrd, E.W., Jr., and Epel, D. 1977. A novel
      procedure for obtaining denuded sea urchin eggs and

observations on the role of the vitelline layer in sperm
reception and egg activation. Exp. Cell Res. 108:365-374.

11. Carroll, E.J., Jr., and Endress, A.G. 1982. Sea urchin
    fertilization envelope: Uncoupling of cortical granule
    exocytosis from envelope assembly and isolation of an envelope
    intermediate from Stronglocentrotus purpuratus embryos. Dev.
    Biol. 94:252-258.

12. Carroll, E.J., Jr., and Epel, D. 1975. Isolation and biological
    activity of the proteases released by sea urchin eggs follow-
    ing fertilization. Dev. Biol. 44:22-32.

13. Carroll, E.J., Jr., and Epel, D. 1975. Elevation and hardening
    of the fertilization membrane in sea urchin eggs: Role of the
    soluble fertilization product. Exp. Cell Res. 90:429-432.

14. Chandler, D.E., and Heuser, J. 1980. Vitelline layer of the sea
    urchin egg and its modification during fertilization:
    Freeze-fracture study using quick-freezing and deep etching.
    J. Cell Biol. 84:618-632.

15. Endo, Y. 1952. The role of the cortical granules in the formation
    of the fertilization membrane in eggs from Japanese sea
    urchins. Exp. Cell Res. 3:406-418.

16. Endo, Y. 1961. Changes in the cortical layer of sea urchin eggs
    at fertilization as studied with the electron microscope. I.
    Clypeaster japonicus. Exp. Cell Res. 25:383-397.

17. Endo, Y. 1961. The role of the cortical granules in the forma-
    tion of the fertilization membrane in the eggs of sea urchins.
    II. Exp. Cell Res. 25:518-528.

18. Epel, D., Weaver, A.M., and Mazia, D. 1970. Methods for removal
    of the vitelline membrane of sea urchin eggs. I. Use of
    dithiothreitol (Cleland's reagent). Exp. Cell Res. 61:64-68.

19. Epel, D., Weaver, A.M., Muchmore, A.V., and Schimke, R.T. 1969.
    β-1,3-glucanase of sea urchin eggs: Release from particles at
    fertilization. Science 163:294-296.

20. Foerder, C.A., and Shapiro, B.M. 1977. Release of ovoperoxidase
    from sea urchin eggs hardens the fertilization membrane with
    tyrosine crosslinks. Proc. Nat. Acad. Sci. USA 74:4214-4218.

21. Gache, C., Niman, H.L., and Vacquier, V.D. 1983. Monoclonal anti-
    bodies to the sea urchin egg vitelline layer inhibit fertili-
    zation by blocking sperm adhesion. Exp. Cell Res. 147:75-84.

22. Glabe, C.G., and Vacquier, V.D. 1977. Isolation and characteriza-
    tion of the vitelline layer of sea urchin eggs. J. Cell Biol.
    75:410-421.

23. Glabe, C.G., and Vacquier, V.D. 1978. Egg surface glycoprotein
    receptor for sea urchin sperm bindin. Proc. Nat. Acad. Sci.
    USA 75:881-885.

24. Gould-Somero, M., and Holland, L. 1975. Oocyte differentiation in
    Urechis caupo (Echiura): A fine structural study. J.
    Morphol. 147:475-506.

25. Gray, J., Justice, R., Nagel, G.M., and Carroll, E.J., Jr. 1986.
    Resolution and characterization of a major protein of the sea
    urchin hyaline layer. J. Biol. Chem., in review.

26. Hall, H.G. 1978. Hardening of the sea urchin fertilization enve-
    lope by peroxidase catalyzed phenolic coupling of tyrosines.
    Cell 15:343-355.
27. Harvey, E.N. 1910. The mechanism of membrane formation and other
    early changes in developing sea urchin eggs as bearing upon
    the problem of artificial parthenogenesis. J. Exp. Zool.
    8:355-376.
28. Hobson, A.D. 1932. On the vitelline membrane of the egg of
    Psammechinus miliaris and Teredo norvegica. Brit. J. Exp.
    Biol. 9:93-106.
29. Inoue, S., and Hardy, J.P. 1971. Fine structure of the fertiliza-
    tion membranes of sea urchin embryos. Exp. Cell. Res. 68:259-
    272.
30. Justice, R.W., and Carroll, E.J., Jr. 1986. Effects of cortical
    granule enzyme inhibitors on development of the ultrastructure
    and polypeptide spectrum of the sea urchin fertilization
    envelope. In preparation.
31. Kay, E., Eddy, E.M., and Shapiro, B.M. 1982. Assembly of the
    fertilization membrane of the sea urchin: Isolation of a
    divalent cation-dependent intermediate and its crosslinking in
    vitro. Cell 29:867-875.
32. Laemmli, U.K. 1970. Cleavage of structural proteins during
    assembly of the head of bacteriophage T4. Nature (London)
    227:680-685.
33. Lillie, R.S. 1911. The physiology of cell division. III. The
    action of calcium salts in preventing the initiation of cell
    division in unfertilized eggs through isotonic solution of
    sodium salts. Am. J. Physiol. 27:289-307.
34. Loeb, J. 1913. In: "Artificial Parthenogenesis and Fertilization"
    (Chicago: University of Chicago Press), p. 208.
35. Loeb, J. 1916. In: "The Organism as a Whole" (New York: G. P.
    Putnam's Sons), p. 108.
36. McBlaine, P.J., and Carroll, E.J., Jr. 1980. Sea urchin egg
    hyaline layer: Evidence for the localization of hyalin on the
    unfertilized egg surface. Dev. Biol. 75:137-147.
37. Motomura, I. 1941. Materials in the fertilization membrane in the
    eggs of echinoderms. Sci. Rep. Tohoku Univ. (4) 16:345-363.
38. Moy, G.W., and Vacquier, V.D. 1979. Immunoperoxidase localization
    of bindin during sea urchin fertilization. Curr. Top. Dev.
    Biol. 13:31-44.
39. Niman, H.L., Hough-Evans, B.R., Vacquier, V.D., Britten, R.J.,
    Lerner, R.A., and Davidson, E.H. 1984. Proteins of the sea
    urchin egg vitelline layer. Dev. Biol. 102:390-401.
40. Rossignol, D.P., Earles, B.J., Decker, G.L., and Lennarz, W.J.
    1984. Characterization of the sperm receptor on the surface of
    eggs of Strongylocentrotus purpuratus. Dev. Biol. 104:308-
    321.
41. Rossignol, D.P., Roschelle, A.J., and Lennarz, W.J. 1981. Sperm-
    egg binding: Identification of a species-specific sperm

receptor from eggs of Stronglylocentrotus purpuratus. J.
Supramol. Struct. Cell Biochem. 15:347-358.

42. Santiago, L., and Carroll, E.J., Jr. 1986. Sea urchin embryo
    fertilization envelope: Immunological evidence that envelope
    hardening involves intermolecular crosslinking of vitelline
    envelope polypeptides. Dev. Biol., in review.

43. Schmell, E., Earles, B.J., Breaux, C., and Lennarz, W.J. 1977.
    Identification of a sperm receptor on the surface of the eggs
    of the sea urchin Arbacia punctulata. Dev. Biol. 72:35-46.

44. Schuel, H., Schuel, R., Dandekar, P., Boldt, J., and Summers,
    R.G. 1982. Sodium requirements in hardening of the fertiliza-
    tion envelope and embryonic development in sea urchins. Biol.
    Bull. 162:202-213.

45. Stevens, R.E., and Kane, R.E. 1970. Some properties of hyalin.
    The calcium-insoluble protein of the hyaline layer of the sea
    urchin egg. J. Cell Biol. 44:611-617.

46. Sugiyama, M. 1938. Effect of some divalent ions upon the mem-
    brane development of sea urchin eggs. J. Fac. Sci. Imp. Univ.
    Tokyo (4) 4:501-508.

47. Sugiyama, M. 1938. Further studies on the development of the
    fertilization membrane in sea urchin egg. Annot. Zool.
    Japon. 17:360-364.

48. Sugiyama, M. 1951. Re-fertilization of the fertilized eggs of the
    sea urchin. Biol. Bull. 101:335-344.

49. Tegner, M.J., and Epel, D. 1973. Sea urchin sperm-egg interac-
    tions studied with the scanning electron microscope. Science
    179:685-688.

50. Veron, M., Foerder, C., Eddy, E.M., and Shapiro, B.M. 1977.
    Sequential biochemical and morphological events during
    assembly of the fertilization membrane of the sea urchin.
    Cell 10:321-328.

51. Villacorta-Moeller, M.N., and Carroll, E.J., Jr. 1982. Sea urchin
    embryo fertilization envelope: Immunological evidence that
    soluble envelope proteins are derived from cortical granule
    secretions. Dev. Biol. 94:415-424.

52. Voller, A., Bidwell, D., and Bartless, A. 1976. Microplate enzyme
    immunoassay for the immunodiagnosis of virus infections. In:
    "Manual of Clinical Immunology," N. Rose and W. Friedman,
    eds., American Society for Micorbiology, Washington D.C.,
    p. 506-512.

53. Yoshida, M., and Aketa, K. 1983. A 225 kdalton glycoprotein is
    the active core structure of the sperm-binding factor of the
    sea urchin Anthocidaris crassispina. Exp. Cell Res. 148:243-
    248.

CHARACTERIZATION OF THE STRONGYLOCENTROTUS PURPURATUS EGG CELL SURFACE RECEPTOR FOR SPERM

Norka Ruiz-Bravo, Daniel P. Rossignol, Glenn L. Decker,
Lawrence I. Rosenberg, and William J. Lennarz

Department of Biochemistry and Molecular Biology
The University of Texas
M. D. Anderson Hospital and Tumor Institute
6723 Bertner Avenue
Houston, Texas 77030

SUMMARY

In earlier studies from our laboratory, the intact sperm receptor was partially purified from Strongylocentrotus purpuratus crude
egg membranes, but due to its insolubility, it was not possible to
purify it to homogeneity. Nonetheless, this receptor preparation
bound with species specificity to acrosome-reacted sperm, thereby
inhibiting fertilization (17). Antibodies against the partially pure
receptor inhibited fertilization in S. purpuratus (but not Arbacia
punctulata) by coating the egg surface, indicating the presence of
binding sites that can be species-specifically recognized by both
sperm and antibody molecules.

Recently we were able to further purify and characterize
the receptor from S. purpuratus eggs. Chaotropic agent solubilization of the receptor prepared from crude egg membranes yielded a very
high molecular weight glycoconjugate that had many of the properties
of a proteoglycan. The receptor interacted with bindin in an in
vitro assay and bound with species specificity to acrosome-reacted
sperm to inhibit fertilization. Unfortunately, this receptor preparation was soluble only in certain chaotropic agents (16).

Exhaustive Pronase digestion of the intact receptor yielded a
soluble high-molecular-weight ($>10^6$) polysaccharide that was virtually devoid of protein. This glycosaminoglycan-like fragment was highly sulfated, and contained fucose, galactosamine, and iduronic acid
(16). The fragment inhibited fertilization, but did not do so with
species specificity.

293

Recently, soluble molecules with receptor activity were
generated by treating intact dejellied eggs with trypsin. These
proteolytically derived molecules contained (on a weight basis)
approximately equal amounts of protein and carbohydrate. Importantly, they inhibited fertilization with species specificity. These
results suggested that the binding activity was conferred by the
polysaccharide component of the receptor and that the intact receptor
and the tryptic fragments contained structural elements in the polypeptide chain necessary for species recognition.

INTRODUCTION

Gamete interaction in sea urchins occurs with a remarkable
degree of species specificity. Previous studies with two sea urchin
species, Strongylocentrotus purpuratus and Arbacia punctulata, have
established that the binding of sperm to egg occurs via the interaction of complementary cell surface ligands (8,18). Bindin, a 30.5-
kilodalton molecule has been isolated from sperm, and there is ample
evidence for its involvement in the adhesion of sperm to egg (24).
Bindin has been localized to both the acrosomal filament and to the
site of sperm and egg binding (15,24). In addition, aggregates of
bindin preferentially bind to eggs of the homologous species, causing
them to agglutinate (9,10). By contrast, the complementary adhesive
molecule on the egg surface has been less amenable to isolation and
characterization, and attempts to purify an intact receptor for sperm
have until recently met with limited success.

MATERIALS AND METHODS

Materials   Aprotonin, 3-amino-1,2,4-triazole, diisopropyl-
fluoro-phosphate, and soybean trypsin inhibitor were obtained from
Sigma Chemical Company, St. Louis, Missouri. Iodogen was purchased
from Pierce Chemical Company, Rockford, Illinois. Guanidine-
thiocyanate was obtained from Fluka, Hauppauge, New York, and
guanidine-HCl from Bethesda Research Laboratories, Inc., Gaithers-
burg, Maryland. Na^{125}I was purchased from Amersham Radiochemicals,
Arlington Heights, Illinois. Pronase was obtained from Calbiochem
Behring Corporation, LaJolla, California, and CsCl from Beckman
Instruments, Inc., Berkeley, California. Bio-Gel gel filtration
matrix was purchased from Bio-Rad Laboratories, Rockville Centre, New
York, and DEAE-cellulose (DE52) from Whatman, Inc., Clifton, New
Jersey. Sepharose and Sephacryl S1000 were obtained from Pharmacia,
Piscataway, New Jersey. All other chemicals were reagent grade or
better. Artificial sea water (ASW) (Instant Ocean) was obtained from
Aquarium Systems, Eastlake, Ohio.

Preparation of bindin and membranes   A. punctulata and S.
purpuratus were maintained at 18 and 7°C, respectively, in aquaria

containing ASW. Gametes were obtained by injection of 0.5 M KCl or
by electric shock. Bindin was isolated from sperm as described by
Vacquier and Moy (24), with the addition of 0.3 M sucrose to the
buffer used to dissociate the acrosomal vesicle. The bindin preparations were greater than 90% pure as assessed by sodium dodecyl
sulfate polyacrylamide gel electrophoresis.

Purification of Intact Receptor and Receptor Fragments   The
jelly coat was removed from eggs by titrating a 10% suspension of
eggs to pH 5 for 2 min and readjusting the pH to 8 with Tris-HCl.
Dejellied eggs were washed in Ca^{2+}-free sea water (8) and membranes
isolated in the presence of 0.1 mM diisopropylfluorophosphate,
Aprotonin (1000 units/ ml), soybean trypsin inhibitor (0.01 mg/ml),
and 3-amino-1,2,4-triazole (2 mM) (16). The membranes were washed
twice in ASW containing 10 mM Tris, pH 8, resuspended by sonication,
and the intact receptor partially purified by detergent extraction
and density gradient centrifugation (17). Further purification was
by solubilization in 4 M guanidine thiocyanate in the presence of 10
mg/ml dithiothreitol (DTT) and gel exclusion chromatography on a
column of Sepharose CL4B or 2B in 4 M guanidine HCl, 10 mM DTT, 10 mM
Tris-HCl, pH 8 (16).

For the generation of Pronase receptor fragments, the high-
molecular-weight intact receptor purified by gel exclusion chromatography was dialyzed against 10 mM Tris-HCl, pH 8, and exhaustively
digested with 1 mg Pronase/100 mg of receptor protein (16). The
digest was heated at 100°C for 20 min, and insoluble material was
removed by centrifugation at 100,000 x g for 1 hr. The soluble
material was chromatographed on a column of Bio-Gel A5m, and the
biologically active fractions pooled, dialyzed against distilled
water, and loaded onto a DEAE-cellulose column. After the column was
washed with 3-4 column volumes of 5 mM NaCl, 1 mM Tris, pH 8, a
linear gradient of 5 mM NaCl, 1 mM Tris-HCl, pH 8 to 750 mM NaCl, 1
mM Tris-HCl, pH 8, and a 2 M NaCl step were used to elute the
material bound to the column. The active fractions were pooled,
dialyzed against distilled water, lyophilized, and resuspended in a
buffer of 50 mM NaCl, 1 mM Tris-HCl, pH 8, 0.02% NaN_3. They were
then chromatographed on a column of Sephacryl S1000 using that same
buffer (16).

For the generation of cell surface tryptic fragments, dejellied
eggs were washed at least twice in ASW and resuspended to 20% (v:v)
in ASW containing trypsin at a final concentration of 100 μg/ml
suspension. The reaction was stopped after 5 min by addition of
soybean trypsin inhibitor (3 mg/mg trypsin) and Aprotonin (12 trypsin
inhibitor units/100 ml suspension). The eggs were allowed to settle
and the supernatant concentrated and spun at 100,000 x g for 1 hr.
The soluble material was chromatographed on a Sepharose CL4B column
(1.6 x 45 cm) in 0.5 M NaCl, 10 mM Tris, pH 8, 0.02% NaN_3. The
active fractions were pooled, dialyzed against 10 mM NaCl, 1 mM Tris-

HCl, pH 8, 0.02% NaN$_3$, and loaded on a DEAE-cellulose column (2.2 x 2.8 cm). After the column was washed with at least 10 column volumes, a linear gradient was constructed using 64 ml each of 10 mM NaCl, 1 mM Tris-HCl, pH 8, 0.02% azide, and 1 M NaCl, 1 mM Tris-HCl, pH 8, 0.02% azide. A 2 M NaCl step and a 5 M NaCl step were used to elute the remaining material bound to the column.

Preparation of Antibodies   Density gradient-isolated intact receptor (17) was emulsified 1:1 with complete Freund's adjuvant. Approximately 2 ml (0.5 mg receptor protein) was injected subcutaneously at six dorsal sites in each of two New Zealand rabbits. An antigen boost of 0.55 mg was similarly injected 3 weeks later, and the rabbits were exsanguinated 2 weeks after the boost.

IgG was isolated from immune serum on a column of Sepharose-protein A as previously described (14). The IgG fraction was concentrated by ammonium sulfate precipitation, and the resolubilized IgG dialyzed first against 0.54 M NaCl, 10 mM KCl, and then against ASW. The preparation was stored at -20°C. Fab fragments were prepared by digestion with mercuripapain as previously described (13). The extent of Fab formation was monitored by gel electrophoresis.

Indirect Immunoperoxidase Staining   Eggs (2% v/v) were fixed for 15 min in 3% formaldehyde in ASW at 0°C. After two washes in ASW containing 10 mM glycine, the eggs were incubated with IgG (final concentration 0.6 mg/ml) for 0.5-1 h at 4°C. Unbound IgG was removed by washing the eggs in ASW. Goat anti-rabbit peroxidase-conjugated antibody was added to a final concentration of 30 µg/ml and allowed to bind for 1 h. Unbound antibody was removed by washing in ASW, and the eggs were treated with 3% glutaraldehyde for 15 min. The eggs were then washed with staining buffer (0.5 M NaCl, 0.05 M Tris-HCl, pH 7.6) and suspended in staining buffer containing 3,3'-diamino-benzidine (0.5 mg/ml) and 0.02% hydrogen peroxide. After 10 min, the samples were again washed in staining buffer, post-fixed in osmium tetroxide, and processed for electron microscopy (3). The above procedures were also followed for immunoperoxidase staining of zygotes.

In Vitro Assay for Receptor-Bindin Interaction   ^{125}I-labeled intact receptor was incubated at room temperature in 12 x 77 mm polypropylene tubes with or without bindin (10 µg) in a final volume of 50 µl, and gently shaken (60-80 rpm) for 20 min. One hundred microliters of Triton X-100 (1.5% final concentration) in ASW was added and the mixture incubated for an additional 5 min. Duplicate 50-µl samples were layered onto 100-µl cushions of 20% sucrose (in ASW) in polypropylene microfuge tubes. The samples were centrifuged for 1 min at 10,000 x g and quick-frozen in dry ice. The bottom 3-5 mm of each tube was cut and both bottom and top of each tube were counted separately in a Packard gamma counter. Competition for binding was assessed by preincubating the competitor with bindin for

20 min in a final volume of 40 µl before addition of ^{125}I-labeled intact receptor.

Fertilization Assays   Fertilization assays to test the activity of receptor preparations were done using acrosome-reacted sperm. Limiting amounts of sperm (i.e., enough to achieve no greater than 90% fertilization) were incubated for 20 sec in ASW containing homologous egg jelly. In the case of A. punctulata sperm, incubation with jelly coat was carried out in ASW containing 30 mM Ca^{2+}. Samples of receptor were added to the jelly coat mixture prior to the addition of sperm, ensuring that sperm were exposed to receptor concomitantly with induction of the acrosome reaction, but prior to their addition to eggs. Fertilization was scored as the percentage of eggs that elevated fertilization envelopes 5 min after addition of acrosome-reacted sperm. The data were normalized to controls that contained sperm at the same dilutions but had been preincubated with jelly coat in the absence of receptor. Sperm were placed on substrate films for electron microscopy, and acrosome reactions quantitated by counting a minimum of 100 sperm (19).

Assays to monitor the effect of anti-receptor IgG or Fab fragments were also done using dejellied eggs and limiting amounts of sperm. A 0.5% suspension of eggs was diluted 1:1 with IgG or Fab fragments in ASW and incubated at room temperature for 30-40 min with gentle, continuous shaking. The eggs were washed twice and resuspended to a final concentration of 0.5% in ASW containing enough homologous crude egg jelly to cause the sperm-acrosome reaction (19). As before, fertilization was scored as the percentage of eggs that elevated fertilization envelopes after addition of a limiting amount of sperm.

Other Methods   Receptor and jelly coat (15 mg protein) were iodinated by the method of Fraker and Speck (7) with 4µg Iodogen and carrier-free ^{125}I (17.26 mCi/g) at 0°C for 10 min. Jelly coat was purified by gel filtration chromatography on Sepharose CL4B and the fucan sulfate polymer by ion exchange chromatography (20). Hexose content was determined by the phenol-sulfuric acid assay (5), and amino sugars were detected by the method of Svennerholm (22). Uronic acid content was determined as described by Dische (4). Iduronic acid content was determined by the method of Brown (1) and confirmed as described by Spiro (21). Sulfate was analyzed according to Terho and Hartiola (23), and saccharides were determined by gas-liquid chromatography (2).

RESULTS AND DISCUSSION

Previous work from this laboratory provided evidence that the S. purpuratus receptor is an egg cell membrane-associated glycoconjugate that binds sperm with species specificity. Fertilization was inhib-

ited by cell membrane preparations from eggs of the same (but not a
different) species (18). The intact sperm receptor was isolated from
crude egg cell membranes by detergent solubilization with 1.5% Triton
X-100 followed by density gradient centrifugation (1) (Fig. 1). Membranes were isolated, radioiodinated, and incubated with or without
bindin. Triton X-100 was added and the mixture loaded onto gradients
consisting of CsCl overlaid with a continuous sucrose density gradient.

After centrifugation, 0.5 ml fractions were collected and
assayed for radioactivity and refractive index. As shown in Fig. 1A,
a labeled fraction sedimented at a position corresponding to the
density of bindin. When the crude ^{125}I-receptor fraction was isolated from membranes by density gradient centrifugation, it retained
the capacity to interact with bindin (Fig. 1B). In addition to its
ability to interact with bindin, this receptor preparation was able
to bind to sperm only after they had undergone the acrosome reaction.
Most important, the partially pure receptor inhibits fertilization of
S. purpuratus but not A. punctulata eggs, showing that it retains its
species specificity (17). Unfortunately, the receptor isolated by
this density gradient procedure was not truly soluble and remained
particulate in a wide variety of detergents and organic solvents.
Its large size and insolubility precluded further analysis by conventional methods.

An immunological approach was therefore taken to examine the
species specificity and binding activity of the receptor. Polyclonal
antibodies were generated in rabbits, using the density gradient
purified receptor as immunogen. Treatment of eggs with either anti-
receptor IgG or Fab fragments inhibited fertilization with species
specificity (Fig. 2). Pretreatment of S. purpuratus eggs with 45
µg/ml IgG inhbited fertilization by 80%; 95% inhibition of fertilization occurred at 270 µg/ml IgG. Preimmune IgG at 270 µg/ml
resulted in less than 25% inhibition of fertilization. Most important, neither preimmune nor anti-S. purpuratus receptor IgG had any
effect on the ability of A. punctulata sperm to fertilize homologous
eggs. Similar results were obtained using Fab fragments from both
anti-receptor IgG and preimmune IgG (Fig. 3).

Previous studies established that Pronase treatment of the
density gradient-isolated receptor yields a high-molecular-weight
carbohydrate-rich fragment that binds to sperm, inhibits fertilization (16), and competes for bindin binding to whole eggs or to the
receptor (8,12,16). It was therefore of interest to determine if
this carbohydrate-rich fragment blocked antibody binding to the
receptor (Fig. 4). The density gradient-purified receptor was
allowed to dry on a microtitre dish, and its ability to bind ^{125}I-
anti-receptor IgG challenged by the Pronase receptor fragment. Two
micrograms (hexose equivalents) of the carbohydrate-rich receptor
fragment showed a maximal antibody-neutralizing activity; 50% of max-

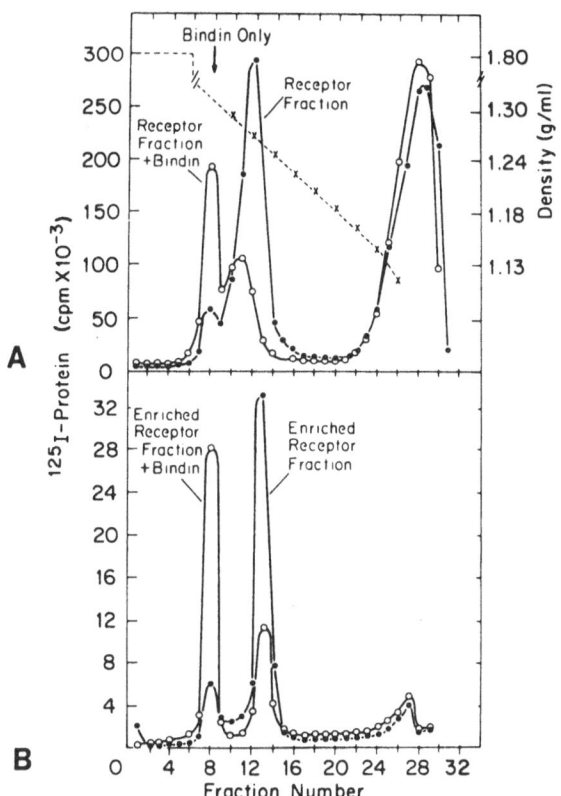

Figure 1. A. Interaction of bindin and components of Triton X-100-
treated membranes from S. purpuratus. Membranes were isolated and radioiodinated as described in Materials and
Methods. Twenty micrograms of membrane protein were incubated with(0—0) or without (●—●) 15 µg bindin for 20 min
with gentle shaking. Triton X-100 in ASW was added to a
final concentration of 1.5%, and the mixture was incubated
with shaking for an additional 5 min. Samples were diluted
to a final detergent concentration of 0.15% and loaded onto
gradients consisting of a 3-ml cushion of CsCl (1.07 g/ml)
in Tris-buffered ASW, overlaid with a continuous sucrose
density gradient of 60% (w/w) sucrose to 28% (w/w) sucrose
in Tris-buffered ASW. After centrifugation for 16 hr at
76,000 x g (average), 0.5 ml fractions were collected and
assayed for radioactivity and refractive index. The arrow
indicates the density at which bindin alone sediments. B.
Interaction of isolated receptor fraction. Receptor fraction was separated from detergent-treated membranes (without bindin) as described in Figure 1A. Fractions 10-14,
containing ^{125}I-labeled receptor, were combined, diluted
with Tris-buffered ASW, and centrifuged at 100,000 x g for

1 h. The pellet was washed in Tris-buffered ASW and
resuspended by sonication. The recovery of ^{125}I label was
greater than 95% by this procedure. This suspension was
incubated in the absence (●—●) or presence (0—0) of
bindin (100 μg), and centrifuged as described above.

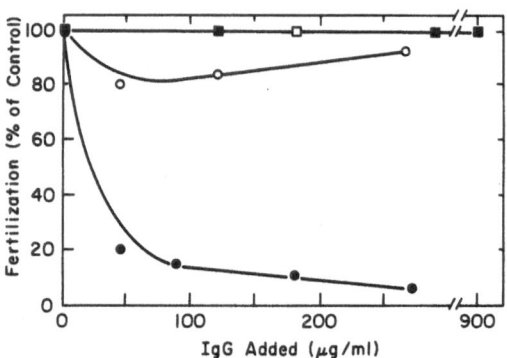

Figure 2. Inhibition of fertilization of anti-sperm receptor IgG.
           Eggs of S. purpuratus (●,0) or A. punctulata (■,□) were
           treated with anti-S. purpuratus receptor (●,■) or pre-
           immune (0,□) IgG at the indicated concentrations. The
           eggs were washed and fertilization was assayed as described
           in Materials and Methods.

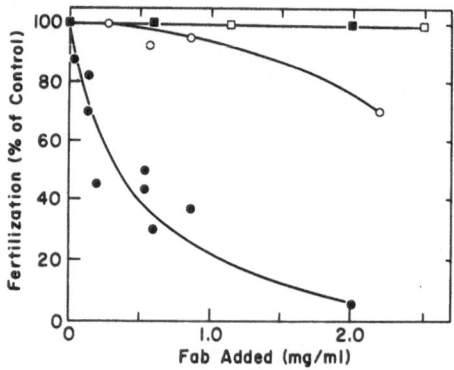

Figure 3. Inhibition of fertilization by anti-receptor Fab. Eggs of
           S. purpuratus (●,0) or A. punctulata (■,□) were treated
           with anti-receptor (●,■) or preimmune (0,□) Fab at the
           indicated concentrations. The eggs were washed and ferti-
           lization was assayed as described in Materials and
           Methods.

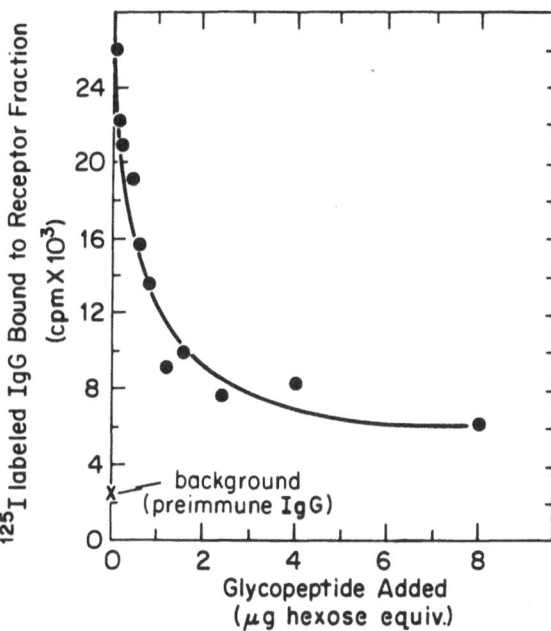

Figure 4. Inhibition of antibody binding to the intact receptor by
the Pronase-derived receptor fragment. Radioiodinated
intact receptor (0.4 µg hexose, 1.4 µg protein) was allowed
to dry overnight in a Linbro microtitre dish. The wells
were washed with 5% bovine serum albumin in phosphate-
buffered saline. IgG (11.8 µg) was added in the presence of
the indicated amounts of the Pronase-derived receptor fragments.

imal neutralization was obtained with 0.6 µg. Since the Pronase-
derived fragment can neutralize up to 85% of the anti-receptor IgG,
the carbohydrate chains of the receptor seem to be a major antigenic
determinant.

Indirect immunoperoxidase staining was used to localize the
receptor on the egg cell surface and to determine its fate during
fertilization envelope elevation. Eggs or zygotes treated with anti-
receptor IgG at a concentration known to completely inhibit fertilization were exposed to immunoperoxidase-conjugated goat anti-rabbit
IgG and stained for peroxidase activity. Anti-receptor antibodies
completely coated the egg surface prior to fertilization (Fig. 5a).
Similar treatment with preimmune IgG revealed no such staining (Fig.
5b). If staining was carried out 1 min after fertilization, the
antigenic material was nearly completely elevated from the egg surface, and the antibodies were bound to the newly formed fertilization
envelope (Fig. 5c). No staining was observed when preimmune IgG was

added to fertilized eggs. This is an important control for the
results in Figure 5c because the ovoperoxidase released at fertilization (6) can, under certain conditions (11), catalyze reaction product formation in the absence of peroxidase-conjugated second antibody.

As previously mentioned, the receptor preparation isolated by
density gradient centrifugation was particulate and remained
insoluble in a wide variety of detergents and organic solvents. This
partially pure receptor was, however, soluble in 4 M guanidine
thiocyanate containing 10 mg/ml DTT. The receptor was therefore
radioiodinated, solubilized, and subjected to gel filtration chromatography on a column of Sepharose CL4B in guanidine hydrochloride and
DTT (Fig. 6). Two peaks of radioactivity were observed. One peak
eluted in the void fraction and a second, broader peak or group of
peaks eluted near the included volume. In the absence of DTT, there
was a 60% increase in the amount of ^{125}I-protein that eluted in the
void fraction of the column. Column fractions were pooled as indicated, dialyzed and assayed for their ability to bind to bindin
particles and to inhibit fertilization with species specificity.

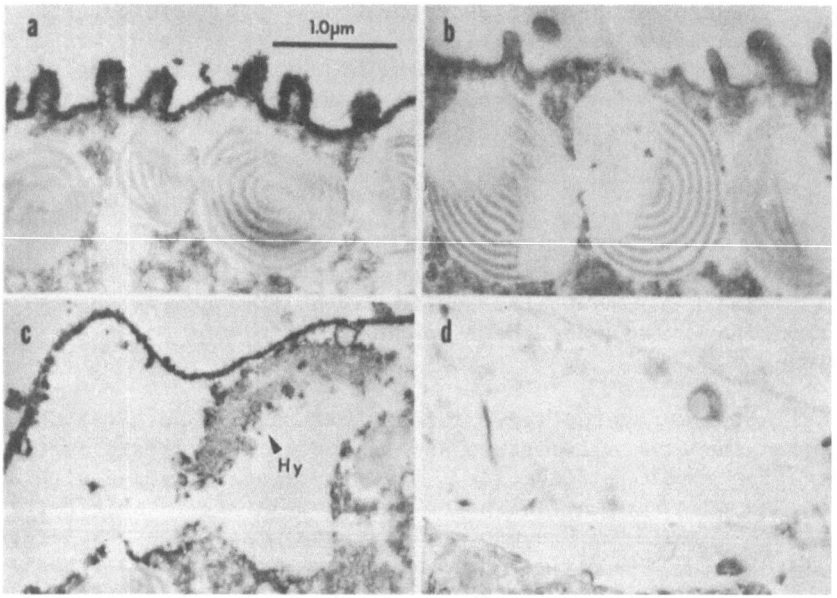

Figure 5. Immunoperoxidase staining of sperm receptor. S. purpuratus
        eggs (a,b) and zygotes 1 min after fertilization (c,d) were
        fixed and treated with anti-receptor antibody (a,c) or
        antibody from preimmune serum (b,d). Immunoperoxidase
        staining was as described in Materials and Methods.

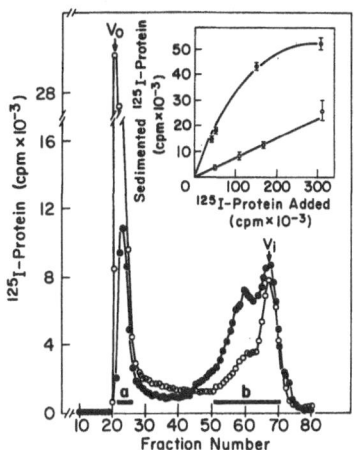

Figure 6. Gel-exclusion chromatography of the ^{125}I-labeled receptor. ^{125}I-labeled receptor (196,000 cpm; 34 µg protein) was solubilized in 4 M guanidine thiocyanate, 10 mM Tris HCl (pH 8) in the absence (O) or presence (●) of dithiothreitol (10 mg/ml), and chromatographed on a column (1 x 55 cm) of Sepharose CL4B in 4 M guanidine HCl, 10 mM Tris-HCl (pH 8) in the absence or presence of 1 mM dithiothreitol. The ^{125}I-labeled proteins were pooled and assayed for their ability to bind to bindin. Inset -- The sedimentation of various amounts of pool a recovered from a column run in the presence of dithiothreitol was measured in the presence (●) or absence (O) of bindin. No bindin-dependent sedimentation of the ^{125}I-labeled protein in pool b was detected.

As previously mentioned, bindin is a low-molecular-weight protein on the tip of acrosome-reacted sperm. A sperm is thought to bind to a receptor(s) on the egg via its bindin-coated acrosomal filament (15). The assay used to monitor receptor interaction with bindin takes advantage of the particulate nature of bindin (24). As shown in Figure 7, sedimentation of ^{125}I-labeled receptor was dependent on the concentration of bindin and on the amount of added receptor. However, no species preference was seen for the binding of homologous over heterologous receptors, indicating that under these conditions the interaction of receptor and bindin was not species specific.

In vitro bindin assays of the material chromatographed on Sepharose CL4B showed that only the material in pool a (Fig. 6) interacted with bindin; no binding activity was detected in pool b. Assays to measure inhibition of fertilization revealed that the pool a material species-specifically inhibited fertilization, indicating that it is the sperm receptor (Fig. 8). S. purpuratus sperm were preincubated with receptor from S. purpuratus or A. punctulata (prepared in a sim-

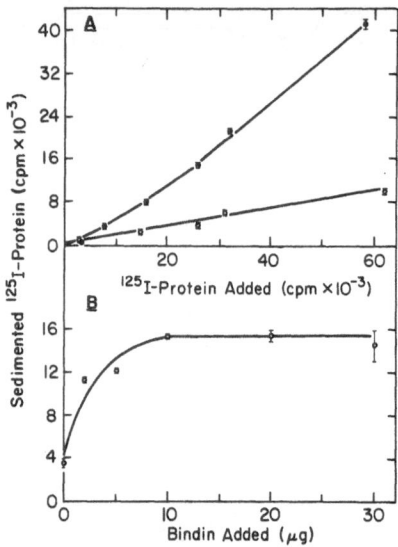

Figure 7. Bindin-mediated sedimentation of the ^{125}I-labeled sperm
          receptor.   A.   Binding as a function of receptor concentra-
          tion. The indicated amount of [125]-I-labeled receptor was
          incubated with (●) or without (0) bindin (10 µg) and
          assayed as described in Materials and Methods. The error
          bars represent the standard deviation (n = 2-4). B. [125]I -
          Receptor binding as a function of bindin concentration.
          The indicated amount of bindin was incubated with the [125]I-
          labeled receptor (150,000 cpm), and binding was assayed as
          described in Materials and Methods.

ilar manner) in the presence of sufficient jelly coat to cause the
acrosome reaction. The acrosome-reacted sperm were then tested for
their ability to fertilize homologous eggs. Nearly complete inhibition of fertilization occurred when 1.6 µg (hexose equivalents) of S.
purpuratus receptor was added to the S. purpuratus sperm. In contrast, preincubation of S. purpuratus sperm with as much as 5 µg
(hexose equivalents) of receptor from A. punctulata did not reduce
their ability to fertilize eggs. In addition, S. purpuratus receptor
affected the ability of A. punctulata sperm to fertilize A. punctulata
eggs only at a very high concentration of hexose equivalents. Ninety
micrograms of hexose equivalents were required for 50% inhibition
(Data not shown).

      Although the receptor was soluble in guanidine hydrochloride, it
did not remain so after removal of the chaotrope by dialysis.
Because of its insolubility and/or high molecular weight, it has not
been possible to determine the number and size of the polypeptide
chains in the receptor preparation. Amino acid analysis did not re-

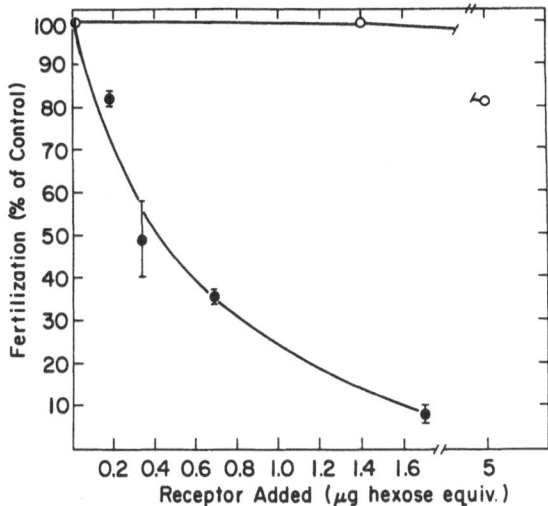

Figure 8. Inhibition of fertilization by isolated sperm receptor.
S. purpuratus sperm (1.5 mg protein/ml) were treated with
receptor prepared from S. purpuratus (●) or A. punctulata
(0) eggs, and their ability to fertilize was assessed as
described in Materials and Methods.

veal unusual composition or the presence of acid-stable cross-linked
amino acids. Amino terminal analysis showed four N-terminal amino
acids, indicating that the receptor was not a gross heterogeneous
mixture of many insoluble proteins. Further analysis of the intact
receptor indicated that carbohydrate accounts for about 12% of its
total mass. Detailed carbohydrate analysis (Table 1) showed that the
receptor contained neutral sugars, amino sugars, uronic acid, and
sulfate. Sialic acid and phospholipid levels were below the limits
of detection. The receptor was very resistant to enzymatic degradation by endo or exoglycosidases. Treatment with mixed sulfatases,
hyaluronidase, chondroitin ABCase, or mixed glycosidases resulted in
no significant decrease in its ability to bind to bindin and did not
result in solubiization.

Because the intact species-specific receptor remained insoluble,
the approach that we took to gain insight into its structure and biological properties was to generate soluble proteolytic fragments and
compare their chemistry and biological properties with those of the
intact receptor. As previously noted, digestion of the isolated
intact receptor or the egg cell surface by Pronase yielded a soluble,
carbohydrate-rich, glycopeptide that inhibited sperm-egg binding,
fertilization, and bindin-mediated egg agglutination, but did so
without species specificity (16). Preliminary experiments also indicated that limited trypsin digestion yielded a soluble receptor fragment that retained species specificity. Efforts in our laboratory,

Table 1. Chemical Composition of the Receptor

Component[a]	Amount (μmoles/mg protein)
Neutral Hexoses	550
Fucose[b]	160
Mannose	170
Galactose	66
Glucose	160
Hexosamine	200
Glucosamine	106
Galactosamine	94
Uronic acid	52
Sialic acid	<20
Sulfate[b]	880
Lipid PO_4	<10

[a] Analyses were performed as described in Materials and Methods.
[b] A portion of the fucose and sulfate content was subsequently found to be due to contaminating fucan sulfate from jelly coat. See text.

therefore, have been directed toward the generation of proteolytic fragments of the receptor using these two proteases. The postulated relationship between the intact receptor, the tryptic fragment, and the Pronase fragment are outlined below:

Intact receptor (polysaccharide–polypeptide)

↓ Trypsin

Tryptic fragments (polysaccharide–smaller polypeptide)

↓ Pronase

Pronase fragments (polysaccharide–little or no polypeptide)

Our working hypothesis is that the carbohydrate portion of the receptor is involved in binding and that the polypeptide is involved in species recognition.

Exhaustive Pronase digestion was used to generate a soluble, carbohydrate-rich fragment (which has almost no detectable protein) from the disulfide-reduced receptor purified by gel filtration. The soluble material after Pronase digestion was applied to a column of Bio-Gel A5m (Fig. 9). Most of the carbohydrate was low-molecular-

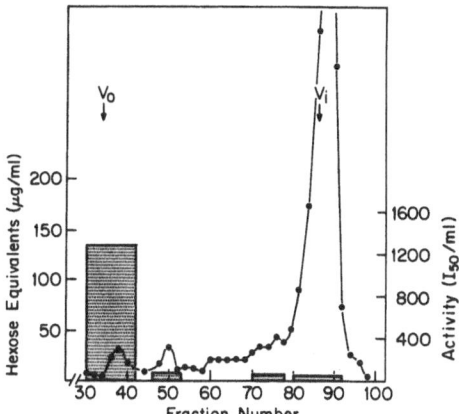

Figure 9. Chromatography of the carbohydrate-rich receptor fragment
on Bio-Gel A5m. The soluble material obtained after Pronase
digestion of the purified receptor was chromatographed on a
column of Bio-Gel A5m (1 x 75 cm) in 0.5 M NaCl, and the
neutral hexose content (O--O) of the fractions determined.
Fractions were pooled as shown and assayed for inhibition
of binding of receptor to bindin (shaded ---). The I_{50},
i.e., the volume of the fraction required to inhibit binding by 50%, was determined and the I_{50}/ml was calculated.

weight glycopeptide. However, most of the ability to inhibit receptor interaction with bindin was found in the carbohydrate peak at V_0.
This material was further purified by ion exchange chromatography
(Fig. 10). Two components that inhibited bindin receptor interaction
eluted starting at approximately 0.5 M and 0.7 M NaCl. The most
active fractions were high in sulfate and hexose, but contained
little or no peptide. The second peak of activity was probably a
contaminating fucan sulfate component of jelly coat, since this
material bound tightly to anion exchange resin (19). Consistent with
this possibility, fucan sulfate prepared from jelly coat and chromatographed under identical conditions eluted in a peak starting at
about 0.7 M NaCl (16). These results indicate that following ion
exchange, the carbohydrate-rich receptor fragment was free of contaminating fucan sulfate from jelly coat.

The receptor fractions were pooled and purified further on
Sephacryl S-1000 (Fig. 11). The broad peak of active material was
pooled as shown and its carbohydrate content determined. The Pronase
receptor fragment was a glycosaminoglycan-like polysaccharide consisting of galactosamine, fucose, and uronic acid. The fragment was
highly sulfated (0.8 mol/mol total carbohydrate), but other common
negatively charged moieties were not detectable. The Pronase receptor fragment was a potent inhibitor of fertilization. The higher

molecular weight fragments in pool a were approximately 10 times more active than the lower molecular weight material in pool b. As expected, because it had little or no detectable protein, the Pronase receptor fragments did not inhibit fertilization species specifically. These data are summarized in Table 2.

Further soluble receptor fragments that interact with bindin and inhibit fertilization with species specificity were generated by treating the cell surface of dejellied eggs with trypsin. The soluble material was applied to a column of Sepharose CL4B as described in Materials and Methods. The cell surface tryptic fragments containing carbohydrate eluted as a broad heterogeneous peak ranging in molecular weight from 1.5×10^5 to 5×10^6 or greater (Data not shown).

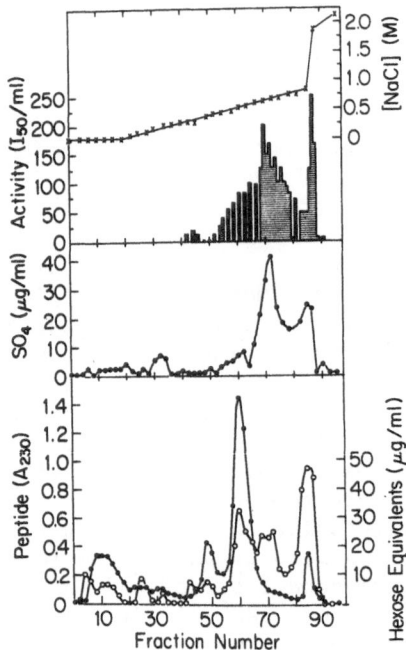

Figure 10. Ion exchange chromatography of the carbohydrate-rich receptor fragment. Active pooled fractions after chromatography on Bio-Gel A5m were applied to a column of DEAE-cellulose (2.7 x 7 cm). Samples were monitored for ability to inhibit binding of receptor to bindin (top panel). The I_{50}, i.e., the volume of the fraction required to inhibit binding by 50%, was determined, and the I_{50}/ml was calculated. Samples were also assayed for sulfate (middle panel) and neutral hexose (●—●) and peptide bond (0—0) content (bottom panel).

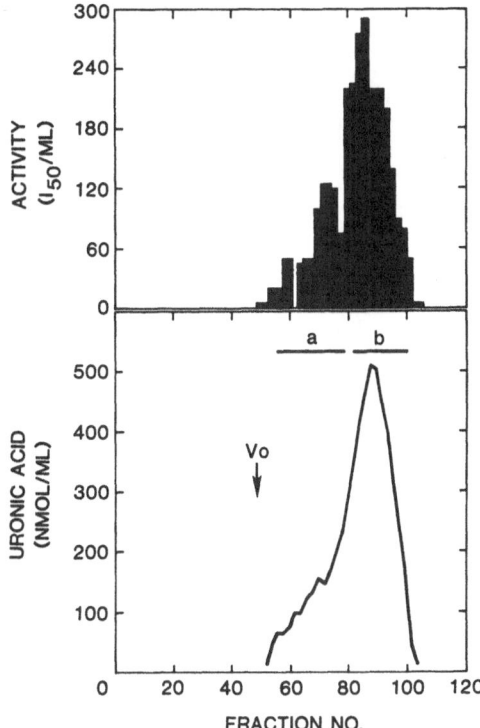

Figure 11. Gel filtration chromatography of the carbohydrate-rich receptor fragment on Sephacryl S1000. The active fractions (67–74) after ion exchange chromatography were pooled and fractionated on a column (1.2 x 113 cm) of Sephacryl S1000. Inhibitory activity and uronic acid content were measured; the peptide bond content (absorbance at 230 nm) was below the limits of detection.

Encouragingly, there were also significant amounts of protein in the high-molecular-weight fractions. This is in contrast to the carbohydrate-containing fragments obtained by Pronase, which were virtually devoid of protein. Fractions were pooled and their ability to bind to bindin tested in vitro. The most active inhibitors of bindin receptor interaction were the high-molecular-weight ($>10^6$) V_0 and slightly included fractions. When tested for their ability to interfere with the fertilizing ability of acrosome-reacted sperm, the high molecular weight fractions were again the most active, requiring approximately 11 µg (hexose equivalents) to inhibit fertilization by 50%.

The most active fractions were loaded onto a DEAE-cellulose column, washed, and eluted with a linear gradient of 10 mM NaCl to 1 M NaCl (data not shown). The hexose-containing material eluted as

Table 2. Biological Activity of Receptor Preparations

	Ratio Hexose:protein	Inhibitory activity[a]	
		S. purpuratus	A. punctulata
Intact receptor pool a from Sepharose CL2B	0.16:1	0.5	NI[b]
Pronase fragment pool DEAE cellulose	no detectable protein	0.2-2.0	0.6

[a] Inhibitory activity is expressed as μg hexose equivalents/reaction necessary to achieve 50% inhibition of fertilization.
[b] NI = no inhibition at equivalent concentrations.

two heterogeneous peaks between 130-300 mM NaCl and 400-610 mM NaCl. Fractions were pooled and tested for their effect on the fertilizing ability of S. purpuratus sperm (data not shown). Interestingly, the slightly charged material in pool a, which had a hexose-to-protein ratio of 0.70:1, was able to inhibit fertilization in S. purpuratus but not in A. punctulata. By contrast, the highly charged material in pool c, which was composed mainly of hexose (hexose-to-protein ratio 8:1), inhibited fertilization in both species of sea urchins at low concentrations. Pool b also inhibited fertilization without species specificity, probably because of lack of resolution from pool c. In a subsequent preparation of cell surface tryptic fragments, the highly charged material (pool c) was found to inhibit fertilization in S. purpuratus but not in A. punctulata. By contrast to the earlier preparation, this fraction had a hexose : protein ratio of 0.9:1, indicating that it was composed of approximately equal amounts of protein and hexose. These results support the hypothesis that the species specificity of the receptor is conferred by the polypeptide chain.

Although the cell surface tryptic fragments obtained in these preliminary studies inhibited fertilization species-specifically, as would be expected of the intact sperm receptor, their identity as the receptor remains to be established. Somewhat puzzling was their comparatively low activity (10 to 20-fold less active than the intact receptor or the Pronase fragment; data not shown). It is possible that, like the Pronase fragment, the tryptic fragments were heterogeneous in size and activity. Alternatively, the tryptic fragments derived from the cell surface may have been functionally related but chemically distinct from the previously isolated receptor. Experiments are in progress to determine whether these fragments bind species-specifically to acrosome-reacted sperm. In addition, the active fractions from DEAE-cellulose will be exhaustively digested with Pronase and their ability to inhibit fertilization species-specifically examined. If species recognition does in fact reside in the polypeptide portion of the receptor, the Pronase-digested tryptic fragments should adversely affect the fertilizing ability of both S. purpuratus and A. punctulata sperm.

CONCLUSIONS

1. A molecule(s) that serves as a species-specific receptor for sperm has been isolated from S. purpuratus eggs. This molecule is believed to be the sperm receptor based on the following criteria:

   a. The isolated receptor binds to bindin in an in vitro assay.

   b. The isolated receptor binds to sperm after (but not before) they have undergone the acrosome reaction.

   c. The isolated receptor inhibits fertilization by direct competition for binding sites on the sperm.

2. The binding activity of the receptor is conferred by the carbohydrate chain(s), whereas species recognition is conferred by the polypeptide chain(s). This conclusion is based on the following observations:

   a. The intact receptor (polysaccharide-polypeptide) inhibits fertilization species-specifically.

   b. Tryptic fragments (polysaccharide-smaller polypeptide) inhibit fertilization species-specifically.

   c. Pronase fragments (polysaccharide-little or no polypeptide) inhibit fertilization without species specificity.

ACKNOWLEDGEMENTS

We would like to thank Ms. Betty J. Earles and Mr. Dana Earles for technical assistance, and Ms. Diana L. Welch for preparation of this manuscript. This work was supported by National Institutes of Health grant HD-18590 to Dr. William J. Lennarz. Dr. Norka Ruiz-Bravo was supported by a National Institutes of Health Postdoctoral Fellowship HD-06477. Dr. William J. Lennarz, who is a Robert A. Welch Professor of Chemistry, gratefully acknowledges the Robert A. Welch Foundation.

REFERENCES

1. Brown, A.H. 1946. Determination of pentose in the presence of
   large quantities of glucose. Arch. Biochem. 2:269-278.
2. Conrad, G.W., Ager-Johnson, P., and Woo, M.L. 1982. Antibodies
   against the predominant glycosaminoglycan of the mammalian
   cornea, keratan sulfate. J. Biol. Chem. 257:464-471.
3. Decker, G.L., and Lennarz, W.J. 1979. Sperm binding and fertilization envelope formation in a cell surface complex isolated
   from sea urchin eggs. J. Cell Biol. 81:92-103.
4. Dische, F. 1947. A new specific color reaction of hexuronic
   acids. J. Biol. Chem. 167:189-198.
5. Dubois, M.K., Giles, A., Hamilton, J.K., Rebers, P.A., and
   Smith, F. 1956. Colorimetric method for determination of
   sugars and related substances. Anal. Chem. 28:350-356.
6. Foerder, C.A., and Shapiro, B.M. 1977. Release of ovoperoxidase
   from sea urchin eggs hardens the fertilization membrane with
   dityrosine crosslinks. Proc. Natl. Acad. Sci. USA 74:4214-
   4218.
7. Fraker, P.J., and Spreck, J.C., Jr. 1978. Protein and cell membrane iodinations with a sparingly soluble chloramide,
   1,3,4,6-tetrachloro-3a6a dipenyl glycoluril. Biochem.
   Biophys. Res. Commun. 80:849-857.
8. Glabe, C.G., and Lennarz, W.J. 1979. Species-specific adhesion in
   sea urchins: A quantitative investigation of bindin-mediated
   egg agglutination. J. Cell Biol. 83:595-604.
9. Glabe, C.G., and Lennarz, W.J. 1981. Isolation of a high molecular weight glycoconjugate derived from the surface of S.
   purpuratus eggs that is implicated in sperm adhesion. J.
   Supramolec. Struct. Cell Biochem. 15:387-394.
10. Glabe, C.G., and Vacquier, V.D. 1977. Species-specific agglutina-
    tion of eggs of bindin isolated from sea urchin sperm. Nature
    267:836-837.
11. Gulyas, B.J., and Schmell, E.D. 1980. Ovoperoxidase activity in
    ionophore treated mouse eggs. I. Electron microscopic
    localization. Gamete Res. 3:267-277.
12. Kinsey, W.J., and Lennarz, W.J. 1981. Isolation of a glycopeptide
    fraction from the surface of the sea urchin egg that inhibits
    sperm-egg binding and fertilization. J. Cell Biol. 91:325-
    331.
13. Kwapinski, G. 1982. In: "The Methodology of Investigative and
    Clinical Immunology", Robert E. Krieger Publishing Co.,
    Florida, pp. 173-174.
14. Lopo, A.C., and Vacquier, V.D. 1980. Antibody to a sperm surface
    glycoprotein inhibits the egg jelly induced acrosome reaction
    of sea urchin sperm. Dev. Biol. 79:325-333.
15. Moy, G.W., and Vacquier, V.D. 1979. Immunoperoxidase localization
    of bindin during sea urchin fertilization. Curr. Top. Dev.
    Biol. 13:31-44.

16. Rossignol, D.P., Earles, B.J., Decker, G.L., and Lennarz, W.J. 1984 Characterization of the sperm receptor on the surface of eggs of <u>Strongylocentrotus purpuratus</u>. Dev. Biol. 104:308-3210.

17. Rossignol, D.P., Roschelle, A.J., and Lennarz, W.J. 1981. Sperm-egg binding: Identification of a species-specific sperm receptor from eggs of <u>Strongylocentrotus purpuratus</u>. J. Supramolec. Struct. Cell Biochem. 15:347-358.

18. Schmell, E., Earles, B.J., Breaux, C., and Lennarz, W.J. 1977. Identification of a sperm receptor on the surface of the eggs of the sea urchin <u>Arbacia punctulata</u>. J. Cell Biol. 72:35-46.

19. SeGall, G.K., and Lennarz, W.J. 1979. Chemical characterization of the component of the jelly coat from sea urchin eggs responsible for induction of the acrosome reaction. Dev. Biol. 71:33-48.

20. SeGall, G.K., and Lennarz, W.J. 1981. Jelly coat and induction of the acrosome reaction in echinoid sperm. Dev. Biol., 86:87-93.

21. Spiro, M.J. 1977. Uronic acid analysis by automated anion exchange chromatography. Anal. Biochem. 82:348-352.

22. Svennerholm, L. 1956. The determination of hexosamines special references to nervous tissue. Acta. Soc. Med. Upsol. 61:287-306.

23. Terho, T.T., and Hartiola, K. 1971. Method for determination of the sulfate content of glycosaminoglycans. Anal. Biochem. 41:471-476.

24. Vacquier, V.D., and Moy, G.W. 1977. Isolation of bindin: The protein responsible for adhesion of sperm to sea urchin eggs. Proc. Natl. Acad. Sci. USA 74:2456-2460.

PEPTIDES ASSOCIATED WITH EGGS: MECHANISMS

OF INTERACTION WITH SPERMATOZOA

David L. Garbers[1,2,3], J. Kelley Bentley[2],
Lawrence J. Dangott[1,2], Chodavarapu S. Ramarao[2],
Hiromi Shimomura[1,2], Norio Suzuki[1,2], and David Thorpe[3]

The Howard Hughes Medical Institute[1]
Departments of Pharmacology[2]
Molecular Physiology and Biophysics[3]
Vanderbilt University School of Medicine
Nashville, TN 37232

SUMMARY

Speract (Gly-Phe-Asp-Leu-Asn-Gly-Gly-Gly-Val-Gly), a peptide
obtained from the culture medium of Strongylocentrotus purpuratus
eggs, stimulates the respiration and motility of S. purpuratus
spermatozoa under appropriate conditions. Resact (Cys-Val-Thr-Gly-
Ala-Pro-Gly-Cys-Val-Gly-Gly-Gly-Arg-LeuNH$_2$), a peptide obtained from
Arbacia punctulata eggs also stimulates the metabolism and motility
of A. punctulata spermatozoa, however, it fails to stimulate S.
purpuratus spermatozoa. Early biochemical responses of the spermatozoa to the egg peptides include a net H$^+$ efflux and elevations of
cyclic AMP and cyclic GMP concentrations. In addition, in A. punctulata spermatozoa, a major plasma membrane protein is modified in
response to resact such that its apparent molecular weight shifts
from 160,000 to 150,000. If cells are incubated with ^{32}P, the
160,000 molecular weight form of the protein becomes radiolabeled;
subsequent addition of resact causes a rapid loss of ^{32}P from the
protein. The plasma membrane protein appears to be the enzyme,
guanylate cyclase; coincident with the shift in apparent molecular
weight, enzyme activity decreases by as much as 90%. Since speract
fails to cause these responses in A. punctulata, it can be concluded
that the events are receptor-mediated.

315

INTRODUCTION

The responses of animal spermatozoa to substances associated
with or released from eggs and/or the female reproductive tract have
been studied by many different investigators (see Table 1). Noted
responses of sperm cells have included the activation (or repression)
of motility and metabolism, and the induction of the acrosome
reaction.

A number of years ago we initiated research aimed at the identification of egg-associated factors which influence spermatozoa. The
special emphasis was on those factors which caused the induction of
the acrosome reaction and the activation of motility and/or metabolism. The sea urchin, an animal used for many years as a model for
studies in fertilization, has been our major animal model although
studies also have been initiated in the bovine.

In initial studies we found that we could separate factor(s)
which caused induction of the acrosome reaction from those which
activated motility and respiration rates. The factor responsible for
the induction of the acrosome reaction was purified based on its
ability to elevate cyclic AMP concentrations (99); the complex was
identified as the fucose-sulfate-rich complex studied by Segall and
Lennarz (169,170). They showed that the complex acted in a species-
specific manner to induce the acrosome reaction; the complex also
retained biological activity after treatment with NaOH, suggestive
that the carbohydrate structure represented the "active" part of the
complex. Later work by Garbers et al. (51) supported these studies
except that it was shown that the amount of protein associated with
the complex markedly affected the ability of the complex to elevate
cyclic AMP concentrations.

Cyclic AMP concentrations were also elevated in isolated sea
urchin sperm heads suggestive that the changes in cyclic AMP were
associated with the acrosome reaction (48).

With respect to the mechanism of action of the factor, our early
studies on the fucose-sulfate-rich complex demonstrated that it
caused a marked increase in $^{45}Ca^{2+}$ accumulation by spermatozoa (99).
Since cyclic AMP concentrations could be dramatically elevated within
seconds, but direct effects of the complex on adenylate cyclase in
the broken cell were not observed, we considered the possibility that
Ca^{2+} was the mediator. In fact, this appeared to be the case since
Nigericin and A23187 increased $^{45}Ca^{2+}$ uptake and elevated sperm
cyclic AMP concentrations, while D-600, verapamil or trifluropera-
zine, blocked the $^{45}Ca^{2+}$ uptake induced by the fucose-sulfate-rich
complex and inhibited the elevations of cyclic AMP (99). Adenylate
cyclase also could be shown to be activated by the fucose-sulfate
complex if intact cells were incubated with it prior to homogeniza-

Table 1. The effect of Egg or Female Reproductive Tract Fluids or Factors on Sperm Motility, Metabolism and Capacitation (Acrosome Reaction)

Animal	Source of fluid or Factors* tested	Motility	Metabolism	Capacitation Acrosome Reaction	References
Nereis	egg°	+			111,112
Holothurian	egg°			+	27
Horseshoe crab	egg°	+			23-25,174,193
Ascidian	egg°,±	+		+	36,104,121
Keyhole limpet	egg°	+	+		194,196
Sea urchin	egg°	+	+	+	22,26,30,32,68,71-73, 98-100,111,112,115, 137,140,152,168-170, 195,198
Mollusc	egg°	+δ		+	33,122
Starfish	egg°	+	+	+	27,58,122,140,197,214
Amphibian	egg°	+		+	91,173,213
Fish	egg°	+	+		3,57,184,185,216
	genital tract	+			77
Bird	genital tract	+	+		47,65,108,139
Hamster[a]	genital tract	+		+	219
	follicular fluid	+		+	63,95,215,218
	cumulus cells	+		+	64
Rat	follicular fluid[b]			+	131
Guinea pig	none[c]			+	81,95,220
Mouse	genital tract	+		+	43,86
	zona pellucida			+	13,165
Sheep	genital tract		+	+	11,12,83,160,206
Cattle	genital tract	+	+	+	143
	follicular fluid	+	+	+	143,191
	cumulus cells	+			15
Rabbit	genital tract	+	+	+	16,29,42,65,66, 130,132,133

(continued)

Table 1. Continued

Animal	Source of fluid or Factors* tested	Motility	Metabolism	Capacitation Acrosome Reaction	References
Rabbit	isthmus-oviduct	−			29,144
Dog	egg°d	+		+	116
Pig	genital tract	+	+		39,166,178
	follicular fluid	+	+		101
Human	follicular fluid	+	+	+	74,75,87,102,131
					150
	genital tract		+		192

*Spermatozoa, in some experiments, were incubated in the female reproductive tract and subsequently studied under in vitro conditions.

°The factors associated with the egg may actually represent female reproductive tract or follicular cell secretions.

±The chorion cells associated with the egg are probably responsible for the observed changes. Mitochondria located in the sperm head region migrate down the tail and are released in response to the chorion cells. Increased motility is associated with the migration of the mitochondria.

δChemotactic response.

aSpermatozoa can be capacitated in the presence of bovine serum albumin.

bHuman follicular fluid and oviduct fluid were used to induce the acrosome reaction.

cNo factors are required at pH 8.3, but at lower pH values (7.4), factors present in serum are required for capacitation.

dSpermatozoa were incubated in the presence of eggs and bovine serum albumin.

tion of the cells (212), although the mechanism of activation has not
been determined. Protein phosphorylation following the addition of
the fucose-sulfate-rich complex seems likely since the cyclic
AMP-dependent protein kinase has been shown to be activated (52).

In 1976, Ohtake (140,141) published reports demonstrating that
the respiration of sea urchin sperm cells could be reproducibly
stimulated by crude mixtures of egg factors if the extracellular pH
was maintained at acidic values. Subsequently, the factor
responsible for the stimulation of respiration was purified from
Strongylocentrotus purpuratus (53,68) and Hemicentrotus pulcherrimus
(182) and was shown to be a peptide (Gly-Phe-Asp-Leu-Asn-Gly-Gly-Gly-
Val-Gly). The peptide, named speract, stimulated sperm respiration
and motility half-maximally at about 30 pM (53,68). Subsequently, in
studies at slightly acidic extracellular pH values, Monensin A (69)
and various other agents capable of elevating intracellular pH (159)
also were shown to stimulate the respiration and motility of sea
urchin spermatozoa. Speract was shown to induce a net H^+ efflux (69)
and to elevate intracellular pH (159); therefore, the possibility
existed that the net proton efflux represented one of the primary
events in response to speract binding. Since speract caused elevations of cyclic AMP and cyclic GMP concentrations, it also remained
possible that one or both of these cyclic nucleotides represented a
primary signal (68,69,100). Of various nucleotides tested, 8-bromo-
cyclic GMP represented an effective analogue of cyclic GMP with
respect to the stimulation of respiration. The physiological
significance, however, was not clear since added cyclic GMP at 10 mM
concentrations failed to increase respiration rates.

In order to determine properties of the receptor for speract,
the Bolton/Hunter adduct of speract was synthesized. With intact
cells, a specific receptor for speract was identified (175). The
^{125}I-Bolton/Hunter speract was bound to the receptor in the absence
of extracellular Mg^{2+} or Ca^{2+}, but deletion of Na^+ reduced binding by
more than 90% (175).

Here, we will describe the purification and sequence of a
peptide from another sea urchin, Arbacia punctulata. The peptide,
named resact, stimulates A. punctulata sperm respiration and motility
and also causes a mobility shift in a plasma membrane protein
subsequently identified as guanylate cyclase (208). Additionally,
the membrane receptor for speract, and methods to prepare membrane
vesicles with retention of speract binding will be described.

MATERIALS AND METHODS

Materials

A. punctulata were obtained from the Marine Biological Laboratory, Woods Hole, MA and from Gulf Speciman Co., Panacea, FL. S.

purpuratus and Lytechinus pictus were purchased from Marinus, Inc.,
Westchester, CA. Seawater was prepared to contain 454 mM NaCl, 9.7
mM KCl, 24.9 mM $MgCl_2$, 9.6 mM $CaCl_2$, 27.1 mM $MgSO_4$, 4.4 mM $NaHCO_3$ and
10 mM Tris (pH 8.0) or 10 mM N-[2-acetamido]-2-aminoethane sulfonic
acid (pH 6.6). Carboxypeptidase Y was from Pierce and ^{14}C-iodoacetic
acid and [α-^{32}P] GTP were from New England Nuclear. Highly purified
Na-dodecyl-SO_4 was from Bio Rad Laboratories. For good resolution of
guanylate cyclase, however, the 95% grade Na-dodecyl-SO_4 from Sigma
Chemical Company was important to use. Acrylamide, bis-acrylamide,
TEMED, ammonium persulfate and molecular weight markers were from Bio
Rad Laboratories. Lubrol PX and Lentil lectin Sepharose 4B were
obtained from Sigma Chemical Company. Na[125]I was purchased from
Amersham. Wheat germ lectin Sepharose was from Pharmacia. Disuc-
cinimidyl suberate was purchased from Pierce Chemical Co. Bis(sul-
fosuccinimidyl)suberate was a generous gift of Dr. James V. Staros,
Department of Biochemistry, Vanderbilt University Medical Center.

Gamete Collection

Gametes were obtained after the injection of 0.5 M KCl. Spermatozoa were collected "dry" at 5°C and were generally stored at 0-2°C
at approximately 400 mg (wet weight)/ml until use. Eggs were collected in seawater and were then treated with acidified (pH 5.5) seawater to rapidly remove the jelly coat. After standing for 20 min at
pH 5.5, the eggs were removed by gentle centrifugation and the supernatant fluid was stored for purification of peptides.

Isolation of Sea Urchin Egg Jelly

The factors associated with the extracellular matrices (jelly)
of the egg were collected by two methods. In some cases, sea urchin
eggs collected in artificial seawater (ASW) were centrifuged at 1500
X g for 10 min at 5°C and the eggs were then resuspended (10% v/v) in
ASW (pH 7.5-7.6). After occasional stirring, the eggs were centrifuged and the supernatant fluid saved. In other cases, eggs were
collected in seawater but were then treated with acidified (pH 5.5)
seawater to rapidly remove the jelly coat. After standing for 20 min
at pH 5.5 the eggs were removed by centrifugation and the supernatant
fluid was saved.

Bioassay of Resact

In general, spermatozoa were incubated in seawater buffered to
pH 6.6 and sperm respiration rates were determined using a Gilson K-
IC Table Top Oxygraph equipped with a 2.2 ml capacity chamber fitted
with a Clark electrode. Sperm (10 mg/30 μl) were added to 2.2 ml of
seawater at 20°C and after basal respiration had remained constant
for 3 min, various agents were added and the new respiration rates
were determined over the next 5 min. The stimulation of respiration
was calculated as the difference between the respiration rates before
and after the addition of factors.

High Pressure Liquid Chromatography (HPLC)

HPLC was generally performed using a Beckman model 322 chromato-graph system with an octyl column (4.6 nm X 25 cm). The column efflu-ent was monitored for absorbance at 235 nm.

Amino Acid and Sequence Analysis

Amino acid analysis was performed using a LKB Amino Acid Analy-zer after a 24-h hydrolysis of peptide in 6N HCl at 110°C under vacuum. Normally, 5-50 nmol of peptide were hydrolyzed. The amino acid sequence was determined by automated Edman degradation (53).

Purification of Guanylate Cyclase

The A. punctulata sperm guanylate cyclase was purified basically as we described in earlier work on the enzyme from S. purpuratus spermatozoa (155). The detergent-solubilized enzyme was applied to a GTP-Sepharose column, followed by DEAE-Sephacel and preparative gel electrophoresis. The resultant enzyme was homogeneous as judged by Na-dodecyl-SO$_4$ polyacrylamide gel electrophoresis.

Gel Scans

Quantitative data on the percent band conversion and on ^{32}P content of the 160,000 molecular weight protein were obtained by the use of a LKB 2202 Laser Densitometer.

Protein

The method of Lowry et al. (114) was used to determine protein content.

Carboxymethylation

Resact (28.0 nmol) was incubated with EDTA (4.5 mM), Tris buffer (50 mM, pH 8.6) and with β-mercaptoethanol (1% v/v) or without β-mer-captoethanol at ambient temperature for 30 min. (1 -^{14}C) Iodoace-tic acid (25 μCi, 2.1 mol) in 0.1 N NaOH was then added and the incubation was continued at ambient temperature for 20 min. The so-lution was applied to an 0.6 cm X 8.2 cm column of Sephadex G-15, eluted with H$_2$0, and freeze-dried.

Amidation

The Leu-amide form of the peptide was prepared by reacting the synthesized peptide with ammonium chloride (5.5 M NH$_4$Cl, pH 4.75) in the presence of 1-ethyl-3-dimethylaminopropylcarbodiimide. The resultant Leu-amide peptide was then purified on a Sephadex G-25 column followed by high pressure liquid chromatography. The amide migrated the same as purified resact on reverse phase columns and on thin layer chromatography.

Peptides

Speract (Gly-Phe-Asp-Leu-Asn-Gly-Gly-Gly-Val-Gly), GGG[Y^2]-speract (Gly-Gly-Gly-Gly-Tyr-Asp-Leu-Asn-Gly-Gly-Gly-Val-Gly), F7-G1 (Phe-Asp-Leu-Asn-Gly-Gly-Gly), G8-G1(Gly-Phe-Asp-Leu-Asn-Gly-Gly-Gly), a carboxy-terminal peptide (Gly-Gly-Gly-Val-Gly) and a tyrosine ana-logue (Tyr-Asp-Leu-Asn-Gly-Gly-Gly-Val-Gly) were custom synthesized by Peninsula Laboratories, Inc., San Carlos, CA or were synthesized in our laboratory as previously described (53).

Guanylate Cyclase Activity of Resact-treated Cells

Spermatozoa were diluted with ASW to give an approximate wet weight of 100 mg/ml. The incubation mixture contained 300 μl of sea-water and 10 μl of H_2O or peptide; the reaction (22°C) was started by the addition of 2.0 ml of a solution containing 20 mM 2[N-morpholino]-ethane sulfonic acid, pH 6.5, 0.5% Lubrol PX, 10 mM KF and 100 μM sodium orthovanadate. The mixture was then placed in an ice bath for 1 h prior to the assay of guanylate cyclase activity.

Guanylate cyclase was assayed at pH 6.5 at 30°C. The reaction typically contained 20 mM 2[N-morpholino] ethane sulfonic acid, pH 6.5, 1 mM 3-isobutyl-1-methylxanthine, 3 mM $MnCl_2$, 10 mM KF, 100 μM sodium orthovanadate, 1 mM nonradioactive cyclic GMP, 500 M GTP and 2-3 X 10^5 dpm of [α-^{32}P]GTP, in a total volume of 100 μl; the time of incubation was generally 10 min. The cyclic [α-^{32}P]GMP formed was separated by $ZnCO_3$ precipittion and neutral alumina column chromatography as described earlier (155), with nucleotide recoveries between 60-70%.

Guanylate Cyclase Gel Electrophoresis

Gels were run essentially as described by Ward and Vacquier (209). The gels were subsequently stained by the method of Morrissey et al. (128).

Radiolabeling of GGG[Y^2]-speract and Binding Studies

Two (2) nmol of GGG[Y^2]-speract were incubated with 1 mCi ^{125}NaI and 150 μg lactoperoxidase (Sigma) in a solution containing 50 mM sodium phosphate (pH 7.0). The reaction was initiated by the addition of H_2O_2 at room temperature and the radiolabeled peptide was then purified by column chromatography on Bio-Gel P-6 equilibrated with a solution containing 10 mM sodium phosphate, 150 mM NaCl, and 0.1% gel-atin, pH 7.2. The specific activity of the iodinated peptide was approximately 100-200 dpm/fmol (34).

Binding studies with intact spermatozoa were performed at 15°C. Cells (20-200 μg wet weight) were incubated with 2 nM ^{125}I-GGG[Y^2]-speract (100-200 dpm/fmol) at pH 6.6 in a total volume of 0.13 ml. The reaction was stopped by dilution with 2 ml of ASW (0-2°C) and subsequent filtration on Whatman GF/C filters. Nonspecific binding was determined by the incubation of ^{125}I-peptide in the presence of 77 nM unlabeled GGG[Y^2]-speract and was typically less than 10% of of total binding.

Crosslinking of ^{125}I-GGGY^2-speract to Intact Spermatozoa

Intact sperm cells were incubated with ^{125}I-GGG[Y^2]-speract for 60 min (equilibrium) at 15°C. The reaction was stopped by the addition of 2 ml of pH 6.6 ASW and the cells were then centrifuged at 5000 X g (4°C) in a JS5.2 rotor (Beckman) for 10 min. The radioactive supernatant fluids were removed and the cells were washed one time with sea water (pH 6.6). Separate dissociation time course experiments with intact sperm cells indicated that less than 10% of the bound ^{125}I-GGG[Y^2]-speract was lost from the cells during the time required for washing under these conditions. The cells were then incubated for 15 min at 15°C with the chemical cross-linking agent, disuccinimidyl suberate (Pierce Chemical Co., Rockford, IL), (50 μM-10 mM) dissolved in dimethyl sulfoxide. The reaction with the disuccinimidyl suberate was stopped by the addition of sea water (pH 6.6) (0-2°C). The cells were again centrifuged and washed as described above.

Electrophoresis and Autoradiography

The cross-linked products were analyzed by Na-dodecyl-SO$_4$ slab gel electrophoresis according to Laemmli (103). Prior to incubation in sample buffer containing 50 mM β-mercaptoethanol the cells were treated with micrococcal nuclease, DNase and RNase according to Garrels (54).

Solubilization of Speract Receptor

Intact spermatozoa were solubilized in a solution containing 5 mM Hepes buffer pH 7.4, and 0.5% Lubrol PX for 1 h at room temperature with occasional vortexing. The solubilized proteins were collected in the supernatant fluids after centrifugation at 100,000 X g for 60 min at 4°C in a Beckman TY 65 rotor, and were then batch adsorbed to Wheat germ lectin Sepharose previously equilibrated with a solution containing 50 mM Tris, pH 7.6, 0.2 M NaCl and 0.1% Lubrol PX by mixing end-over-end for 2 h at 4°C. The agarose beads were centrifuged at 15,000 X g and the supernatant fluids were removed. The beads were washed twice with the above buffer and the proteins bound to the beads were then eluted by incubation of the beads with the above buffer adjusted to contain 0.2 M N-acetylglucosamine or 3 mM diacetylchitotriose.

Cyclic GMP

    The concentration of cyclic GMP in the sperm cells was monitor-
ed as described previously (49,50).

Membrane Preparation

    Spermatozoa were suspended in 550 mM NaCl, 10 mM HEPES, 10 mM
ethylene diamine tetraacetate (EDTA), and 1 mM ethylene glycol-
bis-(βamino ethyl ether)-N,N' tetraacetic acid (EGTA) adjusted to
pH 7.5 with 1 M Tris. The spermatozoa were then centrifuged at 1500
X g in a Beckman TJ-6 centrifuge for 15 min and the resultant pellet
was resuspended and centrifuged again at 4500 X g for 15 min to
obtain spermatozoa. The washed pellet was resuspended in the above
buffer containing 10 mM benzamidine. The cell suspension was subjected to nitrogen cavitation for 20 min at 200 psi in an ice bath,
and the suspension was then centrifuged at 200 X g for 15 min. The
resultant supernatant fluid was removed and centrifuged at 100,000 X
g for 1 h. The pellet was obtained, resuspended in 3 to 5 ml of the
buffer used for cavitation, and was then homogenized with a motor
driven, Teflon-glass homogenizer. The homogenized suspension was
layered over a 25% to 33% sucrose step gradient (the sucrose was
dissolved in the same buffer used for cavitation) and centrifuged at
100,000 X g for 2 h. Particulate fractions were apparent at the
interfaces between 0 to 25% sucrose (Band I) and 25% to 33% sucrose
(Band II). There was also a sedimented particulate fraction. The
fractions within the gradient were removed with a Pasteur pipette
and centrifuged at 200,000 X g for 1 h. The fractionated preparations were suspended with buffer to a protein concentration of about
1 to 5 mg/ml. Band II retained the capacity to specifically bind
speract for at least six months when stored at -70°C.

Electron Microscopy

    Intact or cavitated spermatozoa and isolated fractions were
fixed in a buffer containing 100 mM sodium cacodylate, pH 7.5, 4%
glutaraldehyde, 550 mM NaCl, 10 mM EDTA and 1 mM EGTA. Following
two rinses in buffer, samples were post-fixed in 1% OsO_4 in 100 mM
sodium cacodylate for 1-2 h. Tissue was dehydrated in an ethanol
series and embedded in Epon 812. Thin sections were stained with
uranyl acetate and lead citrate and examined in a Hitachi H-600
electron microscope.

Binding Studies to Plasma Membrane

    [125I]-Bolton/Hunter reagent was used to label speract accord-
ing to the methods of Bolton and Hunter (14). Iodinated speract was
purified, and binding assays were conducted by filtration methods as
previously described (175). All speract binding experiments were
allowed to proceed to equilibrium (1 h, 17°C).

RESULTS

Identification of Speract Receptor

The peptide analogue (GGG[Y^2]-speract) was identical to speract except for the substitution of Tyr for Phe and the extension of the amino terminus by three amino acids (Fig. 1). GGG[Y^2]-speract stimulated sperm respiration rates at the same concentrations as speract and competed for binding against ^{125}I-GGG[Y^2]-speract at the same concentrations as speract (34). Thus, the peptide appeared to be an excellent analogue of speract and was used in studies designed to covalently attach the analogue to the receptor.

The incubation of spermatozoa previously equilibrated with ^{125}I-GGG[Y^2]-speract with disuccinimidyl suberate resulted in a single radiolabeled band when analyzed by Na-dodecyl-SO$_4$ gel electrophoresis (34). The radioactive band had an apparent molecular weight of approximately 77,000 in the presence of β-mercaptoethanol. The high degree of specificity of the interacton was suggested by the ability of an excess of unlabeled peptide to eliminate the binding of ^{125}I-GGG[Y^2]-speract to the 77,000 molecular weight band and the appearance of only one radiolabeled band even at relatively high concentrations of the disuccinimidyl suberate. No other radiolabeled bands were observed under the conditions tested. Identical results were obtained with the membrane impermeant cross-linking reagent Bis(sulfosuccinimidyl) suberate (Fig. 2).

Further evidence for the specificity of the association were experiments where speract analogues, previously shown to compete for speract binding and to exhibit respiration-stimulating activity, were able to prevent the covalent attachment of the iodinated peptide (34). In contrast, speract analogues with little or no biological activity were not capable of competing with the radiolabeled peptide during the chemical reaction. Comparisons of the amount of ^{125}I-GGG[Y^2]-speract bound in the presence of variable concentrations of unlabeled GGG[Y^2]-speract showed that about 1 nM unlabeled peptide reduced ^{125}I-GGG[Y^2]-speract binding by 50%; this was estimated as either specific binding to intact cells or as the reduction in radioactivity found in the 77,000 molecular weight band on autoradiographs after covalent coupling.

| Gly-Phe-Asp-Leu-Asn-Gly-Gly-Gly-Val-Gly | SPERACT |
| Gly-Gly-Gly-Gly-Tyr-Asp-Leu-Asn-Gly-Gly-Gly-Val-Gly | GGGY^2-speract |

Figure 1. The amino acid sequence of speract and of GGG[Y^2]-speract.

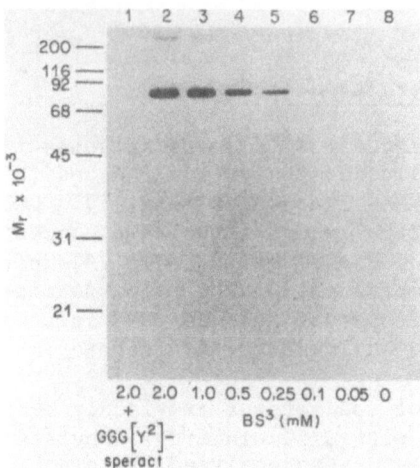

Figure 2. Electrophoretic analysis of the effect of bis(sulfosuccinimidyl)suberate (BS^3) on the covalent coupling of ^{125}I-
GGG[Y^2]-speract to S. purpuratus intact sperm cells. Binding and cross-linking steps were performed as described in
Materials and Methods. Shown is an autoradiographic profile of the cross-linked cells after electrophoresis in a
10% polyacrylamide slab gel containing Na-dodecyl-SO$_4$ and
β-mercaptoethanol. Coupling in the presence of 77 nM
unlabeled GGG[Y^2]-speract is indicated in Lane 1.

Speract cross-reacts with L. pictus spermatozoa but not with
sperm cells from A. punctulata (181). ^{125}I-GGG[Y^2]-speract was
incubated with sperm cells from L. pictus or A. punctulata for 1 h
(15°C), washed, reacted with 10 mM disuccinimidyl suberate and
analyzed by Na-dodecyl-SO$_4$ polyacrylamide gel electrophoresis. A
radiolabeled band with an apparent molecular weight similar to that
seen in S. purpuratus was detected in L. pictus spermatozoa, but no
radiolabeled bands were detected in A. punctulata spermatozoa (34).
Thus, the radiolabeled band appears to be the specific receptor for
speract.

When radiolabeled receptor was solubilized in 0.5% Lubrol PX
and adsorbed to Wheat germ lectin Sepharose 4B under the conditions
described in Materials and Methods, the iodinated material was
specifically eluted from the lectin after incubation with 0.2 M
N-acetylglucosamine or 3 mM diacetylchitotriose, but the detergent-
solubilized radiolabeled receptor was not bound to Lentil lectin
Sepharose 4B under identical conditions (34).

Binding of Speract to Membrane Vesicles

In order to study the mechanism of action of the peptides at the
membrane level the isolation of membrane vesicles which retained the

peptide receptor would be of great advantage. Band II consisted of a membraneous vesicular fraction when examined by electron microscopy, and [^{125}I]-Bolton/Hunter-speract bound specifically to these membranes reaching equilibrium by 1 h. Specific speract binding was linear to 0.5 mg/ml membrane protein and increased in a non-linear fashion at higher protein concentrations. To determine whether the binding of speract was consistent with that of the receptor identified on intact cells, analogues of speract were used. These analogues demonstrated a range of relative potencies from speract (most potent) to completely inactive analogues. Competition against [^{125}I]-Bolton/Hunter-speract binding to the putative receptor was consistent with binding data obtained with intact spermatozoa from S. purpuratus and with the relative potency of the analogues in relation to speract (175). The inactive speract analog, F7-Gl, failed to compete with [^{125}I]-Bolton/-Hunter-speract for binding to these membranes.

The stepwise enrichment of several known membrane-bound enzymes as well as an enrichment of specific binding was observed in the preparation of the Band II vesicles. Guanylate cyclase, adenylate cyclase, and cyclic GMP phospodiesterase were all enriched in Band II along with specific speract binding. Succinic acid-cytochrome C oxidase, a mitochondrial inner membrane marker, in contrast, was not found in this fraction.

Resact Purification and Amino Acid Sequence

A crude mixture of egg-associated factors were obtained by the treatment of A. punctulata eggs with acidified (pH 5.5) seawater for 20 min followed by centrifugation to remove the eggs. Generally, 3 liters of the acidified supernatant fluid were mixed with 6 liters of 95% ethanol and the suspension was then centrifuged at 10,000 X g for 30 min at 4°C. The resultant supernatant fluid was concentrated with a rotary evaporator at 50°C and the concentrated fraction was then applied to a Sephadex G-10 column (2.5 cm X 90 cm) equilibrated with 0.1 M acetic acid. The fractions (7 ml) were assayed for their ability to stimulate sperm respiration and the fractions containing activity were pooled and freeze-dried. The dry residue was dissolved in distilled water and was further purified by high pressure liquid chromatography (reverse phase, octyl). Fractions were assayed for activity and active-fractions were pooled, freeze-dried and again analyzed by high pressure liquid chromatography. The purified, active material eluted as a single peak after the second run (Fig. 3).

Amino Acid Composition and Sequence

The single peptide peak was analyzed for amino acid composition. The following residues (integer values) were identified after 6N HCl hydrolysis: Thr (1), Gly (5), Ala (1), Val (2), Leu (1), Arg (1), and Pro (1).

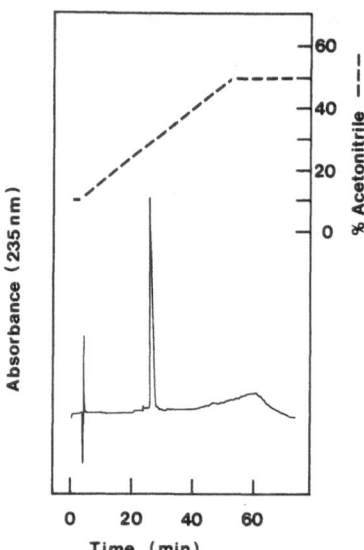

Figure 3. High pressure liquid chromatograph of the purified peptide
(resact) obtained from A. punctulata eggs. Approximately
100 nmol of resact (purified as described in Materials and
Methods) were applied to a 4.6 X 250 mm octyl column
equilibrated with a solution containing 10% acetonitrile
and 0.1% trifluoroacetic acid. A linear acetonitrile
gradient was used to elute the peptide.

Subsequently, the peptide was treated with $[^{14}C]$-iodoacetic
acid in the presence or absence of β-mercaptoethanol. Amino acid
analysis now revealed the presence of carboxymethylcysteine residues
whether the peptide was incubated with β-mercaptoethanol or not.
The nmol of each amino acid recovered were as follows: carboxymethylcysteine (7.49); Thr (3.31); Gly (16.19); Ala (3.72); Val
(6.16); Leu (3.39); Arg (3.32); Pro (2.73). The carboxymethylated
peptide was subjected to sequence analysis and the following
sequence was determined by Edman degradation: CMCys-Val-Thr-Gly-
Ala-Pro-Gly-CMCys-Val-Gly-Gly-Gly-Arg. No Leu was detected suggestive that it existed as the CO_2H-terminal amino acid and/or that it
existed in modified form. To determine the CO_2H-terminal amino
acid, 2.8 nmol of resact were incubated with 10 μg of carboxypeptidase Y. The dried residue was applied to the amino acid analyzer
and Leu was detected as the released amino acid (50-100% yield).
However, the possibility still existed that Leu was actually Leu-
amide since carboxypeptidase Y will also cleave this bond.

The peptide was then synthesized by solid phase methods (53)
with Leu as the carboxyl-terminal amino acid. The synthesized
material had the following amino acid composition: Thr, 1.02 (1);
Gly, 5.14 (5); Ala, 1.15 (1); Val, 1.90 (2); Leu, 1 (1); Arg, 1.05
(1); Pro, 1.09 (1). Cysteine was not determined in these analyses.
The leucine-amide form of the peptide was also prepared and the
amide migrated identically on reverse phase columns and on thin
layer chromatographs with resact while the non-amide form migrated

in a different position. The synthesized peptide containing
Leu-amide also elevated sperm respiration at the same concentrations
as the purified, native resact. The amino acid sequence of resact
and its comparison with speract is shown in Fig. 4.

Effects of Resact on Respiration and Cyclic GMP

Resact stimulated sperm respiration rates and elevated sperm
cyclic GMP concentrations at acidic pH values (183). These effects
were species-dependent in that resact caused these effects in A.
punctulata but not in S. purpuratus or L. pictus spermatozoa. The
peptide was half-maximally active at about 500 pM on respiration and
at about 25 nM on cyclic GMP concentrations. Speract at
concentrations as high as 100 µM failed to increase cyclic GMP
concentrations or to elevate respiration rates of A. punctulata
spermatozoa. Resact had only small effects on respiration rates at
extracellular pH values of 8.0, but it continued to elevate cyclic
GMP concentrations to the same extent at pH 8.0, with half-maximal
effects at about 25 nM peptide.

Stimulation of Respiration at Alkaline Extracellular pH

Speract and other egg-associated peptides (Thr^5-speract; Ser^5-
speract) have been shown to stimulate the respiration of sea urchin
spermatozoa at slightly acidic pH values (53,68,137,182). Physiologically, the stimulation has been suggested to occur because of an
acidic pH of the egg jelly (59,110); it has a high content of sulfate residues (84,85,99,168,170). However, the pH of hydrated egg
jelly is not very acidic [pH 7.4-7.7 in our measurements and pH 8.0
in the studies of Holland and Cross (76)] and thus may not represent
the mechanism by which jelly can inhibit sperm respiration rates.

Since we had pure components from jelly we could now reconsti-
tute the system and study physiological function at normal seawater
pH. To do these studies, we initially determined whether or not
heterologous jelly would inhibit sperm respiration. Jelly from A.
punctulata was tested on L. pictus spermatozoa and vice versa.
Jelly from the heterologous sea urchin inhibited sperm respiration
under conditions where the extracellular pH was maintained constant
(181). The subsequent addition of speract or resact to the homologous spermatozoa greatly stimulated respiration.

| Gly-Phe-Asp-Leu-Asn-Gly-Gly-Gly-Val-Gly | SPERACT |
| Cys-Val-Thr-Gly-Ala-Pro-Gly- Cys-Val-Gly-Gly-Gly-Arg-Leu | RESACT |

Figure 4. Comparison of the amino acid sequence of speract and
          resact. The carboxyl terminal Leu exists as the amide
          form in resact.

The factor(s) in jelly responsible for the respiratory decrease was non-dialyzable and was active between pH 6.6 and pH 7.8. Thus, respiratory inhibition was not necessarily correlated with the occurrence of an acrosome reaction. The activation of jelly-inhibited respiration by speract was dependent on the concentration of the peptide with half-maximal stimulation of respiration occurring at about 10^{-9} M speract. When unfractionated, homologous egg jelly was used, which contains the activating peptide, respiration was not inhibited or inhibited only slightly in either L. pictus or A. punctulata spermatozoa (181).

These results suggested that the function of speract and of activating peptides from other sea urchins is to maintain sperm motility and metabolism while cells traverse the extracellular matrix around the egg. Jelly was then prepared from L. pictus or S. purpuratus to give the fucose sulfate polymer and other jelly components free of speract (181). When the purified jelly [does not contain speract] was added to the spermatozoa, it inhibited respiration (Fig. 5). The addition of speract to the L. pictus spermatozoa under these conditions, however, stimulated sperm respiration.

Whether or not speract still functions to elevate intracellular pH is not known, but Monensin A, an ionophore which catalyzes an electro-neutral Na^{+}/H^{+} exchange and which mimics the effects of speract at pH 6.6 (69), also reproduces the effects of speract, in that it reversed the jelly-inhibited respiration (181). Thus, the mechanism of action of the peptide at normal extracellular pH could be identical to its effects at lower pH.

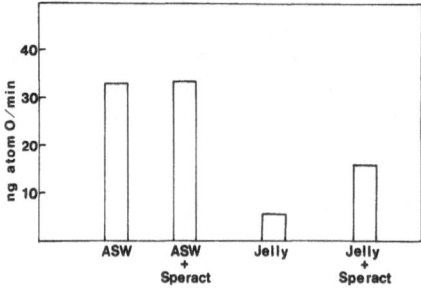

Figure 5. Inhibition of sperm respiration by partially purified egg jelly and partial reversal of the inhibition by speract. Respiration rates were estimated as described in Materials and Methods. The egg jelly was purified free of speract. The addition of such jelly at pH 7.5 caused an inhibition of respiration and speract (1 nM) could partially relieve the inhibition. ASW = artificial seawater at pH 7.5.

Modification of a Major Sperm Plasma Membrane Protei   Crude mixtures of egg factors appear to modify the phosphorylation state and
the electrophoretic properties of a sperm membrane protein (209).
When A. punctulata spermatozoa were treated with various concentrations of resact, one major protein was shifted from about 160,000 to
150,000 in molecular weight (183). Laser scans to determine the
relative proportion of 160,000/150,000 indicated that about 115 nM
resact caused a one half-maximal conversion. Ward and Vacquier
(209) previously established that the 150,000 molecular weight
protein originates from the 160,000 molecular weight form.

The conversion of the 160,000 to 150,000 molecular weight form
occurred completely within 5 s at saturating resact concentrations;
even after the continued incubation of sperm cells for 2 h, no
reversal in molecular weight was apparent (183). Intermediate
concentrations of resact caused partial conversions of the 160,000
dalton protein; again these partial conversions occurred within 5 s.

When spermatozoa were incubated with ^{32}P, the 160,000 dalton
band was radiolabeled and the subsequent addition of resact caused
the loss of ^{32}P in a concentration-dependent manner (183).

Guanylate Cyclase   Ward et al. (208) have reported that 160,000
and 150,000 molecular weight bands represent the enzyme, guanylate
cyclase. With slight modifications of previous procedures used in
our laboratory (155), the enzyme was purified from A. punctulata
spermatozoa (183). Homogenous guanylate cyclase migrated on Na-
dodecyl-SO$_4$ gels identically with the 150,000 molecular weight protein found after resact treatment of cells (183). The 160,000 molecular weight form of the protein, however, has not yet been purified.

Modification of Guanylate Cyclase Activity   The incubation of
intact S. purpuratus or A. punctulata spermatozoa with speract or
resact, respectively, resulted in a marked inhibition of guanylate
cyclase activity when enzyme activity was subsequently measured in
the detergent-extracted cells (Fig. 6). Inhibition was not observed
if the peptide was added to the detergent-extract after cell solubilization (not shown). The response appeared to be receptor-
mediated since peptides of the opposite species failed to alter
guanylate cyclase activity (Fig. 6).

Lubrol-Containing Buffer at pH 6.5   To observe the effects of
the peptides on guanylate cyclase activity, the cell disruption
procedure was of importance. When spermatozoa were incubated with
or without the peptide and then homogenized in hypotonic buffer at
pH 6.0 or 8.0, in the presence or absence of Lubrol PX, the
magnitude of the apparent effect of the peptide on enzyme activity
varied dependent on the homogenization method (156).

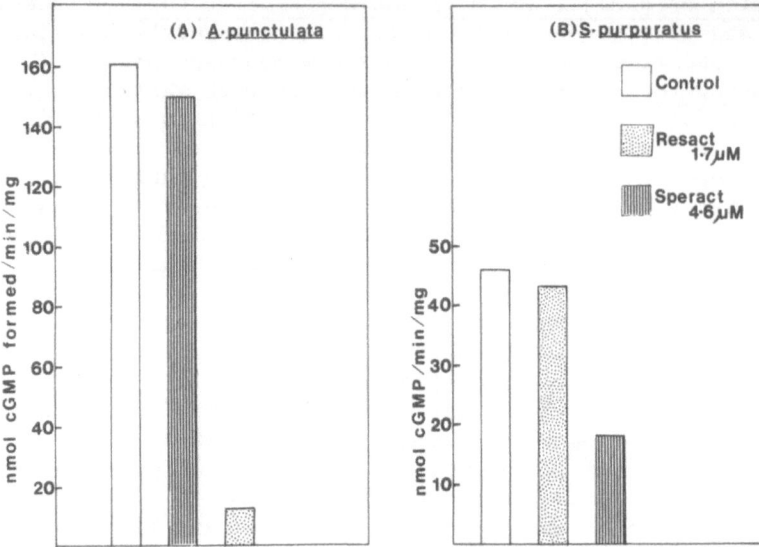

Figure 6. Apparent decrease in guanylate cyclase activity caused by
          the incubation of intact sperm cells with the egg-associated
          peptides. Approximately 10 mg (wet weight) sperm cells were
          incubated in ASW at pH 7.9 with either 4.6 μM speract or 1.7
          μM resact in a final volume of 400 μl. At 1 min, the incu-
          bation was stopped by the addition of 2 ml of 20 mM 2[N-
          morpholino] ethane sulfonic acid buffer, pH 6.5, containing
          0.5% Lubrol PX, 20 mM KF and 100 μM sodium orthovanadate.
          The tubes were allowed to remain at 0-2°C for 1 h and were
          then assayed for guanylate cyclase activity as described in
          Materials and Methods.

        After cells were homogenized in buffers containing either Lubrol
PX (0.5%) or Triton X-100 (0.5%) and were retained at ice-bath temperature, or at 5°C, enzyme activity remained stable with time for at
least 2 h (not shown). In cases where cells were homogenized with
buffer not containing detergent and detergent was added after 1 h of
incubation on ice, the activity at pH 6.0 or 8.0 declined by 50% and
70%, respectively. The activity of enzyme in cells treated with
resact prior to homogenization, however, remained relatively constant
as a function of time after homogenization. The same qualitative
results were obtained in experiments on S. purpuratus spermatozoa.
Because of these results, cells were subsequently solubilized in a
buffer containing 0.5% Lubrol PX at pH 6.5. However, when the enzyme
preparation was stored at 5°C for 24 h, the activity of previously
non-treated cell extracts decreased to the peptide-treated extract
level. The addition of KF (10 mM) and sodium orthovanadate (100 μM)
to the Lubrol-containing buffer resulted in a stabilization of enzyme
activity for at least 24 h, and therefore, the normal extraction

buffer was modified to contain 0.5% Lubrol PX, 10 mM KF and 100 µM sodium orthovanadate (also see Ward et al., this volume).

The enzyme preparation in Lubrol-containing buffer (without KF or vanadate) obtained from cells not treated with peptide and incubated at 22°C showed a time-dependent decrease in the activity of guanylate cyclase (Fig. 7); the activity, in fact, decreased to levels found in the peptide-treated cell extracts. The cells previously treated with peptide, however, did not show any significant decrease in enzyme activity during this time period. The rate of decrease was dependent on the protein concentration suggestive that an enzyme was responsible for the apparent decrease in guanylate cyclase activity.

Time of Inactivation   The effect of the peptide on enzyme activity appeared to be very rapid and irreversible (156). Cells homogenized 30 min after addition of the peptide, had about the same activity as that observed 5 s after addition. In other experiments, enzyme activity measured in cells homogenized 2 h after addition of peptide was also found to be the same as that measured at 5 s. Attempts to decrease the rate of inactivation have been unsuccessful in that the incubation of sperm cells at 0-4°C still results in an apparent complete inactivation within 5 s.

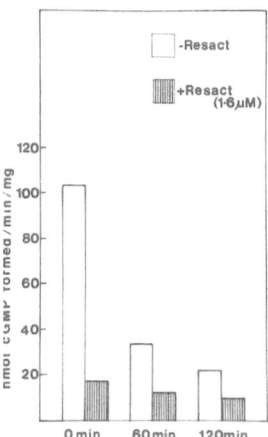

Figure 7. The time-dependent decrease of guanylate cyclase activity when measured at 22°C. A. punctulata sperm cells were incubated with or without resact as described in the legend to Fig. 6. After 1 min, the reaction was stopped with 20 mM 2[N-morpholino] ethane sulfonic acid buffer, pH 6.5, containing 0.5% Lubrol PX. After 1h at 0-1°C, the detergent-extracted cells were incubated at 22° and guanylate cyclase activity was assayed at the indicated time invervals.

Concentration-Response   Half-maximal inhibition of enzyme
activity occurred at about 25 nM speract and 112 nM resact in $\underline{S}$.
purpuratus and $\underline{A}$. punctulata spermatozoa, respectively. Strong evidence in further support of a receptor-mediated effect was the
concentration-response data obtained with various analogues of speract. The relative potencies of the analogues to decrease guanylate
cyclase activity coincided with their previously determined relative
potencies to increase respiration rates, elevate cyclic GMP concentrations and to bind to the receptor of intact cells (156).

Activity as a Function of Protein   Product formation measured
as a function of added enzyme protein resulted in linear plots. Under
all conditions, enzyme activity was measured where the amount of
cyclic GMP formed was linear as a function of time.

Metal Ion Dependence   The sea urchin sperm guanylate cyclase is
known to be highly dependent on Mn^{2+} for activity (155). It was
possible that the peptides would differentially effect the metal-ion
dependence of guanylate cyclase. However, enzyme obtained from
either control or resact-treated cells was markedly dependent on Mn^+.
Activity with Mg^{2+} was very low but it also was decreased by the
previous treatment of intact cells with peptide.

Monensin A and NH_4Cl An early event in response to speract is
the net efflux of H^+ and an increase of intracellular pH (69).
Agents such as Monensin A and NH_4Cl, which elevate intracellular pH,
are also known to stimulate sperm respiration and motility when
tested under the same conditions as speract (159). If intact sperm
cells are incubated with Monensin A, a time-dependent decrease in
guanylate cyclase activity occurs. The activity approaches that of
the resact treated sperm guanylate cyclase activity level but
requires considerably longer time periods of incubation. The effect
of Monesin A appears to be related to ion-transport since it is completely dependent on extracellular Na^+ (156). Approximately 40 mM
extracellular Na^+ was required for one-half maximal effects on guanylate cyclase activity. NH_4Cl also caused time-dependent decreases of
guanylate cyclase activity but the activity appeared to plateau above
the resact-treated level. Neither Monensin A nor NH_4Cl had effects
on guanylate cyclase activity when added after detergent-extraction
of the cells (not shown). Similar results are reported by Ward et
al. (this volume).

DISCUSSION

A summary of known responses of spermatozoa in response to the
binding of speract or resact to their receptor is shown in Fig. 8.
Although the receptors for each peptide fail to recognize the peptide
from the heterologous species, the physiological responses are similar. The interaction of speract with a plasma membrane glycoprotein,
whose apparent molecular weight is 77,000 on Na-dodecyl-SO_4 polyacry-

lamide gels in the presence of dithiothreitol, results in a net H^+ efflux, an increase in intracellular pH, increases in cyclic AMP and cyclic GMP concentrations, a decrease in guanylate cyclase activity as measured in vitro, and increased respiration rates and motility, especially at acidic extracellular pH values. Resact also causes a distinct mobility shift of a plasma membrane protein on Na-dodecyl-SO_4 polyacrylamide gels. All of the responses occur within seconds after addition of the peptides, but the time sequence of the above events is not known.

Whether or not factors similar to those found in the sea urchin exist in mammals is not yet established. As indicated in Table 1, many reports exist to suggest that the female contains substance(s) which can alter sperm motility and metabolism. However, the specificity of the effects remain to be determined. With respect to capacitation and the acrosome reaction, sulfated glycosaminoglycans have been found to have significant positive effects in the bovine (67,-109). The requirement of sulfate is similar to the situation in the sea urchin where a highly sulfated, fucose-rich complex induces the acrosome reaction (169,170).

A hyperactivated form of sperm motility has been reported for various mammals (see Table 2), although the mechanism by which this motility is induced remains to be determined; it does not necessarily occur concomitant with capacitation (acrosome reaction). Taurine and hypotaurine have been reported to stimulate the motility of hamster spermatozoa (120), but it is not yet known whether or not they are natural regulators of sperm motility. We have not observed positive effects of these amino acids on sea urchin sperm cells; however, whether or not sea urchin or other invertebrate spermatozoa can develop a hyperactivated form of motility (change in motility pattern) in response to egg factors has not been reported.

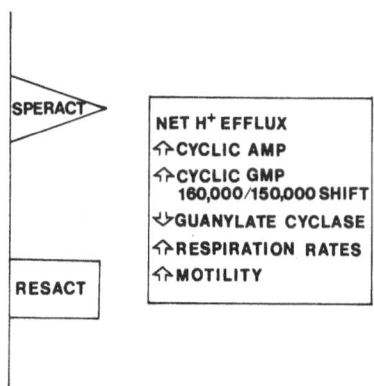

Figure 8. A summary of known responses of spermatozoa to the interaction of speract or resact with their respective receptor.

Table 2. Spermatozoa of Animals Reported to Show Hyperactivated
          Motility

Mice (1,28,43-45,151)

Hamster (92-94,123,217,219)

Guinea Pig (5,6,94,220-222)

Rabbit (29,89,145,146)

Bat (105)

Dog (116)

Sheep (31)

Dolphin (41)

Human, possibly (129)

Guanylate cyclase activity has been reported to be altered by
various agents in many different tissues (see Table 3). The
enclosed Table, for the most part, contains only those reports where
a plasma membrane receptor for a hormone, neurotransmitter or drug
might be expected to mediate the response. In a number of cases,
however, the added agents were tested on the soluble form of guanylate cyclase where a membrane receptor was not involved; these have
been included even though the mechanism of such a response in the
intact cell, if it occurred, would not be clear. Also included are
reports of steroid effects on guanylate cyclase activity where a
cytoplasmic or nuclear receptor could be involved. The many reports
on the direct activation of the soluble form of the enzyme by nitro-
samines, nitric oxide and similar chemicals are not included.

The presentation of the Table represents a biased report since
various reports not listed have observed no effects of many of the
included agents in a number of different tissues. Thus, at least
some of the observed effects must be considered controversial. In
some cases it would appear that reported activations definitely
represent artifacts (82). Some of the reports also can be debated
from a physiological viewpoint. For example, George et al. (55)
reported a stimulation of the soluble form of guanylate cyclase by
acetylcholine, an observation which they thought could explain the
effect of acetylcholine on cardiac cyclic GMP concentrations. They
also reported that atropine could inhibit the effect of acetylcholine

Table 3. Reported Effects of Various Agents on Guanylate Cyclase Activity

Tissue	Agent	Agent Added[1]	Guanylate Cyclase Activity[2]
Brain	Serotonin Secretin	post	+ 154,188
Caudate nucleus	Catecholamines	post	− 46
Neuroblastoma	Muscarinic agonists carbamoylcholine oxotremorine-M	pre	+ 179
Pituitary gland	Bromocriptine Dopamine	post	+ 202
Iris of Eye	Secretin	post	− 188
Lung	GH Secretin	post	+ 188,201
Heart	GH Angiotensin II Acetylcholine	post	+ 55,200,201
Liver	β-antagonists Glucagon	pre	+ 38,142
	Carbamoylcholine	post	+ 106,113,188-190 201,203,204
	Acetylcholine Estrogens Progesterone Secretin Insulin Interferon Testosterone		
Spleen	Interferon	post	+ 203
Pancreas	Secretin GH	post	+ 188,201
Islet of Langerhans	Acetylcholine Cholecystokinin Secretin	post	+ 80,188
	Epinephrine Norepinephrine	post	− 80
Adrenal	ACTH	post	+ 134,135
Adrenal cortical carcinoma	ACTH	post	+ 134,136
Adrenal medulla	ACTH Acetylcholine	----	+ 172
Kidney	Angiotensin II Interferon Testosterone	post	+ 199-201,203

(continued)

Table 3. Continued

Tissue	Agent	Agent Added[1]	Guanylate Cyclase Activity[2]
	GH		
Stomach	Secretin	post	+ 61,188
	Cholecystokinin		
	β–Antagonists	pre	+ 142
Small Intestine	β–Antagonists	pre	+ 118,142,186
	Methylprednisolone		
	Cytotoxin (C. difficile)	post	+ 62,157,158 205
	Heat-stable Enterotoxin (E. coli)		
	Cholera toxin	pre	− 96
Ovary	Bromocriptine	post	+ 202
	Dopamine		
Uterus	Estrogens	post	+ 204
	Progesterone		
Myometrium	Progesterone	pre	− 7,8
Prostate	Testosterone	post	+ 199
Testis	Melatonin	pre	+ 90
Muscle	Testosterone	post	+ 188,199,201
Skeletal	GH		
	Secretin		
Insect	Glutamate	post	+ 161
Cartilage	Somatomedin C	post	+ 180
Fetal osteogenic	PTH	pre	+ 207
Calvarial mesenchyme	Calcitonin		
	Insulin		
Blood Vessels	Angiotensin II	post	+ 200
Dictyostelium discoideum	cAMP	pre	+ 119
Fat	Secretin	post	+ 188
Balb/c 3T3	FGF	post	+ 164
Platelets	Thrombin	pre	+ 4
	ADP		
Burkitt Lymphoma (Daudi)	Interferon-α	pre	+ 162
Sea urchin sperm	FRE (factors released from eggs)	post	− 49,50
	Jelly	pre	− 208

[1]The agent was added to the tissue before (pre) or after (post) homogenization.
[2]The numbers refer to references.

on the enzyme. It seems most doubtful, however, that the soluble form of guanylate cyclase represents a cholinergic receptor.

The first reports to show that a mixture of crude egg factors (FRE) could inhibit guanylate cyclase were those of Garbers and Hardman (49,50). In these studies, particles were used and the factor(s) obtained from eggs which caused inhibition of the enzyme appeared to be non-dialyzable. The extent of inhibition was highly dependent on protein concentration. Later, Ward et al. (208) also showed an inhibition of sperm guanylate cyclase activity by egg-associated factors but under conditions where intact cells were treated with a crude mixture of egg factors and subsequently homogenized. Those effects could be demonstrated under conditions where enzyme inhibition was not observed if the factor(s) were added after homogenization. Here, and in the report of Ramarao and Garbers (156) sperm guanylate cyclase activity appears to decrease in response to an egg peptide (resact or speract). The response is receptor-mediated. Whether or not enzyme activity actually decreases in the cell, however, is not yet known. It is certainly possible that activity actually increases in the intact cell, and that the apparent decrease observed under in vitro conditions reflects a receptor-mediated conformational change. The assay conditions are very different than cellular conditions not only because of the presence of detergent, but also because Mn^{2+} is substituted for Mg^{2+}. The apparent activity was still decreased by the peptides in the presence of Mg^{2+} but possibly other important cofactors were still absent (156). Alternatively, it could be argued that the flux of cyclic GMP is reduced by the peptides. Thus, guanylate cyclase activity decreases but cyclic GMP phosphodiesterase activity decreases to an even greater extent; the result would be an elevation of the steady state concentration of cyclic GMP. Finally, one must consider the possibility that the decrease in guanylate cyclase reflects a rapid inactivation of the enzyme after an initial stimulation. Cyclic GMP concentration are raised by the peptides in only a transient manner.

Intracellular pH has been suggested as an important signal in various cells (see Table 4 and recent review by Busa and Nuccitelli, 19). One response of the sperm cell to speract is an elevation of intracellular pH; it has been suggested that at least under conditions of lower extracellular pH that the effect of the peptide on intracellular pH may represent a primary mechanism by which respiration is increased (159). Others also have suggested that intracellular pH of the sea urchin sperm cell can modulate respiration and motility (21,107). The mobility shift of the 160,000 molecular weight plasma membrane protein can be induced by Monensin A and NH_4Cl suggestive that intracellular pH might also regulate this conversion. However, the response time is very slow relative to that observed with the peptides, while the activation of respiration continues to occur rapidly in response to NH_4Cl or Monensin A. Therefore, the mobility shift caused by the peptide may represent a receptor-mediated response not involving changes in intracellular pH.

Table 4. A Partial Summary of Various Cellular Responses to Agents Where Intracellular pH is Altered

FERTILIZATION

Agent/Event	Species	pH_I Change	Response	Reference
For Sperm:				
Amines	Caenorhabditis elegans	Increased	Spermiogenesis	210
Amines or Nigericin	S. purpuratus, L. pictus	Increased	Increased motility	10,107
Egg Jelly	S. purpuratus	Increased then Decreased	Decreased Fertilizability	22
Speract	S. purpuratus	Increased	Stimulated respiration	159
For Eggs:				
Progesterone or insulin	Xenopus laevis	Increased	Increased protein synthesis and egg maturation	79,177,211
Fertilization	L. pictus	Increased	Increased K^+ Conductance	171
Fertilization	S. purpuratus, L. pictus P. lividus, S. granularis and Xenopus laevis	Increased	Increased protein synthesis and cell division	37,60,88, 138,149

MITOGENESIS AND NEOPLASIA

Agent/Event	Tissue or Cell Line	pH$_i$ Change	Response	Reference
Con A or LPS	T or B Lymphocyte	Increased	Increased DNA Synthesis Increased Mitosis	56
Serum (fetal or newborn calf)	CC1 39 (fibroblast) N1E-115 (neuroblastoma) BSC-1,3T3, NRG, Human Fibroblast	Increased	"	17,20,124, 125,153,163
PDGF	3T3 and NR 6	Increased	"	17,20
EGF	Human Fibroblast and BSC-1	Increased	"	125,163
ADH	3T3	Increased	"	17
Thrombin	CC1 39	Increased	"	153
Thrombin + Insulin	CC1 39	Increased	"	153
PDGF + ADH + Insulin	3T3	Increased	"	167
EGF + Insulin	Human Fibroblast	Increased	"	125
EGF + ADH + Insulin	3T3	Increased	"	18
Phorbol Ester	3T3	Increased	"	17
3-Methyl-4-dimethyl-Amino Azobenzene	liver	Decreased	Carcinogenesis	97

(continued)

Table 4. Continued

PHYSIOLOGIC-PHARMACOLOGIC EFFECTORS

Agent/Event	Species	pH$_I$ Change	Response	Reference
ADH	Bladder (Toad and Frog)	Increased	Increased permeability to water	2,147,148
Nutrients/Glucose	Beta Cells of Islets	Decreased then increased	Increased insulin secretion	35,117,176, 187
Insulin	Skeletal Muscle	Increased	Increased glycolysis	40,126,127
Thrombin	Human Platelets	Increased	Aggregation	78
Quercitin	Ehrlich Ascites Tumor	Decreased	Decreased glycolysis	9

Although the sperm cell is highly differentiated with little, if any, capacity to synthesize new protein, its responses to extracellular signals are similar to those of many other cells. The Tables presented, in part, document some of the similarities. The spermatozoon contains a receptor for peptides associated with eggs and responds to the interaction between the receptor and these peptides with changes in various potential intracellular messengers. Cyclic nucleotides may play an important role as second messengers but other substances (i.e. H^+) also appear to have prominent primary effects. Since the responses of sperm cells are not unlike those of other "activated" cells (whether activated by hormones, mitogens or by agents causing transformation) they may prove to be a less complex and therefore a valuable cell model.

ACKNOWLEDGEMENTS

The authors thank Dr. Victor D. Vacquier and Gary E. Ward for valuable discussions during the course of these studies. The authors are grateful to Penny Stelling and Becky Lawson for excellent typing of the manuscript. Supported by NIH Grants HD 10254 and GM 31362.

REFERENCES

1. Aonuma, S., Okabe, M., Kawaguchi, M., and Kishi, Y. 1980. Studies on sperm capacitation. IX Movement characteristics of sperm in relation to capacitation. Chem. Pharm. Bull. Jpn. (Tokyo) 28:1497-1502.

2. Arruda, J.A.L., Dytko, G., Lubansky, H., Burt, C.T., and Mola, R. 1983. Acid-base metabolism, intracellular pH and water transport by the toad bladder. Arch. Int. Pharmacodyn 265:150-163.

3. Atherton, R.W. 1977. A tentative hypothesis to explain the neurochemical regulation of sperm motility. Adv. Invest. Reprod. 1:120-129.

4. Barber, A.J. 1976. Cyclic nucleotides and platelet aggregation: effect of aggregating agents on the activity of cyclic nucleotide metabolizing enzymes. Biochim. Biophys. Acta 444:579-595.

5. Barros, C., and Berrios, M. 1977. Is the activated spermatozoa really capacitated? J. Exp. Zool. 201:65-72.

6. Barros, C., Berrios, M., and Herrera, E. 1973. Capacitation in vitro of guinea-pig spermatozoa in a saline solution. J. Reprod. Fert. 34:547-549.

7. Beatty, C.H., Bocek, R.M., Herrington, P.T., Young, M.K., and Brenner, R.M. 1979. Effects of estradiol-17 beta and progesterone on cyclic nucleotide metabolism in myometrium of macaques. Biol. Reprod. 21:309-318.

8. Beatty, C.H., Bocek, R.M., Herrington, P.T., Young, M.K., and Brenner, R.M. 1980. Estradiol-17 beta and progesterone: effects on guanylate cyclase activity in the myometrium of macaques. Proc. Soc. Exptl. Biol. Med. 164:292-298.

9. Belt, J.A., Thomas, J.A., Buchsbaum, R.N., and Racker, E. 1979. Inhibition of lactate transport and glycolysis in Ehrlich ascites tumor cells by bioflavonoids. Biochemistry 18:3506-3511.

10. Bibring, T., Baxandall, J., and Harter, C.C. 1984. Sodium-dependent pH regulation in active sea urchin sperm. Dev. Biol. 101:425-435.

11. Black, D.L., Crowley, L.V., Duby, R.T., and Spilman, C.H. 1968. Oviduct secretion in the ewe and the effect of oviduct fluid on oxygen uptake by ram spermatozoa. J. Reprod. Fert. 15:127-130.

12. Black, D.L., Kumar, A., Crowley, L.V., Duby, R.T., and Spilman, C.H. 1970. Composition of oviduct fluid kept at two temperatures during collection. J. Reprod. Fert. 22:597-600.

13. Bleil, J.D., and Wassarman, P.M. 1983. Sperm-egg interactions in the mouse: sequence of events and induction of the acrosome reaction by a zona pellucida glycoprotein. Dev. Biol. 95:317-324.

14. Bolton, A.E., and Hunter, W.M. 1973. The labelling of proteins to high specific radioactivities by conjugation to a ^{125}I-containing acylating agent. Biochem. J. 133:529-539.

15. Bradley, M.P., and Garbers, D.L. 1983. The stimulation of bovine caudal epididymal sperm forward motility by bovine cumulus-egg complexes in vitro. Biochem. Biophys. Res. Comm. 115:777-787.

16. Brown, C.V., and Senger, P.L. 1982. Influence of incubation in utero on motility and head-to-head agglutination of ejaculated rabbit spermatozoa. J. Reprod. Fertil. 66:283-289.

17. Burns, C.P., and Rozengurt, E. 1983. Serum, platelet-derived growth factor, vasopressin and phorbol esters increase intracellular pH in Swiss 3T3 cells. Biochem. Biophys. Res. Commun. 116:931-938.

18. Burns, C.P., and Rozengurt, E. 1984. Extracellular Na^+ and initiation of DNA synthesis: role of intracellular pH and K^+. J. Cell Biol. 98:1082-1089.

19. Busa, W.B., and Nuccitelli, R. 1984. Metabolic regulation via intracellular pH. Am. J. Physiol. 246:R409-R438.

20. Cassel, D., Rothenburg, P., Zhuang, Y.X., Deuel, T.F., and Glasser, L. 1983. Platelet-derived growth factor stimulates Na^+/H^+ exchange and induces cytoplasmic alkalinization in NR6 cells. Proc. Natl. Acad. Sci. USA 80:6224-6228.

21. Christen, R., Schackmann, R.W., and Shapiro, B.M. 1982. Elevation of the intracellular pH activates respiration and motility of sperm of the sea urchin, Strongylocentrotus purpuratus. J. Biol. Chem. 257:14881-14890.

22. Christen, R., Schackmann, R.W., and Shapiro, B.M. 1983. Inter-
    actions between sperm and sea urchin egg jelly. Dev. Biol.
    98:1-14.

23. Clapper, D.L., and Brown, G.G. 1980. Sperm motility in the horse-
    shoe crab, Limulus polyphemus L. I. Sperm behavior near eggs
    and motility initiation by egg extracts. Dev. Biol. 76:341-
    349.

24. Clapper, D.L., and Brown, G.G. 1980. Sperm motility in the horse-
    shoe crab, Limulus polyphemus L. II. Partial characterization
    of a motility initiating factor from eggs and the effects of
    inorganic cations on motility initiation. Dev. Biol. 76:350-
    357.

25. Clapper, D.L., and Epel, D. 1982. Sperm motility in the horseshoe
    crab. III. Isolation and characterization of a sperm motility
    intiating peptide. Gamete Res. 6:315-326.

26. Clowes, G.H.A., and Bachman, E. 1921. A volatile sperm-stimulating
    substance derived from marine eggs. Proc. Exptl. Biol. Med.
    18:120-121.

27. Colwin, L.H., and Colwin, A.L. 1956. The acrosome filament and
    sperm entry in the Thyone briareus (holothuria) and Asterias.
    Bio. Bull. 110:243-255.

28. Cooper, T.G. 1984. The onset and maintenance of hyperactivated
    motility of spermatozoa from the mouse. Gamete Res. 9:55-74.

29. Cooper, G.W., Overstreet, J.W., and Katz, D.F. 1979. The motility
    of rabbit spermatozoa recovered from the female reproductive
    tract. Gamete Res. 2:35-42.

30. Cornman, I. 1941. Sperm activation of Arbacia egg extracts, with
    special reference to echinochrome. Biol. Bull. 80:202-207.

31. Cummins, J.M. 1982. Hyperactivated motility patterns of ram
    spermatozoa recovered from the oviducts of mated ewes. Gamete
    Res. 6:53-63.

32. Dan, J.C. 1952. Studies on the acrosome. I. Reaction to egg water
    and other stimuli. Biol. Bull. 103:54-66.

33. Dan, J.C., and Wada, S.K. 1955. Studies on the acrosome. IV. The
    acrosome reaction in some bivalve spermatozoa. Biol. Bull.
    109:40-55.

34. Dangott, L.J., and Garbers, D.L. 1984. Identification and partial
    characterization of the receptor for speract. J. Biol. Chem.
    259:13712-13716.

35. Deleers, M., Lebrun, P., and Malaisse, W.J. 1983. Increase in
    CO_3H^- influx and cellular pH in glucose-stimulated pancreatic
    islets. FEBS Lett. 154:97-100.

36. deSantis, R., Pinto, M.R., Cotelli, F., Rosati, F., Monroy, A.
    and D'Alessio, G. 1983. A fucosyl glycoprotein component with
    sperm receptor and sperm activating activities from the
    vitelline coat of Ciona intestinalis eggs. Exp. Cell Res.
    148:508-513.

37. Dube, F., and Guerrier, P. 1983. Ca^{2+} influx and stimulation of
    protein synthesis in sea urchin eggs. Exp. Cell Res. 147:209-
    215.

38. Earp, H.S. 1980. The role of insulin, glucagon and cAMP in the regulation of hepatocyte guanylate cyclase activity. J. Biol. Chem. 255:8979-8982.

39. Esbenshade, K.L., and Clegg, E.D. 1980. Acrosome reaction of sperm incubated in the uterus of gilts. Am. J. Vet. Res. 41:1137-1140.

40. Fidelman, M.L., Seeholzer, S.H., Walsh, K.B., and Moore, R.D. 1982. Intracellular pH mediates action of insulin on glycolysis in frog skeletal muscle. Am. J. Physiol. 242:C87-C93.

41. Fleming, A.D., Yanagimachi, R., and Yanagimachi, I. 1982. Spermatozoa of the Atlantic bottlenosed dolphin Tursiops truncatus. J. Reprod. Fert. 63:509-514.

42. Foley, C.W., and Williams, S.L. 1967. Effect of bicarbonate and oviduct fluid on respiration of spermatozoa. Proc. Soc. Exptl. Biol. Med. 126:634-637.

43. Fraser, L.E. 1977. Motility pattern in mouse spermatozoa before and after capacitation. J. Exp. Zool. 202:439-444.

44. Fraser, L.E. 1979. Accelerated mouse sperm penetration in vitro in the presence of caffeine. J. Reprod. Fertil. 57:377-384.

45. Fraser, L.E. 1983. Potassium ions modulate expression of mouse's sperm fertilizing ability, acrosome reaction and hyperactivated motility in vitro. J. Reprod. Fert. 69:539-553.

46. Frey, W.H., Senogles, S.E., Heston, L.L., Tuanson, V.B., and Nicol, S.E. 1980. Catecholamine-sensitive guanylate cyclase from human caudate nucleus. J. Neurochem. 35:1418-1430.

47. Fujihara, N., and Howarth, B., Jr. 1980. Prolonged survival of cock spermatozoa in vitro with fluid removed from tissue cultured oviductal cells. Poultry Sci. 59:164-167.

48. Garbers, D.L. 1981. The elevation of cyclic AMP concentrations in flagella-less sea urchin sperm heads. J. Biol. Chem. 256:620-624.

49. Garbers, D.L., and Hardman, J.G. 1975. Factors released from sea urchin eggs affect cyclic nucleotide metabolism in sperm. Nature 257:677-678.

50. Garbers, D.L., and Hardman, J.G. 1976. Effects of egg factors on cyclic nucleotide metabolism in sea urchin sperm. J. Cyclic Nucleotide Res. 2:59-70.

51. Garbers, D.L., Kopf, G.S., Tubb, D.J., and Olson, G. 1983. Elevation of sperm adenosine 3',5'-monophosphate concentrations by a fucose-sulfate-rich complex associated with eggs. I. Structural characterization. Biol. Reprod. 29:1211-1220.

52. Garbers, D.L., Tubb, D.J., and Kopf, G.S. 1980. Regulation of sea urchin sperm cyclic AMP-dependent protein kinases by an egg-associated factor. Biol. Reprod. 22:526-532.

53. Garbers, D.L., Watkins, H.D., Hansbrough, J.R., Smith, A.S., and Misono, K. 1982. The amino acid sequence and chemical synthesis of speract and of speract analogues. J. Biol. Chem. 257:2734-2737.

54. Garrels, J.I. 1979. Two-dimensional gel electrophoresis and com-

puter analysis of proteins synthesized by clonal cell lines.
J. Biol. Chem. 254:7961-7977.

55. George, W.J., Ignarro, L.J., and White, L.E. 1976. Muscarinic
    stimulation of cardiac guanylate cyclase, in: "Recent
    Advances in Studies on Cardiac Structure and Metabolism," Vol.
    7: Biochemistry and Pharmacology of Myocardial Hypertrophy,
    Hypoxia, and Infarction. P. Harris, R. J. Bing, and A.
    Fleckenstein, eds. (Baltimore: University Park Press),
    pp. 381-390.

56. Gerson, D.F., Kiefer, H., and Eufe, W. 1963. Intracellular pH of
    mitogen-stimulated lymphocytes. Science 216:1009-1010.

57. Ginsburg, A.S. 1963. Sperm-egg association and its relationship
    to the activation of the egg in Salmonid fishes. J. Embryol.
    Exp. Morphol. 11:13-33.

58. Glasser, O. 1914. A qualitative analysis of the egg secretions of
    Arbacia and Asterias. Biol. Bull. 26:367-386.

59. Goldstein, L., Levin, Y., and Katachalski, E. 1964. A water-
    insoluble polyionic derivative of trypsin. II. Effect of the
    polyelectrolyte carrier on the kinetic behavior of the bound
    trypsin. Biochemistry 3:1913-1919.

60. Grainger, J.L., Winkler, M.M., Shen, S.S., and Steinhardt, R.A.
    1979. Intracellular pH controls protein synthesis rate in the
    sea urchin egg and early embryo. Dev. Biol. 68:396-406.

61. Grissen, M. 1978. Effect of Ca^{2+}, Mg^{2+}, NaN_3, cholinergic agents,
    and gastrointestinal hormones on the guanylate cyclase from
    guinea pig gastric mucosa. Gastroenterology 74:263-270.

62. Guerrant, R.L., Hughes, J.M., Chang, B., Robertson, D.C., and
    Murad, F. 1980. Activation of intestinal guanylate cyclase by
    heat-stable enterotoxin of Escherichia coli: studies of
    tissue specificity, potential receptors, and intermediates.
    J. Infect. Dis. 142:220-228.

63. Gwatkin, R.B.L., and Andersen, O.F. 1969. Capacitation of hamster
    spermatozoa by bovine follicular fluid. Nature 224:111-112.

64. Gwatkin, R.B L., Andersen, O.F., and Hutchinson, C.F. 1972. Capa-
    citation of hamster spermatozoa in vitro: the role of cumulus
    components. J. Reprod. Fert. 30:389-394.

65. Hamner, C.E., and Williams, W.L. 1963. Effect of the female
    reproductive tract on sperm metabolism in the rabbit and
    fowl. J. Reprod. Fert. 5:143-150.

66. Hamner, C.E., and Williams, W.L. 1964. Identification of sperm
    stimulating factor of rabbit oviduct. Proc. Soc. Exptl. Biol.
    Med. 117:240-243.

67. Handrow, R.R., Lenz, R.W., and Ax, R.L. 1982. Structural compari-
    sons among glycosaminoglycans to promote an acrosome reaction
    in bovine spermatozoa. Biochem. Biophys. Res. Commun.
    107:1326-1332.

68. Hansbrough, J.R., and Garbers, D.L. 1981. Speract: Purification
    and characterization of a peptide associated with eggs that
    activates spermatozoa. J. Biol. Chem. 256:1447-1452.

69. Hansbrough, J.R., and Garbers, D.L. 1981. Sodium-dependent activation of sea urchin spermatozoa by speract and monensin. J. Biol. Chem. 256:2235-2241.

70. Hansbrough, J.R., Kopf, G.S., and Garbers, D.L. 1980. The stimulation of sperm metabolism by a factor associated with eggs and by 8-bromoguanosine 3'5'-monophosphate. Biochim. Biophys. Acta 630:82-91.

71. Hartmann, M., Kuhn, R., Schartau, O., and Wallenfels, K. 1940. Uber die Wechselwirkung Von Gyno-und Androgamonen bei befructing der Eier des Seeigels. Naturwiss 28:144.

72. Hathaway, R.R. 1960. Stimulation of Arbacia sperm respiration by egg substances. Biol. Bull. 119:318-391.

73. Hathaway, R.R. 1963. Activation of respiration in sea urchin spermatozoa by egg water. Biol. Bull. 125:486-498.

74. Hicks, J.J., Martiniz-Manatou, J., Pedron, N., and Rosado, A. 1972. Metabolic changes in human spermatozoa related to capacitation. Fert. Steril. 23:172-179.

75. Hicks, J.J., Pedron, N., and Rosado, A. 1972. Modifications of human spermatozoa glycolysis by cyclic adenosine monophosphate (cAMP), estrogens, and follicular fluid. Fert. Steril. 23:886-893.

76. Holland, L.Z., and Cross, N.L. 1983. The pH within the jelly coat of sea urchin eggs. Dev. Biol. 99:258-260.

77. Holtz, W., Stoss, J., and Buyukhatipoglu, S. 1977. Beopachtungen zur Aktivierbarkeit von F orellenspermatozoen mit fructwasser, bachwasser und destilliertem Wasser. Zuchtygiene 12:82-88.

78. Horne, W.C., Norman, N.E., Schwartz, D.B., and Simons, E.R. 1981. Changes in cytoplasmic pH and in membrane potential in thrombin-stimulated human platelets. Eur. J. Biochem. 120:295-302.

79. Houle, J.G., and Wasserman, W.J. 1983. Intracellular pH plays a role in regulating protein synthesis in Xenopus oocytes. Dev. Biol. 97:302-312.

80. Howell, S.L., and Montague, W. 1974. Regulation of guanylate cyclase in guinea-pig islets of Langerhans. Biochem. J. 142:379-384.

81. Hyne, R.V., and Garbers, D.L. 1980. Requirement of serum factors for capacitation and the acrosome reaction of guinea pig spermatozoa in buffered medium below pH 7.8. Biol. Reprod. 24:257-266.

82. Ichihara, K., Larner, J., Kimura, H., and Murad, F. 1977. Activation of liver guanylate cyclase by bile salts and contaminants in crude secretin and pancreozymin preparations. Biochim. Biophys. Acta 481:734-740.

83. Iritani, A., Gomes, W.R., and Vandemark, N.L. 1969. The effect of whole, dialyzed and heated female genital tract fluids on respiration of rabbit and ram spermatozoa. Biol. Reprod. 1:77-82.

84. Isaka, S., Hotta, K., and Kurokawa, M. 1970. Jelly coat substan-

ces of sea urchin eggs. I. Sperm isoagglutination and sialopo-
lysaccharide in the jelly. Exp. Cell Res. 59:37-42.

85. Ishihara, K., Oguri, K., and Taniguchi, H. 1973. Isolation and
characterization of fucose sulfate from jelly coat glycoproteins of sea urchin eggs. Biochim. Biophys. Acta 320:628-
634.

86. Iwamatsu, T., and Chang, M.C. 1971. Factors involved in the fertilization of mouse eggs in vitro. J. Reprod. Fert. 26:197-
208.

87. Johnsen, O., and Eliasson, R. 1976. Follicular fluid and succinate oxidation by human spermatozoa. Andrologia 8:283-284.

88. Johnson, J.D., Epel, D., and Paul, M. 1976. Intracellular pH and
activation of sea urchin eggs after fertilization. Nature
262:661-664.

89. Johnson, L.L., Katz, D.F., and Overstreet, J.W. 1981. The movement characteristics of rabbit spermatozoa before and after
activation. Gamete Res. 4:275-282.

90. Kano, T., and Miyachi, Y. 1976. Direct action of melatonin on
testosterone and cyclic GMP production using rat testis
tissue in vitro. Biochem. Biophys. Res. Commun. 72:969-975.

91. Katagiri, C. 1966. Fertilization of dejellied uterine toad eggs
in various experimental conditions. Embryologia 9:159-169.

92. Katz, D.F., and Overstreet, J.W. 1980. Mammalian sperm movement
in the secretions of the male and female genital tracts, in:
"Testicular Development, Structure, and Function," A.
Steinberger, E. Steinberger, Eds., New Nork Raven Press. pp.
481-489.

93. Katz, D.F., and Yanagimachi, R. 1980. Movement characteristics
of hamster spermatozoa within the oviduct. Biol. Reprod.
22:759-764.

94. Katz, D.F., and Yanagimachi, R. 1981. Movement characteristics
of hamster and guinea pig spermatozoa upon attachment to the
zona pellucida. Biol. Reprod. 25:785-791.

95. Katz, D.F., Yanagimachi, R., and Dresdner, R.D. 1978. Movement
characteristics and power output of guinea pig and hamster
spermatozoa in relation to activation. J. Reprod. 52:167-
172.

96. Kiefer, H.C., Atlas, R., and Moldan, D. 1975. Inhibition of
guanylate cyclase and cyclic GMP phosphodiesterase by cholera
toxin. Biochem. Biophys. Res. Commun. 66:1017-1023.

97. Kitagawa, Y., and Kuroiwa, Y. 1976. Change in intracellular pH
of rat liver during azo-dye carcinogenesis. Life Sci.
18:441-450.

98. Kopf, G.S., and Garbers, D.L. 1979. A low molecular weight
factor from sea urchin eggs elevates sperm cyclic nucleotide
concentrations and respiration rates. J. Reprod. Fert.
57:353-361.

99. Kopf, G.S., and Garbers, D.L. 1980. Calcium and a fucose-
sulfate-rich polymer regulate sperm cyclic nucleotide meta-

bolism and the acrosome reaction. Biol. Reprod. 22:1118–1126.

100. Kopf, G.S., Tubb, D.J., and Garbers, D.L. 1979. Activation of sperm respiration by a low molecular weight egg factor and by 8-bromoguanosine 3',5'-monophosphate. J. Biol. Chem. 254:8554–8560.

101. Kozumplik, J. 1978. Effect of follicular fluid on motility of thawed boar spermatozoa. Acta Vet. Brno 47:33–38.

102. Kurzrok, R., Wilson, L., and Birnberg, C. 1953. Follicular fluid. Its possible role in human fertility and infertility. Fert. Steril. 4:479–494.

103. Laemmli, U.K. 1970. Cleavage of structural proteins during the assembly of the head of bacterophage T4. Nature 227:680–685.

104. Lambert, C.C., and Epel, D. 1979. Calcium-mediated mitochondrial movement in ascidian sperm during fertilization. Dev. Biol. 69:296–304.

105. Lambert, H. 1981. Temperature dependence of capacitation in bat spermatozoa monitored by zona-free hamster ova. Gamete Res. 4:525–533.

106. Laurenza, A., Paolisso, G., Choisi, E., Spina, and Illiano, G. 1983. Low insulin concentrations stimulate in vitro the soluble guanylate cyclase activity of rat liver. Biochem. Biophys. Res. Commun. 114:282–288.

107. Lee, H.C., Johnson, C., and Epel, D. 1983. Changes in internal pH associated with initiation of motility and acrosome reaction of sea urchin sperm. Dev. Biol. 95:31–45.

108. Lehrer, A.R., and Schindler, H. 1969. The influence of hen oviduct homogenates on the motility, oxygen uptake and fertilizing capacity of fowl spermatozoa. Poultry Sci. 48:30–36.

109. Lenz, R.W., Az, A.L., Grimek, H.J., and First, N.L. 1982. Proteoglycan from bovine follicular fluid enhances an acrosome reaction in bovine spermatozoa. Biochem. Biophys. Res. Commun. 106:1092–1098.

110. Levin, Y., Pecht, M., Goldstein, L., and Katchalski, E. 1964. A water-insoluble polyanionic derivative of trypsin. I. Preparation and properties. Biochemistry 3:1905–1913.

111. Lillie, F.R. 1912. Studies of fertilization. V. The behavior of the spermatozoa of Nereis and Arbacia with special reference to egg extractives. J. Exp. Zool. 16:515–574.

112. Lillie, F.R. 1919. Problems of fertilization, University of Chicago Press.

113. Lippo de Becembert, I., Herrera, V., Perez Ayuso, V. and Alfonzo, M. 1982. Effects of cholinergic agents on rat liver plasma membrane guanylate cyclase. FEBS Lett. 137:303–306.

114. Lowry, O.H., Rosebrough, N.J., Farr, A.L., and Randall, J.R. 1951. Protein measurement with the Folin phenol reagent. J. Biol. Chem. 193:265–275.

115. Lybing, S., and Hagstrom, B.E. 1957. Isolation of a fertiliza-

tion promoting factor from egg water of Psammechinus milaris. Exp. Cell Res. 13:60-68.

116. Mahi, C.A., and Yanagimachi, R. 1976. Maturation and sperm penetration of canine ovarian oocytes in vitro. J. Exp. Zool. 196:189-196.

117. Malaisse, W.J., Herchulez, A., and Sener, A. 1980. The possible significance of intracellular pH in insulin release. Life. Sci. 26:1367-1371.

118. Marnane, W.G., Tai, Y.H., Decker, R.A., Boedeker, E.C., Charney, A.N., and Donowitz, M. 1981. Methylprednisolone stimulation of guanylate cyclase activity in rat small intestinal mucosa: possible role in electrolyte transport. Gastroenterology 81:90-100.

119. Mato, J.M., and Malchow, D. 1978. Guanylate cyclase activation in response to chemotactic stimulation in Dictyostelium discoideum. FEBS Lett. 90:119-122.

120. Meizel, S., Lui, C.W., Working, P.K., and Mrsny, R.J. 1980. Taurine and hypotaurine: Their effects on motility, capacitation and the acrosome reaction of hamster sperm in vitro and their presence in sperm and reproductive tract fluids of several mammals. Develop. Growth and Differ. 22:483-494.

121. Miller, R.L. 1975. Chemotaxis of the spermatozoa of Ciona intestinalis. Nature 254:244-245.

122. Miller, R.L. 1977. Chemotactic behavior of the sperm of chitons (Mollusca: Polyplacophora). J. Exp. Zool. 20:203-211.

123. Mohri, H., and Yanagimachi, R. 1980. Characteristics of motor apparatus in testicular, epididymal and ejaculated spermatozoa: A study using demembranated sperm model. Exp. Cell Res. 129:191-196.

124. Moolenaar, W.H., Mummery, C.L., van der Saag, P.T., and de Laat, S.W. 1981. Rapid ionic events and the initiation of growth in serum-stimulated neuroblastoma cells. Cell 23:789-798.

125. Moolenaar, W.H., Tsien, R.Y., van der Saag, P.T., and de Latt, S.W. 1983. Na^+/H^+ exchange and cytoplasmic pH in the action of growth factors in human fibroblasts. Nature 304:645-648.

126. Moore, R.D. 1979. Elevation of intracellular pH by insulin in frog skeletal muscle. Biochem. Biophys. Res. Commun. 91:900-904.

127. Moore, R.D., Fidelman, M.L., and S. H. Seeholzer, Correlation between insulin action upon glycolysis and change in intracellular pH. Biochem. Biophys. Res. Commun. 91:905-910.

128. Morrisey, J.H. 1981. Silver stain for proteins in polyacrylamide gels: A modified procedure with enhanced uniform sensitivity. Anal. Biochem. 117:307-310.

129. Mortimer, D., Courtot, A.M., Giovangrandi, Y., and Jeulin, C. 1983. Do capacitated human spermatozoa show an "activated" pattern of motility? in: "The sperm cell: Fertilizing power, surface properties, motility, nucleus and acrosome evolutionary aspects," J. Andre, Ed., Boston, Martinus Nijhoff, pp. 349-352.

130. Mounib, M.S., and Chang, M.C. 1964. Effect of in utero incuba-
     tion on the metabolism of rabbit spermatozoa. Nature
     201:943-944.
131. Mukherjee, A.B., and Lippes, J. 1972. Effect of human follicular
     and tubal fluids on human, mouse and rat spermatozoa in
     vitro. Can. J. Genet. Cytol. 14:167-174.
132. Murdoch, R.N. 1968. The influence of the female genital tract on
     the metabolism of rabbit spermatozoa. I. Direct effect of
     tubal or uterine fluids, bicarbonate, and other factors.
     Aust. J. Biol. Sci. 21:961-972.
133. Murdoch, R.N., and White, I.G. 1967. The metabolism of labeled
     glucose by rabbit spermatozoa after incubation in utero. J.
     Reprod. Fert. 14:213-223.
134. Nambi, P., Aiayar, N.V., and Sharma, R.K. 1982. Adrenocortico-
     tropin-dependent particulate guanylate cyclase in rat adrenal
     and adrenocortical carcinoma: comparison of its properties
     with soluble guanylate cyclase and its relationship with
     ACTH-induced steroidogenesis. Arch. Biochem. Biophys. 217:
     638-646.
135. Nambi, P., and Sharma, R.K. 1981. Adrenocorticotropic hormone-
     responsive guanylate cyclase in the particulate fraction of
     rat adrenal glands. Endocrinology 108:2025-2027.
136. Nambi, P., and Sharma, R.K. 1981. Demonstration of ACTH-
     sensitive particulate guanylate cyclase in adrenocortical
     carcinoma. Biochem. Biophys. Res. Commun. 100:508-514.
137. Nomura, K., Suzuki, N., Ohtake, H., and Isaka, S. 1983. Struc-
     ture and action of sperm activating peptides from the egg
     jelly of a sea urchin, Anthocidaris crassipina. Biochem.
     Biophys. Res. Commun. 117:147-153.
138. Nuccitelli, R., Webb, D.J., Lagier, S.T., and Matson, G.B. 1981.
     ^{31}P NMR reveals increased intracellular pH after fertiliza-
     tion in Xenopus eggs. Proc. Natl. Acad. Sci. USA 78:4421-
     4425.
139. Ogasawara, F.X. and Lorenz, F.W. 1964. Respiratory rate of cock
     spermatozoa as affected by oviduct extracts. J. Reprod.
     Fert. 7:281-288.
140. Ohtake, H. 1976. Respiratory behavior of sea urchin spermatozoa.
     Effects of pH and egg water on the respiratory rate. J. Exp.
     Zool. 198:303-312.
141. Ohtake, H. 1976. Respiratory behavior of sea urchin spermatozoa.
     I. Sperm activating substance obtained from jelly coat of sea
     urchin eggs. J. Exp. Zool. 198:313-322.
142. Okine, L.K.N., Ioannides C. and Parke, D.V. 1983. Effect of some
     beta adrenergic blocking agents on tissue guanylate cyclase
     and cyclic nucleotides in the rat. Toxicol. Lett.
     18:235-240.
143. Olds, D. and Vandemark, N.L. 1957. The behavior of spermatozoa
     in luminal fluids of bovine female genitalia. Am. J. Vet.
     Res. 18:603-607.

144. Overstreet, J.W. and Cooper, G.W. 1975. Reduced sperm motility in the isthmus of the rabbit oviduct. Nature 258:718-719.

145. Overstreet, J.W. and Cooper, G.W. 1979. Effect of ovulation and sperm motility on the migration of rabbit spermatozoa to the site of fertilization. J. Reprod. Fertil. 55:53-59.

146. Overstreet, J.W., Katz, D.F. and Johnson, L.L. 1980. Motility of rabbit spermatozoa in the secretions of the oviduct. Biol. Reprod. 22:1083-1088.

147. Parisi, M. and Bourguet, J. 1984. Effects of cellular acidification on ADH-induced intramembrane particle aggregates. Am. J. Physiol. 246:C157-C159.

148. Parisi, M., Wietzerbin, J. and Bourguet, J. 1983. Intracellular pH, transepithelial pH gradients, and ADH-induced water channels. Am. J. Physiol. 244:F712-F718.

149. Payan, P., Girard J.P. and Ciapa B. 1983. Mechanisms regulating intracellular pH in sea urchin eggs. Dev. Biol. 100:29-38.

150. Perloff, W.H., Schultz, J., Farris, E.J. and Balin, H. 1955. Some aspects of the chemical nature of human ovarian follicular fluid. Fert. Steril. 6:11-17.

151. Phillips, D.W. 1972. Comparative aspects of mammalian sperm motility. J. Cell Biol. 53:561-573.

152. Piatigorsky, J. and Austin, C.R. 1962. Relationship of fertilizin to the acrosome reaction in Arbacia. Biol. Bull. 123:473.

153. Pouyssegur, J., Chambard, J.C., Franchi, A. Paris, S. and Obberghen-Schilling, E. 1982. Growth factor activation of an amiloride-sensitive Na^+/H^+ exchange system in quiescent fibroblasts: coupling to ribosomal protein S6 phosphorylation. Proc. Natl. Acad. Sci. USA 79:3935-3939.

154. Quayle, E.S., Pagel, J., Monti, J.A. and Christian, S.T. 1978. A serotonin sensitive guanylate cyclase associated with specific neurotransmitter binding sites on isolated synaptic membranes from mature rat brain. Life Sci. 23:159-166.

155. Radany, E.W., Gerzer, R. and Garbers, D.L. 1983. Purification and characterization of particulate guanylate cyclase from sea urchin spermatozoa. J. Biol. Chem. 258:8346-8351.

156. Ramarao, C.S. and Garbers, D.L. 1985. Receptor-mediated regulation of guanylate cyclase activity in spermatozoa. J. Biol. Chem. 260:8390-8396.

157. Rao, M.C., Guandalini, S., Smith, P.L. and Field, M. 1980. Mode of action of heat-stable Escherichia coli enterotoxin: tissue and subcellular specification and role of cyclic GMP. Biochim. Biophys. Acta 632:35-46.

158. Rao, M.C., Orellana, S.A., Field, M., Robertson, D.C. and Giannella, R.A. 1981. Comparison of the biological actions of three purified heat stable enterotoxins: effects on ion transport and guanylate cyclase activity in rabbit ileum in vitro. Infect. Immun. 33:165-170.

159. Repaske, D.R. and Garbers, D.L. 1983. A hydrogen ion flux medi-

ates stimulation of respiratory activity by speract in sea
urchin spermatozoa. J. Biol. Chem. 258:6025–6029.

160. Restall, B.J. and Wales, R.G. 1966. The fallopian tubes of the
sheep. IV. The metabolism of ram spermatozoa in the presence
of fluid from the fallopian tube. Aust. J. Biol. Sci.
19:883–893.

161. Robinson, N.L., Cox, P.M. and Greengard, P. 1982. Glutamate reg-
ulates adenylate cyclase and guanylate cyclase activities in
an isolated membrane preparation from insect muscle. Nature
296:354–356.

162. Rochette-Egly, C. and Tovey, M.G. 1982. Interferon enhances
guanylate cyclase activity in human lymphoma cells. Biochem.
Biophys. Res. Commun. 107:150–156.

163. Rothenberg, P., Reuss, L. and Glaser, L. 1982. Serum and epider-
mal growth factor transiently depolarize quiescent BSC-1
epithelial cells. Proc. Natl. Acad. Sci. USA 79:7783–7787.

164. Rudland, P.S., Gospodarowicz, D. and Sifert, W. 1974. Activation
of guanyl cyclase and intracellular cyclic GMP by fibroblast
growth factor. Nature 250:741–742.

165. Saling, P.M., Storey, B.T. and Wolf, D.P. 1978. Calcium-
dependent binding of mouse epididymal spermatozoa to the zona
pellucida. Dev. Biol. 65:515–525.

166. Schul, G.A. II, Foley, C.W., Heinze, C.D., Erb, R.E. and
Harrington, R.B. 1966. Some effects of the porcine female
reproductive tract on metabolism of boar spermatozoa. J.
Animal Sci. 25:406–409.

167. Schuldiner, S. and Rozengurt, E. 1982. Na^+/H^+ antiport in Swiss
3T3 cells: mitogenic stimulation leads to cytoplasmic
alkalinization. Proc. Natl. Acad. Sci. USA 79:7778–7782.

168. Segall, G. and Lennarz, W.J. 1978. The role of jelly coat in the
induction of the acrosomal reaction in sperm. J. Cell Biol.
79:170.

169. Segall, G.K. and Lennarz, W.J. 1979. Chemical characterization
of the component of the jelly coat from sea urchin eggs
responsible for induction of the acrosome reaction. Dev.
Biol. 71:33–48.

170. Segall, G.K. and Lennarz, W.J. 1981. Jelly coat and induction of
the acrosome reaction in echinoid sperm. Dev. Biol. 86:87–
93.

171. Shen, S.S. and Steinhardt, R.A. 1980. Intracellular pH controls
the development of new potassium conductance after fertilization of the sea urchin eggs. Exp. Cell Res. 125:55–61.

172. Shima, S. and Nakao, T. 1974. Studies on cyclic nucleotides in
the adrenal gland V: Adenyl and guanyl cyclase activities in
the cortex and medulla of the adrenal gland. Jap. J.
Pharmacol. 14:48.

173. Shivers, C.A. and James, J.M. 1970. Capacitation of frog sperm.
Nature 227:183–184.

174. Shoger, R.L. and Bishop, D.W. 1967. Sperm activation and fertilization in Limulus polyphemus. Biol. Bull. 133:485.

175. Smith, A.C. and Garbers, D.L. 1983. The binding of an ^{125}I-speract analogue to spermatozoa. In: Biochemistry of Metabolic Processes (D. L. F. Lennon, F. W. Stratman and R. N. Zahlten, eds.) Elsevier/North Holland Biomedical Press, pp. 15-28.

176. Smith, J.S. and Pace, C.S. 1983. Modification of glucose-induced insulin release by alteration of pH. Diabetes 32:61-66.

177. Stith, B.J. and Maller, J.L. 1984. The effect of insulin on intracellular pH and ribosomal protein S6 phosphorylation in oocytes of Xenopus laevis. Dev. Biol. 102:79-89.

178. Stone, R.T., Foley, C.W., Thorne, J.G. and Huber, T.L. 1973. Effect of oviductal fluids on oxidative phosphorylation in spermatozoa. Proc. Soc. Exptl. Biol. Med. 142:64-67.

179. Strange, P.G., Birdsall, N.J.M. and Burgen, A.S.V. 1977. Occupancy of a muscarinic acetylcholine receptor stimulates a guanylate cyclase in neuroblastoma cells. Biochem. Soc. Trans. 5:189-191.

180. Stuart, C.A., Vesely, D.L., Provow, S.A. and Furlanetto, R.W. 1982. Cyclic nucleotides and somatomedin action in cartilage. Endocrinology 111:553-558.

181. Suzuki, N. and Garbers, D.L. 1984. Stimulation of sperm respiration rates by speract and resact at alkaline extracellular pH. Biol. Reprod. 30:1167-1174.

182. Suzuki, N., Nomura, K., Ohtake, H. and Isaka, S. 1981. Purification and the primary structure of sperm-activating peptides from jelly coat of sea urchin eggs. Biochem. Biophys. Res. Commun. 99:1238-1244.

183. Suzuki, N., Shimomura, H., Radany, E.W., Ramarao, C.S., Ward, G.E., Bentley, J.K. and Garbers, D.L. 1984. A peptide associated with eggs causes a mobility shift in a major plasma membrane protein of spermatozoa. J. Biol. Chem. 259:14874-1487.

184. Suzuki, R. 1958. Sperm activation and aggregation during fertilization in some fishes. I. Behavior of spermatozoa around the micropyle. Embryologia 4:93-102.

185. Suzuki, R. 1959. Sperm activation and aggregation during fertilization in some fishes. III. Non species-specificity of stimulating factor. Annot. Zool. Japan 32:105-111.

186. Tai, Y.J., Decker, R.A., Marnane, W.G., Charney, A.N. and Donowitz, M. 1981. Effects of methylprednisolone on electrolyte transport by in vitro rat ileum. Am. J. Physiol. 240:G365-G370.

187. Tamagawa, T. and Henquin, J.C. 1983. Chloride modulation of insulin release, ^{86}Rb$^+$ efflux, and ^{45}Ca^{2+} fluxes in rat islets stimulated by various secretagogues. Diabetes 32:416-423.

188. Thomson, W.J., Johnson, D.G., Lavis, R.W. and Williams, R.H. 1974. Effects of secretin on guanyl cyclase of various tissue. Endocrinology 94:276-278.

189. Thomson, W.J., Williams, R.H. and Little, S.A. 1973. Studies on

the assay and activities of guanyl and adenyl cyclase of rat liver. Arch. Biochem. Biophys. 159:206-213.

190. Thomson, W.J., Williams, R.H. and Little, S.A. 1973. Activation of guanyl cyclase and adenyl cyclase by secretin. Biochim. Biophys. Acta 302:329-337.

191. Triana, L.R., Babcock, D.F., Lorton, S.P., First, N.L. and Lardy, H.A. 1980. Release of acrosomal hyaluronidase follows increased membrane permeability to calcium in the presumptive capacitation sequence of spermatozoa of the bovine and other mammalian species. Biol. Reprod. 23:47-59.

192. Trifunac, N.P. and Berstein, G.S. 1981. Effect of human endometrial secretions on the metabolism of human spermatozoa. Fert. Steril. 35:58-63.

193. Tubb, D.J., Kopf, G.S. and Garbers, D.L. 1979. Starfish and horseshoe crab egg factors cause elevations of cyclic nucleotide concentrations in spermatozoa from starfish and horseshoe crab. J. Reprod. Fert. 56:539-542.

194. Tyler, A. 1940. Sperm agglutination in the keyhole limpet, Megathura crenulata. Biol. Bull. 78:159-178.

195. Tyler, A. 1956. Physio-chemical properties of the fertilizins of the sea urchin Arbacia punctulata and the sand dollar Echinarachinius parma. Exp. Cell Res. 10:377-386.

196. Tyler, A. and Fox, S.W. 1939. Sperm agglutination in the keyhole limpet and the sea urchin. Science 90:517-518.

197. Uno, Y. and Hoshi, M. 1978. Separation of the sperm agglutinin and the acrosome reaction inducing substance in egg jelly of starfish. Science 20:58-59.

198. Vasseur, E. and Hagstrom, B. 1946. On the gamones of some sea urchins from the Swedish west coast. Ark. Zool. 37A:1-17.

199. Vesely, D.L. 1979. Testosterone and its precursors and metabolites enhance guanylate cyclase activity. Proc. Natl. Acad. Sci. USA 76:3491-3494.

200. Vesely, D.L. 1981. Angiotensin II stimulates guanylate cyclase activity in aorta, heart and kidney. Am. J. Physiol. 240:E391-E393.

201. Vesely, D.L. 1981. Human and rat growth hormones enhance guanylate cyclase activity. Am. J. Physiol. 240:E79-E82.

202. Vesely, D.L.1981. Bromocriptine enhances guanylate cyclase activity. Endocrinology, 109:1284-1286.

203. Vesely, D.L. and Cantell, K. 1980. Human interferon enhances guanylate cyclase activity. Biochem. Biophys. Res. Commun. 96:574-579.

204. Velely, D.L. and Hill, D.E. 1980. Estrogens and progesterone increase fetal and maternal guanylate cyclase activity. Endocrinology 107:2104-2109.

205. Vesely, D.L., Straub, K.D., Nolan, C.M., Rolfe, R.D., Finegold, S.M. and Monson, T.P. 1981. Purified Clostridium difficile cytotoxin stimulates guanylate cyclase activity and inhibits adenylate cyclase activity. Infect. Immun. 33:285-291.

206. Wales, R.G. and Restall, B.J. 1966. The metabolism of ram spermatozoa in the presence of genital fluids of the ewe. Aust. J. Biol. Sci. 19:199-209.

207. Walling, M.W., Marvaso, V. and Bernard, G.W. 1978. Stimulation of guanylate cyclase activity in cultured osteogenic murine calvarial mesenchymal cells by PTH, calcitronin and insulin. Biochem. Biophys. Res. Commun. 83:521-527 (1978).

208. Ward, G.E., Garbers, D.L. and Vacquier, V.D. 1985. Effects of extra cellular egg factors on sperm guanylate cyclase. Science 227:7680770.

209. Ward, G.E. and Vacquier, V.D. 1983. Dephosphorylation of a major sperm membrane protein is induced by egg jelly during sea urchin fertilization. Proc. Natl. Acad. Sci. USA 80:5578-5582.

210. Ward, S., Hogan, E. and Nelson, G.A. 1983. The initiation of spermiogenesis in the nematode Caenorhabditis elegans. Dev. Biol. 98:70-79.

211. Wasserman, W.J. and Houle, J.G. 1984. The regulation of ribosomal protein S6 phosphorylation in Xenopus oocytes: a potential role for intracellular pH. Dev. Biol. 101:436-445.

212. Watkins, H.D., Kopf, G.S. and Garbers, D.L. 1978. Activation of sperm adenylate cyclase by factors associated with eggs. Biol. Reprod. 19:890-894.

213. Wolf, D.P. and Hedrick, J.L. 1971. A molecular approach to fertilization. III. Development of a bioassay for sperm capacitation. Dev. Biol. 25:360-376.

214. Woodward, A.E. 1918. Studies on the physiological significance of certain precipitates from the egg secretions of Arbacia and Asterias. J. Exp. Zool. 26:459-500.

215. Yanagimachi, R. 1944. In vitro acrosome reaction and capacitation of golden hamster spermatozoa by bovine follicular fluid and its fractions. J. Exp. Zool. 170:269-280.

216. Yanagimachi, R. 1957. Studies of fertilization of Clupea pallasi. VI. Fertilization of the egg deprived of the membrane. Jap. J. Ichthy 6:41-47.

217. Yanagimachi, R. 1966. Time and process of sperm penetration into hamster ova in vivo and in vitro. J. Reprod. Fert. 11:359-370.

218. Yanagimachi, R. 1969. In vitro capacitation of hamster spermatozoa by follicular fluid. J. Reprod. Fert. 18:275-286.

219. Yanagimachi, R. 1970. The movement of golden hamster spermatozoa before and after capacitation. J. Reprod. Fert. 23:193-196.

220. Yanagimachi, R. 1972. Fertilization of guinea pig eggs in vitro. Anat. Rec. 174:9-20.

221. Yanagimachi, R. and Mahi, C.A. The sperm acrosome reaction and fertilization in the guinea pig: A study in vivo. J. Reprod. Fert. 46:49-54.

222. Yanagimachi, R. and Usui, N. 1974. Calcium dependence of the acrosome reaction and activation of guinea pig spermatozoa. Exp. Cell Res. 89:161-174.

DEPHOSPHORYLATION OF SEA URCHIN SPERM GUANYLATE CYCLASE

DURING FERTILIZATION

Gary E. Ward, Gary W. Moy, and Victor D. Vacquier

Marine Biology Research Division A-002
Scripps Institution of Oceanography
University of California at San Diego
La Jolla, California 92093

SUMMARY

Exposure of Arbacia punctulata spermatozoa to solubilized egg jelly results in the immediate dephosphorylation (within 3 sec) of an abundant 160,000 dalton (160 kDa) sperm membrane protein, and a simultaneous increase in its electrophoretic mobility to 150 kDa. The sperm phosphoprotein has been identified as guanylate cyclase. Correlated with the mobility shift of the cyclase is a decrease in its enzymatic activity. In this paper we will briefly review the work on the sperm guanylate cyclase, present new data on the role of ion fluxes in the control of its dephosphorylation, and discuss what role the dephosphorylation might play in successful sperm-egg interaction.

INTRODUCTION

The sea urchin egg is surrounded by an extracellular coat known as the jelly layer. As the spermatozoon swims through the jelly layer on its way to the plasma membrane of the egg, it undergoes a number of morphological, biochemical, and behavioral changes. These activations include the acrosome reaction, increased respiration, and increased motility (4,8). Two components of egg jelly which play a role in sperm activation have been described: a large fucose-sulfate-rich complex, which induces the acrosome reaction through a mechanism involving altered ionic fluxes, (9,23,24,26) and a sperm activating peptide, (28) or "speract" (12), which has been implicated in the activation of motility and respiration through a sodium-dependent increase in intracellular pH (12,21).

We have previously reported that egg jelly induces the rapid dephosphorylation of sperm guanylate cyclase (34,36). In this paper we review the work on the sperm guanylate cyclase, present new data on the control of its dephosphorylation, and discuss how these findings might help elucidate the sequence of events underlying sperm activation by egg jelly.

MATERIALS AND METHODS

All materials and methods not specifically described below can be found in references 34 and 36. A. punctulata were spawned by intracoelomic injection of 0.5 M KCl. Gametes were collected from the gonopores with a pipette; spermatozoa were stored "dry" on ice and egg jelly was prepared as described (34). Seawater formulations were as follows: MFSW: Millipore-filtered natural seawater, pH 7.9, supplemented with 100 mg l^{-1} each of penicillin-G and streptomycin sulfate. ASW (artificial seawater): 425 mM NaCl, 27 mM $MgCl_2$, 26 mM $MgSO_4$, 10 mM $CaCl_2$, 5 mM KCl, 5 mM MES [2-(N-morpholino) ethanesulfonic acid] , 5 mM HEPES [N-2-hydroxyethylpiperazine-N'-2-ethanesulfonic acid]; the pH was adjusted with 1 M KOH. NaFSW (sodium-free seawater): identical to ASW, except choline chloride was substituted for NaCl. CaFSW (calcium-free seawater): 508 mM NaCl, 27 mM $MgCl_2$, 26 mM $MgSO_4$, 10 mM KCl, 5 mM $NaHCO_3$; the pH was adjusted with Na_2CO_3 to 7.9. For all experiments using CaFSW, both CaFSW and CaFSW + 2 mM EGTA (ethylene glycol-bis(β-aminoethyl ether)N,N,N',N'-tetraacetic acid] were used, with identical results. Preparation of sodium-free jelly was as described for normal jelly (34), with three modifications: prior to spawning the female the test was rinsed in NaFSW; the eggs were pipetted from the gonopores into NaFSW instead of MFSW; and following the removal of the jelly coats at pH 5, the pH was readjusted to 7.9 with 0.1 M KOH instead of 0.1 M NaOH. Calcium-free jelly was prepared exactly as described for sodium-free jelly, except that CaFSW was substituted for NaFSW. Sperm acrosome reactions were scored as described (33).

Monensin (CalBiochem-Behring) was added from a 5 mM stock in DMSO/ethanol (1:1). Nigericin (Sigma) was added from a 2 mM stock in ethanol. NH_4Cl was added from a 5 M stock, pH 7.9, in either ASW or NaFSW. Pronase (Sigma) was preincubated (0.11 mg ml^{-1} in ASW pH 7.9) at 23°C for 60 min prior to use.

Sodium dodecyl sulfate polyacrylamide gel electrophoresis (SDS-PAGE) was as described (34). SDS from Sigma (70% lauryl sulfate: L-5750) was used in all gel solutions; more highly purified sources (Sigma L-4509, BIO-RAD 161-0302) were found to resolve less effectively both the tubulins and the 160/150 kDa proteins. Similar effects of SDS purity on electrophoretic resolution have been seen by others (1,17). Densitometric scans of silver-stained gels were done using the gel scanner accessory on a Varian Techtron 635 spectrophotometer (555 nm).

Intracellular pH was measured at 21°C, using 2.5 μM 9-amino-acridine (Sigma) (2,14); 0.04% Triton X-100 was used to correct for binding to intracellular components (2). For analysis of amino acid composition, the protein was electroeluted from a Coomassie blue-stained gel slice and analyzed as described (19) using a Beckman 121-M amino acid analyzer. Guanylate cyclase was assayed essentially as described (7), with the modifications noted here and in ref. 36. Cyclic GMP levels were measured in 0.5 N perchloric acid extracts of sperm by radioimmunoassay (13).

RESULTS AND DICUSSION

The jelly-induced 160 to 150 kDa electrophoretic mobility shift

When A. punctulata spermatozoa are exposed to homologous egg jelly, a 160 kDa band on SDS-polyacrylamide gels disappears (within 3 sec), and a new band appears at 150 kDa (Fig. 1A). The 160 and 150 kDa bands are the same protein, as shown by a comparison of their cyanogen bromide cleavage patterns (34) and their amino acid composi-tions (Table 1). The 160/150 kDa protein has been shown to be heav-ily enriched in the plasma membrane of the flagellum (34). Recent studies have shown that the protein is iodinated under vectorial ^{125}I-labelling conditions (unpublished) and binds concanavalin A (36), indicating that at least a portion of the protein is externally disposed.

When A. punctulata spermatozoa are incubated in seawater con-taining $^{32}P_i$, they take up the label and incorporate it into the 160 kDa protein (Fig. 1B), in the form of ^{32}P-phosphoserine (ref. 34). When labelled spermatozoa are treated with egg jelly, the label is removed from the 160 kDa protein (Fig. 1B). This observation sug-gests that enzymatic dephosphorylation, proteolysis (perhaps of a portion of the molecule which contains the ^{32}P-phosphoserine), or some combination of the two might be responsible for the 160 to 150 kDa mobility shift.

Two lines of evidence support the idea that the mobility shift occurs as a result of dephosphorylation. First, cyanogen bromide cleavage of the ^{32}P-labelled 160 kDa protein shows that the labelled phosphoserines are not clustered within the amino acid sequence of the protein, but are distributed over a number of distinct 7 to 15 kDa cleavage fragments (Fig. 2). If the 160 kDa protein has been digested to completion under these conditions, these results demons-trate that the 10 kDa mobility shift and loss of label which occur in response to jelly cannot be the result of proteolysis alone. Sec-ondly, we have shown that calf intestinal alkaline phosphatase is able to induce both the near complete mobility shift and loss of label from the ^{32}P-labelled 160 kDa protein in vitro (Fig. 3, lanes c and d). When the phosphatase activity is partially inhibited (by

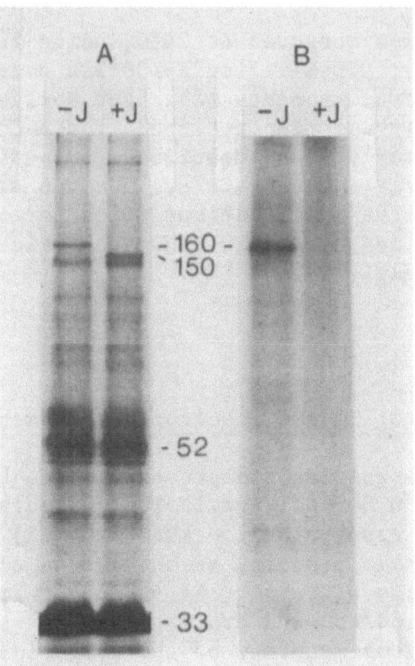

Figure 1. Spermatozoa labelled in vivo with $^{32}P_i^-$ were incubated for
4 min in MFSW (-J) or MFSW containing egg jelly (+J),
extracted with trichloracetic acid, and analyzed by SDS-
PAGE. (A) Coomassie blue staining pattern. (B) Corresponding autoradiogram. The positions of the 160 kDa protein,
the 150 kDa protein, β-tubulin (52 kDa) and histone H1 (33
kDa) are shown. Bands less heavily radiolabelled than the
160 kDa protein are often seen at ∼51 kDa and ∼37 kDa; when
present, the amount of label associated with these bands
does not change in response to egg jelly. (Reprinted from
ref. 34).

pretreatment with EDTA), both loss of label and the mobility shift
are reduced (Fig. 3, lanes e and f). When the phosphatase activity
is completely inhibited (by 20 mM inorganic phosphate), no loss of
label or change in mobility is seen (Fig. 3, lanes g and h).

These results suggest that dephosphorylation alone may be sufficient to cause the 160 to 150 kDa mobility shift. Such a large
effect of phosphorylation state on electrophoretic mobility is not
without precedent (10,37) and is presumably due to an effect of
charged phosphate groups on SDS binding (3,31). However, until it is
demonstrated that the 150 kDa protein can be rephosphorylated and
converted back into its 160 kDa form, a role for proteolysis in the
mobility shift in vivo cannot be conclusively ruled out. The extent

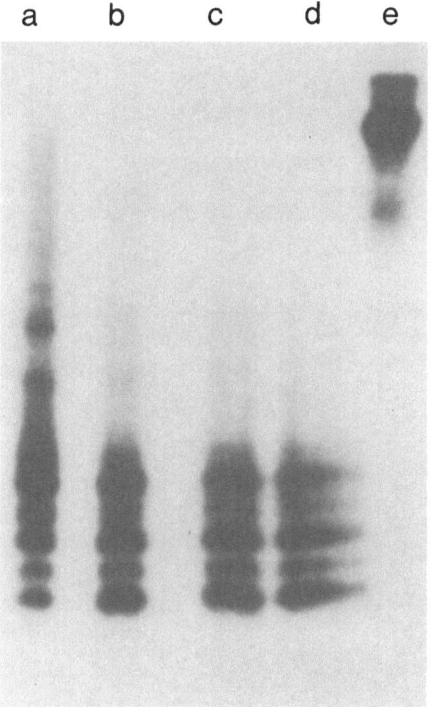

Figure 2. Cyanogen bromide cleavage of the ^{32}P-labelled 160 kDa protein. Sperm were labelled in vivo and subjected to SDS-
PAGE as shown in Fig. 1. The 160 kDa protein was sliced
out of the gel and digested (21°C) for (a) 15 min, (b) 1 h,
(c) 5 h, and (d) 9 h in the presence of approx. 2.5 x 10^8
mol cyanogen bromide per mol protein (34). Very little
protein breakdown is observed when cyanogen bromide is
omitted from the 9 h digestion step (e). The labelled
fragments (b-d) range in size from 7-15 kDa.

to which the 160 kDa protein is dephosphorylated in vivo in response
to jelly is not known; label is apparently removed from the ^{32}P-phos-
phoserine residues, but the total amount of covalently bound phosphate before and after jelly treatment has not yet been determined.

The role of jelly-induced ion fluxes in the 160- to 150-kDa mobility
shift

Changes in ion flux across the sperm plasma membrane are known
to occur in response to egg jelly. These changes include increased
Ca^{2+} influx (23,24), H$^+$ efflux (possibly coupled to Na$^+$ influx; refs.
2,13,14,24,25) and a decrease in the K$^+$ membrane potential (25). The

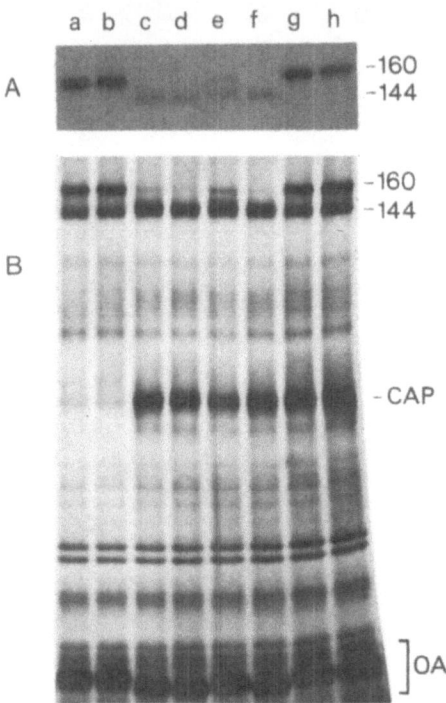

Figure 3. In vitro dephosphorylaton of the 160 kDa protein. An
          extract of Triton X-100 soluble protein from ^{32}P-labelled
          spermatozoa was prepared and incubated for 10 min (a) and
          40 min (b) without calf intestinal alkaline phosphatase,
          10 min (c) and 40 min (d) with phosphatase (0.1 mg ml^{-1}),
          10 min (d) and 40 min (f) with phosphatase (0.1 mg ml^{-1})
          pretreated with 20 mM EDTA, and 10 min (g) and 40 min (h)
          with phosphatase (0.1 mg ml^{-1}) plus 10 mM sodium phosphate.
          (B) Silver-stained gel. (A) Autoradiogram of the 160 to
          144 kDa region of B. Exogenous ovalbumin (OA) was added to
          all extracts as an internal phosphoprotein control; note
          the correlation between altered OA mobility and altered
          mobility of the 160 kDa protein. Phosphatase activity
          under the three incubation conditions was determined to be
          1 unit ml^{-1} (c,d), 0.15 units ml^{-1} (e,f), and 0.04 units
          ml^{-1} (g,h). The positions of phosphatase (CAP) and the
          160 and 144 kDa proteins are shown. (Reprinted from ref.
          34).

increase in Ca^{2+} influx does not appear to be involved in the dephosphorylation: the mobility shift is induced by jelly in calcium-free
conditions, and it is not induced by the ionophore A23187 in seawater
containing 10 mM calcium (5 μM A23187, 5 min; ref. 35).

Table 1. Amino Acid Compositions of the 160 kDa and 150 kDa A. punctulata Proteins[*]

	160 kDal (mol %)	150 kDal (mol %)
Lys	4.7	4.4
Arg	4.0	3.6
His	2.2	2.4
Asx	10.5	10.7
Glx	9.4	8.9
Ala	7.1	7.5
Leu	8.7	8.3
Ile	5.1	5.6
Val	5.8	5.4
Pro	4.4	5.1
Phe	3.2	2.5
Met	1.3	1.8
Gly[†]	19.2	19.6
Ser	6.9	6.3
Thr	4.8	5.0
Tyr	2.7	2.8

[*] Trp and Cys were not determined.

[†] Gly values may be overestimates, due to carryover from the Tris/glycine buffer (34) used in electrophoretic preparation of the protein.

We have previously suggested (35) that jelly-induced changes in H^+ and Na^+ flux might play a role in the mobility shift. The evidence on which this suggestion is based includes: a) At a given jelly concentration, the extent to which the mobility shift occurs shows a strong dependence on the pH of the surrounding seawater (35) (Fig. 4), decreasing the external pH is known to depress intracellular pH (pH_i) in these cells (refs. 2,21 and unpublished results); b) The jelly-induced mobility shift shows an absolute requirement for external sodium (35) (Fig. 5); c) The carboxylic ionophore monensin, which in normal seawater will exchange extracellular Na^+ for intracellular H^+ (ref. 13), induces the mobility shift in the absence of egg jelly (Fig. 6); d) The ionophore nigericin, which in NaFSW supplemented with 100 mM K^+ will exchange extracellular K^+ for intracellular H^+ (ref. 14), also induces the mobility shift (not shown). These results implicate pH_i, and possibly jelly-induced increases in pH_i (refs. 2,14,25), in the control of the mobility shift.

Recent experiments have shown that while the jelly-induced increase in pH_i may play some indirect role, increased pH_i alone is not sufficient to induce the mobility shift. These experiments have made use of the fluorescent dye 9-aminoacridine to directly monitor pH_i (2,14).

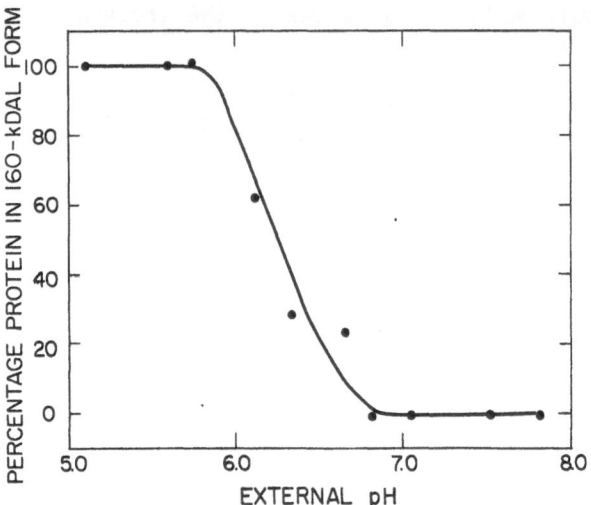

Figure 4. The effect of external pH on the jelly-induced mobility
shift. Aliquots of dry sperm suspension were suspended to 9
x 10^7 cells ml^{-1} in a series of ASW solutions (pH 5.0-7.8).
Egg jelly was added (to 500 ng ml^{-1}), and after 4 min at
21°C the samples were extracted and processed for SDS-PAGE
(34). The silver-stained gel was scanned with a densitometer; percent protein in 160 kDa form was calculated as
the area under the 160 kDa peak divided by the sum of the
areas under the 160 and 150 kDa peaks, x 100.

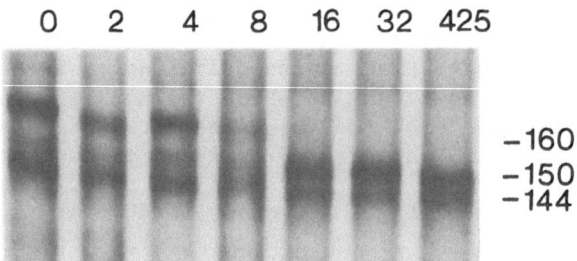

Figure 5. The jelly-induced mobility shift shows an absolute requirement for external Na⁺. Dry spermatozoa were washed once in
200 vol NaFSW (12 min, 3000 x g), and resuspended (2.5 x
10^7 cells ml^{-1}) in ASW/NaFSW mixed in proportions which
yielded the final sodium concentrations indicated (0-425
mM). Sodium-free jelly was added to a final concentration
of 500 ng fucose ml^{-1} (100-fold dilution of 50 µg fucose
ml^{-1} jelly stock). After 4 min at 21°C, the sperm were extracted and subjected to SDS-PAGE/silver stain (34). Positions of the 160, 150, and 144 kDa proteins are shown.

a   b   c   d

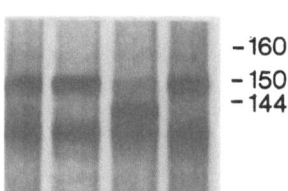

-160
-150
-144

Figure 6. Monensin induces the mobility shift in the absence of
        jelly. Dry spermatozoa resuspended to a final concentration
        of 5×10^7 cells ml^{-1} in (a) ASW pH 6.6 + 20 µM monensin,
        (b) NaFSW pH 8.0 + 20 M monensin , (c) ASW pH 8.0 + 20 M
        monensin, and (d) ASW pH 8.0 + 1% DMSO-ethanol (1:1)
        (equivalent to the amount of DMSO/ethanol added in a-c).
        After 10 min at 21°C, the spermatozoa were extracted and
        subjected to SDS-PAGE/silver stain (34). Positions of the
        160, 150, and 144 kDa proteins are shown.

    A. punctulata spermatozoa diluted in pH 7.9 ASW have a pH_i of
7.23 ± 0.04 (Fig. 7A). Within seconds after jelly treatment, pH_i
increases to 7.36 ± 0.04, a change of 0.13 pH unit (similar to what
has been observed in Strongylocentrotus purpuratus spermatozoa
(2,14). The mobility shift occurs in less than 5 sec under these
conditions. Exposing the spermatozoa to 10 µM monensin results in an
increase in pH_i of 0.25 pH units within 1 min (not shown), yet under
these conditions it takes 8-10 min for the mobility shift to first
become detectable, and 14-16 min for its completion. The weak base
NH_4Cl can also be used to induce increases in pH_i of similar
magnitude and similar kinetics to those induced by jelly, but under
these conditions the mobility shift is not detectable until 30 min
after NH_4Cl addition (Fig. 7B and unpublished results).

    Spermatozoa diluted into NaFSW have a lower pH_i (6.69 ± 0.04)
than spermatozoa diluted into ASW (7.23 ± 0.04). Under sodium-free
conditions, neither 25 mM NH_4Cl, which titrates the depressed pH_i
back to 7.4, nor sodium-free jelly will induce the mobility shift
(Fig. 8A, B). While neither treatment alone has any effect (as
expected from Figs. 5 and 7B), the sodium-free jelly will induce an
immediate mobility shift when added to the spermatozoa in the
presence of 25 mM NH_4Cl (Fig. 8C). This suggests that the inhibition
of the mobility shift by low external sodium (Figs. 5, 8B) is
probably an indirect result of low external sodium depressing the pH_i
below some critical "permissive" value. The same interpretation
would apply to the effects of reduced external pH (Fig. 4).

    Experiments are in progress to determine directly the permissive
pH_i. If it lies between 7.23 and 7.36, then the jelly-induced

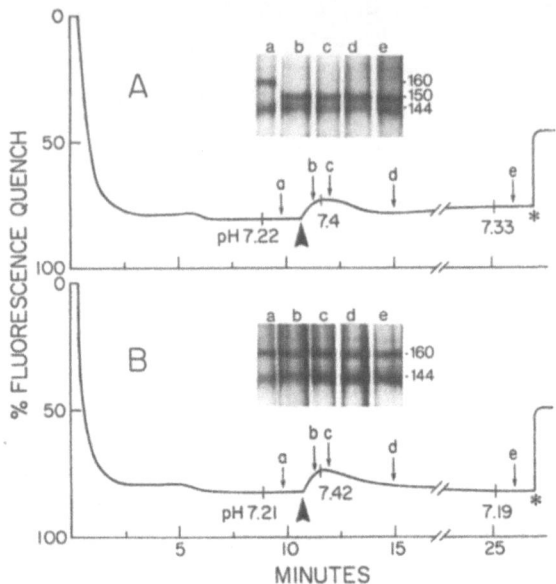

Figure 7. A comparison of the effects of jelly vs. NH_4Cl on pH_i and
the 160 to 150 kDa mobility shift. Dry spermatozoa were
diluted 100-fold into ASW containing 2.5 µM 9-amino-
acridine. Following dye equilibration (monitored spectro-
fluorimetrically; λ_{ex} = 382 nm, λ_{em} = 454 nm), jelly
(A; 200 ng fucose ml^{-1} final) or NH_4Cl (B; 1.25 mM final)
was added (arrowheads) directly to the cuvette. Fluorescence was continuously monitored and samples removed for
SDS-PAGE at the times indicated (a-e). Triton X-100 was
added (asterisk) to a final concentration of 0.04%
(vol/vol), and nominal pH_i values at the indicated times
were calculated as described (2).

increase in pH_i (from 7.23 to 7.36) may play some role in inducing
the mobility shift, although it is not in itself sufficient (Fig. 7).
If the permissive pH_i is below 7.23, then jelly-induced increases in
pH_i are not likely to be in any way involved. It is not clear why
prolonged elevations of pH_i as induced by monensin (Fig. 6) or NH_4Cl
(not shown), eventually cause the mobility shift (10-30 min), but a
mechanism different from that by which jelly normally induces the
change may be responsible. Similar results have been reported by
Garbers et al. (this volume).

A peptide which stimulates A. punctulata sperm respiration has
recently been isolated from A. punctulata egg jelly (29,30). The
effects of the peptide on the respiration and cyclic nucleotide

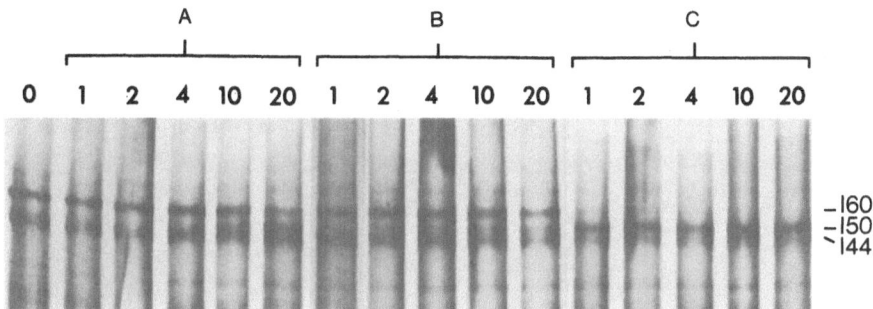

Figure 8. Jelly is able to induce the 160 to 150 kDa mobility shift
         in the absence of external sodium when the pH_i has been
         titrated back to 7.4 with NH_4Cl. Dry spermatozoa were
         diluted 100-fold into NaFSW containing 2.5 µM 9-
         aminoacridine. Following dye equilibration, one aliquot
         (0) was removed for SDS-PAGE (34), and NH_4Cl (A,C; 25 mM
         final) or sodium-free jelly (B; 200 ng fucose ml^{-1} final)
         was then added directly to the cuvette. Sodium-free jelly
         (200 ng fucose ml^{-1} final) was also added to C, 2 min after
         NH_4Cl addition. Fluorescence was continuously monitored,
         and samples removed for SDS-PAGE 1, 2, 4, 10, and 20 min
         after addition of NH_4Cl (A) or jelly (B,C). Intracellular
         pH was calculated as described (2). Spermatozoa diluted in
         NaFSW (0) had a stable pH_i of 6.69 ± 0.04. The pH_i of
         spermatozoa exposed to NH_4Cl (A) increased to 7.4 one min
         after NH_4Cl addition, and remained stable at approximately
         7.4 for 20 min. The addition of jelly had no effect on pH_i
         under these sodium-free conditions; the pH_i of spermatozoa
         exposed to jelly (B) remained at approximately 6.7 for 20
         min, and the pH_i response of spermatozoa exposed to both
         NH_4Cl and jelly (C) was indistinguishable from that of
         spermatozoa exposed only to NH_4Cl (A).

metabolism of A. punctulata spermatozoa are similar to the effects of
speract on S. purpuratus spermatozoa (30). However, the peptide is
different from speract in terms of both its species specificity and
its primary structure, and it has been given the name "resact" (29).
Resact has been identified as the component of egg jelly which
induces the mobility shift (30). While changes in Na^+ and H^+ flux
may occur in A. punctulata spermatozoa in response to resact (as
appears to be the case with speract and S. purpuratus spermatozoa)
(13,21), the above results suggest that some cellular response to
resact other than increased pH_i is directly involved in inducing the
mobility shift.

When intact spermatozoa labelled with $^{32}P_i$ are incubated in
seawater containing 0.1 mg pronase ml^{-1}, the 160 kDa band disappears

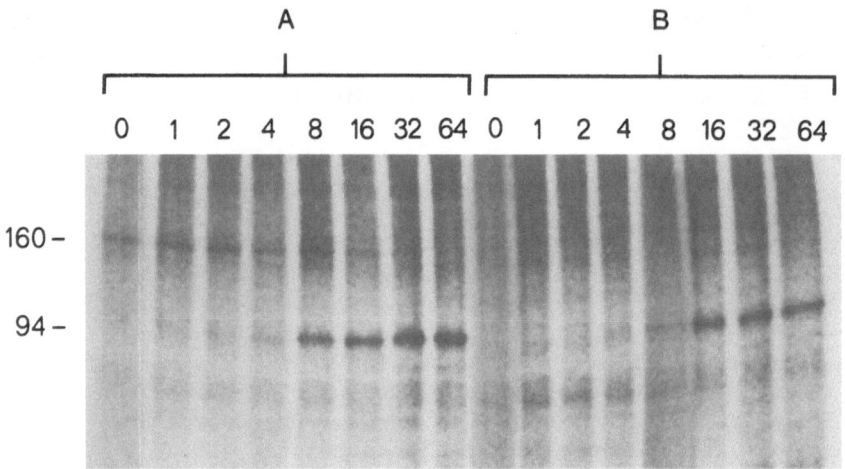

Figure 9. Pronase digestion of the 160 kDa protein in intact spermatozoa. Spermatozoa were labelled with $^{32}P_i$ as described
(34), pelleted (12 min, 3000 xg), diluted into 10 vol 0.11
mg ml^{-1} pronase in ASW, and incubated at 23°C for 0-64 min
as indicated. At each time point, an aliquot was removed
and divided into 2 lots; jelly (500 ng fucose ml^{-1} final)
added to one lot (B) and an equivalent amount of ASW pH 7.9
to the other lot (A). Following a 2 min incubation (23°C),
the spermatozoa were extracted and processed for SDS-PAGE
and autoradiography (34); a portion of the autoradiogram is
shown here. The positions of the 160 kDa protein and its
94 kDa cleavage fragment are shown.

over the course of 10-15 min. As the 160 kDa band disappears, a new
^{32}P-labelled band appears at 94 kDa (Fig. 9A). The ^{32}P-labelled 94
kDa band is a digestion product of the ^{32}P-labelled 160 kDa protein
as shown by a comparison of their cyanogen bromide cleavage fragments
(not shown). Interestingly, label is not removed from the 94 kDa
fragment when the proteolyzed spermatozoa are exposed to jelly (Fig.
9B). This result suggests that like speract (27), resact may interact with the spermatozoon (including induction of the dephosphorylation) through a surface receptor-mediated mechanism.

The 160/150 kDa protein is guanylate cyclase

A 135 kDa particulate guanylate cyclase has been purified to
homogeneity from spermatozoa of S. purpuratus (20). When a monospecific polyclonal antiserum against the S. purpuratus guanylate
cyclase (20) is used in immunoblots of A. punctulata spermatozoa
membranes, a single cross-reacting band is observed, at

160 kDa (Fig. 10a). The cross-reacting band undergoes a mobility
shift to 150 kDa in jelly-treated spermatozoa (Fig. 10c). This
antiserum also inhibits more than 97% of the guanylate cyclase
activity in homogenates of A. punctulata spermatozoa (Fig. 11).

A procedure for isolating the 150 kDa form of the protein has
been developed, based on its ability to bind to concanavalin A (it
has not yet been possible to maintain the protein in its 160 kDa form
during the isolation) (36). The 150 kDa protein and guanylate
cyclase activity copurify throughout the isolation; in the final

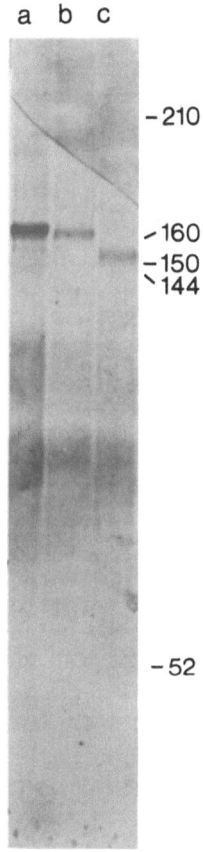

Figure 10. Immunoblots showing specific cross-reaction between polyclonal antiserum to S. purpuratus sperm guanylate cyclase
(20) and A. punctulata (a) sperm membranes (10 µg), (b)
whole spermatozoa (50 µg), (c) whole jelly-treated spermatozoa (50 µg). The numbers on the right denote relative
molecular mass in kDa. (Reprinted from ref. 36; copyright
1985 by the American Association for the Advancement of
Science).

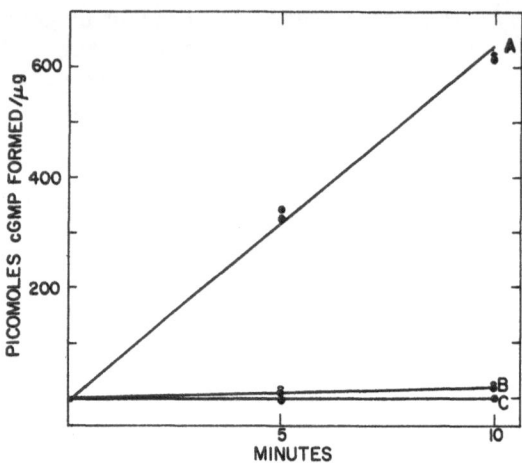

Figure 11. The antiserum which cross-reacts specifically with the
160/150 kDa protein (Fig. 10) inhibits 97% of the
guanylate cyclase activity in homogenates of A. punctulata
spermatozoa. Dry spermatozoa were diluted into 10 ml
0.25% (wt/vol) Lubrol WX, 20 mM Tris pH 6.5, 10 mM NaF,
100 μM Na vanadate to a final concentration of ∿7.5 x 10⁶
cells ml⁻¹ (2°C). The sperm suspension was subjected to
nitrogen cavitation at 300 psi for 10 min (2°C). An
aliquot of the cavitate was removed for determination of
protein (16), and the remainder divided into 2 lots. One
lot (B) was incubated with antiserum and the other (A)
with non-immune serum (5 h at 19°C, both sera diluted
50-fold). Each was then assayed for guanylate cyclase
activity. The non-immune serum had no detectable
endogenous guanylate cyclase activity (C). Final assay
conditions (21°C) were as follows: 0.06% Lubrol WX, 25 mM
Tris pH 6.5, 1 mM IBMX (3-isobutyl-1-methylxanthine), 3 mM
MnCl₂, 8 mM Na₃N, 20 mM NaF, 200 μM Na vanadate, 4 mM
dithiothreitol, 1 mM cGMP, 0.5 mM GTP (including 10⁶ cpm
³H-GTP per 200 μl assay tube). Samples were diluted such
that product formation was linear with both protein
concentration and time.

150 kDa fraction, guanylate cyclase specific activity is enriched
1760-fold from the starting material (Fig. 12 and Table 2),
confirming that the 150 kDa protein is guanylate cyclase.

The specific activity of the purified 150 kDa guanylate cyclase
from A. punctulata spermatozoa (1.2 μmol cGMP formed min⁻¹ mg⁻¹) is
comparable to that reported for the 135 kDa guanylate cyclase from S.
purpuratus spermatozoa (15 μmol min⁻¹ mg⁻¹) when the different assay
conditions are taken into account (20,36). The enzymes from these
two species are comparable in many other respects.

Table 2. Copurification of the 150 kDa Protein with Guanylate Cyclase
         Activity

Purification step[*]	Guanylate Cyclase Specific Activity[†]	Fold Enrichment of Specific Activity
Whole spermatozoa (b)	0.7	(1.0)
Membranes (c)	2.8	4.0
Solubilized membranes (d)	4.4	6.3
Concanavalin A eluate (e)	1231	1760

[*] Letters in parentheses correspond to gel lanes in Fig. 12 in which
    assayed fractions are displayed.
[†] Expressed as nmol cGMP formed min^{-1} mg^{-1} (reprinted from ref. 36).

Their amino acid compositions are very similar (Table 1 and ref. 20),
they share antigenic determinants (Fig. 10), and both are externally
disposed membrane proteins which bind concanavalin A (20,36). Since
S. purpuratus spermatozoa do not take up exogenous $^{32}P_i$ (ref. 32), it
is not clear at this point whether the S. purpuratus guanylate
cyclase is also dephosphorylated in response to egg jelly.

What role does the dephosphorylation of guanylate cyclase play in
sperm activation?

    When cGMP levels are measured in intact A. punctulata sperma-
tozoa, the response to jelly is similar to that previously described
(13) for S. purpuratus spermatozoa (Fig. 13). An initial rapid
increase in cGMP (to more than 6 times basal levels) is followed by a
more gradual return to basal and sub-basal levels (Fig. 13). The
dephosphorylation of guanylate cyclase could play a role in these
changes in cGMP metabolism.

    Homogenates containing the guanylate cyclase in various propor-
tions of 160:150 kDa can be prepared as previously described (34).
When these homogenates are assayed for guanylate cyclase activity, a
striking correlation is observed: homogenates containing the enzyme
in its 160 kDa form have the highest guanylate cyclase specific
activity, homogenates containing the 150 kDa form have the lowest
specific activity, and homogenates containing both forms have intermediate activities (Fig. 14). The decrease in specific activity of
homogenates containing only the 160 kDa form to those containing
only the 150 kDa form is 38-fold (average of five separate experiments) (36). The decrease in activity does not result from inhibition by residual egg jelly, since adding jelly back into the assay
tubes has no effect (Fig. 14F).

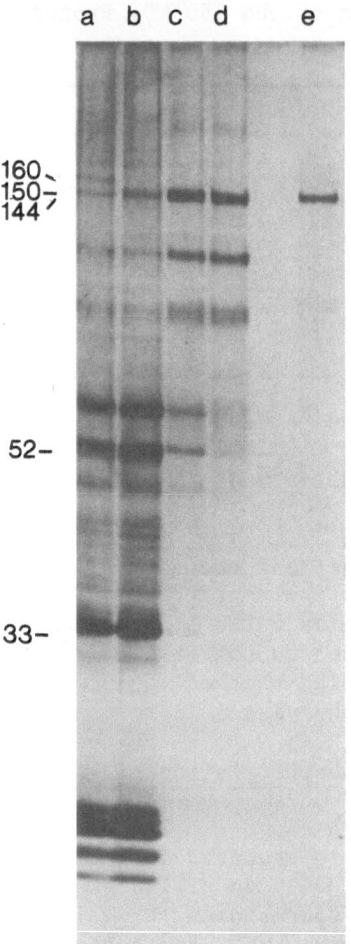

Figure 12. Purification of the 150 kDa protein: (a) untreated con-
          trol spermatozoa, (b) jelly-treated spermatozoa, (c)
          membranes, (d) solubilized membranes, (e) eluate from
          concanavalin A affinity column. A silver-stained gel is
          shown; numbers on left denote apparent molecular mass in
          kDa. Guanylate cyclase specific activity at each step of
          the purification is shown in Table 2. (Reprinted from
          ref. 36; copyright 1985 by the AAAS).

     Monensin, which is able to induce the dephosphorylation of the
enzyme in the absence of egg jelly (Fig. 6), has a similar effect on
guanylate cyclase activity; homogenates prepared from monensin-
treated spermatozoa show large decreases in guanylate cyclase activity when compared to control spermatozoa (10-fold in Fig. 15).
Adding monensin back into the assay tubes has no effect on activity.

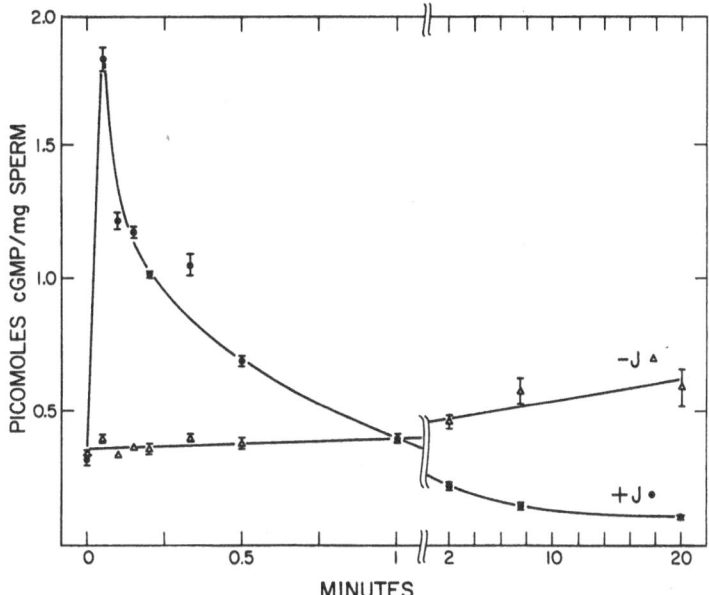

Figure 13. Egg jelly causes a rapid increase in cGMP levels in A̲. punctulata spermatozoa. Dry spermatozoa were diluted to 10^8 cells ml^{-1} in MFSW and divided into two lots. After 5 min preincubation at 21°C, jelly (+J) was added to one lot (to 900 ng fucose ml^{-1}) and an equivalent volume of MFSW (-J) to the other lot (t=0). At each subsequent time point, three aliquots were quenched in 0.5 N perchloric acid for cGMP determination (13), and one aliquot was removed for analysis by SDS-PAGE (32). Gel analysis showed that the 160 to 150 kDa mobility shift was complete by the first time point (3 sec) in the +J samples.

Similar effects on guanylate cyclase activity have recently been observed in response to purified resact (Garbers et al., this volume).

These results show that there is a strong correlation between the specific activity of the enzyme and whether it is in its 160 kDa or 150 kDa form. This observation introduces a paradox; if dephosphorylation of guanylate cyclase in response to egg jelly decreases its enzymatic activity (Fig. 14), then why do cGMP concentrations increase in response to jelly (Fig. 13)? It is possible that the guanylate cyclase activity measurements made in vitro on whole cell homogenates do not accurately reflect what is happening in the intact cell. What is seen in vitro as a decrease in activity may be a manifestation of some quite different change which the enzyme undergoes upon dephosphorylation in vivo. On the other hand, a decrease in guanylate cyclase activity could be easily reconciled with the transient increase in cGMP levels if cGMP phosphodiesterase activity were

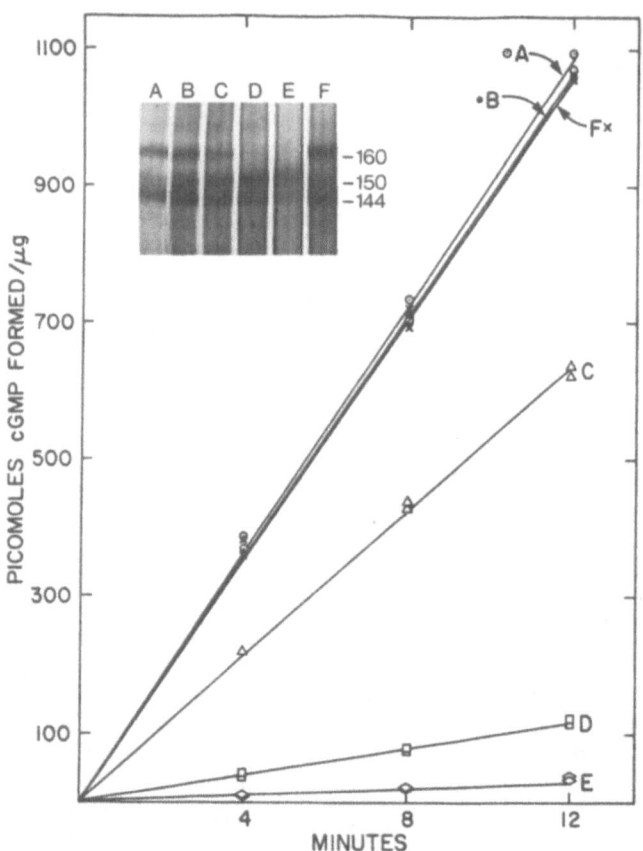

Figure 14. Spermatozoa were exposed to various concentrations of egg
jelly; (B) 112 ng fucose ml^{-1}, (C) 224 ng ml^{-1}, (D) 560 ng
ml^{-1}, (E) 1400 ng ml^{-1} (A = no jelly control); homogenized, and assayed for guanylate cyclase activity. The
assay was carried out at lower pH and temperature (36)
than are conventionally used for guanylate cyclase (7,20),
in order to minimize spontaneous conversion of the 160 kDa
form of the enzyme to 150 kDa during the assay. Curve F
is a control in which egg jelly, at the maximum possible
concentration it could be present in homogenate E, was
added back into homogenate A. At the end of the 12 min
assay, aliquots were removed for analysis by SDS-PAGE
(inset). The numbers on the right of the inset denote
relative molecular mass in kDa. (Reprinted from ref. 36;
copyright 1985 by the AAAS).

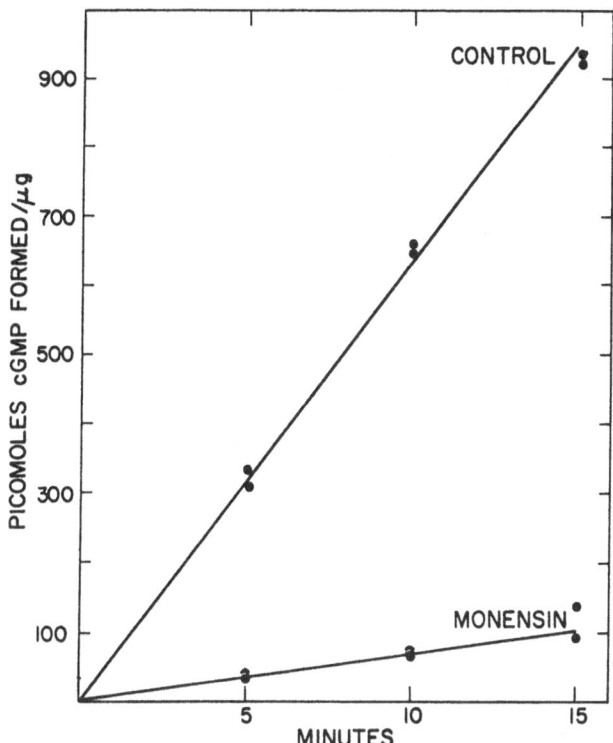

Figure 15. Guanylate cyclase activity of monensin-treated vs. control
spermatozoa. Dry spermatozoa were diluted to 9 x 10[7]
cells ml^{-1} in ASW and divided into two lots; monensin was
added to one lot (40 μM final), and an equivalent amount
of DMSO/ethanol to the other (control). After 7 min at
21°C, the cells were washed, disrupted by nitrogen
cavitation, and assayed for guanylate cyclase activity as
described in Fig. 11. Electrophoretic analysis showed no
change in mobility of the 160 kDa protein in the control
sample, while the mobility shift was complete in the
monensin-treated sample by 7 min. In other experiments,
20 min incubation in 10 μM monensin was found to have a
similar effect on guanylate cyclase activity and the
mobility shift. Monensin had no detectable inhibitory
effect on guanylate cyclase activity when added back to
the control assay tube (100 μM).

to also change in response to egg jelly. The resolution to this
question will only come when more is known about what happens to the
rate of cGMP turnover in vivo following jelly treatment.

Extraction of whole A. punctulata egg jelly with 75% ethanol separates the activity which induces the acrosome reaction from the factor responsible for dephosphorylation of the guanylate cyclase (Fig. 16). The ethanol precipitate is as effective as whole jelly (on a per mg fucose basis) at inducing the acrosome reaction, but is no longer able to induce the dephosphorylation of the enzyme (even at concentrations 80-fold higher than the amount of control jelly required to induce the complete dephosphorylation). The ethanol precipitate is therefore able to cause 100% of the spermatozoa to acrosome react, in the absence of any detectable change in phosphorylation of the cyclase, demonstrating that the dephosphorylation is not an obligatory step in the sequence of events leading to the acrosome reaction.

CONCLUSIONS

In summary, we have shown that egg jelly induces the rapid electrophoretic mobility shift of an abundant membrane protein of A. punctulata spermatozoa. The mobility shift is probably the direct result of dephosphorylation of the protein. A single peptide of known sequence isolated from A. punctulata egg jelly, resact, is responsible for the dephosphorylation (30). The phosphoprotein has been identified as guanylate cyclase. Since resact induces not only the dephosphorylation, but changes in cGMP levels as well (30), it seems likely that the dephosphorylation of guanylate cyclase is somehow involved in the altered cGMP metabolism of jelly-treated spermatozoa. Initial in vitro experiments (Figs. 14,15) suggest that there is in fact a correlation between the activity of the enzyme and whether it is in its 160 kDa ("phospho") or its 150 kDa ("dephospho") form.

Although sea urchin spermatozoa contain extremely high levels of guanylate cyclase activity (8), the function of cGMP in these cells remains unknown. Guanylate cyclase activity is heavily enriched in the plasma membrane of the flagellum (11,22,34), and it has been shown that the sperm head contains very little cGMP (6). These observations suggest that cGMP metabolism may be involved in some aspect of motility regulation, or some function which requires the large surface area of the flagellum.

In most cells thus far studied, cGMP has been shown to exert its effects through the activation of cGMP-dependent protein kinase (15), but this activity is undetectable in sea urchin spermatozoa (8). Cyclic GMP-binding proteins other than the kinase have been reported in various cells (15), including one report of a non-hydrolytic cGMP-binding site on sea urchin sperm phosphodiesterase (5). A third possibility, currently generating much excitement in the visual cell literature, is that the absolute levels of cGMP may not be as important as the rate of cGMP turnover; some byproduct of cGMP formation and hydrolysis (possibly H^+), rather than cGMP itself, may be what

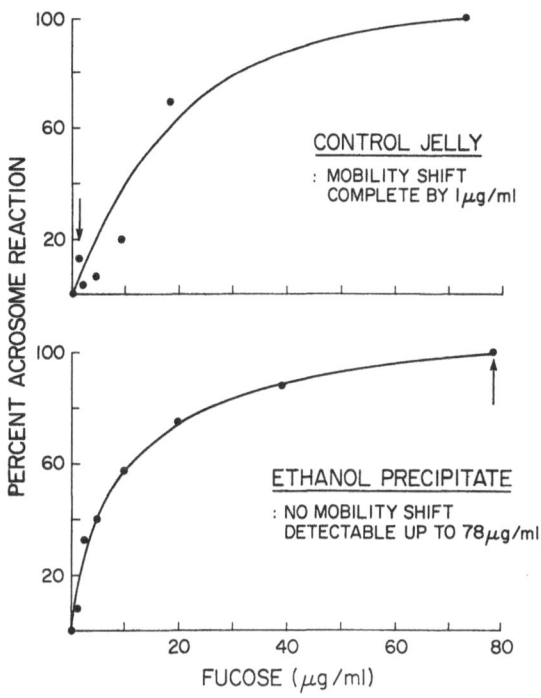

Figure 16. An ethanol precipitate of whole egg jelly induces the
acrosome reaction without the mobility shift. Ethanol (3
vol, 95% vol/vol) was added to 1 vol egg jelly, stirred
for 15 min (23°C), and the precipitate pelleted (10 min,
16000 xg). The pellet was resuspended in 70% (vol/vol)
ethanol, stirred for 15 min, and repelleted (this
resuspend/pellet sequence was repeated 3 times after the
initial precipitation). The final pellet was lyophilized
and dissolved in 1 vol ASW pH 8.0; the conductivity of the
dissolved pellet was identical to the conductivity of the
original whole jelly. Various dilutions of control jelly
and the ethanol precipitate solution (expressed in fucose
equivalents ml^{-1}) were assayed for their ability to induce
the 160 to 150 kDa mobility shift (34) and the acrosome
reaction (33). With control jelly, 1 µg fucose ml^{-1} was
sufficient to induce the complete mobility shift. With
the ethanol precipitate, no 160 to 150 kDa conversion was
detectable even at 78 µg fucose ml^{-1}. A value of one
hundred percent acrosome reaction was assigned to the
maximum number of spermatozoa in a given sperm batch which
acrosome reacted in response to saturating amounts of
jelly, minus spontaneous (0 jelly) acrosome reactions.

underlies the release of Ca^{2+} from intracellular stores in the rod outer segment (18). Further studies on the dephosphorylation of sea urchin sperm guanylate cyclase may lead to new approaches to the question of how cGMP exerts its effects in the spermatozoon, and what role cGMP metabolism plays in successful sperm-egg interaction.

ACKNOWLEDGEMENTS

We thank Drs. D. L. Garbers, N. Suzuki, and G. Schatten, and D. C. Porter for many stimulating discussions and the sharing of unpublished data, Dr. Garbers and D. J. Tubb for help with the cGMP assays, and Valerie Lavergne for technical assistance. This work was supported by NIH Grant HD-12986 to VDV.

REFERENCES

1. Best, D., Warr, P.J. and Gull, K. 1981. Influence of the composition of commercial sodium dodecyl sulfate preparations on the separation of α-and β-tubulin during polyacrylamide gel electrophoresis. Anal. Biochem. 114:281-284.
2. Christen, R., Schackmann, R.W. and Shapiro, B.M. 1982. Elevation of intracellular pH activates sperm respiration and motility of sperm of the sea urchin Strongylocentrotus purpuratus. J. Biol. Chem. 257:14881-14890.
3. Dunker, A.K. and Rueckert, R.R. 1969. Observations on molecular weight determinations on polyacrylamide gel. J. Biol. Chem. 244:5074-5080.
4. Epel, D. 1978. Mechanisms of activation of sperm and egg during fertilization of sea urchin gametes. Curr. Topics Dev. Biol. 12:185-246.
5. Francis, S.H., Lincoln, T.M. and Corbin, J.D. 1980. Characterization of a novel cGMP-binding protein from rat lung. J. Biol. Chem. 225:620-626.
6. Garbers, D.L. 1981. The elevation of cyclic AMP concentrations in flagella-less sea urchin sperm heads. J. Biol. Chem. 256:620-624.
7. Garbers, D.L. and Murad, F. 1979. Guanylate cyclase assay methods. Adv. Cyclic Nucleotide Res. 10:57-67.
8. Garbers, D.L. and Kopf, G.S. 1980. The regulation of spermatozoa by calcium and cyclic nucleotides. Adv. Cyclic Nucleotide Res. 13:251-306.
9. Garbers, D.L., Kopf, G.S., Tubb, D.J. and Olson, G. 1983. Elevation of sperm adenosine 3':5'-monophosphate concentrations by a fucose-sulfate-rich complex associated with eggs: I. Structural characterization. Biol. Reprod. 29:1211-1220.
10. Gates, R.E. and King, L.E. 1982. The EGF receptor-kinase has multiple phosphorylation sites. Biochem. Biophys. Res. Comm. 105:57-66.

11. Gray, J.P. and Drummond, G.I. 1976. Guanylate cyclase of sea urchin sperm: subcellular localization. Arch. Biochem. Biophys. 172:31-38.

12. Hansbrough, J.R. and Garbers, D.L. 1981. Speract; Purification and characterization of a peptide associated with eggs that activates spermatozoa. J. Biol. Chem. 256:1447-1452.

13. Hansbrough, J.R. and Garbers, D.L. 1981. Sodium-dependent activation of sea urchin spermatozoa by speract and monensin. J. Biol. Chem. 256:2235-2241.

14. Lee, H.C., Johnson, C. and Epel, D. 1983. Changes in internal pH associated with initiation of motility and acrosome reaction of sea urchin sperm. Develop. Biol. 95:31-45.

15. Lincoln, T.M. and Corbin, J.D. 1983. Characterization and biological role of the cGMP-dependent protein kinase. Adv. Cyclic Nucleotide Res. 15:139-192.

16. Lowry, O.H., Rosebrough, N.J., Farr, A.L. and Randall, J.R. 1951. Protein measurements with the folin-phenol reagent. J. Biol. Chem. 193:265-275.

17. Margulies, M.M. and Tiffany, H.L. 1984. Importance of sodium dodecyl sulfate source to electrophoretic separations of thylakoid polypeptides. Anal. Biochem. 136:309-313.

18. Miller, W.H. Does cyclic GMP hydrolysis control visual signal transduction in rods? Trends Pharmacol. Sci. (Dec. 1983) 509-511.

19. Moore, S. and Stein, W.H. 1963. Chromatographic determination of amino acids by the use of automated recording equipment. Meth. Enz. 6:819-831.

20. Radany, E.W., Gerzer, R. and Garbers, D.L. 1983. Purification and characterization of particulate guanylate cyclase from sea urchin spermatozoa. J. Biol. Chem. 258:8346-8351.

21. Repaske, D.R. and Garbers, D.L. 1983. A hydrogen ion flux mediates stimulation of respiratory activity by speract in sea urchin spermatozoa. J. Biol. Chem. 258:6025-6029.

22. Sano, M. 1976. Subcellular localizations of guanylate cyclase and 3',5'-cyclic nucleotide phosphodiesterase in sea urchin sperm. Biochim. Biophys. Acta 428:525-531.

23. Schackmann, R.W., Eddy, E.M. and Shapiro, B.M. 1978. The acrosome reaction of Strongylocentrotus purpurtatus sperm; Ion requirements and movements. Develop. Biol. 65:483-495.

24. Schackmann, R.W. and Shapiro, B.W. 1981. A partial sequence of ionic changes associated with the acrosome reaction of Strongylocentrotus purpuratus. Develop. Biol. 81:145-154.

25. Schackmann, R.W., Christen, R. and Shapiro, B.M. 1981. Membrane potential depolarization and increased intracellular pH accompany the acrosome reaction of sea urchin sperm. Proc. Nat. Acad. Sci. USA 78:6066-6070.

26. SeGall, G.K. and Lennarz, W.J. 1979. Chemical characterization of the component of the jelly coat from sea urchin eggs responsible for induction of the acrosome reaction. Develop. Biol. 71:33-48.

27. Smith, A.C. and Garbers, D.L. 1983. The binding of an [125]I-
    speract analogue to spermatozoa, in: "Biochemistry of Meta-
    bolic Processes," D. L. F. Lennon, F. W. Stratmen, and R. N.
    Zahlten, eds., pp. 15–28, Elsevier Biomedical, New York.
28. Suzuki, N., Nomura, K., Ohtake, H. and Isaka, S. 1981. Purifica-
    tion and the primary structure of sperm-activating peptides
    from the jelly coat of sea urchin eggs. Biochem. Biophys.
    Res. Comm. 99:1238–1244.
29. Suzuki, N. and Garbers, D.L. 1984. Stimulation of sperm respira-
    tion rates by speract and resact at alkaline extracellular pH.
    Biol. Reprod. 30:1167–1174.
30. Suzuki, N., Shimomura, H., Radany, E.W., Ramaro, C.S., Ward,
    G.E., Bentley, J.K. and Garbers, D.L. A peptide associated
    with eggs causes a mobility shift in a major plasma membrane
    protein of spermatozoa. J. Biol. Chem. (in press).
31. Swank, R.T. and Munkres, K.D. 1971. Molecular weight analysis of
    oligopeptides by electrophoresis in polyacrylamide gel with
    sodium dodecyl sulfate. Anal. Biochem. 39:462–477.
32. Swarup, G. and Garbers, D.L. 1982. Phosphoprotein phosphatase
    activity of sea urchin spermatozoa. Biol. Reprod. 26:953–
    960.
33. Vacquier, V.D. Rapid immunoassays for the acrosome reaction of
    sea urchin sperm utilizing antibody to bindin. Exp. Cell Res.
    (in press).
34. Ward, G.E. and Vacquier, V.D. 1983. Dephosphorylation of a major
    sperm membrane protein is induced by egg jelly during sea
    urchin fertilization. Proc. Nat. Acad. Sci. USA 5578–5582.
35. Ward, G.E. and Vacquier, V.D. 1983. Sea urchin egg jelly induces
    the dephosphorylation of a 160 kDa sperm flagellar membrane
    protein: the role of Na^+/H^+ exchange. J. Cell Biol. 97:250a.
36. Ward, G.E., Garbers, D.L. and Vacquier, V.D. Effects of extra-
    cellular egg factors on sperm guanylate cyclase. Science (in
    press).
37. Wegenar, A.D. and Jones, L.R. 1984. Phosphorylation-induced
    mobility shift in phospholamban in sodium dodecyl sulfate-
    polyacrylamide gels. J. Biol. Chem. 259:1834–1841.

STEPS IN THE FERTILIZATION PROCESS: UNDERSTANDING AND CONTROL

Michael G. O'Rand

Laboratories for Cell Biology
Department of Anatomy
University of North Carolina
Chapel Hill, NC 27514

SUMMARY

The fertilization process begins when spermatozoa are released from the male and enter into the oocyte's environment (external fertilization) or are deposited into the female (internal fertilization) and it ends when the zygote is formed. In this report the fertilization process is discussed in the context of a hierarchy of specificities leading to successful fertilization. In 1919 F.R. Lillie conceptualized the specific interaction and fusion of sperm and egg in the "receptor terminology" that is in use today. Sperm and egg receptors are discussed in this context and the rabbit sperm membrane protein, RSA-1 is presented as an example of a sperm zona binding protein. Lillie's sperm-egg receptor paradigm is still not fully understood, yet with a better understanding of each level in the hierarchy will come the possibility of control not only to inhibit fertilization but also to treat infertility.

Attempts at defining the fertilization process have varied over the last seventy years, from the very narrow spermatozoon-oocyte interaction definition to the broadest of definitions which includes both sperm and egg maturation. Currently, defining the process as one of spermatozoon-female interaction would seem to be the most meaningful. This definition would encompass all the changes that the spermatozoon must undergo in association with female tissue in order to fertilize the egg. The fertilization process begins when spermatozoa are released from the male and enter into the oocyte's environment (external fertilization) or are deposited into the female (internal fertilization) and it ends when the zygote is formed.

This definition is appropriate for every phylum. Moreover, it incorporates the functional change that many species of spermatozoa must undergo within the female tract before fertilization (26,38), namely capacitation. Most phyla have representatives with internal fertilization. In sponges, the spermatozoon penetrates a choancyte which fuses with the egg (12), and in coelenterates (cnidarians) fertilization may occur inside the female gonophore (26). In the Platyhelminthes, where hermaphroditism and internal fertilization are probably characteristic of the turbellarian archetype (1), internal fertilization occurs in the complex hermaphroditic reproductive systems. In the Mollusca it occurs in the neritacea, neogastropods, mesogastropods, opisthobranchs, pulmonates and some octopod cephalopods (11). Many arthropods, particularly insects (see 8 for a general discussion) have internal fertilization and the spermatozoa are often maintained within the female for long periods of time. In all of these phyla, however, capacitation has been documented in only insects and cnidarians (see 28 for review), yet it should also be pointed out that serious attempts to look for capacitation in other invertebrate groups have not been made.

Conceptually, it would seem quite clear that species with internal fertilization should evolve mechanisms to regulate the time and place of the sperm's acrosome reaction in order to ensure successful fertilization. Consequently, capacitation would not logically seem to be an exclusive preserve of mammalian gametes as some authors would have us believe.

If the fertilization process is put into the larger context of animal reproduction, one can better examine and understand the specificities that exist at different levels of the entire reproductive process. A hierarchy of specificities can be made which is essentially a list of steps leading to successful fertilization (Table 1). The first step, with the strictest of specificities, would be geographic isolation and the second step would be the mating behavior of species. Both of these serve in large part to maintain species' diversity. If mating does occur, then there are several steps which are generally unfriendly to heterologous spermatozoa. In species with internal fertilization, reproductive tract passage is the next level of specificity, while in those with external fertilization specificity would pass directly to the cumulus matrix/egg jelly step. Thus, if a particular step in the heirarchy does not exist for a given species then the specificity simply defaults to the next level.

Within the female reproductive tract, sperm passage may not occur through the heterologous cervix, nor may passage to the site of fertilization be appropriate for the timing of ovulation. Capacitation, the functional change in spermatozoa that allows them to fertilize the egg, is the next level within the tract. While specificity does not reside in species specific environmental factors (36), it is rather more likely to be in the timing of the sperm's acrosome reac-

Table 1

HIERARCHY OF SPECIFICITIES

Geographic Isolation

Mating Behavior

Reproductive Tract Passage

Cervix
Uterus
Oviduct

Capacitation

Cumulus Matrix/Egg Jelly

Zona Pellucida/Vitelline Layer

Egg Plasma Membrane

tion in the presence of the freshly ovulated egg. Having completed successful passage of the female reproductive tract, the spermatozoon faces the three final levels which are analogous in internally and externally fertilizing species (see, for example, Table 1 in 22). These final three steps will now be considered in more detail and in the context of fertilization as it was first described just after the turn of the century.

Until the first decade of the twentieth century, the importance of fertilization as a process upon which evolution depends was not yet established. The reduction-division controversy had to be resolved first, giving the fertilizing spermatozoon an equal genetic contribution to the zygote. Following the observations of Boveri (7) and Hertwig (15) on oogenesis and spermatogenesis, Weismann (42) made it quite clear that the differentiation and development of the germ cells was important for an understanding of the importance of fertilization. He emphasized that the result of fertilization was a mixing of the maternal and paternal hereditary units. With the additional discovery of the centrosome, the sperm cell provided not only a nucleus but also "a centrosome which by its own division leads the way in the embryonic development" (7). Thus, the spermatozoon moved from being a physiochemical stimulus and a mere footnote to fertilization, to an important and necessary component in the reaction.

By 1910, F. R. Lillie had begun a series of experiments to investigate the mechanism of sperm penetration of the egg. Being influenced to a large extent by Ehrlich's work on immunity, Lillie formulated the first theoretical basis for the fertilization reaction (18,19). These classical experiments have been reviewed numerous times, most recently by Metz (24; see also C. Metz, this volume) and they led Lillie to postulate that "fertilization is a reaction between three bodies of which one is born by the sperm and one by the egg; the third body, which is secreted by the egg, reacts with both the others" (19). This model conceptualized the specific interaction and fusion of two cells in the "receptor terminology" that is in use today. The third body, which Lillie called "fertilizin", is primarily seen today as the initiator of the acrosome reaction, although it also agglutinates spermatozoa. The agglutination was seen by Lillie (20) as not essential for fertilization and is still regarded (24) "as an artifact of laboratory conditions which however provides an assay procedure of unusual analytical value in terms of sperm-egg interacting receptor substances."

Lillie's paradigm for fertilization has been subsequently expanded and refined (e.g., 40) and numerous workers have contributed to the scientific literature of sperm-egg interaction (for review see 23,24,4,44). However, Lillie's essential elements remain of interest to workers today in both vertebrate and invertebrate species. These essential elements are: a sperm receptor, an egg receptor and an extracellular "third body" receptor which binds sperm. Each of these will be discussed below.

The egg receptor "element" on spermatozoa is perhaps better conceptualized as a group of molecules which function in binding to the egg's sperm receptor and/or to the egg's extracellular receptor, i.e., the zona pellucida or vitelline envelope. The sperm's binding components are most probably proteins and these may function at two different levels in sperm-egg interaction. First, in an intact spermatozoon, capable of fertilizing the egg, glycoproteins may exist in the sperm's plasma membrane which respond to the egg's extracellular coats and initiate the acrosome reaction.

In sea urchins, the egg jelly (Lillie's fertilizin) binds to the sperm surface in a species-specific manner to initiate the acrosome reaction (38). The homologous jelly coat which initiates the acrosome reaction also initiates the simultaneous uptake of Ca^{++} whereas the heterologous jelly does not. SeGall and Lennarz (38) suggest that a specific jelly binding component may be present on spermatozoa. That such a sperm surface component exists was also suggested by Lopo and Vacquier (21) from studies using an antibody against an 84,000 dalton sperm glycoprotein. Egg jelly could not induce the acrosome reaction in the presence of anti-sperm 84K. The idea of a jelly binding protein on sperm is further strengthened by the report of Eckberg and Metz (10) that a 68,000 dalton Arbacia sperm membrane glycoprotein neutralizes the anti-sperm antibody (Fab fragment) action which inhibits the acrosome reaction. The differences in molecular weight may reflect the species-specificity: Strongylocentrotus purpuratus,

84,000 and _Arbacia_ 68,000, and _Arbacia_ jelly does not induce the S. purpuratus acrosome reaction (38).

In eutherian mammals, the ovulated oocyte is not surrounded by egg jelly but by a complex of granulosa cells and extracellular matrix. Proteoglycans constitute the bulk of the extracellular matrix and their constituent glycosaminoglycans, heparin sulfate, chondroitin sulfates A, B, C, and hyaluronic acid all enhance the bovine (16) and rabbit sperm (17) acrosome reaction. As with sea urchin egg jelly (38), the sulfation of the matrix was found to be an important parameter in the initiation of the bovine sperm acrosome reaction (14). It will be of interest to see if a sperm membrane glycoprotein which binds glycosaminoglycans and initiates the acrosome reaction can be found.

The second level in sperm-egg interaction at which the sperm's binding component(s) may interact with the egg's extracellular coats is following the acrosome reaction. The acrosome reacted spermatozoon binds to the vitelline envelope (sea urchins) or zona pellucida (mammals). In some laboratory experiments dejellied or cumulus-free eggs may be used and the acrosome intact sperm may bind to the vitelline envelope or zona pellucida. However, these in vitro situations will not be considered further here (for further discussion on this point, see 3).

In sea urchins the sperm protein responsible for binding the sperm to the vitelline layer is bindin (41). Bindin is present in the sperm acrosomal vesicle as a non-glycosylated protein of 30,500 daltons which is localized on the acrosomal process following the acrosome reaction (25). Bindins from two different species of Strongylocentrotus are very similar biochemically and specific antisera to bindin from Strongylocentrotus will cross-react with all the species within the class Echinoidea (25,22). Although bindin-like molecules have been identified in the Molluska and Echiuroidea (see 22 for details), their existence in mammals has not yet been established.

However, it has now been established (31) that mammalian sperm have a set of zona binding proteins (ZBP), each species with its own characteristic set, and each set also able to bind heterologous zonae to some extent. That spermatozoa from a variety of mammalian species will bind to a heterologous zona has been pointed out by numerous workers (see 2,37,44 for review). Thus, binding of spermatozoa to the zona pellucida is neither species specific nor class specific. Rather it would seem that the specificity must lie in the spermatozoon's ability to penetrate the zona pellucida (43,13).

In the rabbit, for example, four major ZBP have been identified by electrophoresing washed ejaculated sperm on SDS-polyacrylamide gels, blotting the protein bands onto nitrocellulose and incubating the blots in ^{125}I-heat solubilized rabbit zonae (^{125}I-HSRZ; 31). The

sperm proteins which bind zona have molecular weights of 32 Kd, 18
Kd, 16 Kd and 14 Kd. Additional evidence now indicates that the rabbit sperm autoantigen, RSA-1 (33,34), when isolated from the testis,
will bind ^{125}I-HSRZ and is the 14 Kd molecule identified in ejaculated sperm.

RSA-1 is a sialoglycoprotein of 13,000 ± 1200 daltons which is
an integral sperm plasma membrane component (33). RSA-1 first appears
on the surface of pachytene spermatocytes and increases in amount
throughout spermatogenesis (35). Two additional low molecular weight
intrinsic plasma membrane glycoproteins which are similar in amino
acid and carbohydrate composition to RSA-1 have also been identified
(34,32, Table 2). On ejaculated spermatozoa the RSA family appears
to be bound to a complex of extrinsic proteins, one of which has a
molecular weight of 84,000 ± 10,000 daltons (35), and which is
removed during capacitation (30).

That the RSA family has a role in sperm-egg interaction has
become apparent through the use of antibody studies and ^{125}I-HSRZ
experiments. Specific polyclonal Fab antibody fragments to RSA-1 can
reduce in vivo fertility and inhibit sperm penetration of the zona in
vitro (29). Moreover, monoclonal antibodies to RSA-1 inhibit in vitro
fertilization at the level of zona penetration but do not block the
sperm's acrosome reaction (32). Anti-RSA-1 antibodies will bind to
acrosome intact spermatozoa bound to the zona pellucida of rabbit
eggs but not to acrosome-reacted spermatozoa in or on the intact rabbit zona (29). From these experiments it was hypothesized that the
zona must be blocking the antibody's access to RSA-1 in some manner.
This now seems to be the case since ^{125}I-HSRZ will bind to isolated
RSA-1 (O'Rand, unpublished). RSA-1 has also been shown to be incorporated into the egg plasma membrane after fertilization (32), sub-
stantiating the observation that sperm membrane proteins become an
integral part of the plasma membrane after gamete fusion (27).

Thus, RSA-1 is an example of a sperm specific plasma membrane
molecule which arises during spermatogenesis and which is localized
in a particular region of the mature sperm's surface to perform a
particular function, namely to bind zona pellucida. When spermatozoa
approach the egg and its surrounding investments and presumably
during their preceeding time of capacitation in the female tract, the
RSA-1 becomes available to bind zona (Fig. 1). This is currently
thought to occur mostly in the postacrosomal region of the rabbit
sperm surface, although other regions such as the plasma membrane
over the apical acrosome undoubtedly also have zona binding proteins.
Additional ZBP may include the galactosyltransferase identified by
Shur and Hall (39; see also B. Shur, this volume).

If the specificity at the level of the zona pellucida/vitelline
layer lies in the sperm's ability to penetrate, then penetration must
depend in some manner upon the affinity of the zona for more than one

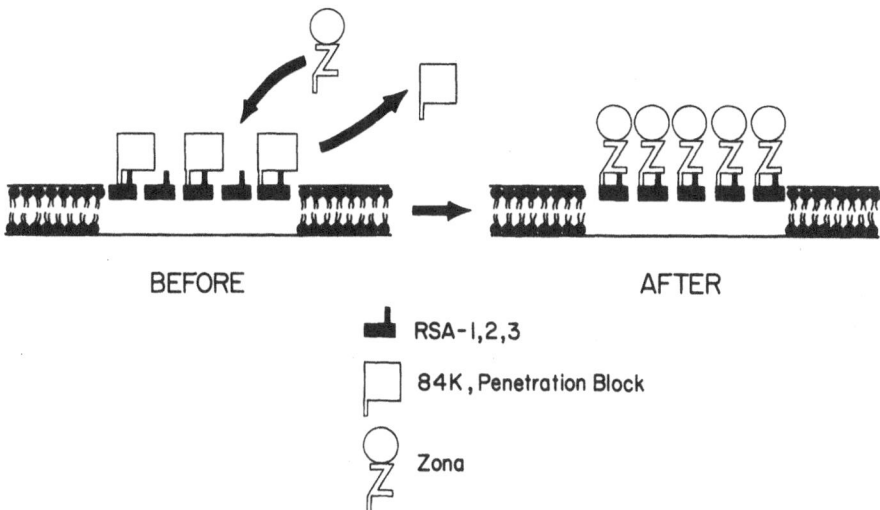

Figure 1. Model of the spermatozoon plasma membrane depicting the
relationship between RSA, 84K and the zona. As the spermatozoon becomes capacitated, extrinsic proteins are lost
from the surface, including the 84,000 dalton protein shown
here. As the 84K is removed, the RSA family becomes available to bind zona components. After the acrosome reaction
(right-hand side of figure), the zona bound to RSA prevents
anti-RSA antibodies from binding to RSA. If anti-RSA is
added to spermatozoa before the acrosome reaction, the
spermatozoa may still bind zona via other ZBP and will
undergo an acrosome reaction, but will not penetrate the
zona.

of the sperm's ZBP. The sum of all binding events will perhaps
determine whether or not penetration can be accomplished.

On the other side of the sperm-zona interaction equation is the
zona's sperm receptor. The sperm receptor in the vitelline layer of
sea urchins which binds to the sperm's bindin has been characterized
as a large molecular weight ($> 10^6$ daltons) glycoprotein containing
sulfate, but distinct from egg jelly (22; see also N. Ruiz-Bravo,
this volume). Similarly the sperm receptor in the zona pellucida of
mice has been characterized as an 83,000 dalton glycoprotein (ZP3, 5,
6; see also P. Wasserman, this volume). Recently, East et al. (9;
see also J. Dean, this volume), have presented evidence based on
monoclonal antibodies to the mouse zona components, ZP2 and ZP3, that
both anti-ZP2 and anti-ZP3 inhibit fertilization. Therefore, both
ZP2 and ZP3 may function as sperm receptors.

Having traversed the hierarchy of specificities, including the barriers which surround the egg, the spermatozoon is finally faced with one last step: fusion with the egg plasma membrane. This step seems in many ways the least specific because a variety of different spermatozoa will fuse with a heterologous egg which is free of its outer investments (43,44 - cf. Table 2). Currently, nothing is known about the existence of sperm receptors on the egg plasma membrane. Even the need for such receptors may be questioned because the fusion of two lipid bilayers may occur under the "correct" environmental conditions without the necessity for specific receptors or alternatively the spermatozoon may have a fusigenic protein. Ultimately, of course, if fertilization is successful between heterologous gametes the genome will determine compatibility and embryonic survival.

Having examined each of the steps in the fertilization process, it should be apparent that although considerable progress has been made, Lillie's sperm-egg receptor paradigm is still not fully understood. At each level in the hierarchy of specificities control is exercised by the difficulty imposed upon heterologous gametes. With a better understanding of each level in the human reproductive system will come the possibility of control not only to inhibit fertilization but also to treat infertility.

Table 2

Characteristics of the RSA Family of

Intrinsic Sperm Membrane Glycoproteins

	RSA-1	RSA-2	RSA-3
M_r	13,000	12,000	10,000
pI	4.2, 4.3, 4.5	4.2	4.2
% CHO	9.1	10.3	N.D.
Relative amount of sialic acid	+	++	+++
Amino sugar	glucosamine	glucosamine	N.D.
Amino acid residues	105	104	N.D.
Absorption maximum	258λ	258λ	258λ
% non-polar amino acids	59	57	N.D.
Hydrophobicity	646	N.D.	N.D.

ACKNOWLEDGEMENTS

     This work was supported by NIH grant HD 14232. The author would
like to thank Ms. Emily White for the typing of this manuscript.

REFERENCES

 1. Ax, P. 1963. Relationships and phylogeny of the Turbellaria, in:
       "The Lower Metozoa," E. C. Dougherty, ed., University Califor-
       nia Press, Berkeley, CA, pp. 191-224.
 2. Bedford, J.M. 1977. Sperm/egg interaction: The specificity of
       human spermatozoa. Anat. Rec. 188:477-488.
 3. Bedford, J.M. 1983. Significance of the need for sperm capacita-
       tion before fertilization in eutherian mammals. Biol. Reprod.
       28:108-120.
 4. Bedford, J.M., and Cooper, G.W. 1978. Membrane fusion events in
       the fertilization of vertebrate eggs, in: "Membrane Fusion-
       Cell Surface Reviews, 5", G. Poste and G. Nicholson, eds.,
       North Holland Publ., Amsterdam, pp. 65-125.
 5. Bleil, J.D., and Wassarman, P.M. 1980a. Structure and function of
       the zona pellucida: Identification and characterization of
       the proteins of the mouse oocyte's zona pellucida. Dev. Biol.
       76:185-202.
 6. Bleil, J.D., and Wassarman, P.M. 1980b. Mammalian sperm-egg
       interaction: Identification of a glycoprotein in mouse zonae
       pellucida possessing receptor activity for sperm. Cell
       20:873-882.
 7. Boveri, T. 1887. Zellenstudien I: Die Bildung der Richtungs
       Korper bei Ascaris megalocephala and Ascaris lumbricoides,
       Jena.
 8. Davey, K.G. 1965. Reproduction in Insects, W. H. Freeman and Co.,
       San Francisco, CA, pp. 51-61.
 9. East, I.J., Mattison, D.R., and Dean, J. 1984. Monoclonal anti-
       bodies to the major protein of the murine zona pellucida:
       Effects on fertilization and early development. Dev. Biol.
       104:49-56.
10. Eckberg, W.R., and Metz, C.B. 1982. Isolation of an Arbacia sperm
       fertilization antigen. J. Exptl. Zool. 221:101-105.
11. Fretter, V., and Graham, A. 1964. Reproduction, in: "Physiology
       of Mollusca," K. M. Wilbur and C. M. Yonge, eds., Academic
       Press, N.Y., Vol. 1, pp. 127-164.
12. Gatenby, J.B. 1919. Germ cells, fertilization and early develop-
       ment of Grantia (Sycon) compressa. Linn. Soc. London J. Zool.
       34:261-297.
13. Hanada, A., and Chang, M.C. 1978. Penetration of zona-free or
       intact eggs by foreign spermatozoa and the fertilization of
       deer mouse eggs in vitro. J. Exptl. Zool. 203:277-286.
14. Handrow, R.R., Lenz, R.W., and Ax, R.L. 1982. Structural compari-
       sons among glycosaminoglycans to promote an acrosome reaction

in bovine spermatozoa. Biochem. Biophys. Res. Comm. 107:1326–1332.

15. Hertwig, O. 1890. Vergleich der Ei-und Samenbildung bei Nematoden: Eine Grundlage fur cellulare Stretifragen. Arch mikro Anat. 36:61.

16. Lenz, R.W., Ball, G.D., Lohse, J.K., First, N.L., and Ax, R.L. 1983a. Chondroitin sulfate facilitates an acrosome reaction in bovine spermatozoa as evidenced by light microscopy, electron microscopy and in vitro fertilization. Biol. Reprod. 28:683–690.

17. Lenz, R.W., Bellin, M.E., and Ax, R.L. 1983b. Rabbit spermatozoa undergo an acrosome reaction in the presence of glycosaminoglycans. Gamete Res. 8:11–19.

18. Lillie, F.R. 1913. The mechanisms of fertilization. Science 38:524–528.

19. Lillie, F.R. 1914. Studies on fertilization. VI. The mechanism of fertilization in Arbacia. J. Exp. Zool. 16:523–590.

20. Lillie, F.R. 1915. Sperm agglutination and fertilization. Biol. Bull. 28:18–33.

21. Lopo, A.C., and Vacquier, V.D. 1980. Antibody to a sperm surface glycoprotein inhibits the egg jelly-induced acrosome reaction of sea urchin sperm. Develop. Biol. 79:325–333.

22. Lopo, A.C., and Vacquier, V.D. 1981. Gamete interaction in the sea urchin, in: "Fertilization and Embryonic Development In Vitro." L. Mastroianni and J. D. Biggers, eds. Plenum Press, N.Y., pp. 199–232.

23. Metz, C.B. 1967. Gamete surface components and their role in fertilization. In: "Fertilization," vol. 1, C.B. Metz and A. Monroy, eds., Academic Press, N.Y., pp. 163–236.

24. Metz, C.B. 1978. Sperm and egg receptors involved in fertilization. Curr. Top. Devel. Biol. 12:107–147.

25. Moy, G.M., and Vacquier, V.D. 1979. Immunoperoxidase localization of bindin during sperm-egg interaction. Curr. Top. Develop. Biol. 13:31–44.

26. O'Rand, M.G. 1972. In vitro fertilization and capacitation-like interaction in the hydroid Campanularia flexuosa. J. Exp. Zool. 182:299–306.

27. O'Rand, M.G. 1977. The presence of sperm-specific surface isoantigens on the egg following fertilization. J. Exptl. Zool. 202:267–273.

28. O'Rand, M.G. 1979. Changes in sperm surface properties correlated with capacitation, in: "The Spermatozoon," D. Fawcett and J. M. Bedford, eds., Urban and Schwarzenberg, Inc., Baltimore, pp. 195–204.

29. O'Rand, M.G. 1981. Inhibition of fertility and sperm-zona binding by antiserum to the rabbit sperm membrane autoantigen, RSA-1. Biol. Reprod. 25:611–618.

30. O'Rand, M.G. 1982. Modification of the sperm membrane during capacitation, Ann. N.Y. Acad. Sci. 383:392–402.

31. O'Rand, M.G., Emery, J.J., Welch, J.E., and Fisher, S.J. 1984a.

Zona binding proteins of mammalian spermatozoa. J. Cell Biol. 99:395a.

32. O'Rand, M.G., Irons, G.P., and Porter, J.P. 1984b. Monoclonal antibodies to rabbit sperm autoantigens. I. Inhibition of in vitro fertilization and localization on the egg. Biol. Reprod. 30:721-729.

33. O'Rand, M.G., and Porter, J.P. 1979. Isolation of a sperm membrane sialoglycoprotein autoantigen from rabbit testes. J. Immunol. 122:1248-1254.

34. O'Rand, M.G., and Porter, J.P. 1982. Purification of rabbit sperm auto-antigens by preparative SDS gel electrophoresis: Amino acid and carbohydrate content of RSA-1. Biol. Reprod. 27:713-721.

35. O'Rand, M.G., and Romrell, L.J. 1981. Localization of a single sperm autoantigen (RSA-1) on spermatogenic cells and spermatozoa. Devel. Biol. 84:322-331.

36. Saling, P.M., and Bedford, J.M. 1981. Absence of species specificity for mammalian sperm capacitation in vivo. J. Reprod. Fertil. 63:119-123.

37. Schmell, E.D., and Gulyas, B.J. 1980. Mammalian sperm-egg recognition and binding in vitro. I. Specificity of sperm interactions with live and fixed eggs in homologous and heterologous inseminations of hamster, mouse and guinea pig oocytes. Biol. Reprod. 23:1075-1085.

38. SeGall, G.K., and Lennarz, W.J. 1981. Jelly coat and induction of the acrosome reaction in echinoid sperm. Develop. Biol. 86:87-93.

39. Shur, B.D., and Hall, N.G. 1982. A role for mouse sperm surface galactosyltransferase in sperm binding to the egg zona pellucida. J. Cell Biol. 95:574-579.

40. Tyler, A., and Tyler, B. 1966. Physiology of fertilization and early development, in: "Physiology of Echinodermata," R. A. Boolootian, ed. John Wiley and Sons, N.Y., pp. 683-741.

41. Vacquier, V.D., and Moy, G.W. 1977. Isolation of bindin: The protein responsible for adhesion of sperm to sea urchin eggs. Proc. Natl. Acad. Sci. USA 74:2456-2460.

42. Weismann, A. 1891. Amphimixis, or the essential meaning of conjugation and sexual reproduction. Essays upon Heredity 2:121.

43. Yanagimachi, R. 1977. Specificity of sperm egg interaction, in: "Immunobiology of Gametes," M. Edidin and M. H. Johnson, eds. Cambridge Univ. Press, London, pp. 255-295.

44. Yanagimachi, R. 1981. Mechanisms of fertilization in mammals, in: "Fertilization and Embryonic Development In Vitro," L. Mastroianni and J. D. Biggers, eds. Plenum Press, N.Y., pp. 81-182.

CONTROL OF FERTILIZATION BY IMMUNIZATION

WITH PEPTIDE FRAGMENTS OF SPERM SPECIFIC LDH-C$_4$

Erwin Goldberg and Joyce A. Shelton

Northwestern University
Department of Biochemistry, Molecular and Cell Biology
Evanston, Illinois

SUMMARY

The use of a well-defined, synthetic antigen is essential to progress in developing a vaccine for fertility control and to establish with certainty the utility of such technology. The studies with LDH-C$_4$ are consistent with the feasibility of using a synthetic antigen in a vaccine to control fertility. LDH-C$_4$ remains the best-developed candidate for future studies. Immunization with LDH-C$_4$ does suppress fertility, and the complete biochemical characterization of this isozyme is well underway. These studies will provide a wealth of synthetic antigens to substitute for the natural product in experiments to determine whether immunologic contraception is appropriate and desirable for wide-scale application in human beings.

INTRODUCTION

The antigenicity of spermatozoa is well documented (6,12,25) and forms the basis for immunization strategies to manipulate reproduction. The rationale for development of a contraceptive vaccine is that fertilization can be blocked by sperm-agglutinating and cytotoxic antibodies in the female reproductive tract. Presumably, such antibodies would recognize a target antigen on the sperm surface where an immune reaction could occur.

Antigens associated with the sperm surface are, for the most part, integral components of the plasma membrane. As such, their isolation and purification in amounts sufficient for characterization present special problems. One notable exception is the sperm-specific form of lactate dehydrogenase, LDH-C$_4$. This isozymic form

395

of LDH is a cytosolic protein that diffuses through the plasma
membrane of the spermatozoan and becomes associated with the surface.
Thus, it can serve as a potential target antigen for immunocontraception.

LDH-C$_4$ is a useful model for anti-sperm vaccine development. It
is absolutely sperm-specific and therefore unlikely to cause patho-
logical auto-immune reactions with somatic tissues. It can be purified to homogeneity in sufficient quantity both for experimental
manipulation and to ensure specificity of the immune response. It
elicits an immune response that impairs fertility. Antibodies to
LDH-C$_4$ have been demonstrated in certain sera from infertile patients
(Shelton and Goldberg, J. Reprod. Immunol., in press).

Fertility Reduction by Sperm Antigens

Immunization of females with whole sperm or with various crude,
semi-purified, and purified extracts of mammalian spermatozoa can
impair fertility. Experimental data in animal models, and observation of circulating anti-sperm antibodies in infertile women support
the concept of immunocontraception. Selected studies performed with
a variety of species over the past twenty years are summarized in
Table 1 and detailed in the following discussion. These experiments
reveal that sperm antigens, or more correctly, antibodies to antigenic constiuents of spermatozoa reduce fertility.

Fertility reduction from 60 to 85% was observed after immunization of female mice with whole spermatozoa (19,29). This level was
increased to 100% suppression of in vitro fertilization by rabbit
anti-mouse epididymal spermatozoa (28). Saling and O'Rand (28) provided preliminary identification, based on antibody recognition, of
at least eight polypeptides that might be considered candidates for
specific functions during the fertilization process. LDH-C$_4$, which
will be discussed in detail below, reduced fertility in mice by 32%
after systemic immunization (17) and by 73% after local immunization
via uterine implants of LDH-C$_4$ in agarose (Shelton and Goldberg, in
preparation).

FA-1, a monoclonal antibody to an antigen in the human germ cell
membrane, did not agglutinate or immobilize mouse sperm but did block
murine fertilization in vitro. FA-1 also inhibited binding and penetration of zona-free hamster ova by human sperm (23). The antigen
was identified as a 23Kd glycoprotein from detergent-solubilized
human testis.

Immunization of the guinea pig with homologous testis extract
reduced fertility 71% (11); however, the purified fractions of guinea
pig testes "S", "P", and "T" (1) had no effect on pregnancy rate.

Table 1. Summary of Fertility Reduction by Sperm Antigens

Species	Antigen	% Fertility Reduction	Reference
Mouse	Whole sperm	60	McLaren, '64
	Whole sperm	85	Tung et al., '79
	Whole sperm[a]	100	Saling & O'Rand, '82
	Soluble sperm extract	37	Bell & McLaren, '69
	LDH-C_4	32	Lerum & Goldberg, '74
	LDH-C_4	73	Shelton & Goldberg, '84
	"23Kd glycoprotein"[a]	46	Naz et al., '84
Guinea Pig	Testis Extract	71	Isojima et al., '58
	S, P, T	0	Almeida & Voisin, '79
Rabbit	Whole Sperm; Supernatant; Pellet	99	Munoz & Metz, '78
	Semen	90	Menge, '71
	Sperm	98	Kummerfeld & Foote, '76
	SE-LIS, SP-LIS, TE-LIS, TP	64-100	Menge et al., '79
	"63 Kd glycoprotein"[b]	97	Naz et al., '81
	RSA-1^2	68	O'Rand, '81
	LDH-C_4	67	Goldberg, '73
Baboon	LDH-C_4	63	Goldberg et al., '81

[a]In vitro fertilization

[b]AI with AS-treated sperm

Many studies have been reported in which the rabbit was used as the experimental model. Munoz and Metz (21) reported virtually complete suppression of fertility after immunization with either homologous, intact sperm or the 98,000xg supernatant, or with the insoluble pellet of $MgCl_2$ disrupted spermatozoa. Menge et al. (20) have performed a series of experiments in which rabbits immunized with lithium diiodosalicylate (LIS) soluble extracts of sperm and testis showed reduced fertility. Some of these preparations led to post-fertilization inhibition of fertility. Kummerfeld and Foote (14) injected rabbits with washed ejaculated, epididymal or β-amylase treated rabbit sperm. All treatments resulted in 98% suppression, apparently by blocking fertilization. Naz et al. (22) obtained mono-clonal antibodies to antigens released by detergent-solubilized rabbit epididymal sperm. Insemination of female rabbits with sperm pre-incubated with these antibodies resulted in post-fertilization infertility. The antigen has been identified as a 63Kd glycoprotein (23), immunogenic in mice, but not in rabbits and therefore possibly of somatic tissue origin. A 68% reduction in fertility was reported by O'Rand (24) upon insemination of female rabbits with sperm pre-incubated with antiserum to RSA-1. RSA-1 is a 13Kd sialoglycoprotein and is intrinsic to the rabbit sperm plasma membrane (26). Whether immunization with RSA-1 (26) or with the 63Kd glycoprotein (23) will provoke an immune response that suppresses fertility remains to be established when sufficient quantities of these antigens become available. It is already known that immunization of female rabbits with purified mouse $LDH-C_4$ results in fertility suppression of about 67%. The immune response blocks sperm transport (11) and fertilization (5).

Female baboons immunized with the $LDH-C_4$ preparation, exhibited a fertility reduction of 63% that was antibody titer related (7). These results support the potential of $LDH-C_4$ as a contraceptive vaccine.

Development of an Immunocontraceptive Technology: Sperm Specific $LDH-C_4$

Isozymes of lactate dehydrogenase are found in all vertebrate tissues. The tetrameric enzyme is assembled randomly from two types of subunits, A and B, in somatic cells. A third subunit type, C, is synthesized only in testes during active spermatogenesis, and is genetically and immunologically distinct from the A and B peptides (4). $LDH-C_4$ provides a remarkable example of spatial and temporal regulation of gene expression. In the mature testes, this isozyme is restricted to the germinal epithelium (Fig. 1). $LDH-C_4$ can be local-ized by immunofluorescence to primary spermatocytes and spermatids, while the spermatogonia and the non-germinal elements, such as interstitial and Sertoli cells, are clearly devoid of this isozyme. The first appearance of $LDH-C_4$ in mid-pachytene primary spermatocytes

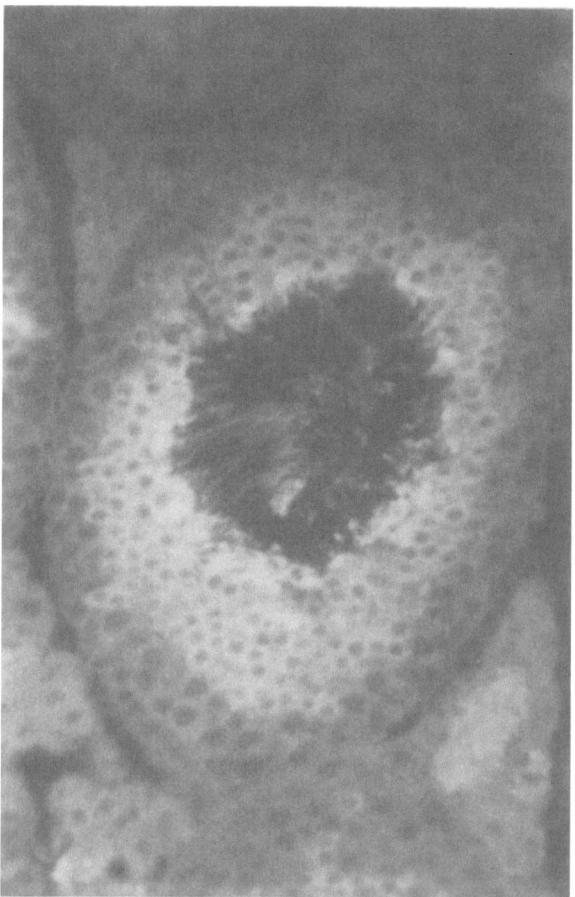

Figure 1. Localization of LDH-C_4 in germinal epithelium of adult
          mouse testis by immunofluorescence. The spermatogonia,
          Sertoli cells and Sertoli cell cytoplasm are devoid of
          specific fluorencence in this cross section.

is illustrated in the immature testis section (Fig. 2). The most
advanced stage of spermatogenesis seen in this section is the mid-
pachytene primary spermatocyte, the only cell type that contains LDH-
C_4. Synthesis of this isozyme continues, so that it becomes the most
abundant form of LDH in spermatozoa. The unique developmental and
biochemical properties of LDH-C_4 are of considerable significance to
testes and sperm metabolism (30).

Since LDH-C_4 is not present in prepubertal testes or in any
other tissue of the male and is completely absent from the female, it

is antigenic in both sexes of mice and other mammalian species. Rabbit antiserum to mouse LDH-C$_4$ reacts with this isozyme from all mammalian species, but does not react with isozymes composed of A and B subunits (4). This immunological specificity has been challenged by Wright and Swofford (32). However, their data contain internal inconsistencies that could be explained by non-specific antibody-protein interaction due to the high concentrations of serum in the dilutions tested. In fact, we have been unable to duplicate their results, but have confirmed antibody specificity by the inability to competitively inhibit anti LDH-C$_4$ binding by LDH-A or B in a radio-immunoassay (Liang et al., submitted).

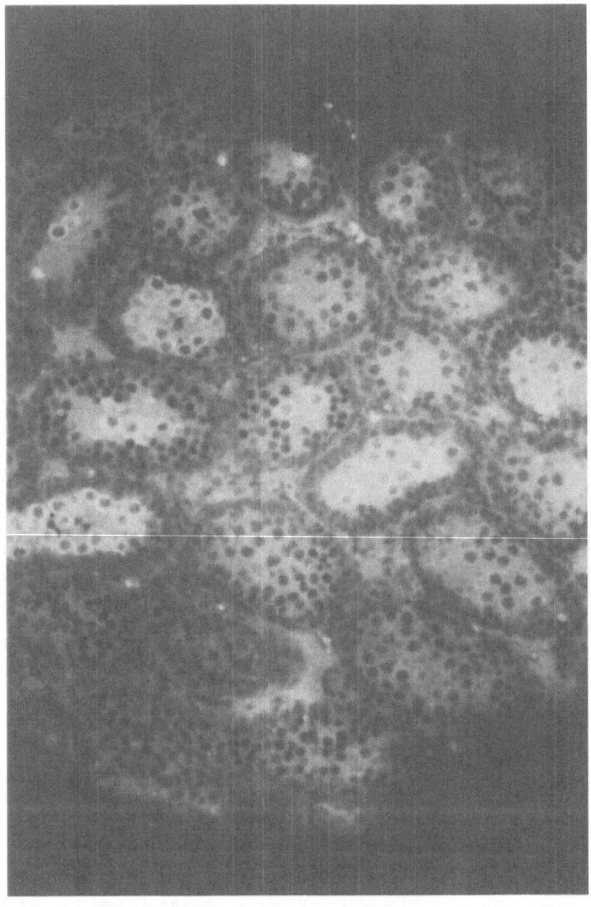

Figure 2. Localization of LDH-C$_4$ in testis tubules from an immature (21 day) mouse. The most prominent cell type with specific immunofluorescence is a primary spermatocyte in midpachytene.

The availability of relatively large quantities of mouse LDH-C_4 (15,31) facilitates rigorous analysis, not only of its biochemical and immunological properties but also of the effect of immunization with LDH-C_4 on fertility. However, use of LDH-C_4, or any natural product antigen, as the basis for a practical immunological contraceptive, presents problems of both supply and homogeneity. A synthetic substitute for the natural product antigen must be identified. Our strategy in developing such a vaccine, therefore, has been to define the antigenic determinants of LDH-C_4 (8). The procedure essentially involves chemical or enzymic digestion of LDH-C_4 and isolation of the fragments. Those peptides that bind anti-LDH-C_4 are chemically characterized. Appropriate peptides are synthesized and conjugated to carrier molecules for immunization. The anti-peptide antibodies are tested for binding to LDH-C_4. Finally, the immunocontraceptive properties of these synthetic determinants are assessed.

The primary structure of mouse LDH-C_4, illustrating tryptic peptides and side chain environment is presented in Figure 3. Those peptides that are immunologically active are underlined and can be considered to contain intact antigenic domains, or antibody combining sites. These fragments range in size from 8 to 35 residues. Parts of several immunologically active peptides are not available for antibody binding because they are buried in the subunit or obscured by an adjacent subunit. It is therefore useful to consider the side chain environment of each residue. First, however, it is important to note that this representation is a very simplistic view of antigenic domains and by no means necessarily reveals all or even the immunodominant epitopes of this subunit. For example, alternative stragies to fragment the molecule might allow identification of domains disrupted by trypsin. Nevertheless, useful information is available from the present model (Fig. 3).

There are 127 residues on the surface of the LDH-C subunit, and 71 of these amino acids are in immunologically active peptides. The tryptic peptides that are probably too small to bind antibody and therefore excluded by our assay include 30 surface residues. Two relatively long superficial sequences may contain antigenic domains that are destroyed by trypsin. The carboxyl terminal helix extending from residue 318 through 330 is cleaved at position 326. Superficial residues 274 and 276 are separated from the peptide containing 277, 278, 280 and 282. In this context it is interesting to note that peptides 98-104 and 105-110 do not bind antibody while synthetic 97-110 is active. From these data, we can estimate, conservatively, that 56% of the surface of LDH-C is antigenic. These antigenic domains have been described in detail (Wheat and Goldberg, Molec. Immunol.).

Four peptides that contain antigenic domains have been prepared synthetically. These are MC5-15, MC97-110, MC148-155 and MC211-220. The antigenicity of these domains was confirmed by their ability to

Primary Structure of Mouse LDH-C_4: Tryptic Peptides and Side Chain Environment

```
1                                  10                                        20
Ser-Thr-Val-Lys Glu-Gln-Leu-Ile-Gln-Asn-Leu-Val-Pro-Glu-Asp-Lys Leu-Ser-Arg Cys-Lys Ile-
 E   E   E   E   E   E   I       E   E       E   E   E       E   E   E   E   I   E   I
22                                 30                                    40
Thr-Val-Val-Gly-Val-Gly-Asp-Val-Gly-Met-Ala-Cys-Ala-Ile-Ser-Ile-Leu-Leu-Lys Gly-Leu-Ala-
 I   I   I   I   I   I   E   I   I   I   I   I   I       I   I   I   I   C   C   I   I
44                      50                                    60
Asp-Glu-Leu-Ala-Leu-Val-Asp-Ala-Asp-Thr-Asp-Lys Leu-Arg Gly-Glu-Ala-Leu-Asp-Leu-Leu-His-
 C   E   I       I   I       E   E   E   E   E   I   E   I       E           I   I   E
66                  70                                   80
Gly-Ser-Leu-Phe-Leu-Ser-Thr-Pro-Lys Ile-Val-Phe-Gly-Lys Asp-Tyr-Asn-Val-Ser-Ala-Asn-Ser-
 I   E   E           I   C   I   E   E   I   E       E   E   E       E   I   I   E   E   I
88      90                                   100
Lys Leu-Val-Ile-Ile-Thr-Ala-Gly-Ala-Arg Met-Val-Ser-Gly-Gln-Thr-Arg Leu-Asp-Leu-Leu-Gln-
 E   I   I   I   I   I   I   E       E   E   E   E   E   E   E   E   I   E   E   I
110                                    120                                         130
Arg Asn-Val-Ala-Ile-Met-Lys Ala-Ile-Val-Pro-Gly-Val-Ile-Gln-Asn-Ser-Pro-Asp-Cys-Lys Ile-
 E   E       E   E   I   E   E   I   I   E   E   I   I   E   E   I   E   E   E       I
132                              140                                     150
Ile-Val-Val-Thr-Asn-Pro-Val-Asp-Ile-Leu-Thr-Tyr-Val-Val-Trp-Lys Ile-Ser-Gly-Phe-Pro-Val-
 I   I   I   I   I   `I  I   I   I   I   I           I       E   .       E       E   E
154                    160                                     170
Gly-Arg Val-Ile-Gly-Ser-Gly-Cys-Asn-Leu-Asp-Ser-Ala-Arg Phe-Arg Tyr-Leu-Ile-Gly-Glu-Lys
 E   E   I   I   I   I   I   I   C   I       C   C   I   I   I       I   I       E   E
176             180                                    190
Leu-Gly-Val-Asn-Pro-Thr-Ser-Cys-His-Gly-Trp-Val-Leu-Gly-Glu-His-Gly-Asp-Ser-Ser-Val-Pro-
 I           C   C   C   C   I   I   I   I   I   I   I               E   E   I   I   I
198     200                                  210
Ile-Trp-Ser-Gly-Val-Asn-Val-Ala-Gly-Val-Thr-Leu-Lys Ser-Leu-Asn-Pro-Ala-Ile-Gly-Thr-Asp-
 I   I   I   I   I   I   I   I               I       E       E   E   E   I   E   E   E
220                                    230                                         240
Lys Asn-Lys Gln-His-Trp-Lys Asn-Val-His-Lys Gln-Val-Val-Glu-Gly-Gly-Tyr-Glu-Val-Leu-Asp-
 E   E   E   E   E   I   E   E   I       E   E   I   I   E   E   I       E   I   I   E
242                              250                                        260
Met-Lys Gly-Tyr-Thr-Ser-Trp-Ala-Ile-Gly-Leu-Ser-Val-Thr-Asp-Leu-Ala-Arg Ser-Ile-Leu-Lys
 I       E       I   I   I   I   I   I   I   C   I   I   C   I   I   C   I   I   I   C
264                  270                                      280
Asn-Leu-Lys Arg Val-His-Pro-Val-Thr-Thr-Leu-Val-Lys Gly-Phe-His-Gly-Ile-Lys Glu-Glu-Val-
     I   E   C   I   C   I   I   I       E   I   E   E   E       E       E   E   E   I
286             290                                    300
Phe-Leu-Ser-Ile-Pro-Cys-Val-Leu-Gly-Glu-Ser-Gly-Ile-Thr-Asp-Phe-Val-Lys Val-Asn-Met-Thr-
 I   I   I   I   I   I   I   I   E       E   I   I   E   E   I   I   E   I   E   I   E
308     310                                  320                                         330
Ala-Glu-Glu-Glu-Gly-Leu-Leu-Lys Lys Ser-Ala-Asp-Thr-Leu-Trp-Asn-Met-Gln-Lys Asn-Leu-Glu-Leu
 E   E   E   E   E   E   I   E   E   I   E   E   E   I   E   E   I   E   E   E   I   E   E
```

Figure 3. Amino acid sequence of mouse LDH-C. Side chain environment data from Eventoff et al. (3). Sequence from Pan et al. (27) and Wheat and Goldberg (Molec. Immunol., in press). The tryptic peptides that contain antigenic domains are underlined.

elicit an immune response to the native protein (8,10; Wheat and Goldberg, NY Acad. Sci., Ann., in press; Wheat et al., submitted). Each peptide, except MC148-155, was prepared with an N-terminal cysteine to effect conjugation to the carrier. These peptides were then conjugated to carrier proteins, either bovine serum albumin (BSA) or diphtheria toxoid (DT), with the heterobifunctional reagent 6-maleimido caproic acyl N-hydroxysuccinimide (16). Rabbits immunized with peptide-carrier conjugates all responded by synthesizing antibodies specific for native LDH-C_4 (Wheat et al., Molec. Immunol., in press) immunogenicity of these peptides therefore supports our strategy to produce antibodies to a sperm specific antigen with a synthetic epitope.

From a biological standpoint, the anti-peptide antibodies bind to both mouse and human spermatozoa (2) and also inhibit in vitro fertilization of mouse ova by mouse sperm. Furthermore, in an ongoing fertility trial, female baboons were immunized with DT-MC5-15. These animals all made an immune response to LDH-C_4 and, by comparison, there were significantly fewer pregnancies in immune females than in control females immunized with carrier only.

While these data certainly confirm the feasibility of using synthetic peptides as substitutes for native LDH-C_4 in an immunocontraceptive, the most effective antigenic peptide for such a vaccine remains to be determined. The many factors that contribute to the overall efficacy of the immune response to these peptides must be dissected. For example, the interactions between the carrier proteins and the peptides, which affect their structural conformation in solution, must be examined and exploited. Effective but benign adjuvants need to be identified. In addition, the sequence homologies of these epitopes in mouse and human LDH-C must be determined more precisely. For example, the region 97-110 differs by two amino acid residues in mouse and rat LDH-C. Binding of a monoclonal antibody that recognizes this sequence in the mouse is 50% less to the rat LDH-C, with the effect apparently due to the Thr → Ser substitution at position 103 (Goldman-Leikin and Goldberg, unpublished). Similarly, anti-MC 211-220 binds mouse but not rat LDH-C (Hogrefe and Goldberg, unpublished). The mouse and rat isozymes differ at position 218 by a Thr → Ser substitution (18).

In summary, the use of a well-defined, synthetic antigen is essential to progress in devloping a vaccine for fertility control and to establish with certainty the utility of such technology. The studies with LDH-C_4 are consistent with the feasibility of using a synthetic antigen in a vaccine to control fertility. LDH-C_4 remains the best-developed candidate for future studies. Immunization with LDH-C_4 does suppress fertility, and the complete biochemical characterization of this isozyme is well underway. These studies will provide a wealth of synthetic antigens to substitute for the natural product in experiments to determine whether immunologic contraception is appropriate and desirable for wide-scale application in human beings.

ACKNOWLEDGEMENT

Research from this laboratory was supported by a grant from the National Institutes of Health (NIH) and by contracts with the Northwestern University Program for Applied Research on Fertility Regulation (USAID/DPE-0546-A-00-1003-00).

REFERENCES

1. Almeida, M., and Voisin, G. 1979. Resistance of female guinea pig fertility to efficient iso-immunization with spermatozoa

autoantigens. J. of Reprod. Immunol. 1:237-247.

2. Beyler, S.A., Wheat, T.E., and Goldberg, E. 1985. Binding of antibodies against antigenic domains of murine lactate dehydrogenase-C_4 to human and mouse spermatozoa. Biol. Reprod. 32:1201-1210.

3. Eventoff, W., Rossmann, M.G., Taylor, S.S., Torff, H.J., Meyer, H., Keil, W., and Kiltz, H.H. 1977. Structural adaptations of lactate dehydrogenase isozymes. Proc. Natl. Acad. Sci. USA 74:2677-2681.

4. Goldberg, E. 1971. Immunochemical specificity of lactate dehydrogenase-X. Proc. Natl. Acad. Sci. 68:349-352.

5. Goldberg, E. 1973. Infertility in female rabbits immunized with lactate dehydrogenase X. Science 181:458-459.

6. Goldberg, E. 1983. Current status of research on sperm antigens: potential applications as contraceptive vaccines, in: "Research Frontiers in Fertility Regulation," G.I. Zatuchni, ed., Chicago, IL, Program for Applied Research on Fertility Regulation, Vol. 2(6), pp. 1-11.

7. Goldberg, E., Gonzales-Prevatt, V., and Wheat, T.E. 1981. Immuno-suppression of fertility by injection of sperm-specific LDH-C_4 (LDH-X): Prospects for development of a contraceptive vaccine, in: "Human Reproduction," Proc. 3rd World Congress, Semm, K., Mettler, L., eds, Amsterdam, Elsevier-North Holland, pp 360-364.

8. Goldberg, E., Wheat, T.E., and Gonzales-Prevatt, V. 1983. Development of a contraceptive vaccine based on synthetic antigenic determinants of lactate dehydrogenase C_4, in: "Reproductive Immunology," Gill, T. J. III, Wegmann, T. G., eds, New York, Oxford University Press, pp 492-504.

9. Goldman-Leikin, R., and Goldberg, E. 1983. Characterization of monoclonal antibodies to sperm-specific lactate dehydrogenase isozyme. Proc. Natl. Acad. Sci. USA 80:3774-3778.

10. Gonzales-Prevatt, V., Wheat, T.E., and Goldberg, E. 1982. Identification of an antigenic determinant of mouse lactate dehydrogenase C_4. Molec. Immunol. 19:1579-1585.

11. Isojima, S., Graham, R.M., and Graham, J.B. 1959. Sterility in female guinea pigs induced by injection with testis. Science 129:44.

12. Jones, W.R. 1982. Immunological fertility regulation, Blackwell Scientific Publications, Victoria, Australia.

13. Kille, J.W., and Goldberg, E. 1980. Inhibition of oviducal sperm transport in rabbits immunized against sperm-specific lactate dehydrogenase (LDH-C_4). J. Reprod. Immunol. 2:15-21.

14. Kummerfield, H.L., and Foote, R.H. 1976. Infertility and embryonic mortality in female rabbits immunized with different sperm preparations. Biol. Reprod. 14:300-305.

15. Lee, C-Y, Pegoraro, B., Topping, J.L., and Yuan, J.H. 1977. Purification and biochemical studies of lactate dehydrogenase X from mouse. Molec. Cell Biochem. 18:49-57.

16. Lee, A.C.J., Powell, J.E., Tregear, G.W., Niall, H.D., and Stevens, V.C. 1980. A method for preparing βHCG COOH peptide-carrier conjugates of predictable composition. Molec. Immunol. 17:749-756.

17. Lerum, J.E., and Goldberg, E. 1974. Immunological impairment of pregnancy in mice by lactate dehydrogenase-X. Biol. Reprod. 11:108-115.

18. Li, S.S.-L., Fitch, W.M., Pan, Y.-C.E., and Sharief, F.S. 1983. Evolutionary relationships of vertebrate lactate dehydrogenase isozymes, A_4 (muscle), B_4 (heart), and C_4 (testis). J. Biol. Chem. 258:7029-7032.

19. McLaren, A. 1964. Immunological control of fertility in female mice. Nature 201:582-585.

20. Menge, A.C., Peegel, H., and Riolo, M.L. 1979. Sperm fractions responsible for immunologic induction of pre- and post-fertilization infertility in rabbits. Biol. Reprod. 20:931-937.

21. Munoz, M.G., and Metz, C.B. 1978. Infertility in female rabbits isoimmunized with subcellular sperm fractions. Biol. Reprod. 18:669-678.

22. Naz, R.K., Saxe, J.M., and Menge, A.C. 1983. Inhibition of fertility in rabbits by monoclonal antibodies against sperm. Biol. Reprod. 28:249-254.

23. Naz, P., Alexander, N., and Isahakia, M. 1984. Monoclonal antibody to a human germ cell membrane glycoprotein that inhibits fertilization. Science 225:342-344.

24. O'Rand, M.G. 1981. Inhibition of fertility and sperm-zona binding by antiserum to the rabbit sperm membrane autoantigen RSA-1. Biol. Reprod. 25:621-628.

25. O'Rand, M.G. 1980. Antigens of spermatozoa and their environment, in: "Immunological Aspects of Infertility and Fertility Regulation," Dhindsa, D.S., Schumacher, G.F.B., eds, New York, Elsevier-North Holland, pp. 155-171.

26. O'Rand, M.G., and Porter, J.P. 1982. Purification of rabbit sperm autoantigens by preparative SDS gel electrophoresis: Amino acid and carbohydrate content of RSA-1. Biol. Reprod. 27:713-721.

27. Pan, Y.C.E., Sharief, F.S., Okabe, M., Huang, S., and Li, S. S.-L. 1983. Amino acid sequence studies on lactate dehydrogenase C_4 isozymes from mouse and rat testes. J. Biol. Chem. 258:7005-7016.

28. Saling, P.M., and O'Rand, M.G. 1982. Fertility inhibition in vitro and preliminary antigen identification. J. Androl. 3:434-439.

29. Tung, K.S.K., Goldberg, E.H., and Goldberg, E. 1979. Immunological consequences of immunization of female mice with homologous spermatozoa: Induction of infertility. J. Reprod. Immunol. 1:145-158.

30. Wheat, T.E., and Goldberg, E. 1983. Sperm-specific lactate dehydrogenase C_4: Antigenic structure and immunosuppression

of fertility, in: "Isozymes: Current Topics in Biological and
Medical Research," Rattazzi, M.C., Scandalios, J.G., Whitt,
G.S., eds, Vol. 7, pp. 113-130.

31. Wheat, T.E., and Goldberg, E. 1977 Isolation of the sperm-
specific lactate dehydrogenase from from mouse, rabbit and
human testes and human spermatozoa, in: "Immunological Influ-
ences on Human Fertility," Boettcher, B., ed., Australia
Harcourt Brace Jovanovich, pp. 221-227.

32. Wright, L.L., and Swofford, J.H. 1984. Mouse lactate dehydrogenase (LDH)C_4 (testis) is immunochemically cross-reactive
with LDH-A_4 (muscle) and LDH-B_4 (heart). Scand. J. Immunol.
19:247-254.

PATHWAYS TO IMMUNOCONTRACEPTION: BIOCHEMICAL AND IMMUNOLOGICAL

PROPERTIES OF GLYCOPROTEIN ANTIGENS OF THE PORCINE ZONA PELLUCIDA

Edward C. Yurewicz, Anthony G. Sacco, and
Marappa G. Subramanian

Department of Obstetrics and Gynecology
C.S. Mott Center for Human Growth and Development
Wayne State University School of Medicine
Detroit, Michigan

SUMMARY

The zona pellucida (ZP) is an extracellular coat which surrounds
the mammalian oocyte and whose macromolecular composition is unique
for each species. Two-dimensional gel electrophoresis of porcine ZP
under reducing conditions resolves four major acidic, charge-hetero-
geneous glycoproteins with apparent molecular weights of 82,000
(ZP1), 61,000 (ZP2), 55,000 (ZP3) and 21,000 (ZP4). Fractionation of
heat-solubilized porcine ZP under nondissociative, nonreducing conditions resulted in a preparation, termed purified pig zona antigen
(PPZA), substantially enriched in ZP3 and lacking ZP1 and ZP2. PPZA
was highly immunogenic in both rabbit and squirrel monkey. Anti-PPZA
sera exhibited contraceptive potential as assessed by in vitro
inhibition of homologous sperm-zona binding following antibody treatment of human, pig and squirrel monkey oocytes. Alternatively,
fractionation of sodium dodecyl sulfate (SDS)-dissociated porcine ZP
under nonreducing conditions resulted in isolation of two electrophoretically homogeneous antigens with apparent molecular weights of
82,000 (82K antigen) and 55,000 (55K antigen) as judged by non-
reducing SDS-polyacrylamide gel electrophoresis. The 82K antigen
exhibited an estimated molecular weight of 345,000 by gel filtration
chromatography in nondissociative solvent and was comprised of ZP1
and disulfide-linked heterodimers of ZP2+ZP4 and ZP3+ZP4. The 55K
antigen was comprised of ZP3 which was highly aggregated in nondissociative solvent. Each antigen had a characteristic carbohydrate
composition and proved to be immunogenic in the rabbit. The effects
of active heteroimmunization of squirrel monkeys with 55K antigen on
fertility are under investigation.

INTRODUCTION

Methods for safe and effective contraception are crucial for control of human fertility. Research aimed at development of new contraceptive methodologies with improved safety, efficacy and duration of effect is therefore of fundamental significance to the health and welfare of the world's reproductive age population. One approach under study is immunocontraception, which is defined as an inhibition of fertility by immunological methods. Major prerequisites for an immunologic method of fertility regulation would include:

1. Identification of a physiologically active antigen specific to the reproductive system; 2. Antigen must be tissue-specific but not species-specific; 3. Antigen must be highly immunogenic; 4. Antigen must be available in sufficient quantities in a homo-geneous state or be capable of being produced synthetically; 5. Antibodies produced against the antigen must be capable of reaching the antigen in situ; and 6. Antibodies must inhibit the reproductive function of the antigen.

Several gamete, embryonic and hormonal antigens have been identified as potential antifertility vaccine candidates (1). Especially promising are antigens localized within the zona pellucida (ZP) of the mammalian oocyte (4,21). The ZP of several species have been demonstrated to possess tissue-specific antigens. Both in vitro and in vivo studies in several species have demonstrated that antibodies to ZP can block fertility, presumably by interfering with sperm receptor sites on the surface of the ZP or by preventing sperm pene-tration through the ZP (7,10,11,26,28). In theory, therefore, active heteroimmunization of humans with a ZP macromolecule isolated from another species and possessing shared antigenic determinants with the human ZP should result in an immunologically mediated reduction in fertility.

A major requirement for immunocontraception in humans will be the use of a homogenous and chemically defined immunogen, i.e. a pur-ified and physicochemically characterized ZP macromolecule or synthetic derivative thereof. Within this context, several laboratories have focused attention on the porcine ZP as the most promising source of such an antigenic entity. Earlier studies have shown that porcine ovaries and ZP contain highly immunogenic tissue-specific antigens and that these antigens elicit antibodies which cross-react with ZP of other species, including those of human oocytes (19,20,22). The past several years have seen an impressive growth in our knowledge of the composition and structure of the porcine ZP and biochemical and immunochemical properties of individual porcine ZP glycoproteins. This presentation will highlight recent advances demarcating the pathway to immunocontraception with porcine ZP, including:

1. Development of methods for large scale isolation of porcine ZP;
2. Elucidation of the macromolecular composition of porcine ZP;
3. Development of purification protocols for isolation of porcine ZP antigens; and
4. Analysis of the macromolecular specificity and in vitro contraceptive potential of antibodies to purified porcine ZP antigens.

ISOLATION OF PORCINE ZONAE PELLUCIDAE

A prime advantage for the study of porcine ZP antigens is the relative abundance in which porcine ovaries can be obtained. A single worker from our laboratory can collect 1000-1500 pig ovaries in a time span of approximately 2 h at a local slaughterhouse. Once collected, the ovaries are returned to the laboratory and stored frozen unless used immediately. Depending on numbers required, two methods are used for collection of zona-encased oocytes. Sperm binding studies and radioiodination for radioimmunoassay (RIA) and electrophoretic studies require collection of limited numbers of oocytes. For such experiments, zona-encased oocytes are retrieved manually from either follicular aspirates or minced ovaries with the aid of a micropipette and a stereomicroscope (23). For isolation of ZP, the oocytes are forced through a restricted bore micropipette and the fractured ZP are collected, washed and solubilized by heating (70°C, 20 min) in Tris-EDTA buffer (50 mM Tris, pH 8.0, 150 mM NaCl, 4 mM EDTA, 0.02% sodium azide).

Isolation of large numbers of zona-encased porcine oocytes, as required for purification of ZP antigens, is readily accomplished by sieving techniques. The use of either stainless steel mesh filters (18) or nylon mesh screens (6,28) for isolation of zona-encased porcine oocytes from ovary homogenates has been described in detail. In our laboratory, ovaries are homogenized by passage through a meat grinder, and the oocytes collected in Tris-zona buffer (50 mM Tris, pH 8.0, 150 mM NaCl, 4 mM EDTA, 2 mM sodium citrate, 0.02% sodium azide) after sieving through nylon screens with meshes of 500, 350, 200 and 175 μm. For isolation of ZP, oocytes are disrupted with a teflon-glass homogenizer and the ZP ghosts collected on a 40 μm screen. After washing in Tris-zona buffer, the ZP are resuspended into 20 mM Tris buffer, pH 8.0, and solubilized by heating for 20 min at 73°C. Typically, one person can process 1000 ovaries in 5-7 h with a yield of 15-20 mg protein as heat-solubilized porcine ZP.

Porcine ZP isolated by sieving techniques contain only minor amounts of nonzonae contaminants. Dunbar et al. (6) reported greater than 93% purity as assessed by chemical, enzymatic and microscopic analyses. The use of collagenase during the isolation procedure, as described by Gwatkin et al. (9), is neither necessary nor desirable

and may result in protelysis of one or more of the porcine ZP macromolecules.

MACROMOLECULAR COMPOSITION OF THE PORCINE ZONA PELLUCIDA

The macromolecular composition of porcine ZP dissociated by
exposure to the detergent sodium dodecyl sulfate (SDS) has been
evaluated by both one-dimensional (14) and two-dimensional (17) gel
electrophoresis. Analogous to the murine ZP (2), the electrophoretic
profiles observed with porcine ZP were dependent upon whether disulfide bonds were disrupted (reduced) or were maintained intact (nonreduced). As previously described (29), SDS-polyacrylamide gel
electrophoresis (PAGE) of nonreduced porcine ZP yielded two major,
characteristically diffuse bands with apparent molecular weights (Mr)
of 82,000 and 55,000 as visualized with Coomassie blue stain (Fig. 1,
lane c). These will be subsequently referred to as 82K antigen and
55K antigen and represent structural entities which may be isolated
from porcine ZP under conditions where integrity of disulfide bonds
is preserved (30).

SDS-PAGE of mercaptoethanol-treated porcine ZP (Fig. 1, lane b)
yielded a major diffuse band at 55,000 Mr (6,29). A discrete band at
82,000 Mr was no longer observed and the diffuseness of the 55,000 Mr
band was enhanced (Mr range = 40,000-70,000). On the basis of SDS-
PAGE studies, it may be concluded that the component(s) migrating at
82,000 Mr in nonreduced porcine ZP apparently migrate within the
55,000 Mr band following reduction of disulfide bonds.

A more detailed evaluation of the macromolecular composition of
the porcine ZP was obtained by high resolution two-dimensional gel
electrophoresis (2-D GE) technique of O'Farrell (17). Isoelectric
focusing in the first dimension coupled with SDS-PAGE in the second
dimension yielded separation by both charge and size. 2-D GE of
mercaptan-treated porcine ZP, either radioinert or ^{125}I-labeled,
resolved three families of proteins as detected by Coomassie blue
staining or autoradiography, respectively (5,27). Adopting the
nomenclature of Subramanian et al. (27), these are designated ZP1,
ZP2 and ZP3 for families with Mr values of 82,000, 61,000 and 55,000,
respectively (Fig. 2, plate A). Each family was composed of multiple
charge isomers. Within each family, the progressively increasing
values of Mr of individual charge isomers toward the acidic end of
the gel yielded diagonal patterns which are characteristic of charge
heterogeneous glycoproteins and account for the presence of only a
single diffuse band following one-dimensional SDS-PAGE of reduced
porcine ZP (Fig. 1, lane b).

Two points of caution must be made regarding nomenclature of
porcine ZP proteins. First, it must be emphasized that ZP1-ZP3 refer
to protein families resolved by 2-D GE of porcine ZP under <u>reducing</u>

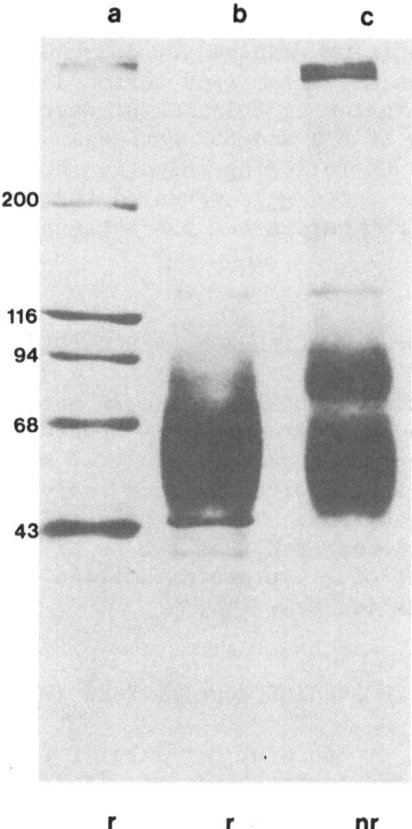

Figure 1. SDS-PAGE of reduced (r) versus nonreduced (nr) heat-solubi-
lized porcine ZP. Gels were stained with Coomassie blue.
Lane a) Protein standards. Lane b) Reduced porcine ZP.
Lane C) Nonreduced porcine ZP. (Modified from 29).

conditions. It cannot be assumed that ZP1 (Mr = 82,000) and ZP3 (Mr
= 55,000) are equivalent to the 82K and 55K antigens resolved by SDS-
PAGE of porcine ZP under non-reducing conditions. As will be discussed later, the 82K antigen and ZP1 are in fact separate and distinct entities although the 55K antigen and ZP3 are indistinguishable. Second, the Mr values reported here for ZP1 and ZP2 differ
from those reported by other laboratories (5,12). The differences
are due to variations in electrophoresis conditions. The porcine ZP
macromolecules are glycoproteins and estimation of their apparent
molecular weights relative to protein standards will be influenced by
experimental conditions, such as use of uniform versus gradient
gels.

Further experiments were undertaken to define the relationships of the 82K and 55K antigens resolved by SDS-PAGE on nonreduced porcine ZP with the three major families (ZP1-ZP3) resolved by 2-D GE of reduced porcine ZP. Following SDS-PAGE of nonreduced ^{125}I-labeled porcine ZP, the bands of 82K and 55K antigens were eluted and subjected separately to 2-D GE following reduction of disulfide bonds. Autoradiography of the dried gels revealed that the 82K antigen was composed of ZP1 and ZP2 whereas the 55K antigen yielded only ZP3 (29).

COMPARATIVE ASPECTS OF MAMMALIAN ZONA PELLUCIDA STRUCTURE

In a comparative study of mammalian ZP structure, the macromolecular composition of ZP from several species was evaluated by 2-D GE. ZP from porcine, rabbit, mouse, squirrel monkey and human oocytes were radioiodinated with ^{125}I and analyzed by 2-D GE under standardized conditions (26). The 2-D gels (Fig. 2, plates A,C,D,E,F) clearly demonstrated that the ZP of each species was comprised of a unique set of 2-4 protein families, each comprised of multiple acidic charge isomers.

ISOLATION OF A PURIFIED PORCINE ZONA ANTIGEN (PPZA)

A major research emphasis of our laboratories has been the development of purification protocols for isolation of porcine ZP macromolecules for subsequent biochemical, immunochemical and immunocontraceptive studies. When these studies were initiated, several important questions remained unanswered, including:

1. Was purification of individual families (ZP1-ZP3) from heat solubilized porcine ZP technically feasible, or were strongly associated supramolecular complexes present which would make such attempts futile?

2. Were antibodies elicited against porcine ZP equally reactive with all three major families, or would one family prove to be immunodominant?

The experimental approach to the first question was to subject heat-solubilized porcine ZP to classical column chromatographic procedures in nondissociative buffers in an attempt to isolate one or more of the porcine ZP families in purified form. In response to the second question, a RIA was developed to quantitate zona antigen activity (ZAA) associated with porcine ZP.

RIA for zona antigen activity   Details of a competitive binding RIA for quantitation of ZAA have been previously reported (19). Briefly, a rabbit antiserum was raised against intact manually-

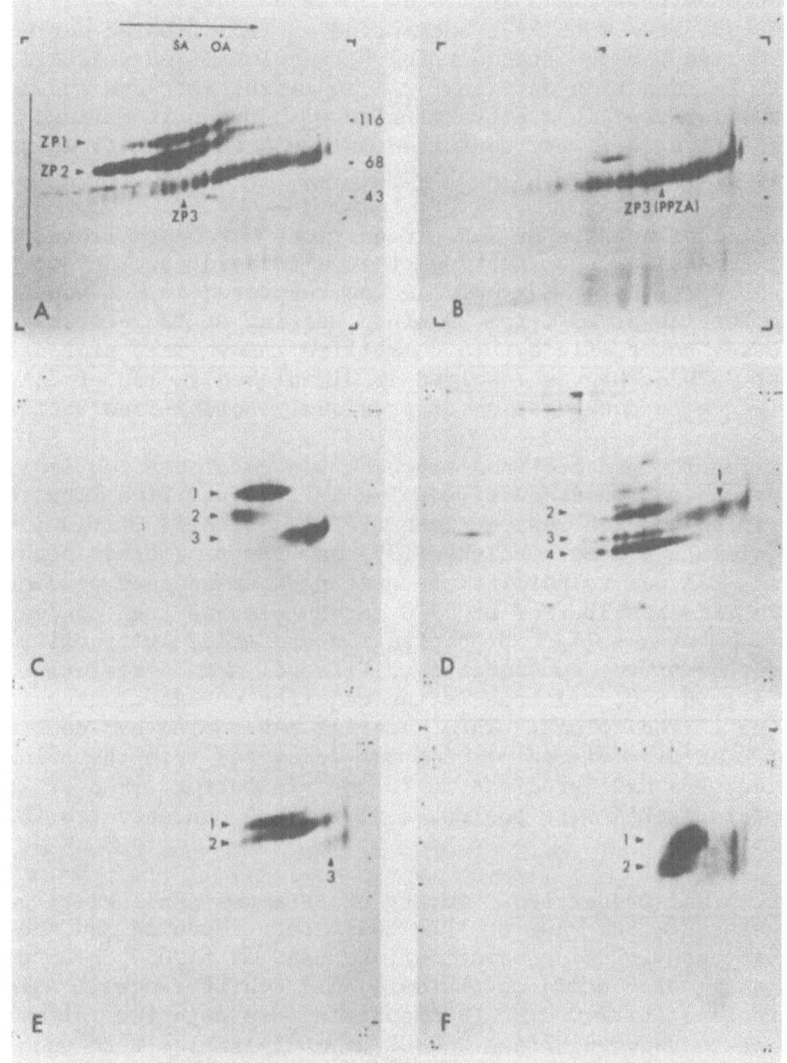

Figure 2. Macromolecular compositions of mammalian ZP. Heat-
        solubilized radioiodinated ZP from (A) porcine, (C) human,
        (D) squirrel monkey, (E) rabbit and (F) mouse oocytes and
        (B) radioiodinated PPZA were analyzed by two-dimensional
        gel electrophoresis. Isoelectric focusing is represented
        in the horizontal dimension (acidic end to the right) and
        SDS-PAGE is represented in the vertical dimension (migra-
        tion from top to bottom). Proteins were visualized by
        autoradiography (from 26).

isolated porcine ZP and subsequently rendered specific to porcine ZP
by absorption with follicular fluid. Heat-solubilized porcine ZP
were radioiodinated with ^{125}I and served as radiolabeled antigen.
Standard curves were generated using heat-solubilized unlabeled
porcine ZP as competing antigen. ZAA in tissue extracts and column
eluates was reported in arbitrary units with one unit defined as the
amount of antigen activity contained in one heat-solubilized porcine
ZP capable of reacting with antibody.

Heat solubilization of ZAA   Conditions were established for
maximal yield of ZAA upon heating of zona-encased porcine oocytes in
Tris-zona buffer (29). Maximal ZAA was recovered in the supernate
following heating at 73°C for 20 min. Heating at temperatures less
than 60°C was not sufficient to solubilize ZAA whereas prolonged
heating at 75°C or higher resulted in diminished yields of ZAA,
presumably due to denaturation of previously solubilized antigen.

Purification protocol   A protocol was developed for isolation
of 55K antigen from nonreduced porcine ZP in a purified form referred
to as purified porcine zona antigen (PPZA). Details of the purification protocol have been published (29) and are presented in summarized form. ZAA was solubilized by heating zona-encased porcine
oocytes in Tris-zona buffer at 73°C for 20 min and then concentrated by ammonium sulfate precipitation and Amicon ultrafiltration.
Gel filtration of the concentrate on Ultrogel AcA 34 resin in 20 mM
Tris buffer, pH 8.0, resulted in the majority of ZAA eluting in the
void volume of the column. This material was pooled and applied to a
column of DEAE Bio-Gel resin. ZAA was recovered from the column by
application of a NaCl gradient in 20 mM Tris buffer, pH 8.0. Fractions containing ZAA were pooled, dialyzed and concentrated to yield
PPZA.

Biochemical properties   Purity of PPZA was assessed by gel
electrophoresis. SDS-PAGE of PPZA under both reducing and nonreducing conditions yielded a major diffuse band at 55,000 Mr as detected
with Coomassie blue stain or periodic acid Schiff reagent. Significantly, no band at 82,000 Mr (82K antigen) was detected following
SDS-PAGE of nonreduced PPZA. 2-D GE of ^{125}I-labeled PPZA under
reducing conditions yielded a family of charge isomers at 55,000 Mr
corresponding to ZP3 (Fig. 2, plate B). Minor contaminants were
visible as discrete spots and areas of smearing. More importantly,
2-D GE failed to detect any material migrating in the positions of
ZP1 and ZP2. Electrophoretic analyses thus established that PPZA,
although not homogeneous, was composed primarily of a 55,000 Mr
glycoprotein (ZP3) and was free of detectable ZP1 and ZP2.

Further experiments confirmed that the ZAA of PPZA was associated with the 55,000 Mr glycoprotein. Nonreduced PPZA was subjected
to SDS-PAGE after which the gel was cut into 2 mm slices, each of
which was subsequently eluted. RIA analysis of the gel slice eluates

yielded a single peak of ZAA coincident with the 55,000 Mr band
visualized with Coomassie blue stain.

Amino acid and saccharide analyses indicated the presence of
approximately 25% carbohydrate by weight of PPZA. Fucose, mannose,
galactose, N-acetylglucosamine, N-acetylgalactosamine and sialic acid
were present in a molar ratio of 1.5 : 3.0 : 9.6 : 11.9 : 1.9 : 1.8,
respectively. No unusual amino acids were detected.

ANTIBODIES TO PPZA

The immunologic properties of PPZA were probed with rabbit and
squirrel monkey anti-PPZA sera. Rabbit anti-PPZA serum was produced
following immunization of a male rabbit with 1 mg of PPZA via
multiple intradermal injections (27). Squirrel monkey anti-PPZA sera
were elicited in a group of six animals following multiple intradermal injections of from 0.25 mg to 1 mg of PPZA (25). The specificity and in vitro immunocontraceptive potential of the resultant
antisera were studied.

Specificity of rabbit anti-PPZA serum   The macromolecular
specificity of the rabbit anti-PPZA serum was evaluated. Nonreduced
porcine ZP were subjected to SDS-PAGE after which the gel was cut
into 2 mm slices and eluted. Gel slice eluates were analyzed for
ability to compete with ^{125}I-labeled PPZA for binding with rabbit
anti-PPZA antibodies. A single peak of antigen activity was detected
at 55,000 Mr. The absence of detectable antigen activity co-
migrating with the 82K antigen (Mr = 82,000) suggested that the
rabbit anti-PPZA serum was monospecific for the 55,000 Mr glycoprotein (ZP3) of porcine ZP.

The species specificity of the rabbit anti-PPZA serum was evaluated by a direct binding RIA. Binding of ^{125}I-labeled heat-
solubilized ZP of homologous and heterologous species to rabbit anti-
PPZA serum was quantitated (26). As might be anticipated, the greatest percentage binding was obtained with the homologous antigen, i.e.
^{125}I-labeled heat-solubilized porcine ZP (Fig. 3). Intermediate
levels of binding were observed with radioiodinated heat-solubilized
ZP of rabbit, squirrel monkey and human oocytes. No binding could be
detected using ^{125}I-labeled heat-solubilized mouse ZP. It was concluded that rabbit antiserum to PPZA recognized shared antigenic
determinants localized within the ZP of pig, rabbit, squirrel monkey
and human oocytes. In contrast, PPZA and mouse ZP appear to be
immunologically unrelated.

Immunocontraceptive potential of rabbit anti-PPZA serum   In
order to assess immunocontraceptive potential, the effects of rabbit
anti-PPZA serum on homologous sperm-egg interactions were examined in
vitro (26). Incubation of zona-encased mouse, rabbit, pig, squirrel

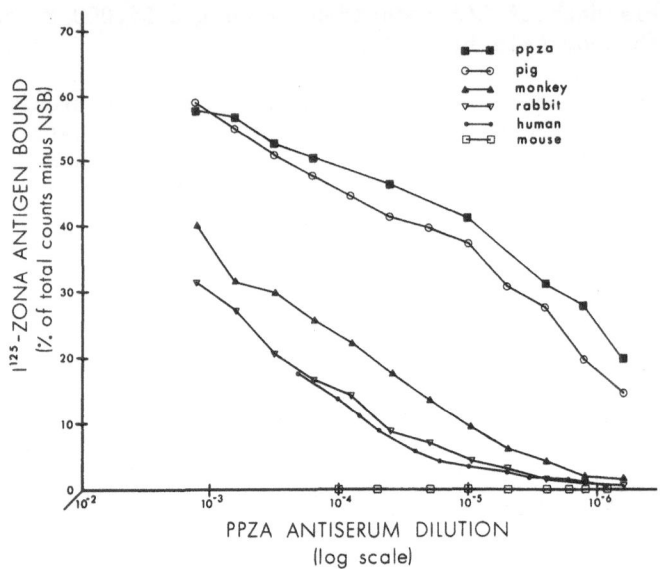

Figure 3. Species specificity of rabbit anti-PPZA serum. The binding
of radioiodinated PPZA and radioiodinated heat-solubilized
ZP of different species to anti-PPZA serum is presented
(From 26).

monkey and human oocytes with rabbit anti-PPZA serum resulted in each
instance in an altered appearance of the outer surface of the ZP,
i.e. the formation of a "precipitation" layer resulting from the
interaction of antibody and antigen. Subsequent addition of homologous sperm preparations to antibody-treated pig, squirrel monkey and
human oocytes resulted in an inhibition of sperm-ZP binding relative
to that observed with oocytes pretreated with control serum. In
contrast, adherence of homologous sperm was not reduced with zona-
encased mouse or rabbit oocytes pretreated with anti-PPZA serum.
Such results suggested that rodents would not be suitable animal
models for studies of the in vivo immunocontraceptive efficacy of
PPZA. In fact, it has been reported that fertility in the mouse
(24), as well as the rat (3), is not reduced following active heteroimmunization with porcine ZP. Rather, the experiments presented
strongly supported use of the squirrel monkey as a promising animal
model for immunocontraceptive studies using purified porcine ZP antigens.

     Squirrel monkey anti-PPZA sera   Active heteroimmunization of
squirrel monkeys (Saimiri sciureus) with PPZA resulted in production
and maintenance of antisera with high antibody titers (25). Equivalent antibody titers were obtained with immunizing doses of 0.25, 0.5

and 1.0 mg PPZA indicating a high degree of immunogenicity. The resultant anti-PPZA sera were tested for ability to inhibit homologous sperm-zona attachment in vitro using pig, squirrel monkey and human gametes. For all three species, preincubation of zona-encased oocytes with squirrel monkey anti-PPZA serum, but not preimmune control serum, resulted in formation of a precipitation layer on the outer surface of the ZP (Fig. 4, plates A,C,E) and a corresponding inhibition of adherence by homologous sperm (Fig. 4, plates B,D,F). These experiments were the first to demonstrate the contraceptive effects of antibodies elicited in a nonhuman primate via active heteroimmunization with a purified macromolecule of porcine ZP.

Immunochemical properties of PPZA   Rabbit anti-PPZA serum was used to probe structural features of antigenic determinants of PPZA. The consequences of various chemical and enzymatic treatments on the antigenicity of PPZA were evaluated by a competitive binding assay utilizing rabbit anti-PPZA serum and ^{125}I-labeled PPZA (29). Digestions with proteases suggested a major contribution of the protein moiety to antigenicity. Overnight incubation of PPZA with either pronase, trypsin or elastase resulted in loss of 50% to 70% of original antigen activity concomitant with degradation of the 55,000 Mr glycoprotein. The residual antigen activity suggested that carbohydrate structures or protease-resistant glycopeptide domains also contributed to antigenicity. Losses of 40% to 80% of antigen activity were observed following exposure of nonreduced PPZA to strongly dissociative agents such as 5 M urea, 6 M guanidine hydrochloride or 1% SDS. In contrast, loss of antigen activity was quantitative following exposure of PPZA to denaturing agents in the presence of reagents capable of cleaving disulfide bonds. It was concluded that intramolecular disulfide bonds are important in stabilization of antigenic determinants of PPZA.

ISOLATION AND PROPERTIES OF 82K AND 55K ANTIGENS FROM SDS-DISSOCIATED NONREDUCED PORCINE ZONAE PELLUCIDAE.

Several major conclusions were derived from our studies of PPZA, including:

1. The principal macromolecular and antigenic component of PPZA is the 55,000 Mr glycoprotein (ZP3) of porcine ZP.
2. Rabbit and squirrel monkey antibodies to PPZA cross-react with ZP of pig, rabbit, squirrel monkey and human oocytes.
3. Rabbit and squirrel monkey antibodies to PPZA block homologous sperm-zona binding in vitro with pig, squirrel monkey and human gametes.
4. The 55,000 Mr glycoprotein of porcine ZP is identified as a target antigen for a potential antifertility vaccine in the human female.

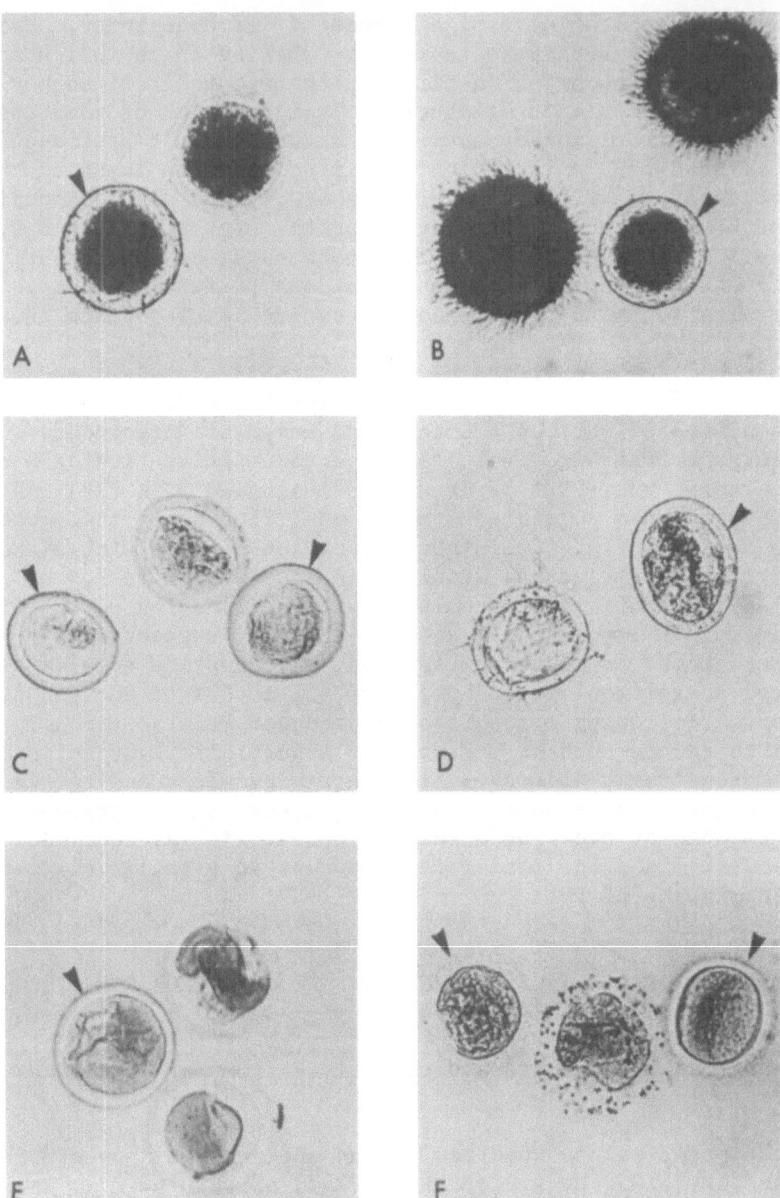

Figure 4. Inhibition of sperm-zona binding by squirrel monkey anti-
PPZA serum. Oocytes obtained from (A,B) porcine, (C,D)
squirrel monkey or (E,F) human ovaries were pretreated with
either squirrel monkey anti-PPZA serum (arrows) or preimmune control serum (unlabeled). Plates A,C,E, demonstrate the formation of a precipitation layer on the
surface of the ZP of oocytes pretreated with anti-PPZA

serum but not on the ZP of oocytes pretreated with preimmune control serum. Plates B,D,F, demonstrate the inhibition of adherence of homologous sperm to the ZP of
oocytes pretreated with squirrel monkey anti-PPZA serum
versus the oocytes pretreated with preimmune control serum.
Only nonliving human oocytes obtained from frozen ovaries
recovered from cadavers were used for these experiments
(From 25).

5. The squirrel monkey is identified as an appropriate nonhuman
   primate animal model for evaluation of the immunocontraceptive potential of purified porcine ZP antigens.

However, the inability to obtain the 55,000 Mr glycoprotein in
homogeneous form was considered inconsistent with the use of PPZA for
continued biochemical and immunocontraceptive studies. This prompted
attempts to devise an alternative purification protocol for isolation
of the 55,000 Mr glycoprotein from nonreduced porcine ZP. As a
result, a purification scheme was developed which yielded immunologically active, electrophoretically homogeneous preparations of the 55K
antigen, and also 82K antigen, of porcine ZP.

Purification protocol   The purification scheme, outlined in
Fig. 5, utilized column chromatographic procedures for isolation of
82K and 55K antigens from SDS-dissociated nonreduced porcine ZP.
Zona-encased porcine oocytes were recovered from homogenates of pig
ovaries by screening techniques as described earlier. The oocytes
were homogenized and the fractured ZP collected and solubilized by
heating (73°C) in 20 mM Tris buffer, pH 8.0. The solubilized zona
proteins were concentrated by ultrafiltration and then treated with
SDS and heat under nonreducing conditions to disrupt noncovalent
intermolecular interactions. Gel filtration chromatography of the
SDS-dissociated porcine ZP on Sephacryl S-400 resin in buffer containing 0.1% SDS resulted in separation of 82K and 55K antigens.
Further purification of each was achieved by chromatography on hydroxylapatite resin in SDS-containing buffer. Elution was achieved by
application of linear salt gradients. Fractions containing homogeneous 82K or 55K antigen, as monitored by SDS-PAGE, were pooled,
dialyzed and lyophilized. Following removal of residual SDS by ion
pair extraction (13), the purified antigens were dialyzed from a
solution of 6 M guanidine hydrochloride under conditions promoting
renaturation (16). Typically, 0.5-1 mg of 82K antigen and 4-7 mg of
55K antigen could be recovered from 4000 pig ovaries.

Electrophoretic analyses   The macromolecular composition of the
purified 82K and 55K antigens was studied by SDS-PAGE (Fig. 6 and 7)
and 2-D GE (Yurewicz et al., manuscript in preparation). The electrophoretic behavior of the 82K antigen was radically affected by
exposure to reagents capable of disrupting disulfide bonds. In the

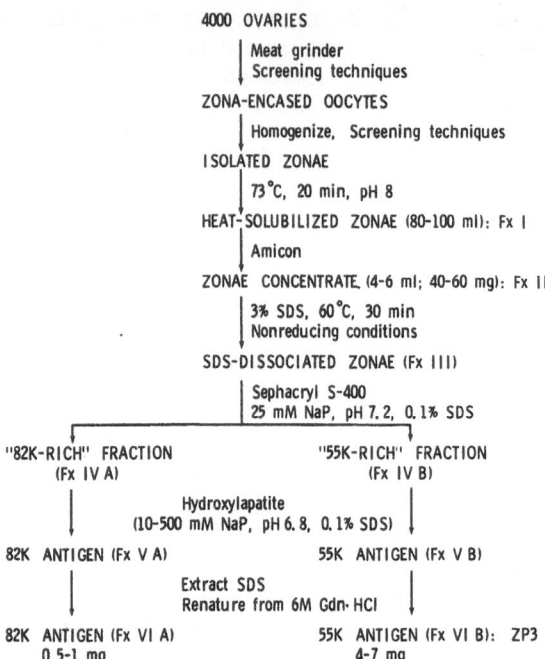

Figure 5. Scheme for the purification of 82K and 55K antigens from
nonreduced SDS-dissociated porcine ZP.

absence of mercaptans, SDS-PAGE yielded a single 82,000 Mr band (Fig.
6, lane c) and 2-D GE resolved a single family comprised of greater
than 25 acidic charge isomers. To avoid confusion, it must be emphasized that the number of charge isomers, their pI range (4.5-6) and
spatial distribution were clearly distinct from those of ZP1, the
82,000 Mr family resolved by 2-D GE of reduced porcine ZP. In
contrast, SDS-PAGE (Fig. 6, lane g) and 2-D GE of 82K antigen preincubated with dithiothreitol resolved bands at 82,000 Mr, 61,000 Mr
and 55,000 Mr - identified as ZP1, ZP2 and ZP3, respectively - and a
previously unreported porcine ZP macromolecule of 21,000 Mr, termed
ZP4. Collectively, the electrophoretic analyses suggested that 82K
antigen isolated from nonreduced SDS-dissociated porcine ZP is
comprised of a complex of ZP1 and disulfide-linked heterodimers of
ZP2+ZP4 and ZP3+ZP4 (Table 1).

Purified 55K antigen yielded a single characteristically diffuse
55,000 Mr band when examined by SDS-PAGE under both reducing and
nonreducing conditions (Fig. 7, lanes c and g). Likewise, a single
family of greater than 20 charge isomers, clearly identified as ZP3,
was observed following 2-D GE of either reduced or nonreduced

antigen. On the basis of the data presented, it was concluded that
55K antigen is comprised exclusively of ZP3. In contrast to 82K
antigen, no evidence for the presence of intermolecular disulfide
bonding was obtained (Table 1).

    Gel filtration studies The elution behavior of purified 82K
and 55K antigens on a calibrated column of TSK G4000SW resin was
examined. A 50 mM sodium phosphate buffer, pH 6.8, containing 300 mM
sodium chloride was employed as mobile phase. The 82K antigen eluted
as a single broadened peak with a calculated apparent molecular
weight of 345,000. This result suggests that purified 82K antigen
may exist in nondissociating solvent as a stoichiometric complex of
noncovalently associated ZP1 monomers and ZP2+ZP4 heterodimers. In
contrast, the majority of the purified 55K antigen applied to the
column was eluted in the void volume (exclusion limit = 10^6 daltons).
A minor peak of 185,000 apparent molecular weight was also detected
and appeared to exist in apparent equilibrium with the material eluting in the void volume. The data suggest that the purified 55K antigen possesses strong self-associative properties resulting in a
propensity for formation of high molecular weight aggregates. Within
this context, it was noted during isolation of PPZA that the zona
antigen activity (i.e. 55,000 Mr glycoprotein) of heat-solubilized
porcine ZP was excluded from columns of Ultrogel AcA 34 resin (exclusion limit = 400,000 daltons). The self-associative properties of
55K antigen suggest a major role of this glycoprotein in formation
and maintenance of porcine ZP ultrastructure.

Table 1. Proposed model: structure and composition of the porcine
          zona pellucida

Intact porcine ZP	SDS–PAGE (nonreduced)	2–D GE (reduced)
ZP1		ZP1 (7%, 82,000 Mr)
ZP2+ZP4 dimer	82K antigen (30%)[a]	ZP2 (20%, 61,000 Mr)
ZP3+ZP4 dimer		ZP3 (1%, 55,000 Mr)
		ZP4 (2%, 21,000 Mr)
ZP3	55K antigen (70%)	ZP3 (70%, 55,000 Mr)

[a]Values in parentheses are approximate percentages of total porcine
ZP protein as estimated by Coomassie blue staining.

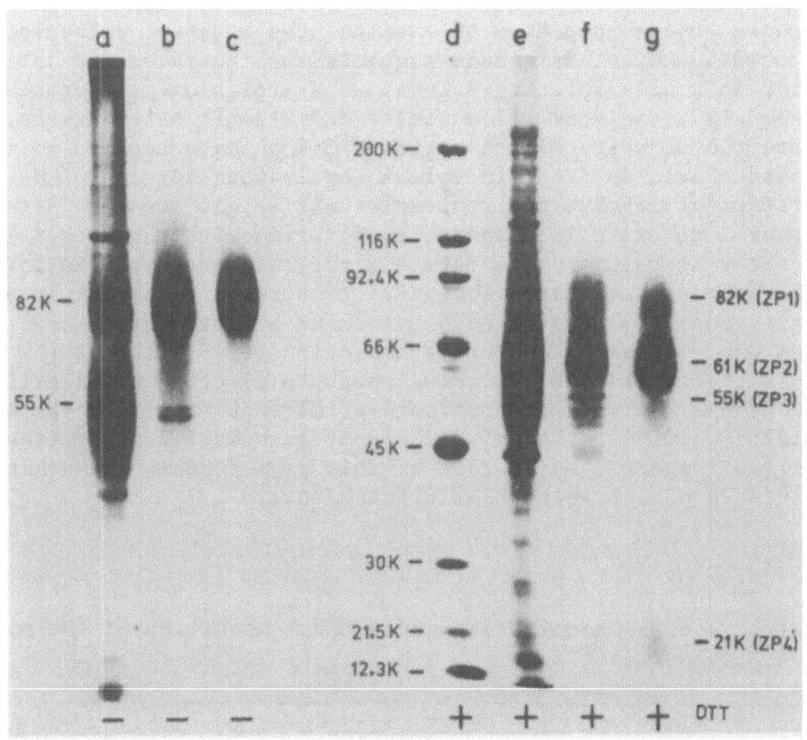

Figure 6. SDS–PAGE analysis of 82K antigen purification. Gels were
stained with Coomassie blue. Lanes a,e) Heat-solubilized
porcine ZP (Fx II). Lanes b,f) Sephacryl S-400 (Fx IV A).
Lanes c,g) Hydroxylapatite (Fx VI A). Lane d) Molecular
weight standards. Samples were pretreated with sample
buffer in the presence and absence of dithiothreitol as
indicated.

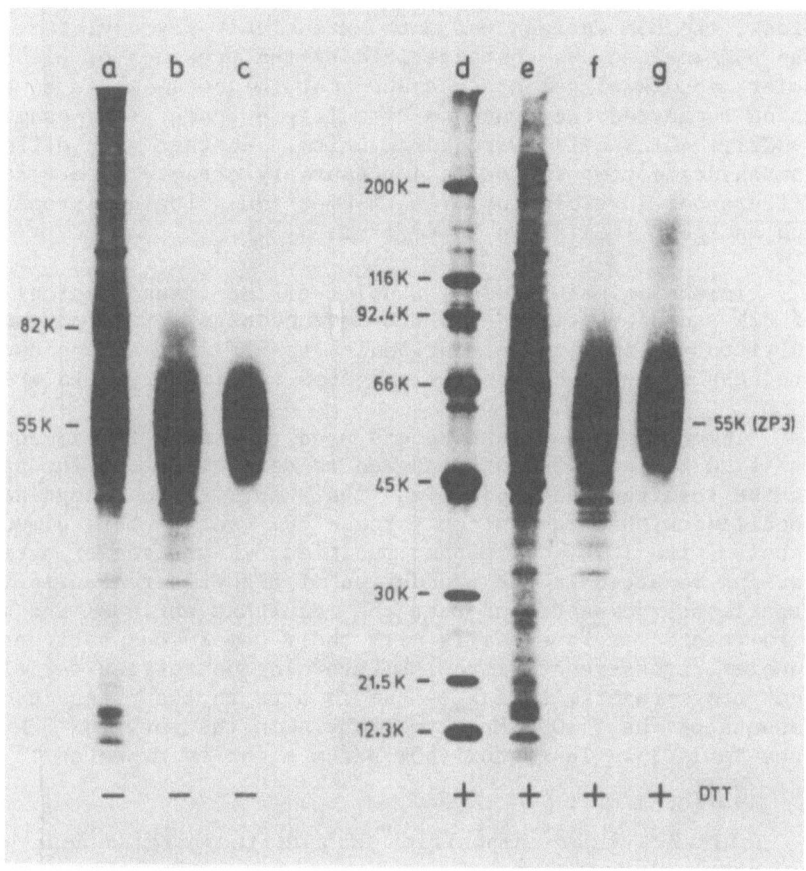

Figure 7. SDS-PAGE analysis of 55K antigen purification. Gels were
stained with Coomassie blue. Lanes a,e) Heat-solubilized
porcine ZP (Fx II). Lanes b,f) Sephacryl S-400 (Fx IV B).
Lanes c,g) Hydroxylapatite (Fx VI B). Lane d) Molecular
weight standards. Samples were pretreated with sample
buffer in the presence and absence of dithiothreitol as
indicated.

Chemical composition   The 82K and 55K antigens are composed of protein and carbohydrate. Sugar analyses revealed the presence of fucose, mannose, galactose, N-acetylglucosamine, N-acetylgalactosamine and sialic acid in each (Table 2). Two differences were noted. First, the 55K antigen was more extensively glycosylated. Second, the 55K antigen was characterized by the presence of elevated, equimolar amounts of galactose and N-acetylglucosamine, a composition which suggested the presence of multiple N-acetyllactosamine ($\rightarrow$3Galβ1$\rightarrow$4GlcNAcβ1$\rightarrow$) structural units. Whether such differences in carbohydrate composition (and presumably structure) are related to differences in biological function or immunological properties of the two antigens remains to be determined.

Immunological studies Studies of the immunochemical properties of 82K and 55K antigens and the immunocontraceptive potential, both in vitro and in vivo, of antibodies to 55K antigen are currently in progress in our laboratories and some preliminary data are available.

Immunization of rabbits was used to assess the immunogenicity of purified 82K and 55K antigens and to determine both the specificity of the resultant antibodies and their ability to recognize antigen localized within the intact porcine ZP. Active heteroimmunization of rabbits with 100 μg of either purified 82K antigen or purified 55K antigen resulted in the production of high-titered antisera. RIA experiments demonstrated that the resultant anti-82K and anti-55K sera reacted preferentially with their homologous antigens (Table 3). However, cross-reactivity with heterologous antigen was also noted. Such cross-reactivity may be due in part to the presence of minor amounts of the 55,000 Mr glycoprotein in the purified 82K antigen (see Table 1). Thus, anti-55K serum might be expected to recognize

Table 2. Sugar composition of purified porcine zona pellucida antigens.

Sugar	82K antigen		55K antigen	
	Molar ratio	%(w/w)	Molar ratio	%(w/w)
Fucose	0.9	0.7	1.2	0.8
Mannose	3.0	2.5	3.0	2.3
Galactose	4.8	4.0	10.7	8.3
GlcNAc	5.3	5.6	10.4	10.2
GalNAc	0.6	0.6	1.6	1.6
Sialic acid	0.8	1.2	1.7	2.4
		14.6		25.6

the 55,000 Mr glycoprotein component of the 82K antigen. Conversely, immunization with 82K antigen might be expected to yield a subpopulation of antibodies directed against the 55,000 Mr glycoprotein and thus be cross-reactive with 55K antigen. Finally, incubation of zona-encased porcine oocytes with either antiserum resulted in formation of a precipitation layer on the outer surface of the ZP. It was concluded that each antiserum is capable of recognizing antigen as it exists within the intact porcine ZP.

The effects of active immunization of female squirrel monkeys with purified 55K antigen are currently under investigation in our laboratories. These studies, as yet incomplete, will test the in vivo contraceptive potential of the purified porcine ZP macromolecule. Among the questions we seek to answer are the following:

1. What dosage of purified antigen is required to elicit an effective immune response?
2. Will immunization with 55K antigen result in an inhibition of fertility?
3. Will the presence of circulating antibodies to 55K antigen result in disruption of normal ovarian function, as has been previously reported to occur in rabbit (28) and cynomologus monkeys (8) following immunization with porcine ZP and in dogs (15) following immunization with PPZA?
4. What is the chemical nature (protein versus carbohydrate) and structure (sequence) of antigenic determinants recognized by squirrel monkey anti-55K sera?
5. What are the long-term effects of active immunization with 55K antigen, i.e. for how long a period of time will antibody titers remain elevated and fertility be blocked?

Table 3. Specificity of rabbit antisera to purified porcine zona pellucida antigens.

Radioimmunoassay	50% Inhibition (ng)[a]	
	82K antigen	55K antigen
Rabbit anti-82K serum (1:25,000)[b] ^{125}I-labeled 82K antigen	3.5	550
Rabbit anti-55K serum (1:75,000)[b] ^{125}I-labeled 55K antigen	60	3.3

[a] Amount of unlabeled antigen required to achieve 50% inhibition of binding of radioiodinated antigen to antiserum.

[b] Dilution of antiserum necessary to achieve 30% binding of homologous radioiodinated antigen in the absence of competing unlabeled antigen.

REFERENCES

1. Anderson D.J., and Alexander, N.J. 1983. A new look at antifertility vaccines. Fertil. Steril. 40:557-571.
2. Bleil, J.D., and Wasserman, P.M. 1980. Structure and function of
   the zona pellucida: characterization of the proteins of the
   mouse oocyte's zona pellucida. Dev. Biol. 76:185-202.
3. Drell, D.W., Wood, D.M., Bundman, D., and Dunbar, B.S. 1984.
   Immunological comparison of antibodies to porcine zonae pellucidae in rats and rabbits. Biol. Reprod. 30:435-444.
4. Dunbar, B.S. 1983. Antibodies to zona pellucida antigens and
   their role in fertility, in: "Immunology of Reproduction", T.
   G. Wegmann, T. J. Gill, C. D. Cumming and E. Nisbet-Brown,
   eds., pp. 507-534, Oxford University Press, New York.
5. Dunbar, B.S., Liu, C., and Sammons, D.W. 1981. Identification of
   the three major proteins of porcine and rabbit zonae pellucidae by high resolution two-dimensional gel electrophoresis:
   comparison with serum, follicular fluid, and ovarian cell
   proteins. Biol. Reprod. 24:1111-1124.
6. Dunbar, B.S., Wardrip, N.J., and Hedrick, J.L. 1980. Isolation,
   physico-chemical properties and macromolecular composition of
   zona pellucida from porcine oocytes. Biochemistry 19:356-
   365.
7. East, I.J., Mattison, D.R., and Dean, J. 1984. Monoclonal antibodies to the major protein of the murine zona pellucida:
   effects on fertilization and early development. Dev. Biol.
   104:49-56.
8. Gulyas, B.J., Gwatkin, R.B.L., and Yuan, L.C. 1983. Active
   immunization of cynomolgus monkeys (Macaca fascicularis) with
   porcine zonae pellucidae. Gamete Res. 4:299-307.
9. Gwatkin, R.B.L., Andersen, O.F., and Williams, D.T. 1980. Large
   scale isolation of bovine and pig zonae pellucidae: chemical,
   immunological, and receptor properties. Gamete Res. 3:217-
   231.
10. Gwatkin, R.B.L., and Williams, D.T. 1978. Immunization of female
    rabbits with heat-solubilized bovine zonae: production of
    anti-zona antibody and inhibition of fertility. Gamete Res.
    1:19-26.
11. Gwatkin, R.B.L., Williams, D.T., and Carlo, D.J. 1977. Immuniza-
    tion of mice with heat-solubilized hamster zonae: production
    of antibody and inhibition of fertility. Fertil. Steril.
    28:871-877.
12. Hedrick, J.L., and Wardrip, N. 1980. The macromolecular composi-
    tion of the porcine zona pellucida. Fed. Proc. 39:2081.
13. Henderson, L.E., Oroszlan, S., and Konigsberg, W.A. 1979. A
    micromethod for complete removal of dodecyl sulfate from pro-
    teins by ion-pair extraction. Anal. Biochem. 93:153-157.
14. Laemmli, U.K. 1970. Cleavage of structural proteins during the
    assembly of the head of bacteriophage T4. Nature 227:680-
    685.

15. Mahi-Brown, C.A., Huang, T.T.F., Jr., and Yanagimachi, R. 1982. Infertility in bitches induced by active immunization with porcine zonae pellucidae. J. Exp. Zool. 222:89-95.

16. McCoy, L.F. Jr., and Wong, K.P. 1981. Renaturation of bovine erythrocyte carbonic anhydrase B denatured by acid, heat and detergent. Biochemistry 20:3062-3067.

17. O'Farrell, P.H. 1975. High resolution two-dimensional electrophoresis of proteins. J. Biol. Chem. 250:4007-4021.

18. Oikawa, T. 1978. A simple method for the isolation of a large number of ova from pig ovaries. Gamete Res. 1:265-267.

19. Palm, V.S., Sacco, A.G., Syner, F.N., and Subramanian, M.G. 1979. Tissue specificity of porcine zona pellucida antigen(s) tested by radioimmunoassay. Biol. Reprod. 21:709-713.

20. Sacco, A.G. 1977. Antigenic cross-reactivity between human and pig zona pellucida. Biol. Reprod. 16:164-173.

21. Sacco, A.G. 1981. Immunocontraception: consideration of the zona pellucida as a target antigen, in: "Obstetrics and Gynecology Annual", R. M. Wynn, ed., Appelton-Century Crofts, New York, Vol. 10, pp. 1-26.

22. Sacco, A.G., and Palm, V.S. 1977. Heteroimmunization with isolated pig zonae pellucidae. J. Reprod. Fertil. 51:165-168.

23. Sacco, A.G., and Shivers, C.A. 1978. Immunologic inhibition of development, in: "Methods in Mammalian Reproduction", J. C. Daniel, ed., Academic Press, New York, pp. 203-228.

24. Sacco, A.G., Subramanian, M.G., and Yurewicz, E.C. 1981. Active immunization of mice with porcine zonae pellucidae: immune response and effect on fertility. J. Exp. Zool. 218:405-418.

25. Sacco, A.G., Subramanian, M.G., Yurewicz, E.C., DeMayo, F.J., and Dukelow, W.R. 1983. Heteroimmunization of squirrel monkeys (Saimiri sciureus) with a purified porcine zona antigen (PPZA): immune response and biologic activity of antiserum. Fertil. Steril. 39:350-358.

26. Sacco, A.G., Yurewicz, E.C., Subramanian, M.G., and DeMayo, F.J. 1981. Zona pellucida composition: species cross-reactivity and contraceptive potential of antiserum to a purified pig zona antigen (PPZA). Biol. Reprod. 25:997-1108.

27. Subramanian, M.G., Yurewicz, E.C., and Sacco, A.G. 1981. Specific radioimmunoassay for the detection of a purified porcine zona pellucida antigen (PPZA). Biol. Reprod. 24:933-943.

28. Wood, D.M., Liu, C., and Dunbar, B.S. 1981. Effect of alloimmunization and heteroimmunization with zonae pellucidae on fertility in rabbits. Biol. Reprod. 25:439-450.

29. Yurewicz, E.C., Sacco, A.G., and Subramanian, M.G. 1983. Isolation and preliminary characterization of a purified pig zona antigen (PPZA) from porcine oocytes. Biol. Reprod. 29:511-523.

30. Yurewicz, E.C., Sacco, A.G., and Subramanian, M.G. 1984. Purification and characterization of glycoprotein antigens of the porcine zona pellucida. Fed. Proc. 43:2039.

RESEARCH FRONTIERS IN HUMAN IN VITRO FERTILIZATION

Don P. Wolf

Department of Obstetrics, Gynecology
 and Reproductive Sciences
The University of Health Science Center at Houston
Houston, Texas 77030

SUMMARY

The acceptance of IVF-ET as a treatment modality for certain types of infertility in the human has triggered a flurry of activity in providing a patient service, in applied research endeavors designed primarily to improve clinical pregnancy rates and in basic research in reproductive endocrinology and physiology. This paper outlines the state of the art in technologic and research frontiers associated with these endeavors.

INTRODUCTION

In considering research frontiers in human in vitro fertilization, it is appropriate to highlight several of the milestones that mark the development and ultimate application of this technology to the human. Significant attempts at human in vitro fertilization were initiated in the 1960's with the studies of Edwards and co-workers (4) and the fertilization, after in vitro maturation, of human oocytes recovered from excised ovaries. Some of these oocytes could be fertilized, but it wasn't until laparoscopic aspiration techniques were perfected that acceptable fertilization levels were realized (16). These pioneering efforts set the stage for Steptoe and Edwards in the early 1970's to establish a pregnancy by the transplantation of an in vitro fertilized human oocyte. Their first clinical success was a tubal ectopic pregnancy (17). By the mid 1970's, Steptoe and Edwards temporarily abandoned the use of ovarian stimulatory drugs in favor of oocyte retrieval in spontaneous menstrual cycles. This approach culminated in the achievement of four pregnancies, the most

notable resulting in the birth of Louise Brown on July 25, 1978 (5).
In the late 1970's, similar work was progressing in Australia, and a
second baby was born in 1979 (11). The Australians established that
clinical pregnancies could result following the application of follicular recruitment drugs. This finding contributed substantially to
the spread of IVF-ET as the development of several preovulatory follicles provided the basis for an increase in the number of oocytes
recovered per patient, with considerably less time and effort expend-
ed. The application of IVF-ET technology was introduced in this
country by the program in Norfolk, Virginia initiated by Howard and
Georgeanna Jones which also pioneered the exclusive use of human
menopausal gonadotropin (hMG) in follicular recruitment. After this
relatively short developmental time period, tremendous growth in the
use of IVF-ET as a treatment modality for the infertile couple has
been realized. By the end of 1984, it has been estimated that as
many as 500 groups around the world will be involved in IVF-ET with
more than 750 live births resulting from this technology. Progress
can also be appreciated by examining clinical pregnancy rates which,
when expressed relative to the number of patients undergoing embryo
transfers, have increased from well under 5% in the late 1970's and
early 1980's to as high as 40% in selected phases of some programs at
the present time.

The current technology of IVF-ET will undoubtedly be expanded in
the near future to include ovum collection by ultrasound-guided per-
cutaneous aspiration. This approach carries the advantage of avoid-
ing general anesthesia and provides new hope for women with ovaries
that are inaccessible to laproscopic aspiration. Other possibilities
include the superimposition of embryo cryopreservation on IVF-ET
protocols with embryo transfer delayed to a subsequent menstrual
cycle that is not disrupted by follicular recruitment and surgery.
Recently, the birth of the first baby resulting from an embryo frozen
(for two mon) prior to transfer was reported (18). The types of
infertility treated by in vitro fertilization have increased markedly
from the original application - for patients that were hopelessly
infertile because of irrepairably damaged or absent fallopian tubes,
to the present inclusion of oligospermia, endometriosis, unexplained
infertility, and immunologic infertility. Other landmarks in the
expansion of the technology are listed in Table 1.

Research frontiers in human IVF can conveniently be subdivided
into three major areas: endocrinology, gamete biology, and embryonic
development and implantation. While each of these will be treated in
due course, because of our focus in this symposium, gamete biology
will be emphasized.

Endocrine Aspects
─────────────────

An intimate knowledge of the endocrinology of the menstrual
cycle has played a critical role in the development of in vitro

Table 1.   Chronology of milestones in the application of in vitro
            fertilization-embryo transfer technology in the human

---

7/25/78   England. First baby born after in vitro
           fertilization-embryo transfer.

1/13/84   Australia. First baby born to a woman
           with non-functioning ovaries, using in vitro
           fertilization of a donor egg and hormone
           supplementation.

2/3/84   California. First baby born after embryo
           transfer from embryo donor.

3/24/84   Australia. First baby born from an in vitro
           fertilized egg cryopreserved as a 8-cell
           stage embryo before transfer.

---

fertilization technology, for improvements in clinical pregnancy
rates can be attributed to the increased numbers of embryos available
for transfer. The latter, in turn, is dependent upon adequate
follicular recruitment protocols. Following the initial success of
Steptoe and Edwards in the unstimulated cycle, the development of
follicular recruitment protocols utilizing clomiphene citrate or hMG
occurred. The use of gonadotropin releasing hormone and pure follicle stimulating hormone (FSH) in follicular recruitment represent
areas of ongoing research in non-human primates. The most commonly
employed follicular recruitment protocol presently in use is a combination regimen involving clomiphene citrate starting on either day
three or day five of the menstrual cycle in combination with one
ampule of hMG (75 IU ea of LH and FSH) for five days. Following this
combination therapy, hMG alone is administered at a dose depending
upon the individual patient's response, but usually varying from one
to three ampules per day until the patient receives human chorionic
gonadotropin (hCG) to induce preovulatory oocyte maturation.
Adequacy of the ovarian response is normally monitered by direct
measurements of follicular size and number ultrasonographically and
by assessments of follicular function employing peripheral estradiol
quantitation. As experience accumulates, ranges for normal responses
in these categories have been defined. As an example, when the peripheral estradiol levels are low (300pg/ml or less), the prognosis is
poor and the treatment cycle may be discontinued. Similarly, a high
responder category exists where the clinical outcome is also likely
to be jeapordized. It is refinements in our ability to define
normalcy within the confines of inherent biologic variability and
differing numbers of recruited follicles that will allow improvement
in outcome at minimal risk and expense to the patient. Knowledge of

the endocrine events associated with the normal menstrual cycle has
allowed the establishment and maintenance of pregnancy with exogenous
hormones in an anovulatory patient with primary ovarian failure (12).
A discussion of research frontiers in endocrinology associated with
human IVF would not be complete without mention of follicular fluid.
Follicular fluid and associated granulosa cells represent major
resources generated by follicular aspiration, and rather extensive
efforts are being devoted towards correlating changes in follicular
fluid composition with changes in oocyte maturational state, or with
the subsequent ability of the oocyte to undergo normal fertilization,
development and implantation (2). The steroid composition of follicular fluid has been examined; progesterone levels may be useful prog-
nostic indicators of follicular and/or oocyte maturation (6).

Gamete Biology

     Protocols for the in vitro maturation and fertilization of human
ovarian oocytes have improved substantially such that fertilization
levels are high (approximately 80%) perhaps even maximal. Oocytes
that are clearly immature, germinal vesicle intact, are recovered on
occasion and can be fertilized if incubated for 24 to 36 h in vitro
prior to insemination (Fig. 1). Such in vitro matured oocytes have
supported clinical pregnancies in a limited number of cases (20). In
optimizing the fertilization of preovulatory oocytes, a four to six
hour in vitro maturation prior to insemination represents the most
commonly employed protocol, the rational being that oocytes recovered
by follicular aspiration are somewhat immature and in vitro incubation allows completion of maturation prior to insemination. Premature insemination of preovulatory oocytes has led to unacceptably
high levels of polyspermy which may be associated with an inadequate
cortical response (19). The sperm concentration required to effect
fertilization has been optimized (21); fertilization success, in our
hands, is independent of the follicular recruitment protocol employed, and with preovulatory eggs is inversely related to sperm concentration over the range of 2.5 to 50 x 10^4 motile sperm/ml. Maximum
fertilization (80.8%) occurs at a concentration of 2.5 x 10^4 motile
sperm/ml. The instance of polyspermic fertilization in this series
was directly related to sperm concentration, decreasing from 5.5% at
10 x 10^4 to 1.8% at 5 x 10^4 and to 0% at 1-2.5 x 10^4 motile sperm/
ml.

     One of the research areas that has received impetus from the
development of human IVF-ET is sperm physiology including sperm
capacitation, the induction of an acrosome reaction, sperm-zona
interaction and the role of changes in specific parameters of flagellar motility. Although these frontiers are not uniquely associated
with the human, the application of IVF-ET in the treatment of male
infertility provides additional incentive to examine these events in
humans. Improved conception rates have already been reported from

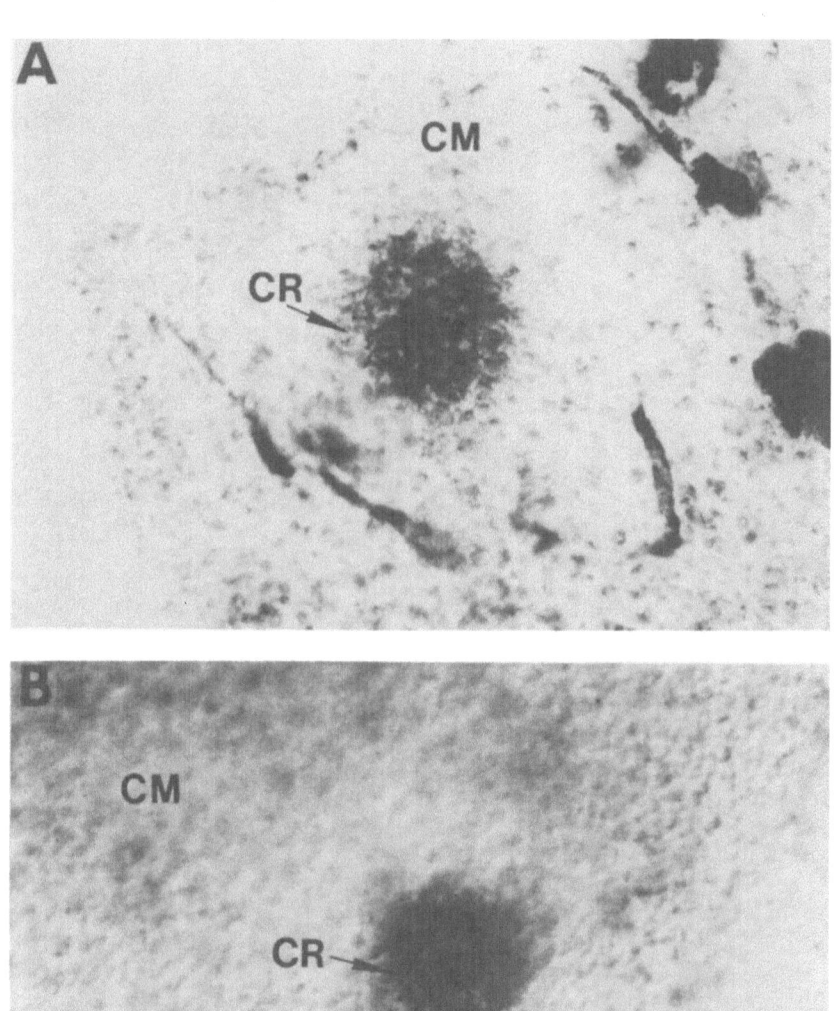

Figure 1. Light microscopy of human ovarian oocytes; A) a preovulatory oocyte aspirated from 20 mm follicle following follicular recruitment with hMG; B) a preovulatory oocyte
recovered from a 21 mm follicle after clomiphene citrate
stimulation (50 mg/day); C) an immature oocyte aspirated
from a 12 mm follicle and D) the immature oocyte from

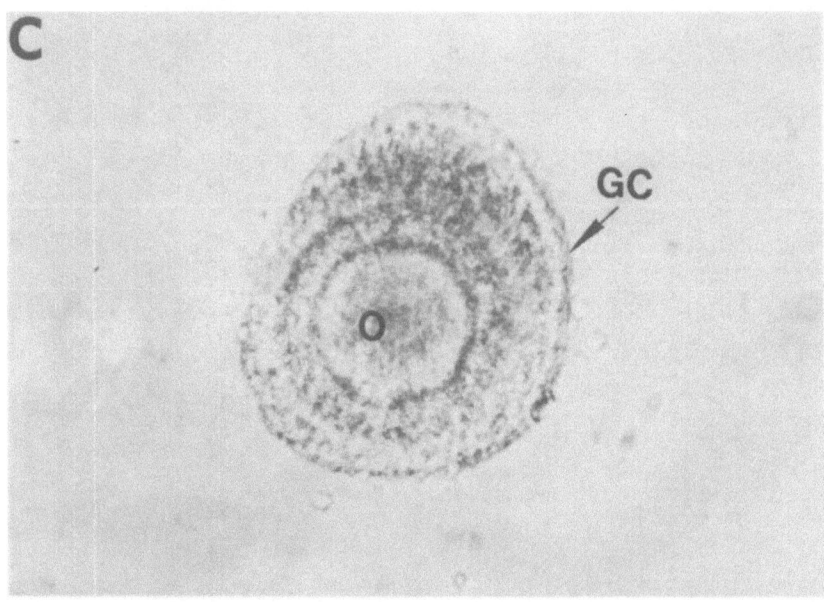

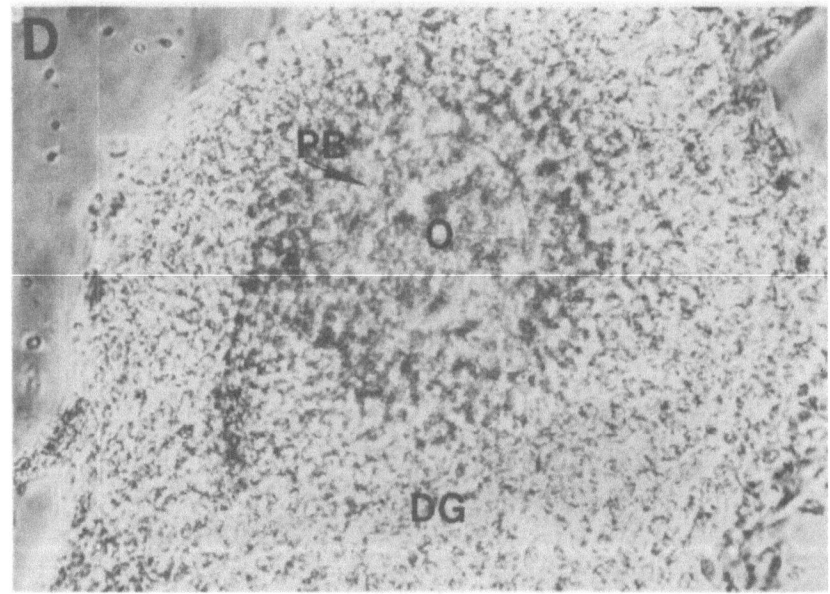

Cont. Figure 1C after culturing for 24 h. Legends: CR = corona
      radiata; CM = cumulus mass; O = oocyte; GC = granulosa cells;
      PB = polar body; DG = dispersed granulosa. Magnification:
      2A,B = 19X, 2C,D = 47X. (Reproduced with permission from Human
      In Vitro Fertilization and Embryo Transfer, D.B. Wolf and M.M.
      Quigley, Eds., Plenum Press, New York, 1984).

sperm samples washed in vitro and inseminated directly into the uterus (9). Sperm capacitation and the induction of an acrosome reaction are phenomena under active investigation in my laboratory. Capacitation, which undoubtedly occurs at different rates in different cells, includes a series of poorly defined biochemical and biophysical changes in sperm prerequisite to successful sperm/egg interaction. In molecular terms, changes in the sperm surface are involved which in turn may result in intracellular calcium ion changes. The activity of a calcium regulatory protein present in seminal plasma has been described; addition of this protein to epididymal sperm inhibits their ability to transport extracellular calcium (15). While sperm capacitation has been the subject of extensive investigation in a number of rodents, it is important to point out that epididymal sperm may not always represent good models for the ejaculated cell. Despite the limited availability of methods for quantitating human sperm capacitation, the protocols for effecting capacitation in vitro are well established, usually involving sperm incubation with stock salt solutions fortified with energy sources and the protein albumin, either as purified serum albumin or as heat-inactivated serum. Capacitation culminates in the occurrence of an acrosome reaction and the exposure of acrosomal enzymes adherent to the inner acrosomal membrane and of new sperm surface. The in vitro trigger for the acrosome reaction has been the subject of study in various rodent systems; in some cases, at least, the reaction can be triggered by specific components of the egg's zona pellucida. In vivo, in the presence of an intact cumulus mass, very little is known about acrosome reaction induction. Recently we have investigated chemical induction of the acrosome reaction in human ejaculated sperm. As with capacitation, methods for quantitating the occurrence of an acrosome reaction in humans are limited, in the latter case, because of the relatively small size of the acrosome. Monoclonal antibody technology has, however, provided reagents that allow the rapid quantitation of acrosome status. Several groups, in addition to my own (10), have generated monoclonal antibodies that recoginze antigens localized exclusively in the acrosomal cap of human sperm and the loss of these antigens, measured by direct or indirect immunofluorescence microscopy is a convenient method for quantitating acrosomal status (Fig. 2). This development has supported efforts to correlate acrosomal status with motility, viability and the in vitro fertilization of human eggs, to quantitate the kinetics of the acrosome reaction in human sperm incubated under capacitating conditions in vitro, and to assess whether the egg or its cumulus plays a role in the induction of an acrosome reaction. We are also interested in whether the rapid quantitative assay of acrosomal status may ultimately be employed in quantitating the state of sperm capacitation (3). Our working hypothesis is that capacitated sperm show an enhanced tendency or ability to undergo a chemically induced acrosome reaction. Our study of human sperm capacitation and induction of the acrosome reaction, represents a component of an effort to define the characteristics of fertile human sperm. Another prerequisite for fertility is motility, and unique motility

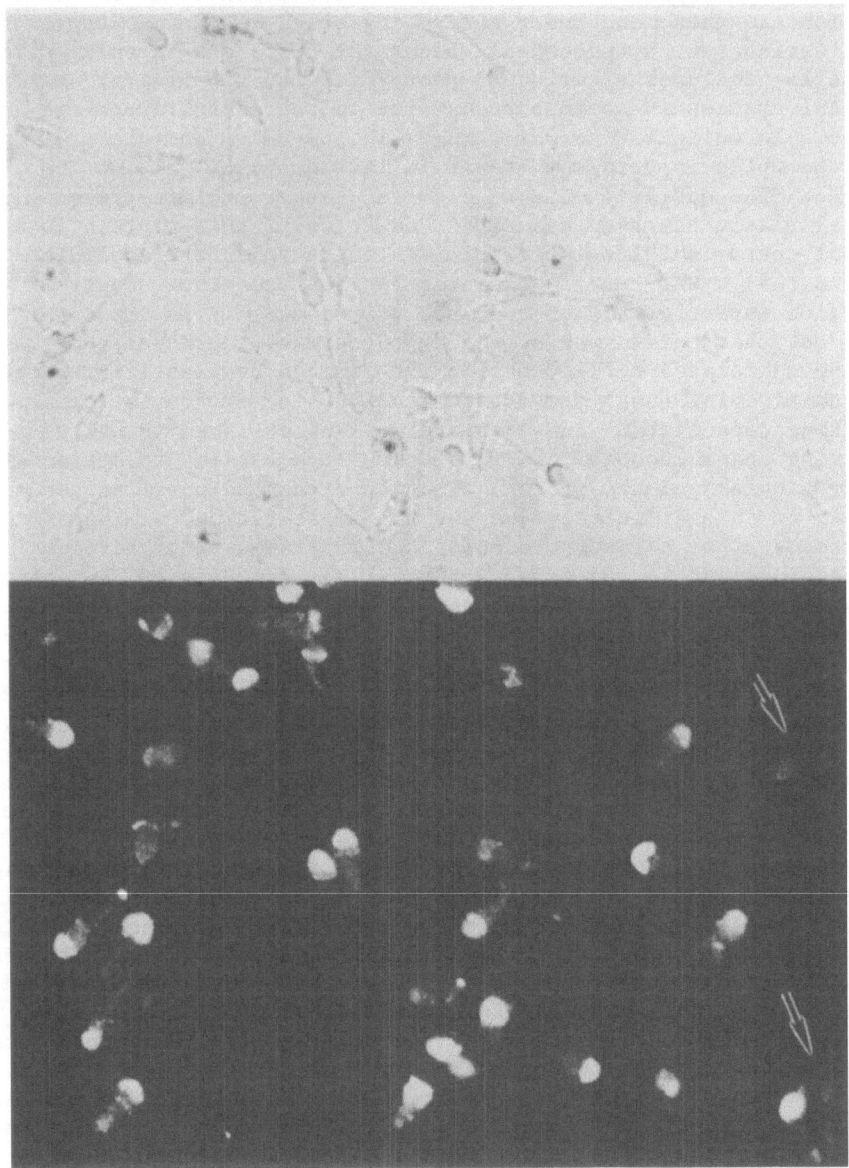

Figure 2. Indirect immunofluorescence staining patterns for human
            ejaculated sperm exposed to a specific monoclonal antibody
            and FITC-antimouse Ig. The figure depicts corresponding
            epifluorescence and light micrographs and the arrows mark
            sperm that do not have acrosomes.

charactertistics termed "hyperactivated motility" have been
associated with capacitation in laboratory and domestic animals. In
the human, motility characteristics have been assessed primarily only
in sperm swimming in whole semen (non-capacitating conditions), and
only recently has evidence been presented that human sperm display
hyperactivated motility (13).

In human gamete biology unlike sperm, egg availability represents a major practical limitation. For this reason, our knowledge
base and our ability to conduct relevant research is limited. From
the perspective of an IVF-ET program, conditions have been optimized
for handling and maturing eggs in vitro as well as for achieving very
high levels of fertilization. Most fertilized eggs undergo cleavage
in vitro, and the rate at which they do so, at least through the
first two or three divisions, is now well established (Fig. 3). As
an inevitable correlate of any IVF-ET program, a limited number of
oocytes fail to fertilize and/or respond abnormally. This provides a
limited source of material for studies concerning sperm/zona binding,
zona penetration and biochemical characterization of the zona.

Embryonic Development and Implantation

At the present time, a major factor limiting clinical pregnancy
in IVF-ET is the inability of transferred concepti to implant.
Potential explanations for this inefficiency are related to the
quality of the endometrium and the embryo or both. Synchrony in time
between the developing embryo and the endometrium is undoubtedly
important. It is already somewhat unusual that the uterine environment in the human is tolerant of an embryo present there approximately two days premature. The quality of the endometrium may be
influenced by the follicular recruitment protocol employed; in the
case of high clomiphene citrate regimens, antiestrogenic effects on
the endometrium may account for the relatively low clinical pregnancy
rates (14). Such protocols may advance or retard endometrial growth
by affecting the interplay of endocrine events. Endometrial dating
or staging following biopsy has been done on a few IVF patients who
for various reasons did not undergo embryo transfer (8). In one
study, biopsy results were reported in patients undergoing transfer
(1). Early indications are that endometrial development can be
advanced or retarded and that, pregnancy rates are highest when
endometrial development is normal. Endometrial development may also
be influenced in IVF-ET patients by inadequate ovarian production of
progesterone since granulosa cells are often aspirated in large
number during laparoscopic recovery of oocytes. In many cases, this
possibility has lead to the administration of exogenous progesterone
during the luteal phase of the menstrual cycle.

An evaluation of the normalcy or quality of the developing
preimplantation stage embryo in the human is problematic. Two
parameters are employed; the rate of development and the appearance
of the embryo. Because embryos are usually transferred back to the

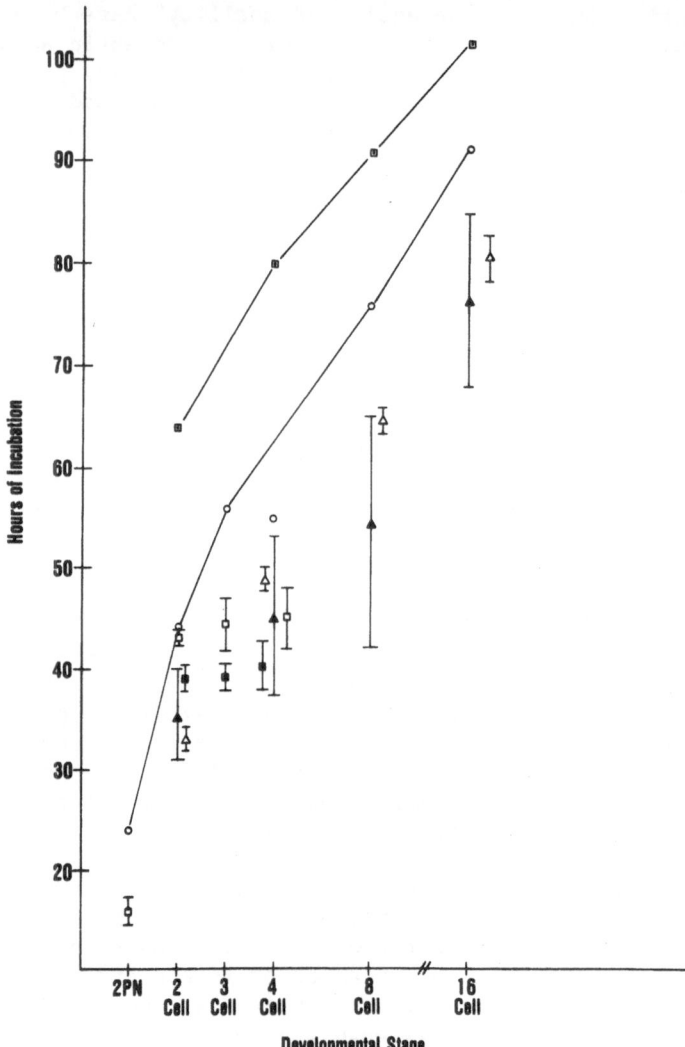

Figure 3. Preimplantation development rates for embryos produced by
          IVF. Legends: (⊡) = Lauritsen, 1982; (O) = Sundstrom et
          al., 1981; (△) = Edwards et al., 1981; (▲) = Trounson et
          al., 1982; (■) = U.T. Houston, 1982-1983; (⊔) = Wortham
          et al., 1983. (Reproduced with permission from Human In
          Vitro Fertilization and Embryo Transfer, D.P. Wolf and M.M.
          Quigley, Eds., Plenum Press, New York, 1984).

donor at the 2-8 cell stage, (40-48 h post insemination), these
parameters are not very discriminating. Additionally, low quality
and/or slowly developing embryos do support pregnancies although

Table 2. The relationship between the number of embryos transferred and pregnancy rates in various IVF-ET programs. (Reproduced with permission from Human In Vitro Fertilization and Embryo Transfer, D.P. Wolf and M.M. Quigley, Eds., Plenum Press, New York, 1984)

	Pregnancy Success[a]				
	Number of Embryos Transferred				
Reference	1	2	3	4	5
Fishel et al., 1983	19% N=120	33% N=36	–	–	–
Feichtinger and Kemeter, 1983	12% N=34	30% N=30	36% (c,e) N=11	–	–
Houston IVF-ET[f] 5/82 - 6/83	10% N=31	19% N=27	10% N=10	33% (b) N=6	50% N=2
Johnston et al., 1983	7% N=135	16% N=82	31% N=13	50% N=6	–
Jones et al, 1983	17% N=107	26% N=102	41% N=29	36% N=11	–
Kerin et al., 1983	0 N=29	31% (d) N=32	39% (b) N=18	50% (c,e) N=6	–
Speirs et al., 1983	10% N=60	22% N=49	19% N=37	35% (c) N=23	0 N=1
Trounson, 1983	7% N=43	25% N=44	23% N=44	0 N=9	–
Mean pregnancy Success	12% N=559	24% N=402	33% N=162	33% N=61	33% N=3
Percent of Transfers Resulting in Multiple Pregnancies	0	1%	3%	10%	0

[a] Total pregnancies reported, b) 1 set of twins, c) 2 sets of twins, d) 3 sets of twins, e) 1 set of triplets, f) clinical pregnancies progressing beyond first trimester and live births only (clomiphene 50 ± hMG trials)

presumably at lower rates. Although clinical pregnancy rates upon transferring one embryo varies between programs, the rate (for singleton pregnancies) doubles when two embryos are transferred, and increases to nearly three-fold when three embryos are transferred (Table 2). This implies that not all embryos are equal and supports the contention that other factors contribute to pregnancy establishment. One of the approaches that undoubtedly will help unravel this dilemma involves embryo cyropreservation and transfer during subsequent cycles when the endometrium should be entirely normal.

ACKNOWLEDGMENTS

Appreciation is extended to my colleagues participating in the human IVF-ET program here in Houston, to Drs. Jeffrey Boldt, E. William Byrd, and Denise Ochs and Ms. Eve Shen for their contributions to the research activities of this laboratory and to Ms. Gail Alexander for her excellent secretarial assistance. The study of human sperm capacitation was supported in part by a grant from the National Science Foundation.

REFERENCES

1. Abate, V., Call, A., and Stanchi, A. 1984. Endometrial biopsy at the time of embryo transfers: Correlation of histologic diagnosis with therapy and pregnancy rates. Fertil. Steril. 41:43S.
2. Botero-Ruiz, W., Laufer, N., DeCherney, A.H., Polan, M.L., Haseltine, F.P., and Behrman, H.R. 1984. The relationship between follicular fluid steroid concentration and successful fertilization of human oocytes in vitro. Fertil. Steril. 41:820-826.
3. Byrd, W., and Wolf, D.P. 1984. Evaluation of capacitation in human sperm using monoclonal antibodies against acrosomal cap antigens. Biol. Reprod. 30:90.
4. Edwards, R.G., Bavister, B.D., and Steptoe, P.C. 1969. Early stages of fertilization in vitro of human oocytes matured in vitro. Nature 221:632.
5. Edwards, R.G., Steptoe, P.C., and Purdy, J.M. 1980. Establishing full-term human pregnancies using cleaving embryos grown in vitro. Brit. J. Obstet. Gynaecol. 87:737.
6. Fishel, S.B., Edward, R.G., and Walters, D.E. 1983. Follicular steriods as a prognosticator of successful fertilization of human oocytes in vitro. J. Endocr. 99:335-344.
7. Isahakia, M. and Alexander, N.J. 1984. Interspecies cross-reactivity of monoclonal antibodies directed against human sperm antigens. Biol. Reprod. 30:1015-1026.
8. Jones, G.S. 1984. Update on in vitro fertilization. Endocrine Reviews 5:62-75.

9. Kerin, J.F.P., Peek, J., Warnes, G.M., Kirby, C., Jeffrey, R., Matthews, C.D., and Cox, L.W. 1982. Improved conception rate after intrauterine insemination of washed spermatozoa from men with poor quality semen. Lancet 1:533-534.

10. Lee, C-Y.G., Huang, Y-S., Huang, C-H., Hu, P-C., and Menge, A.C. 1982. Monoclonal antibodies to human sperm antigens. J. Reprod. Immunol. 4:173-181.

11. Lopata, A., Johnston, I.W.H., Hoult, I.J., Speirs, A.I. 1980. Pregnancy following interuterine implantation of an embryo obtained by in vitro fertilization of a preovulatory egg. Fertil. Steril. 33:117.

12. Lutjen, P., Trounson, A., Leeton, J., Findlay, J., Wood, C., and Renou, P. 1984. The establishment and maintenance of pregnancy using in vitro fertilization and embryo donation in a patient with primary ovarian failure. Nature 307:174-175.

13. Mortimer, D., Courtot, A.M., Giovangrandi, Y., Jeulin, C., and David, G. 1984. Human sperm motility after migration into, and incubation in, synthetic media. Gamete Research 9:131-144.

14. Quigley, M.M., Maklad, N.F., and Wolf, D.P. 1983. Comparison of two clomiphene citrate dosage regimens for follicular recruitment in an in vitro fertilization program. Fertil. Steril. 40:178-182.

15. Rufo, G.A., Jr., Singh, J.P., Babcock, D.F., and Lardy, H.A. 1983. Purification and characterization of a calcium transport inhibitor protein from bovine seminal plasma. J. Biol. Chem. 257-4627.

16. Steptoe, P.C., and Edwards, R.G. 1970. Laparoscopic recovery of preovulatory human oocytes after priming of ovaries with gonadotropins. Lancet 1:683.

17. Steptoe, P.C., and Edwards, R.G. 1976. Reimplantation of a human embryo with subsequent tubal pregnancy. Lancet 1:880.

18. Trounson, A., and Mohr, L. 1983. Human pregnancy following cryopreservation, thawing and transfer of an eight-cell embryo. Nature 305:707-709.

19. Trounson, A.O., Mohr, L.R., Wood, C., and Leeton, J.F. 1982. Effect of delayed insemination on in vitro fertilization, culture and transfer of human embryos. J. Reprod. Fert. 64:285.

20. Veeck, L.L., Wortham, J.W.E., Witmyer, J., Sandow, B.A., Acosta, A.A., Garcia, J.E., Jones, G.S., Jones, H.W., Jr. 1983. Maturation and fertilization of morphologically immature human oocytes in a program of in vitro fertilization. Fertil. Steril. 39:594-602.

21. Wolf, D.P., Byrd, W., Dandekar, P., and Quigley, M.M. 1984. Sperm concentration and the fertilization of human eggs in vitro. Biol. Reprod. Biol. Reprod. 31:837.

INDEX

Row 1. J. Dean, P. Betancourt, M. Betancourt/ E. Wassarman, P.
Wassarman, D. Katz, D. Phillips. Row 2. A. Hino, H. Sasaki, K.
Utsumi/ J. Meizel, E. Goldberg, D. Garbers. Row 3. E. Oliphant, E.
Yurewicz/A. Wyrobeck, C. Metz, B. Gledhill, C. Birr, J. Biggers/P. Jego
F. Dube, R. Sullivan. Row 4. A. Lopo, G Gerton, D. Wolf/P. Thomas,
S. Meizel, T. Oikawa, K. Turner/R. Waibel, P. Saling, A. Fleming,
J. Tezon. Row 5. D. Micelli, S. Sepsenwol, P. Jego/C. Katagiri,
A. Hino, H. Sasaki.

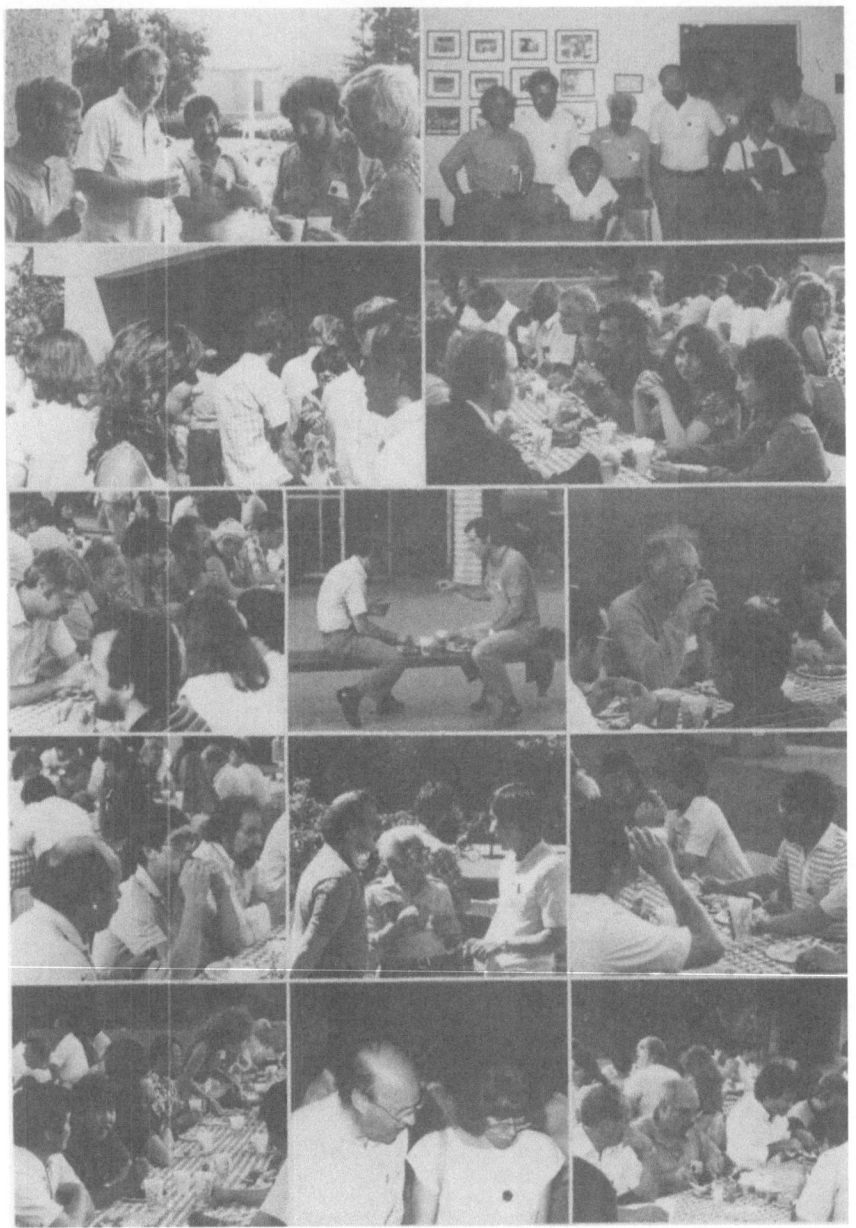

Row 1. A. Yudin, W. Clark, S. Vijayarghavan, G. Cherr, G. Hinsch/
C. Katagiri, P. Jego, D. Micelli, A. Monroy, J. Hedrick, U. Urch, C.
Birr, G. Gerton. Row 2. P. Cuasnicu, M. O'Rand/ Bar-B-Que Dinner.
Row 3. G. Gerton, C. Katagiri, G. Cherr, D. Katz, G. Hinsch, S.
Stolzenberg/ D. Wolf, C. Brown/ M. Betancourt, D. Phillips, E.
Drobnis. Row 4. U. Urch, E. Carroll, A. Smith/ P. Wassarman, A.
Monroy, M. Hoshi/ N. Cross, H. Sasaki, K, Utsumi/ Row 5. S. Hochi,
K. Nishimura, A. Lopo, T. Oikawa, L. Schlichter/ J. Biggers, P.
Saling/ D. Hoskins, E. Goldberg, C. Carron